AF327644

Natural Computing Series

Series Editors: G. Rozenberg
Th. Bäck   A.E. Eiben   J.N. Kok   H.P. Spaink

Leiden Center for Natural Computing

For further volumes:
http://www.springer.com/series/4190

Nataša Jonoska
Masahico Saito

Editors

Discrete and Topological Models in Molecular Biology

 Springer

Editors
Nataša Jonoska
Masahico Saito
Department of Mathematics & Statistics
University of South Florida
Tampa FL, Florida
USA

Series Editors
G. Rozenberg (Managing Editor)

Th. Bäck, J.N. Kok, H.P. Spaink
Leiden Center for Natural Computing
Leiden University
Leiden, The Netherlands

A.E. Eiben
Vrije Universiteit Amsterdam
The Netherlands

ISSN 1619-7127 Natural Computing Series
ISBN 978-3-642-40192-3   ISBN 978-3-642-40193-0 (eBook)
DOI 10.1007/978-3-642-40193-0
Springer Heidelberg New York Dordrecht London

Library of Congress Control Number: 2013957136

The Cre-loxP recombination system has been used as a tool to characterize DNA tertiary structure, for example in difference-topology experiments that are based on tangle analysis. Difference topology depends on the formation of a looped recombinase-DNA intermediate; however, kinetic details of the Crerecombination pathway have not been sufficiently characterized for the recombination system to be used quantitatively. By analyzing the synapsis steps in Cre recombination, Shoura and Levene showed that the free energy of DNA-loop formation can be directly measured using a novel fluorescence resonance energy transfer (FRET)-based reporter technique. The featured method uses special DNAs bearing loxP recombination sites that contain fluorophore tags (in red and green) at specific nucleotide positions. Addition of Cre protein (yellow) to the reaction generates a looped synaptic complex and recombination products (linear and circular DNA). Innovative applications of their technique include superhelical, knotted, or catenated DNA substrates in vitro and in living cells. See the chapter by Shoura and Levene for more details. (Conceptual design and artwork is by Massa Shoura and Udayana Ranatunga. Supercoiled DNA PDB are from MD simulations by Sarah Harris. Other DNA structures were made using VMD and the GraphiteLifeExplorer software. Final image was rendered using PovRay.)

Preface

Commonly used models in mathematical biology involve dynamical systems, differential equations, and statistics. These fields often study the general behavior and dynamics of biological systems either at the population level or in terms of cell-to-cell interactions, tissue development, or organ function, but they are less often used in the study of biomolecular processes such as genetics, biomolecular structures and interactions.

With the explosion of research in molecular biology, and in particular the enormous experimental data generated in the last couple of decades, new mathematical tools are being developed using graph theory, algebra, combinatorics, discrete stochastic processes, and topology. These new methods allow us to "zoom-in" to the cell to better understand spatial macromolecular arrangements and molecular interactions within the cell, or a portion of the cell.

With this volume, we wish to introduce several aspects of these contemporary approaches in mathematical (molecular) biology that contain a variety of models, covering a wide spectrum of problems in molecular biology.

The chapter authors are experts in their own fields and have diverse scientific background ranging from biology to biophysics, physics, computer science, and mathematics. The collection of their experiences gives different perspectives on sometimes similar biological problems and, we hope, will help in understanding the mathematical tools as well as the biological process.

The book is divided into five parts devoted to general biological themes, while the mathematical methods introduced in each theme differ dramatically.

The first part of the book concentrates on data analysis, including genetic data and data related to brain activities. The chapters by Franco and Angeleska et al. deal with short nucleotide segments. While Franco describes methods from formal language theory to learn about the functionality of genetic segments, Angeleska et al. survey methods, mainly based on graph theory, for parsing and annotating sequencing data for genome assembly. The next two chapters, by Carbone and Bonizzoni et al., describe methods to understand gene or protein (co)evolutions through analysis of distances in phylogenetic trees (Carbone) or graph theoretical methods (Bonizzoni et al.). The last two chapters deal with different types of data analyses. Based on MRI scans, Daley describes a method to construct a graph for the brain neural network through threshold functions, while Nanda and Sazdanović show how to use algebraic topology to analyze a variety of data types, including neural connections in the brain.

The second part of the book deals with biomolecular spatial arrangements. Chapters in this part use techniques from combinatorics and graph theory as well as purely algebraic methods using groups of symmetries. This segment starts with problems about RNA secondary structures (Heitch and Poznanović); continues with rigid and flexible regions in a protein ternary structure (Fox and Streinu), supramolecular assembly of viral capsids with clusters of proteins (Sitharam), and transitions of dodecahedral and icosahedral symmetries in viral capsid expansions (Cermelli et al.); and concludes with three-dimensional synthetic DNA structures (Ellis-Monaghan et al.).

DNA rearrangements have been observed on both developmental and evolutionary scales, and some of the most extensive shuffling of genetic material has been observed in certain species of single-cell organisms (ciliates). Goldman et al. start the third part of the book with a brief survey of the biological process, while the next two chapters cover mathematical methods that describe the rearrangement process using matrix algebras (Brijder and Hoogeboom) and topological aspects of graphs (Dolzhenko and Valencia).

The fourth part on spatial embeddings of biomolecules starts with an introduction to DNA topology by Darcy et al., providing basic biological and topological background of the subject. Buck's exposition on methods from knot theory capturing enzymatic actions that control topological embeddings of DNA is followed by Baker's mathematical development of these methods. This part of the book ends with a chapter by Ishihara et al. applying the described methods to a specific experimentally observed biological process.

The fifth and last part of the book deals with the kinetics and dynamics of molecular interactions. It starts with analyzing looping of DNA through enzyme kinetics (Shoura and Levene) and moves into reaction networks using the quasi-steady-state assumption by differential and matrix equations (Pantea et al.) and a survey of algebraic methods in systems biology (Laubenbacher et al.). This part, and the book, concludes with chapters by Savageau and Lomnitz, who use dynamical systems to study phenotype development, and Rejniak, who shows a computational model that can capture spatial tissue development including mutant morphologies.

We hope this volume will be suitable as a reference book for researchers in mathematics and theoretical computer science who are interested in modeling molecular and biological phenomena using discrete methods as well as for biologists looking for available mathematical tools in discrete models. It may also serve as a guide and supplement for a graduate course in mathematical biology or bioinformatics, to introduce discrete aspects of mathematical biology. Many chapters end with open problems that can serve as the basis for research developments.

We wish to thank all the contributing authors for their work in producing these chapters. Every contribution was reviewed by at least two researchers, whose valuable comments and suggestions helped in composing the final product. We express our deep gratitude to them for their time and effort. Many of the contributors were also participants of a workshop on Discrete and Topological Models in Molecular Biology held at the University of South Florida in March 2012. The workshop received generous support from the National Science Foundation through the grant DMS-1157242. This allowed for a very successful workshop and provided opportunities for the research community to meet and exchange ideas that spurred the development of this volume. Finally, we wish to acknowledge support for this project, in part, by the NSF grant DMS-0900671 and CCF-1117254.

Tampa, FL, USA   Nataša Jonoska
June 2013   Masahico Saito

Contents

Part I
Discrete and Graph-Theoretic Models for Data Analysis

Perspectives in Computational Genome Analysis

Giuditta Franco

Abstract DNA segments which together cover a genome may be collected together to form a *genomic dictionary* of specific words, which may be annotated either by biological information (according to the functional role they may play in regulatory mechanisms), or by numerical information (such as the position in the genome, the total number of occurrences, the occurrences lying inside or outside genic sequences, the CpG content, and more sophisticated informational indexes of text analysis). In this chapter, two analogous and complementary dictionary-based approaches to genome analysis are reviewed. We give a sketch of some of the relevant knowledge about the (human) genome, in terms of structure and functional role of its parts, and an informational view based on a mathematical analysis of k-mer dictionaries, with the aim of opening the way to the formulation of a model. Basic notions about genomic regulatory activity, where the underlying mechanisms of information exchange are far from understood, are given. A description of an initial attempt at computational modeling of genomes, seen as a new language to be deciphered, concludes the chapter.

1 Introduction

Biological and computational human genome analysis is one of the most important and intriguing research challenges we are currently facing. It involves efforts from numerous countries all over the world, and any result towards an understanding of the basic mechanisms underlying genome structure and functioning could have high impact on major problems in medicine, such as the understanding and control

G. Franco (✉)
Computer Science Department, University of Verona, Strada le Grazie 15, I–37136 Verona, Italy
e-mail: giuditta.franco@univr.it

N. Jonoska and M. Saito (eds.), *Discrete and Topological Models in Molecular Biology*,
Natural Computing Series, DOI 10.1007/978-3-642-40193-0_1,
© Springer-Verlag Berlin Heidelberg 2014

of (often incurable) genetic diseases such as multiple sclerosis, lupus, rheumatoid arthritis, Crohn's disease, celiac disease, and various kinds of cancer. Advances in next-generation sequencing (NGS) technology have boosted studies of both population genomics and genomic sequence analysis. However, it is still like having at hand an encrypted book, written in an unknown language that we have to decipher.

After the revolutionary Human Genome Project, the retrieval of any sort of genomic text is nowadays possible online, by freely accessible databases (for example, the NCBI, UCSC, and EMBL-EBI websites[1]), containing genes, chromosomes, and whole genomes. Unfortunately, there are many regions of the human genome which, for a variety of reasons, are still not well characterized or have not been characterized at all. However, at the end of the project, we had additional evidence that only a small part of the genome (less than 2 %) is genic (an initial estimate of 35,000 traditional protein-coding genes was whittled down to about 21,000 genes, with an average length of 3,000 bases, whereas the longest human gene known is that for dystrophin, with a length of 2.4 million bases); the remaining part (98 %) was denoted *junk DNA* (and often referred to as *dark matter*). Indeed, even including the (about 20,000) RNA genes, encoding RNA strands with myriad roles, a large, noncoding portion of the genome still appeared useless.

Discovered in 1977 [23], more than 18,000 *pseudogenes* were dismissed as junk DNA as well. By definition, these derive from gene duplication (most of them are associated with a few abundantly expressed gene families). They are located either on the same chromosome as the genes from which they originate or on a different chromosome, but have some alterations in the sequence structure (loss of promoter sequences, premature stop codons, frameshift mutations, or alterations in splice sites) which prevent them from being transcribed or translated (see Fig. 1 for a simple sketch of the traditional central dogma). However, they are not functionally disabled: they can promote or inhibit the expression, or enhance the function, of the parental gene. In some cases, mutations acquired by RNA pseudogenes allow them to perform functions unrelated to those of their parental genes, and even to produce (truncated) proteins. Interestingly, they provide a mechanism whereby genomes can evolve new functions from existing sequences (and represent a reservoir of protein diversity): pseudogenic proteins are produced under different conditions or in different cell types from the proteins derived from their parental genes. In recent research, pseudogenes have turned out to be modulators of expression of parental or unrelated genes [33]. Often, cancer-related genes (such as the PTEN tumor suppressor gene and the oncogenic KRAS) possess "biologically active" pseudogenes, which means that they regulate coding-gene expression [36]. Other recent advances in the precise annotation of pseudogene loci and in pseudogene statistics developed using computational approaches may be found in [32], within the framework of the GENCODE project.

---

[1]See   http://www.ncbi.nlm.nih.gov/sites/genome,   http://hgdown\load.cse.ucsc.edu/downloads. html, and http://www.ebi.ac.uk/genomes/, respectively.

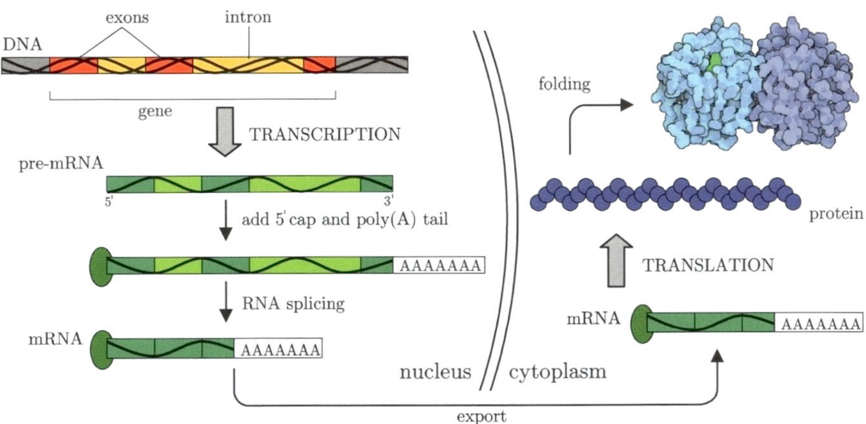

Fig. 1 The central dogma of molecular biology (so named by Francis Crick) states that genomic information guides the formation of proteins. DNA is *transcribed* (copied) into single RNA strands, called *messenger RNA* (mRNA), since it carries the information originally present in the genome outside the nucleus, to the *ribosomes*, machines which synthesize proteins. Transcription consists of two internal steps: the strand first assembled by the transcription process is called *pre-mRNA*, which is then processed in the cell nucleus to become (mature) mRNA. This last process consists of three phases: *post-transcriptional capping*, where a modified G is attached to the $5'$-end, step-by-step removal of introns present in the pre-mRNA, that is, *splicing* of exons (done by *splicesomes*), and a final *post-transcriptional polyadenylation* (elongation by a polyA) at the $3'$-end. The resulting protected information is moved to the cytoplasm, where *translation* into a protein is performed by ribosomes, according to the *genetic code*, which associates three-letter words, called *codons*, to single amino acids (the basic protein constituents). Different codons may code for the same amino acid

Our current knowledge about the structure and function of the human genome has been strongly influenced by recent results from the ENCODE (Encyclopedia of DNA Elements) project, which are changing our previous characterizations of genic regions. Several traditional computational (namely, machine-learning-based) methods have supported this project, and genome informational analysis has been emerging from the computational and linguistic investigation of genomic dictionaries.

1.1 The ENCODE Project

Several revolutionary computational and experimental results about the crucial informational and regulatory role of junk DNA have recently been published in about 40 papers, mainly in *Nature*, *Genome Research*, and *Genome Biology*, and some also in *Biological Chemistry* and *Science*. A selection of these papers are listed in the references. The decade-long joint project ENCODE, involving 440 scientists from 32 laboratories around the world (at MIT, Harvard, Stanford, and

SUNY in the USA, and at universities in Germany, the UK, Spain, Switzerland, Singapore, China, and Japan) provided integrated evidence that about 80 % of the human genome is covered by active regulatory elements (landing spots for proteins in order to promote, inhibit, or silence gene activity) [10, 11], although if has a lot of redundancy. More than half of it is transcribed into the direct output of genetic information, that is, into different RNA types for synthesis, processing, transport, modification, and translation activities [8, 10, 33]. In an initial effort, an ENCODE pilot project focused on just 1 % of the genome, and its results (published in 2007) indicated that the list of human genes (and corresponding intergenic regions) was incomplete. Recent advances in low-cost, rapid DNA sequencing technology have allowed investigators to scale up this research, from a more specifically driven analysis to the whole genome, supported by massive data analysis work that has been performed in specialized centers. NHGRI has invested about US$300 million in ENCODE, including the pilot project, technology development, and preliminary studies on the genomes of mice, nematodes, and fruit flies [34].

There are now 1,640 publicly available genome-wide datasets for different types of cell,[2] such as a complete catalogue of annotated human transcripts (identifying different types of RNAs) and functional elements (such as promoters,[3] switches, transcription factors, protein-binding regions, transcription start sites (TSS), and transcriptional repressors such as CTCF,[4] also known as 11-zinc finger protein or CCCTC-binding factor[5]). These data have enabled us to assign biochemical functions to 80 % of the junk genomic portion. The newly identified elements also show a statistical correspondence with sequence variants linked to human diseases, and can thereby guide the interpretation of such variations [10].

The ENCODE project has systematically mapped regions of transcription, transcription factor association, chromatin structure, and histone modification,[6] providing new insights into the mechanisms of gene regulation [9, 30, 40]. The creation of an encyclopedia of DNA elements was made possible by a special enzyme (discovered by Vogelstein and Gillespie in 1979) called DNase, whose hypersensitive sites (DHSs) result in markers for regulatory DNA. In fact, all classes

---

[2]Since genomic sequences are expressed differently in different kinds of cells and tissues, the project first started work on three types of cell (an immature white blood cell line, a leukemia line called K562, and a human embryonic stem cell line) and then extended the analysis to 147 cell types (including the liver cancer cell line HepG2, the laboratory cancer cell line HeLa S3, and human umbilical cord tissue) [34].

[3]See, for example, http://epd.vital-it.ch/.

[4]The human genome contains from 15,000 to 40,000 CTCF-binding sites, depending on cell type.

[5]This protein plays a major role as an enhancer, and was found to bind to three regularly spaced repeats of the core sequence CCCTC.

[6]There is a helpful explorer at http://www.nature.com/ENCODE, which allows one to access papers containing the results of the project and the investigate various topics, thematically organized in to threads. For example, thread 3 is about the definition of a gene, threads 5 and 6 are about RNAs, thread 9 is about long-range looping, and thread 10 is about computational methods.

of *cis*-regulatory elements, including enhancers, promoters, insulators, silencers, and locus control regions, are attached to regulatory factors, which protect genomic DNA (along with the underlying sequence) from cleavage by DNaseI (which leaves nucleotide-resolution footprints of regulatory elements). About 2.9 million DHSs were found and experimentally validated in the ENCODE project.

Annotating functional attributes has revealed novel relationships between chromatin accessibility, histone modification, gene expression, DNA methylation, and regulatory factor occupancy patterns [42]. The DHSs in pluripotent and immortalized cells exhibit higher mutation rates than those in highly differentiated cells, so exposing an unexpected link between chromatin accessibility, proliferative potential, and patterns of human variation.

Of course, none of the discoveries above would have been possible without the computational support of powerful software to process data and a data warehouse. Ad hoc algorithms and specific computational approaches were often helpful for cataloguing and analyzing datasets. As an example, in [17], a machine learning approach was used to systematically identify locations of transcription factors. There are plenty of genomic data analysis tools in the literature, such as genome browsers and visual software for sequencing data. As an example, we shall mention only one of the most recent of these tools, called ggbio [45], which is used to visualize and explore genomics annotations and high-throughput data. The plots produced by ggbio provide views of genomic regions, sequence alignments, splicing patterns, and genome-wide overviews. The methods used by the tool include a combination of statistical functionalities from the R software package[7] and a grammar of graphics.

1.2 Open Questions

The majority of the elements of the human genome have been annotated by ENCODE in terms of their biochemical function. Also, from an informational viewpoint, new insights into the mechanisms of gene regulation have been obtained, with new data about the correspondence between promoter sequences, the binding of specific protein combinations and encoding regions. However, there are still parts of the genome that are not understood, and there is a clear lack of a model which could explain how the major informational processes work to keep the cell alive by means of an interplay of its metabolism, growth, and duplication.

The genome appears to be a stunningly complex system, with many open questions. Namely, it is not clear *how* dark matter communicates with both the close and the far genes that it affects. Promoters and distal elements engage in more than 1,000 long-range looping interactions (up to 120,000 bases upstream of the TSS), which are not well understood [37]. The differences between proximal and distal

---

[7]The ggbio R package is available at http://tengfei.github.com/ggbio/.

regulation (as well as the commensurately larger number of distal binding sites) seem to be a unique feature of human regulation [17]; possibly they reflect the much larger intergenic space in humans than in other organisms.

Furthermore, millions of genetic switches have been found packed into the dark matter. They control which genes are active in a cell, thus determining when those genes are transcribed and the type of cell. There are so many switches, to control only 21,000 genes, and many complex diseases appear to be caused by tiny changes (point mutations or polymorphisms) in hundreds of gene switches. It may happen that one individual gets a disease (such as cancer or depression) while an identical twin sibling remains perfectly healthy. This is related to environmental causes, which may induce genome modification (and DNA damage). A key point for attacking diseases is to control the activation process, that is, the genomic language of signals that lead promoters to activate the corresponding genes, and the whole gene regulation machinery.

In next section, a rather simplified description of the structure and functionality the human genome is given, including the most recent results, in order to better contextualize the main goals of computational genome analysis methods. A specific approach in which genomes are represented by means of dictionary-based indexes is presented in Sect. 3, along with a review of recent results and open problems. This is based on a recent view, where the aim is to analyze collections of genomic k-mers (factors of length k) to discover some sort of "genomic code" underlying the communication between genes and their respective promoters.

2 Genomic Sequence

DNA molecules are chains of nucleotides of four types (A, T, C, G), chemically concatenated by strong covalent phosphodiester bonds and paired by weak hydrogen bonds with complementary strands (according to the Watson–Crick complementation C–G and A–T) having opposite (or antiparallel) orientations. The reading direction of the sequence goes from a dangling phosphoric group, at the $5'$-end, to a dangling hydroxyl group, at the $3'$-end. It may be argued that such a bilinear, antiparallel complementarity in the structure has a logical motivation in the efficiency of DNA replication algorithm [15]. The double strand is twisted into the familiar DNA double helix, which is in turn wrapped around barrels called *histones* (the cylindrical basic proteins of chromatin), so forming a *nucleosome* (a structure about 30 nm long). The nucleosomes are assembled into higher-order structures, called *chromosomes* (the human genome is organized into 46 chromosomes), which are especially visible during cell division. This complex three-dimensional agglomeration of DNA, about 3 m (and three billion bases) long when stretched out, is stuffed into the microscopic nucleus of a cell (about 10^{-5} m in diameter), tightly wound and coiled around itself, just like a ball of wool.

While floating, genomic DNA locally opens up its stitches to allow DNA replication, gene activation, and regulation of genomic activity. In fact, the chromatin is induced to expose a portion of one of its single strand sequences for transcription by means of (cascades of) external signals, which may be activated, inhibited, or promoted by binding of (possibly chemically modified) proteins. The secondary and tertiary structures play an important functional role, because portions which are far apart in the sequence may become close in the 3D structure (and therefore have a reciprocal influence). For example, looping of chromosomes that brings enhancers close to promoters (and promoters close to other promoters) is a mechanism to ensure the expression (or inhibition) of groups of genes that must perform together.

In addition to genes, which are copied into messenger RNA (mRNA), the genome contains large segments which are transcribed into RNA-based regulatory elements, and another portion (including *centromeres* and *telomeres*, located at the middle and the ends of chromosomes) is never transcribed (see Fig. 1). Of this RNA, only mRNA codes for polypeptides; all the other classes are regulatory RNA. Whenever genes in prokaryotes are contiguous segments, eukaryotic genes are segmented into *exons* and *introns* (representing coding and noncoding parts, respectively), which are cut off and reassembled in the nucleus in a splicing phase, in such a way that only a concatenation of (some) exons (called an *exome*) is then translated into a protein. One of the main results achieved in the ENCODE project was the annotation of isoforms for human genes, i.e., the annotation of those genes whose product is due to *alternative splicing*, a key mechanism in which only some of the exons in a gene are assembled to be translated into protein (so that one gene may possibly code for several different proteins). Human genes include *untranslated regions* (UTRs), which are transcribed into mRNA but never translated into protein. The 5′-flanking regions of genes contain a specific TSS sequence where *RNA polymerase* starts the transcription (unlike DNA polymerase, it does not need any primer to start, but just recognizes some specific sites at which it stars and stops). Transcription is activated by molecular signals (such as hormones), which interact with the regions where promoters, switchers, enhancers, and protein-binding sites are located. *Enhancers* may be located upstream, or downstream of the gene they control, or even within it, and increase the rate of transcription.

A typical form of regulation is that due to *transcription factors*, which are (not necessarily site-specific) binding proteins that may assume several chemical states, capable of playing the role of activators or repressors. When repressor or activator proteins bind, then expression of the gene is repressed or activated, respectively. Transcription factors bind in a combinatorial fashion to specify the on and off states of genes; the set of these binding events forms a cell regulatory network. In [17], it was shown that distinct combinations of transcription factors bind at specific genomic locations (the combinatorial coassociation of transcription factors is thus highly context-specific), which are patterns in gene-proximal or gene-distal regions. It is especially interesting to study the unique combination of promoter sites (and transcription factors) for a single gene, since different genes may share the same set of transcription factors. However, the principles that define clearly the relationship between the regulatory elements and (distal) target genes remain unknown.

2.1 The Role of RNA

Eukaryotic cells make many types of primary and processed RNAs, which are found either in specific subcellular compartments or throughout the cell. A complete catalogue of these RNAs is not available yet, and their characteristic subcellular localizations are also poorly understood [8]. There are 8,800 small RNA molecules and 9,600 long noncoding RNA molecules that have recently been defined, all at least 200 bases long, working in different compartments of the cell [34]. Some regulatory RNAs, both long (up to a few thousand bases) and short (less than 200 bases), also have been identified [33], but much remains to be learned about the function(s) of some of them. Here we would like to give a short (absolutely not exhaustive) list of RNAs of interest, all synthesized in the nucleus of eukaryotic cells, at very different rates.

Nuclear gene transcription relies on the work of *small nuclear RNA* (snRNA), which catalyzes the splicing process to remove introns (as a component of the *spliceosomes*), and *small nucleolar RNA* (snoRNA), molecules 60–300 bases long, which produce large precursor molecules ("primary transcripts") that must be processed within the nucleus to form the functional molecules for export to the cytosol.

The expression of mRNA is regulated by tiny RNA molecules (about 22 nucleotides long), called *microRNAs* (miRNAs), by means of several mechanisms, including the famous pathway employing the Dicer protein. In one of these mechanisms, it inhibits the translation of several mRNAs by binding to the UTR region.[8] Since miRNAs bind to RNA, they most likely possess a regulatory role that relies on their ability to compete to bind, independently of their protein-coding function [41]. Promising laboratory experiments have show that miRNAs that inhibit genes needed for metastasis suppress the metastasis of treated human breast cancer cells.[9]

The translation process in the cytoplasm is based on the work of both *ribosomal RNA* (rRNA) to build ribosomes, which process mRNA to assemble proteins, and 32 kinds of *transfer RNA* (tRNA). These are molecules 73–93 nucleotides long that carry one of the 20 amino acids by means of a 3D structure with one loop containing three unpaired bases called an *anticodon*. Most of the amino acids have more than one tRNA responsible for them. Base pairing between an anticodon on a tRNA molecule and the complementary codon allows us to bring the correct amino acid to the ribosome. Each tRNA is the product of a separate gene, called a structural gene as it is not involved in regulation. These genes are needed for morphological or functional traits of the cell.

---

[8]This phenomenon is called *gene silencing*. It is performed by means of *small interfering RNAs* (siRNAs), which are stable when in a double structure, and which activate enzymatic destruction of pre-RNAs.

[9]More details of various roles of RNA may be found at http://users.rcn.com/jkimball.ma.ultranet/BiologyPages/T/Transcription.html#types.

Our current knowledge of genome functioning seems to necessitate a redefinition of the concept of a gene, where the basic unit of heredity should be the transcript – RNA decoded from DNA – rather than, as in the traditional view, a portion of the genome that finally codes for a protein [34]. The genome is exactly the same in all cells of a human individual, but cells differentiate from each other on the basis of their gene activation network, that is, of which RNA is transcribed, on the one hand, and of which proteins are produced, with different functions, on the other hand.

The challenges for the future include the design of a model for the dynamic changes in the regulatory landscape during specific developmental pathways. To understand how regulation processes work, we could observe what happens in the gene activation network during the differentiation from a stem cell to a specific cell. Repression of gene expression by miRNAs appears to be a key mechanism to ensure regulated and coordinated gene expression as cells differentiate along particular paths. Thus miRNAs play an important role as transcription factors in regulating and coordinating the expression of multiple genes, in a particular type of cell, at particular times.

3   Dictionary Based Approaches

In the ENCODE project, the concept of a *functional element* is central, and is defined as a genomic segment that codes for a defined product or displays a reproducible biochemical signature. Furthermore, the distinct distribution of transcribed RNA species across segments suggests that underlying biological activities are captured in segmentation [10].

In an initial, simplified approach, a genome may be linearized, and then analyzed by synthetic methods as just a long sequence of letters. Along with this natural approach, alignment-free methods have recently been developed [43], where systemic views replace local sequence analyses [2]; these methods are based on empirical studies of the frequencies of DNA k-mers in whole genomes [7, 38, 39]. Regularity properties of long strings may be discovered by the application combinatorial algorithms to biological strings [14], seen as collections of k-mers, or of genomic words with different lengths, that may be studied as codes using information theory methods [5]. Once we have a DNA dictionary, we may annotate words by a biochemical description (as has been done in ENCODE project), by statistical information [13, 21], or in terms of formal languages [26, 35]. In order to get an understanding of the complexity of genomes, some work was done on modeling biological sequences by means of formal languages [2, 20, 38, 39], and by approaches inspired by linguistic semantics [18]. A recent development has been the introduction of context-free grammars to formalize design principles for new genetic constructs, by starting from a library of genetic fragments already organized according to their biological function [3, 4]. Related previous dictionary-based studies of genomes may be found in [19, 22], as well as in [29], where entropy measures were employed to estimate the randomness or repeatability of DNA sequences; this has even been done as a function of the organism's "biological complexity" [44].

Among the numerous mathematical and computational genome analysis methods in the literature, we would like here to present one developed as the Infogenomics project. We report some of the results presented in [5], where the definition, computation, and analysis of well-characterized dictionary-based genomic indexes have identified some phenomena of genomic regularity and specificity.

3.1 Infogenomics

Infogenomics is a research project initiated in 2010 at the University of Verona in Italy, where a methodology for genome analysis has been developed [5] for computational investigations of k-mer distributions integrated with informational, dictionary-based indexes. Here we report the main features of this approach, which currently has several different trends in the ongoing research.

We assume an alphabet $\Gamma = \{a, t, c, g\}$, over which a given genome G is a formal string, that is, $G \in \Gamma^*$, and $D_k(G) \subseteq \Gamma^k$ is the k-*genomic dictionary* of all k-mers occurring in the genome G. As in [5], we refer to k-words that appear once as *hapaxes*, to k-words that appear more than once as *repeats*, and to k-words which do not appear in the genome G as *forbidden*.

Cardinalities of the corresponding dictionaries $D_k(G)$, $H_k(G)$, $R_k(G)$, and $F_k(G)$, have been computed and analyzed by varying both the word length k and the genome G. These dictionaries have also been analyzed with respect to other more sophisticated informational indexes (such as k-lexicality, or k-dictionary selectivity, which take into account the number of occurrences, i.e., the *multiplicity*, of single repeats).

In Fig. 2, the variations of the number of genomic k-words (top chart), hapax k-words (second chart), repeated k-words (third chart), and forbidden k-words (bottom chart) in *Homo sapiens* chromosome 19, for $k = 1, \dots, 18$, are shown. The effect of evolutive pressure may be revealed by observing the top chart in Fig. 2, where the number of genomic words longer than 11 is smaller than the number of factors (of the same length) in randomly permuted sequences. Since, by definition, the genomic dictionary is partitioned into hapax and repeat dictionaries for each value of k, the curves in the top chart show values equal to the sum of the values for the curves in the second and third charts. We may notice that the human chromosome 19 contains fewer hapaxes and many more repeats longer than 14 than random permutations do; similar results were found in [5] for a series of variegate, whole genomes. It may be deduced that the words in genomic dictionaries are the result of a selection process, and that genomic repeats longer than 14 seem to be significant from an informational viewpoint.

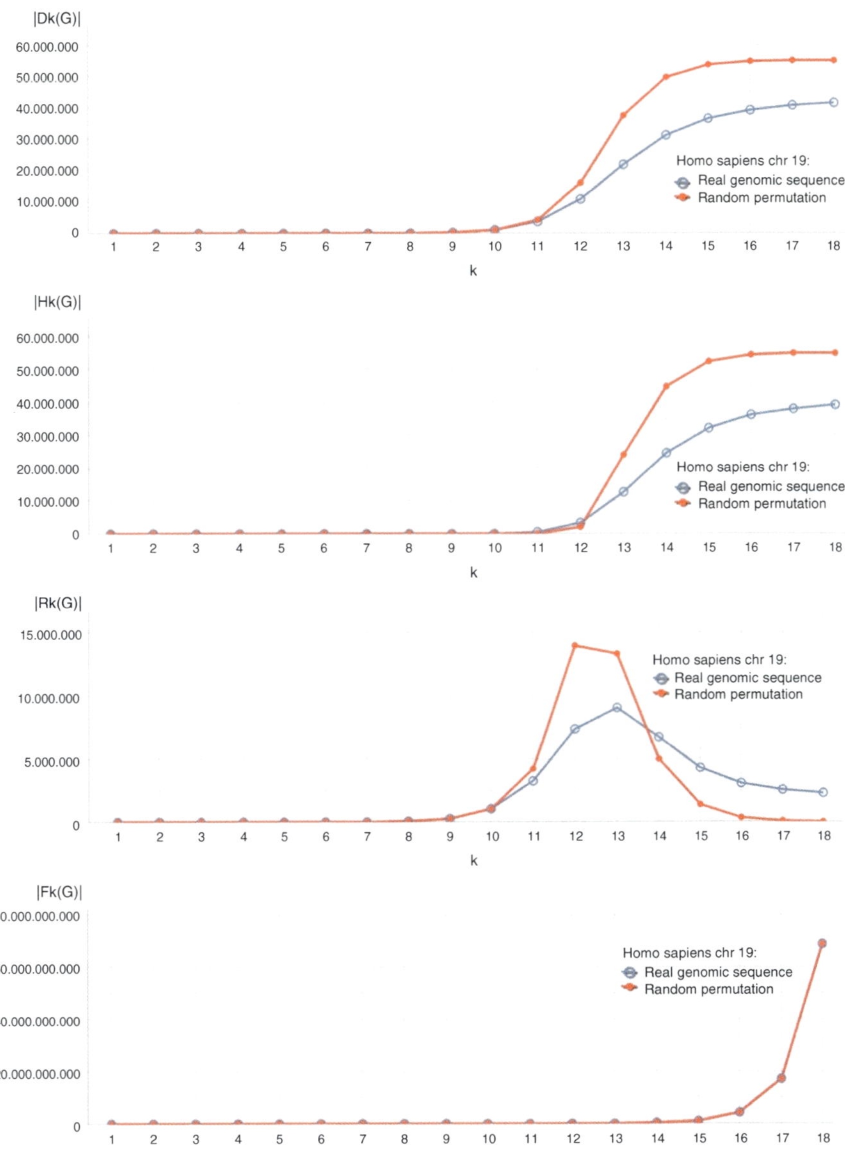

Fig. 2 Variations of $|D_k(G)|$ (chart at *top*), $|H_k(G)|$ (*second chart*), $|R_k(G)|$ (*third chart*), and $|F_k(G)|$ (*bottom chart*), where G is the *Homo sapiens* chromosome 19, for $k = 1, \ldots, 18$. The *blue lines* and *large, empty circles* represent the dictionary size variation for the real genomic sequence, and the *red lines* and *small, filled circles* represent the dictionary size variation for the average of several random permutations of the original genomic sequence

The curves describing the variation of such cardinalities show a similar pattern for all the genomes investigated in [5], whose supplementary material[10] presents other diagrams that represent genomic information, including the k-multiplicity–comultiplicity profiles (i.e., multiplicities of k-words versus amount of k-words having given multiplicities). Quite intuitive genomic indexes, such as the *minimal hapax length* (if we think that the genome itself is an hapax, and that any word including an hapax is itself an hapax, it is reasonable to look for the minimal length), *maximal repeat length* (any subword of a repeat is obviously a repeat), *minimal forbidden length*, and k-*repeat positions* in a genome for a given repeat length k, constitute a starting point of the infogenomic research we developed on 12 different genomic sequences, ranging from *Nanoarchaeum equitans* genome to human chromosomes. The concept of minimal forbidden words, as finite words which are not factors and are such that every proper factor of them appears in the genome, was also studied in [22] and [27], where a linear-time algorithm was given to reconstruct a finite word from a set of its factors (either with a fixed known length and single occurrences [28], or being forbidden factors with a minimum length [12]).

3.2 Some Results

The minimum (forbidden) length of nonappearing factors in a genome G tells us the value of k such that the k-genomic dictionary $D_k(G)$ does not contain Γ^k, that is, the length for which the dictionary contains only some of the possible k-words. It was found empirically that this k is smaller than 12 for all the genomes investigated in [5].

As it can be seen in Table 1, from [5], the cardinalities $|H_{12}(G)|$ and $|D_{12}(G)|$ appear not to be correlated to the length of G, while those of $D_{18}(G)$ and $H_{18}(G)$ increase with the genome length, as expected. The relative number of k-repeats, measured by the value $RD_k = |R_k(G)|/|D_k(G)|$, increases with the genome size and decreases noticeably with the word length from $k = 12$ to $k = 18$. This is due to the fact that 12-repeat words constitute a considerable portion of the 12-genomic dictionaries; in fact, the percentage increases with the genome length (from 11 to 90 %). A surprising result though, is that the 18-repeat-factor ratio is firmly fixed (over all of the genomes) in a very small portion of the 18-genomic dictionary, mostly ranging from 0.01 to 0.07, independently of the genome length. On the other hand, we can observe that the 12-hapax-repeat ratio $HR_{12} = |H_{12}(G)|/|R_{12}(G)|$ is a roughly decreasing function of the genome length, whereas the 18-hapax-repeat ratio (which has much greater values) does not show any evident correlation with it.

---

[10]www.cbmc.it/external/Infogenomics3.

Table 1 Indexes related to $D_{12}(G)$ and $D_{18}(G)$. Genomes are listed in increasing order of in the length (reported in Table 2). The main result here is the transition phase discovered in [5] in the distributions of repeats having a length between 12 and 18

Genome G	$|D_{12}|$	$|H_{12}|$	$|R_{12}|$	RD_{12}	HR_{12}	$|D_{18}|$	$|H_{18}|$	$|R_{18}|$	RD_{18}	HR_{18}
N. equitans	431,046	385,146	45,900	0.11	8.39	489,465	488,802	663	0.001	737.25
M. genitalium	496,194	435,502	60,692	0.13	7.175	569,202	563,045	6,157	0.01	91.44
M. mycoides	646,965	442,836	204,129	0.32	2.169	987,645	913,599	74,046	0.07	12.33
H. influenzae	1,495,701	1,256,043	239,658	0.17	5.240	1,795,492	1,775,531	19,964	0.01	88.93
E. coli	3,478,923	2,675,846	803,077	0.24	3.331	4,557,590	4,518,585	39,005	0.008	115.84
P. aeruginosa	2,949,852	1,799,637	1,150,215	0.39	1.564	6,183,215	6,117,968	65,247	0.01	93.76
S. cerevisiae	6,597,259	3,977,392	2,619,867	0.40	1.518	11,499,795	11,307,098	192,697	0.01	58.67
S. cellulosum	3,863,399	1,924,969	1,938,430	0.51	0.993	12,640,960	12,340,846	300,114	0.02	41.12
H. sapiens chr19	10,735,683	3,359,705	7,375,978	0.69	0.455	41,529,106	39,256,297	2,272,809	0.05	17.27
C. elegans	13,929,915	3,099,744	10,830,171	0.78	0.286	89,444,661	85,157,627	4,287,034	0.04	19.86
D. melanogaster	15,891,212	1,632,045	14,259,167	0.9	0.114	116,446,627	112,977,046	3,469,581	0.02	32.56

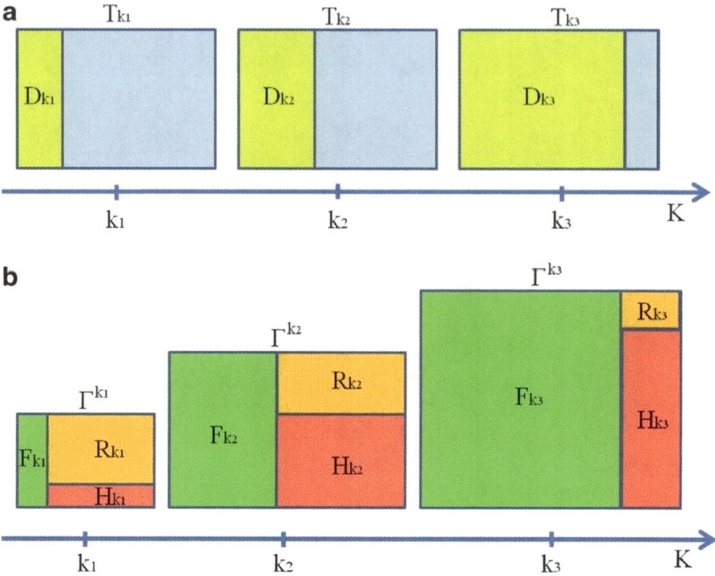

Fig. 3 (**a**) Ratio of the cardinality of all distinct k-words, $|D_k(G)|$, to that of all the words counted with their multiplicity, $|T_k(G)|$, estimated for all the genomes listed in Table 1 and three increasing values of k. This ratio is called the k-lexicality index, and is defined as $L_k(G) = |D_k(G)|/|T_k(G)|$. For all the genomes, this value increases with the word length k. This result is confirmed by the relative increase in the number of k-hapax words in the genomic dictionary. The set of repeats becomes smaller with increasing word length – see (**b**), where rough proportions are visualized among the sizes of k-hapax, k-repeat, and k-forbidden words, which together make up the set Γ^k of all possible k-words

The computational results reported in Table 1 are visualized in Fig. 3. From these data we may conclude that, in general, for greater lengths k, the number of k-repeats is relatively smaller (compared with the dictionary); for $k = 18$, this number is about 0.05 % of the genomic dictionary, independently of the genome size. The fact that genomic words with a length of about 20 are almost all hapax gives an intuitive explanation for the reliability of DNA microarrays.

Interestingly enough, in Fig. 4 one may observe the (very slow) exponential decay of the number of genomic k-repeats with the word length k. Moreover, *all* of the genomes reported in Table 1 proved to have *only one* repeat with the maximum length (and multiplicity 2), and the distance between the two positions (in proportion to the genome length) is reported in Table 2. We may notice that there is no apparent correlation between the genome length and the maximum repeat length (denoted by MR). For all the genomes analyzed in [5], we found that $|R_{\mathrm{MR}}| = 1$ (the maximal repeat is unique, and occurs twice). The occurrences of the maximal repeat turned out to be relatively close together (note that *N. equitans* is a prokaryote, with a circular genome).

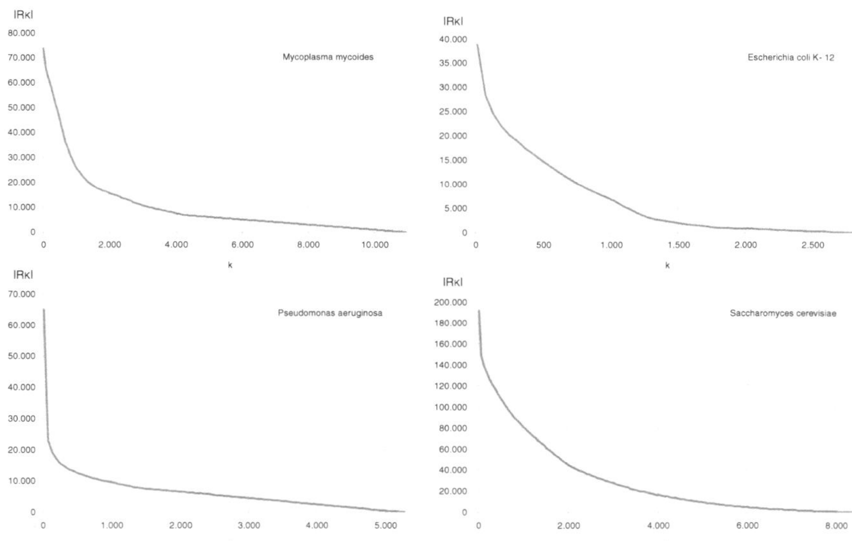

Fig. 4 Length–cardinality repeat distributions of *Mycoplasma mycoides, Escherichia coli, Pseudomonas aeruginosa*, and *Sorangium cellulosum*. For all the genomes in Table 1 an exponential decay of the number of repeats was observed when the MR value was computed [5]

Table 2 MR index (i.e., the maximum repeat length) and MRD, which is the distance between the two occurrences of the maximal repeat, normalized with respect to the genome length

Genome G	Genome length (bp)	MR	MRD/$|G|$ (%)
N. equitans	490,885	139	96.95
M. genitalium	580,076	243	0.15
M. mycoides	1,211,703	10,963	0.019
H. influenzae	1,830,138	5,563	8.05
E. coli	4,639,675	2,815	0.89
P. aeruginosa	6,264,404	5,304	12.37
S. cerevisiae	12,070,898	8,375	0.07
S. cellulosum	13,033,779	2,720	27.68
H. sapiens chr19	63,800,000	2,247	0.02
C. elegans	100,267,632	38,987	0.10
D. melanogaster	129,663,327	30,892	0.02

According to preliminary results, presented in [16] and discovered from computational analysis of the genomes of three specific organisms, the longest (i.e., most significant) repeats seem to be located in genic regions, which is a good, basic motivation to investigate the gene networks defined in the next section.

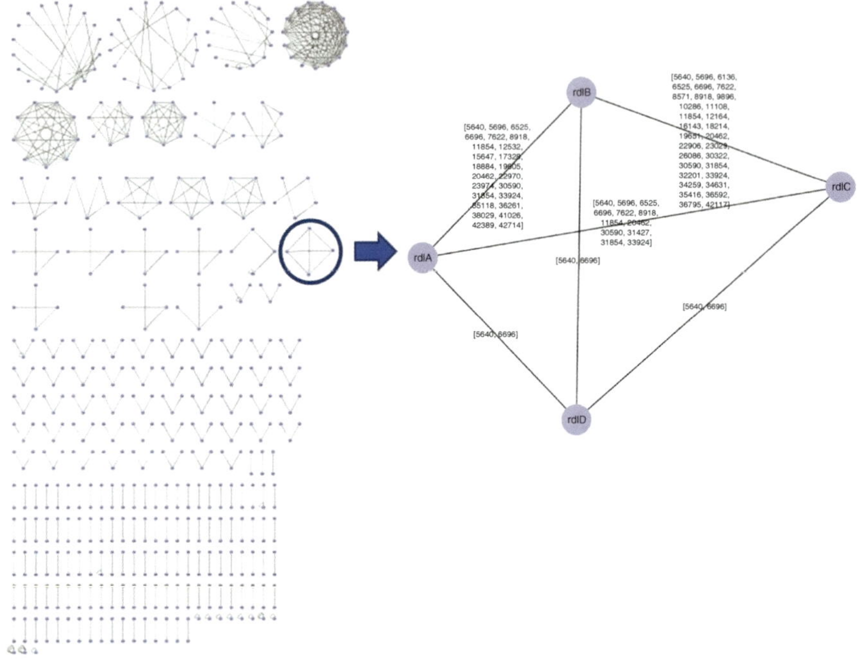

Fig. 5 Repeat-sharing gene network N_{18} of *E. coli*. This genome has a high percentage (89 %) of genes. The four genes in the figure on the right are all connected (thus forming a clique) by a few repeats labeling half of the connections and quite a high number of common repeats labeling the others [5]

3.3 *Repeat-Sharing Gene Networks*

Cells have evolved an intricate system of interconnections, which is often modeled by biological networks. All retrieved networks, whether experimentally tested or just predicted *in-silico*, are stored in databases such as KEGG (Kyoto Encyclopedia of Genes and Genomes) [31].

Here we give a definition of genetic network which was inspired by our repeats analysis and by the biological evidence of "communication among genes" observed in recent experiments, where common substrings seem to compete for short (around 20 bp) miRNA sequences [36, 41]. We consider all genes of an organism as a set of (labeled) nodes, and we connect each pair of nodes by an edge if the associated genes contain at least one k-word as a common substring. We label each edge by the set of all the k-words shared by the two genes, and delete from the network all the gene-nodes which are not connected by any edge. We call this (k-parameterized) graph a *k-repeat-sharing gene network,* $N_k(G)$. As an example, the gene network for *Escherichia coli* is reported in Fig. 5.

When k is increased, the nodes and edges of N_k decrease in number, until the network disappears. However, the network persists to high values of k, because the longest repeats are often located in genes [16]. There is a break point at $k = 18$, where the observed networks pass from having a large connected component to being a set of clusters, which vanishes as k is increased. Some work in progress [6] focuses on the analysis of these gene networks, starting from the computation of the maximum degree, maximum label weight, and number of complete clusters. According to preliminary results, for *N. equitans*, *E. coli*, and *S. cerevisiae*, the genes which are present in the networks N_k for $k > 30$ are all paralogs or pseudogenes, and cliques in the network appear to have both informational relevance and a biological meaning. Namely, highly connected genes are involved in important biological pathways, such as DNA repair and replication.

4 Conclusion and Open Problems

The ENCODE and the Infogenomics projects share a dictionary-based view of genome analysis and an effort to gather global, holistic information contained in genomes, in order to understand how genomes orchestrate coordinated processes to maintain cellular metabolism, growth, and replication. Recent advances in computational genomics in these contexts have opened up new perspectives on genome analysis, focused on the understanding of a genomic code which explains how promoters communicate with their corresponding (proximal or distal) genes; the aim is to generate a sequence-based dynamical model for describing mechanisms of gene activation and regulation.

We expect that systemic alignment-free methods will be helpful in cases where alignment methods fail. For example, sequence similarities among coding regions which produce proteins having very similar functions could be revealed by a dictionary-based score, even in those cases (well known in the literature) where the sequence alignment score is very low.

In future work, we would like to apply the Infogenomics methodology: to genomes of patients affected by some disease (as individual phenotypic variations are influenced by sequence variation across the entire genome), to the transcriptome (via RNA sequence data), and to better understand the intricate links between promoters and the corresponding genes. A systematic analysis of transcripts is essential to identify regulatory regions, since the transcript may be considered as the basic unit of heredity. Moreover, the bipartition of a genomic dictionary into hapax and repeat words corresponds to an interesting genomic representation, by which a string reconstruction problem may be seen as a graph problem. Some preliminary results have been obtained from a variant of this approach where, given a genome, we look for a characterization of other genomes that have the same set of k-hapaxes.[11]

---

[11]Manuscript in preparation, available on request from the first author.

Repeat-sharing gene networks (where genes are linked if they have common factors) may be of interest for recognizing similarities among genes, such as conserved miRNA binding sites; these could be investigated by further graph theory methods [1] or defined differently. For example, we could associate nodes to the exomes of genes, that is, sequences of the corresponding mature mRNA (where the splicing of exons has been already performed, see Fig. 1). On the other hand, a study of the networks described above, defined on generic nucleotidic sequences, could include all cases of possible genomic variations, such as single-nucleotide polymorphisms, which are of great biomedical interest. Indeed, understanding how alterations in noncoding regions contribute to human diseases may result in new DNA engineering interventions, or personalized drugs [24, 25]. The first challenge in this context relates to the mechanism connecting such changes to cancer growth.

A further application of the Infogenomics methodology will concern the computation of intersections of genomic dictionaries, to find evolutionarily conserved motifs among genomes. As a test for functionality of DNA, one could evaluate which sequences are conserved between species: DNA sequences which are not conserved among multiple populations suggest that these regions are no longer functional.

Acknowledgements The author is very grateful to Vincenzo Manca, who designed the Infogenomics project, for very stimulating discussions on the subject of the manuscript, as well as to Alessio Milanese and Alberto Castellini for providing her with Figs. 1 and 3, respectively. These people carefully reviewed the chapter, together with Vincenzo Bonnici, Ricardo Henrique Guiraldelli, Zsuzsanna Lipták, and Luca Marchetti. The author is thankful to all of them and to the anonymous referees for relevant comments, pointers to the literature, and specific suggestions that have much improved the initial version of this work.

References

1. T. Aittokallio, B. Schwikowski, Graph-based methods for analysing networks in cell biology. Brief. Bioinform. **7**(3), 243–255 (2006)
2. V. Brendel, H. Busse, Genome structure described by formal languages. Nucleic Acids Res. **12**(94), 2561–2568 (1984)
3. Y. Cai, B. Hartnett, C. Gustafsson et al., A syntactic model to design and verify synthetic genetic constructs derived from standard biological parts. Brief. Bioinform. **23**(20), 2760–2767 (2007)
4. Y. Cai, M. Lux, L. Adam et al., Modeling structure-function relationships in synthetic DNA sequences using attribute grammars. PLoS Comput. Biol. **5**(10), e1000529 (2009)
5. A. Castellini, G. Franco, V. Manca, A dictionary based informational genome analysis. BMC Genomics **13**(1), 485 (2012)
6. A. Castellini, G. Franco, A. Milanese, A genome analysis based on repeat sharing gene networks (Submitted)
7. B. Chor, D. Horn, N. Goldman et al., Genomic DNA k-mer spectra: models and modalities. Genome Biol. **10**, R108 (2009)

8. S. Djebali, C. Davis, A. Merkel et al., Landscape of transcription in human cells. Nature **489**, 101–108 (2012)
9. X. Dong, M. Greven, A. Kundaje et al., Modeling gene expression using chromatin features in various cellular contexts. Genome Biol. **13**, R53 (2012)
10. I. Dunham, A. Kundaje, S. Aldred et al. (the ENCODE Project Consortium), An integrated encyclopedia of DNA elements in the human genome. Nature **489**, 57–74 (2012)
11. J.R. Ecker, W. Bickmore, I. Barroso et al., Genomics: ENCODE explained. Nature **489**, 52–55 (2012)
12. G. Fici, F. Mignosi, A. Restivo et al., Word assembly through minimal forbidden words. Theor. Comput. Sci. **359**, 214–230 (2006)
13. Y. Fofanov, Y. Luo, C. Katili et al., How independent are the appearances of n-mers in different genomes? Bioinformatics **20**(15), 2421–2428 (2008)
14. G. Franco, Biomolecular computing – combinatorial algorithms and laboratory experiments. Doctoral thesis, Department of Computer Science, University of Verona, 2006
15. G. Franco, V. Manca, An algorithmic analysis of DNA structure. Soft Comput. **9**, 761–768 (2005)
16. G. Franco, A. Milanese, An investigation on genomic repeats. LNCS **7921**, Springer, 149–160 (2013)
17. M. Gerstein, A. Kundaje, M. Hariharan et al., Architecture of the human regulatory network derived from ENCODE data. Nature **489**, 91–100 (2012)
18. M. Gimona, Protein linguistics – a grammar for modular protein assembly? Nature **7**, 68–73 (2006)
19. G. Hampikian, T. Andersen, Absent sequences: nullomers and primes. Pac. Symp. Biocomput. **12**, 355–366 (2007)
20. T. Head, Formal language theory and DNA: an analysis of the generative capacity of specific recombinant behaviors. Bull. Math. Biol. **9**(6), 737–759 (1987)
21. J. Herold, S. Kurtz, R. Giegerich, Efficient computation of absent words in genomic sequences. BMC Bioinform. **9**, 167 (2008)
22. H. Hoogeboom, W. Kosters, Substring differences in genomes, in *Proceedings of the Benelux Bioinformatics Conference (BBC)*, Maastricht, ed. by R. Armañanzas, Y. Saeys, I. Inza, M. García-Torres, Y. Van de Peer, C. Bielza, P. Larrañaga 2008, p. 62
23. C. Jacq, J.R. Miller, G.G. Brownlee, A pseudogene structure in 5S DNA of Xenopus laevis. Cell **12**(1), 109–20 (1977)
24. J. Leja, H. Dzojic, E. Gustafson et al., A novel chromogranin-A promoter-driven oncolytic adenovirus for midgut carcinoid therapy. Clin. Cancer Res. **13**, 2455–2462 (2007)
25. J. Leja, B. Nilsson, D. Yu et al., Double-Detargeted oncolytic adenovirus shows replication arrest in liver cells and retains neuroendocrine cell killing ability. PLoS One **5**(1), e8916 (2010)
26. M. Lynch, *The Origins of Genome Architecture* (Sinauer Associates, Sunderland, 2002)
27. F. Mignosi, A. Restivo, M. Sciortino, Forbidden factors and fragment assembly. RAIRO Theor. Inform. Appl. **35**(6), 565–577 (2001)
28. F. Mignosi, A. Restivo, M. Sciortino, Words and forbidden factors. Theor. Comput. Sci. **273**, 99–117 (2002)
29. G. Navarro, V. Mäkinen, Compressed full-text indexes. ACM Comput. Surv. **39**(1), article 2 (2007)
30. S. Neph, J. Vierstra, A. Stergachis et al., An expansive human regulatory lexicon encoded in transcription factor footprints. Nature **489**, 83–90 (2012)
31. H. Ogata, S. Goto, K. Sato et al., KEEG: Kyoto encyclopedia of genes and genomes. Nucleic Acids Res. **27**(1), 29–34 (1999)
32. B. Pei, C. Sisu, A. Frankish et al., The GENCODE pseudogene resource. Genome Biol. **13**, R49 (2012)
33. L. Poliseno, Pseudogenes: newly discovered players in human cancer. Sci. Signal. **5**(242), re5 (2012). doi:10.1186/gb-2012-13-8-r77
34. E. Pennisi, ENCODE project writes eulogy for junk DNA. Science **337**(6099), 1159–1161 (2012)

35. J. Percus, *Mathematics of Genome Analysis*. Cambridge Studies in Mathematical Biology (Cambridge University Press, Cambridge, 2007)
36. L. Poliseno, L. Salmena, J. Zhang et al., A coding-independent function of gene and pseudogene mRNAs regulates tumour biology. Nature **465**(7301), 1033–1038 (2010)
37. A. Sanyal, B. Lajoie, G. Jain et al., The long-range interaction landscape of gene promoters. Nature **489**, 109–113 (2012)
38. D. Searls, String variable grammar: a logic grammar formalism for the biological language of DNA. J. Logic Program. **24**, 73–102 (1995)
39. D. Searls, The language of genes. Nature **420**, 211–217 (2002)
40. M. Spivakov, J. Akhtar,P. Kheradpour et al., Analysis of variation at transcription factor binding sites in Drosophila and humans. Genome Biol. **13**, R49 (2012)
41. Y. Tay, L. Kats, L. Salmena et al., Coding-independent regulation of the tumor suppressor PTEN by competing endogenous mRNAs. Cell **147**, 344–357 (2011)
42. R. Thurman, E. Rynes, R. Humbert et al., The accessible chromatin landscape of the human genome. Nature **489**, 75–82 (2012)
43. S. Vinga, J. Almeida, Alignment-free sequence comparison – a review. Bioinformatics **19**(4), 513–523 (2003)
44. S. Vinga, J. Almeida, Local Renyi entropic profiles of DNA sequences. BMC Bioinform. **8**, 393 (2007)
45. T. Yin, D. Cook, M. Lawrence, ggbio: an R package for extending the grammar of graphics for genomic data. Genome Biol. **13**(8), R77 (2012)

The Sequence Reconstruction Problem

Angela Angeleska, Sabrina Kleessen, and Zoran Nikoloski

Abstract Despite recent advances, assembly of genomes from the high-throughput data generated by the next-generation sequencing (NGS) technologies remains one of the most challenging tasks in modern biology. Here we address the sequence reconstruction problem, whereby, for a given collection of subsequences or factors, one has to determine the set of sequences compliant with the collection. First, we give a brief review of sequencing technologies, along with an exposition of the advantages and shortcomings of the existing algorithmic approaches to sequence assembly. In addition, we enumerate some properties of subsequences, which have been overlooked in the existing heuristic solutions despite their effect on the quality of the assembly. We then give an overview of the sequence reconstruction problem from a language-theoretic perspective, and present a comprehensive review of theoretical results that may prove relevant to the genome assembly problem. Finally, we outline a new optimization-based formulation which casts the sequence reconstruction problem as a quadratic integer programming problem.

1 Introduction

Next-generation sequencing (NGS) technologies of ever-increasing quality offer the possibility to use the plethora of sequence data that results from them to assemble whole genomes. Currently, tremendous amounts of time, money, and computational resources are being invested in genome sequencing. The existing

A. Angeleska (✉)
The University of Tampa, Tampa, FL, USA
e-mail: aangeleska@ut.edu

S. Kleessen · Z. Nikoloski
Systems Biology and Mathematical Modeling Group, Max Planck Institute for Molecular Plant Physiology – Golm, Potsdam, Germany
e-mail: Kleessen@mpimp-golm.mpg.de; Nikoloski@mpimp-golm.mpg.de

N. Jonoska and M. Saito (eds.), *Discrete and Topological Models in Molecular Biology*, 23
Natural Computing Series, DOI 10.1007/978-3-642-40193-0_2,
© Springer-Verlag Berlin Heidelberg 2014

NGS technologies cannot be used to sequence entire genomes at once; in fact, only short reads (i.e., substrings) are sequenced, and these must then be assembled to form the original genome. The resulting reads differ in length, depending on the technology used. To increase the coverage, the genome to be investigated is first copied several times. These copies are broken up randomly into fragments, which are then sequenced to produce reads. The aim of genome assembly is then to obtain the entire genome sequence with the help of the overlaps of all the reads while taking into account the particularities of the NGS technology used.

The contribution of this chapter is threefold: (1) to provide a brief review and history of existing methods for genome assembly, while critically comparing and contrasting them with existing results from language theory; (2) to offer a source for interdisciplinary understanding of the genome assembly problem, by presenting a comprehensive overview of the relevant mathematical results and pointing out problems where mathematical endeavor may improve the existing solutions; and (3) to formulate an optimization-based approach that provides a different framework for the problem at hand.

2   A Brief Review of Sequencing Technologies

In this section, we present a short background on DNA and whole-genome sequencing. In addition, we present a brief historical overview of the commonly used sequencing technologies, including their advantages and drawbacks. For a detailed review of next-generation sequencing technologies, we refer the reader to [16].

2.1   The Basics of Sequencing

DNA sequencing is a process used to determine the order of nucleotides (adenine, A; guanine, G; cytosine, C; and thymine, T) in a strand of a DNA molecule. Knowing exact DNA sequences is necessary for research in molecular biology, and such knowledge has resulted in numerous breakthrough applications in medicine, forensics, and other fields. The process of DNA sequencing of the full genome of an organism is referred to as complete or *whole-genome* sequencing. Whole-genome sequencing includes the sequencing of chromosomal, mitochondrial, and (in plants) chloroplast DNA.

Owing to the limited power of even the most modern technologies, a whole genome or a long DNA strand cannot be sequenced affordably as a whole piece in a reasonable time and with the desired quality. Therefore, most of the existing techniques shatter the molecule into millions of smaller pieces (anywhere between 20 and 2,000 bp), amplify each of them by the polymerase chain reaction (PCR), and then run them through sequencers. The resulting "small" DNA sequences are called *reads*, and are then assembled into longer segments and, eventually, genomes.

This process is known as *genome assembly* and is performed by the use of various algorithms known as *assembly algorithms* (with their respective implementations referred to as *assemblers*).

The sequencing can be de novo, where new sequences are assembled without a reference sequence, or *mapping-based*, where the method relies on a reference sequence. There are two types of assembly strategies, depending on the type of sequencing. In the de novo assembly approach, sequence reads are compared with each other, and then assembled into longer segments called *contigs* by using the overlaps of the sequences. The reference-based assembly approach involves mapping each read to a reference genome sequence [27].

To ensure that all nucleotides from the sequenced DNA are read, the sequencing process must have a certain *depth*. The average number of reads that contain a given nucleotide is called the *sequencing coverage or the depth*. It is denoted by C, and can be calculated from $C = NL/G$, where N is the total number of reads, L is the average read length, and G is the size of the whole genome. A higher-depth sequencing provides greater accuracy of the assembly [25].

2.2 The History of Sequencing Technologies

The ultimate goal of the genome sequencing and assembly process is to correctly determine the complete genome sequence of an organism and to characterize and annotate the protein-coding genes.

An understanding of organismal genomes is expected to revolutionize molecular medicine, pharmaceutics, and environmental studies [11]. The first genome to be completely sequenced was bacterial, and this was done in 1995 [8]. The human genome was sequenced de novo in the Human Genome Project, which started in 1990, and was completed in 2003. The total cost of the project has been estimated at $3 billion [38]. To illustrate the advancement of genome assembly techniques since then, we may mention that at the beginning of 2012, Life Technologies announced a sequencer designed to sequence an individual human genome in 1 day for a cost of $1,000 [9]. Despite the fact that the latter example is not de novo sequencing, it shows a significant improvement in sequencing technologies, leading to a decrease in the time invested and the cost.

Historically, sequencing technologies can be divided into several categories. Before 1980, the sequencing procedure was performed manually [1]. Then, the Sanger sequencing method was adopted in laboratories around the world and was the prevalent method used for two decades. The capillary sequencer machine incorporated Sanger's sequencing method, for which Sanger won a Nobel Prize in 1980 [32]. This method was the main method used in the Human Genome Project.

The "Next-generation sequencing" (NGS) technologies are widely used today. Some of the sequencing methods commonly used are based on the Roche/454, Illumina, and SOLiD platforms [24]. More details of these platforms can be found in recent reviews [19, 31].

The Illumina reads are 50–150 bp long with up to six billion reads per run, where as 454 can manage 400 bp but with lower throughput. Therefore, these technologies are fast and cheap, and have relatively high coverage depth. Current second-generation sequencing technologies produce read lengths ranging from 35 to 400 bp. The shortcomings of the NGS technologies are short reads and high data volume, which often lead to difficulties in assembly. Correct assembly and mapping are computationally challenging, as short segments create ambiguities in alignment and in genome assembly, which, in turn, can produce errors when the results are interpreted.

Recently, new technologies such as "single-molecule real-time technology" (SMRT) and the nanopore sequencing method [19] have been proposed. The general characteristics of these methods are a shorter DNA preparation time before sequencing, capturing the nucleotide signal in real time, and longer reads. The error rates of single-molecule reads are high (less than 85 % nucleotide accuracy). To overcome this problem, a new hybrid method that combines these technologies with NGS and yields very good accuracy for long reads has been developed [14].

3   Three Categories of Assembly Algorithms

Although NGS technologies provide the possibility of fast and cheap sequencing, they result in a computationally challenging assembly problem. There are three basic approaches to designing assembly algorithms: the greedy-algorithm approach; overlap–layout–consensus methods, relying on overlap graphs; and de Bruijn methods, based on de Bruijn graphs.

3.1   Greedy Assembly

Two reads are considered to overlap if a prefix of one nucleotide sequence is the same as or very similar to a suffix of the other. The quality of an overlap is quantified by the size of the overlapping sequence and the percentage of matching base pairs in the overlapping region. All of the greedy assemblers use a variant of the greedy-algorithm approach. Some of greedy algorithms start by iteratively joining the reads with the best overlaps, forming multiple contigs. Others extend a given read to a contig by consecutively attaching the read that has the best overlap with the previous one. This is performed at both the 3′ and the 5′ end of the read until no further extensions are possible. An unassembled read is then chosen to be the start of a new contig. In this selection process, priority is given to reads of better quality. The greedy approach was first used on Sanger data sets by assemblers such as TIGR [35] and CAP3 [10]; a second approach was more recently applied to short-read data assemblers (e.g., SSAKE [37], VCAKE [13], and SHARCGS [5]).

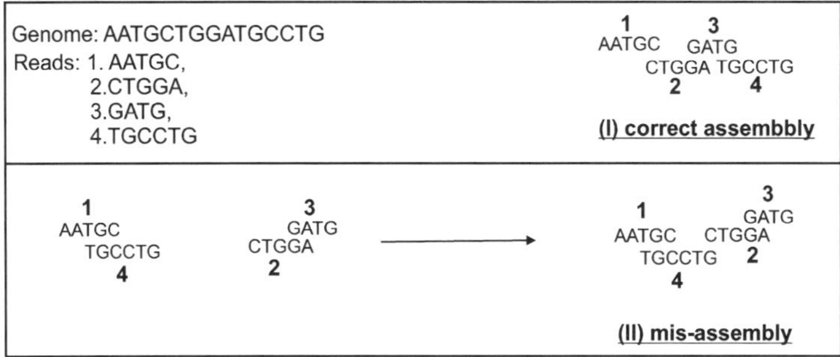

Fig. 1 Genome misassembly by greedy algorithm. (I) A small genome is to be assembled from four reads, labeled 1, 2, 3, and 4. The correct assembly is 1–2–3–4. (II) The greedy algorithm assembles 1 and 4 first, since they have the best overlap, of three nucleotides, and then 2 and 3 are assembled. As a result, the misassembled genome 1–4–2–3 is obtained

These assembly algorithms, like every other greedy algorithm, aim at determining the optimal global assembly by finding locally optimal assemblies at each step. This is a simple assembly strategy and very easy to implement, but it does not necessarily result in the optimal solution. In addition, a greedy assembly might also lead a misassembly, as illustrated in Fig. 1.

3.2 Overlap–Layout–Consensus Assembly

The *overlap–layout–consensus (OLC)* assembly approach is graph-based. A *graph* is a structure composed of a set of *vertices* connected by *edges*. If the edges are directed, the graph is called *directed*. If a vertex u is an endpoint of an edge e, we say that e is an incident edge on u. A *path* in a graph is an alternating sequence of vertices and edges $u_1, e_1, u_2, e_2, \ldots, e_{k-1}, u_k$ such that each edge e_i in the sequence is incident on both u_i and u_{i+1}. A *Hamiltonian path* is a path that visits every vertex of the graph exactly once. Examples of a graph, a path, and a Hamiltonian path are given in Fig. 2 (I), (II), and (III), respectively.

The assembly of a genome by the OLC approach method can be seen as a mathematical problem of finding a Hamiltonian path in a directed graph. The graph structure used by OLC assemblers is called an overlap graph. The vertices of the overlap graph represent the reads, so that the graph has as many vertices as there are reads. The first (overlap) step in an OLC assembly is the identification of overlapping reads by pairwise comparison. The second (layout) step is the construction of the graph, such that two vertices are connected with an edge if the corresponding reads overlap. The direction of the edge is from the vertex that contains the overlap as a suffix towards the vertex that contains the overlap as a prefix. Each path in such a graph constructed in this way corresponds to an assembled contig, and each Hamiltonian path corresponds to a completely

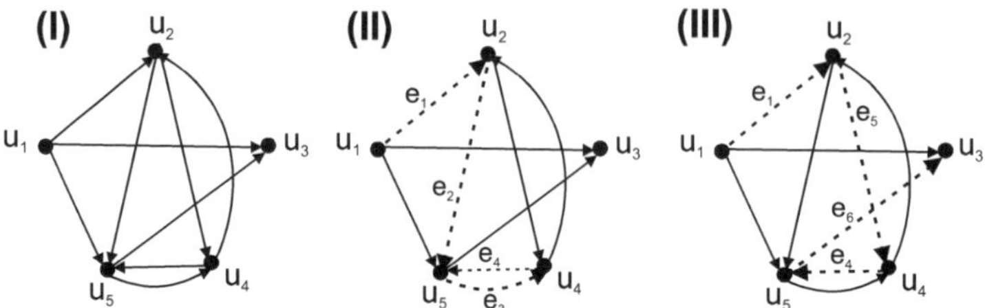

Fig. 2 (I) A directed graph with five vertices, labeled u_1, u_2, u_3, u_4, u_5. (II) The alternating sequence of vertices and edges $u_1 e_1 u_2 e_2 u_5 e_3 u_4 e_4 u_5$ is a path. (III) The alternating sequence of vertices and edges $u_1 e_1 u_2 e_5 u_4 e_4 u_5 e_6 u_3$ is a Hamiltonian path

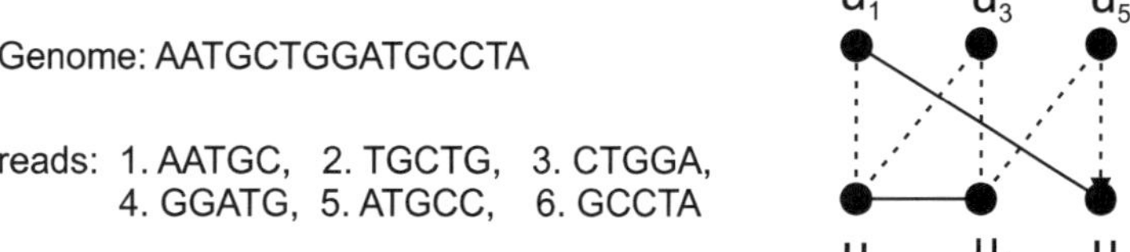

Fig. 3 Overlap graph. The graph corresponds to the overlap graph of the sequence used in Example 1

assembled genome. Finding a Hamiltonian path is then the third step in the process. A simplified example to illustrate the OLC assembly process is given in Fig. 3.

Example 1. This example refers to the overlap graph shown in Fig. 3. A small genome $\Gamma = AATGCTGGATGCCTA$ is to be assembled from reads 1, 2, 3, 4, 5, and 6. Each vertex of the graph corresponds to a read. Two vertices are connected by a directed edge if the corresponding reads have an overlap of at least two nucleotides. There is a Hamiltonian path 1–2–3–4–5–6 in the overlap graph. By listing the reads corresponding to each vertex on this path (including only a single copy of each overlap) in consecutive order as they appear in the path, one obtains the assembled genome.

Two well-known assemblers that use the OLC method are Newbler [22] and Celera Assembler [26]. The OLC assembly method is exact, and therefore more accurate than the greedy approach. However, finding Hamiltonian paths in a graph is, in general, an NP-hard problem, for which there exists no efficient algorithm. It should also be pointed out that the resulting graphs include millions of vertices, so even efficient heuristics need to scale linearly with the order and size of the graph. Therefore, this approach might be convenient for smaller number of reads (i.e., larger read lengths). As the reads become shorter, as in the case of NGS data, the use of OLC assembly techniques becomes prohibitive. Some of the difficulties can be resolved by considering Eulerian instead of Hamiltonian paths and creating de Bruijn instead of overlap graphs, as discussed in the following section.

Fig. 4 Eulerian path. The
sequence $u_1e_1u_2e_2u_4e_3u_5e_4u_4e_5u_2e_6u_5e_7u_3e_8u_1$
is an Eulerian path in the
graph

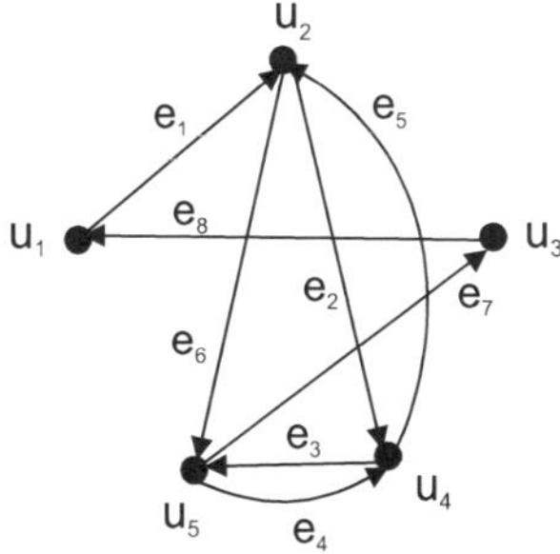

3.3 De Bruijn Assembly

De Bruijn graphs were introduced in the 1940s by the Dutch mathematician
Nicholas Govert de Bruijn long before their application in genome assembly. De
Bruijn was interested in the following "string reconstruction problem": Find the
shortest superstring that contains as substrings all possible strings of a given length k
over an arbitrary alphabet. He solved the problem by encoding it in directed graphs,
later known as de Bruijn graphs.

For every sequence of length $k - 1$, there is a vertex in the de Bruijn graph. Two
vertices u_1 and u_2 are connected by an edge directed from u_1 to u_2 if there is a k-mer
that has the $(k - 1)$-mer u_1 as its prefix and the $(k - 1)$-mer u_2 as its suffix. The
k-mer composed of u_1 and u_2 is the label of the edge connecting them. Therefore,
traversing each edge in the graph yields a path whose edge labels give the smallest
sequence that has all possible k-mers as subsequences.

A path that contains every edge of the graph exactly once is called an *Eulerian
path*. Hence, finding an Eulerian path in the de Bruijn graph solves the sequence
reconstruction problem. An example of an Eulerian path is depicted in Fig. 4.

De Bruijn graphs can readily be used to solve the genome assembly problem (see
[12, 28]). If we consider an alphabet $\{A, C, G, T\}$ and, instead of using all possible
k-mers as edges, we use only those generated from the reads, we can construct a de
Bruijn graph as described above. An Eulerian path in such a graph corresponds to
an assembled DNA sequence.

This approach has been used in many assemblers, such as Velvet [39],
ABySS [34], and AllPaths [2]. For more details of how the de Bruijn graph was used
in these "DBG" assemblers, see [25]. A simple example of DBG genome assembly
is illustrated in Fig. 5 and described as follows.

Example 2. Consider the genome $\Gamma = AATGCTGGATGCCTA$. The set of reads is

$$\{AATGC, TGCTG, CTGGA, GGATG, ATGCC, GCCTA\}.$$

We construct a graph with a vertex set composed of every 3-mer obtained from the
reads

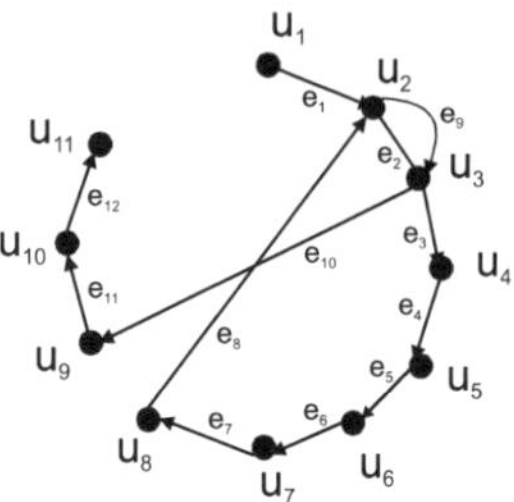

Fig. 5 De Bruijn graph. The graph on the *right* is the de Bruijn graph that is used to assemble the genome *AATGCTGGATGCCTA*

$$\{u_1 = AAT, u_2 = ATG, u_3 = TGC, u_4 = GCT, u_5 = CTG, u_6 = TGG,$$

$$u_7 = GGA, u_8 = GAT, u_9 = GCC, u_{10} = CCT, u_{11} = CTA\}.$$

A directed edge from vertex v to vertex w is included if v is a prefix and w is a suffix of a 4-mer that belongs to a read. The de Bruijn graph depicted in Fig. 5 is obtained. The path $u_1e_1u_2e_2u_3e_3u_4e_4u_5e_5u_6e_6u_7e_7u_8e_8u_2e_9u_3e_{10}u_9e_{11}u_{10}e_{12}u_{11}$ is an Eulerian path which corresponds to the assembled genome Γ.

Finding an Eulerian path in a graph is a computationally easy problem, rendering the de Bruijn approach more applicable than the Hamiltonian-path approach used in the OLC method. On the other hand, one of the drawbacks of the de Bruijn approach is the loss of information caused by decomposing a read into a path of k-mers [33].

3.4 Summary of Advantages and Disadvantages of the Popular Assembly Methods

The quality of DNA sequencing and assembly can be quantified by a few characteristics, including the speed, accuracy, and cost of the sequencing, the computational complexity of the assembly, and the accuracy of the assembly. All of these categories are interlaced with each other, and also all of them depend on the sequencing method, the biotechnology used, the computational approach, and the software. Unfortunately, there is no solve-it-all assembly or sequencing technique for the genome assembly problem. All of the methods described above might perform very well in terms of one characteristic, but not another.

The Sanger sequencing method produces reads ranging from 800 to 900 bp in length, which are much longer than the NGS read lengths (50–150 bp). Therefore, the assembly process is less complex, and an approach such as the OLC method might yield reasonable computational complexity. But the cost of generating Sanger data is much higher, and this technology is more time-consuming. Next-generation

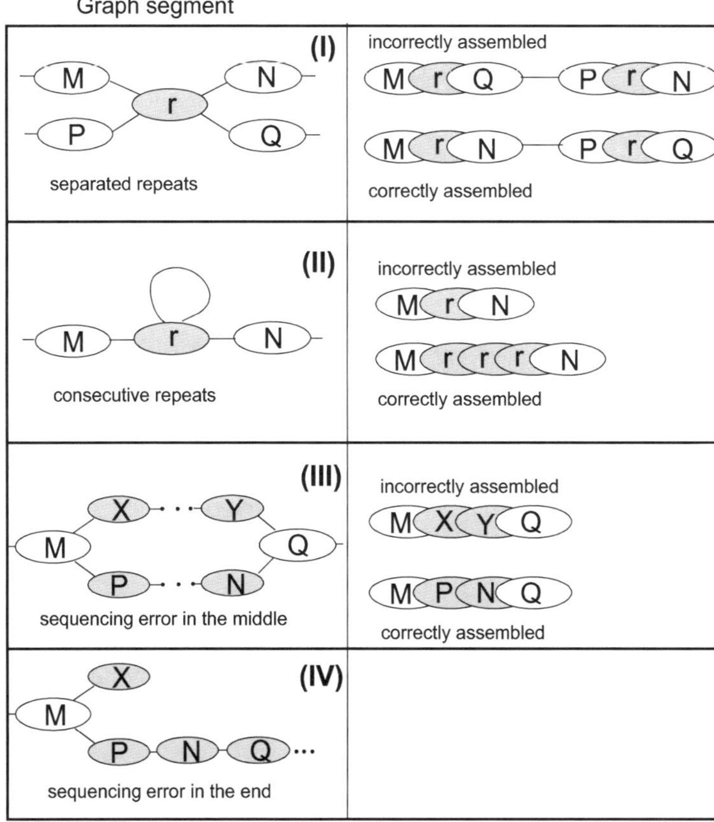

Fig. 6 Errors caused by repetition. On the *left-hand side*, a portion of an assembly graph that contains a repetitive segment is shown. On the *right-hand side*, the correct assembly is depicted, along with a possible incorrect assembly

sequencers can read base pairs at a thousandth of the cost of Sanger sequencers and faster, but the size of the reads makes the computation very laborious and error-prone. Therefore, there is no single assembler that will perform well on any type of data set. For example, although the greedy approach is computationally feasible and might work well on small genomes, its output may be far from the optimal. On the other hand, the robustness of the OLC method comes at the price of its being more computationally intensive, especially when the number of reads is large, which is the case for the NGS technologies. The de Bruijn approach can easily be implemented, but suffers from loss of information when the reads are chopped up into k-mers [30].

All existing assemblers and assembly approaches face the same problems, which are largely due to sequencing errors and repeats in the genome. Sequencing errors in the middle of the set of reads might produce "bubbles" (see Fig. 6 (III)) in the graph structure used for assembly. This implies that there is no Hamiltonian path that traverses all vertices in the region and no Eulerian path that includes the edges in both branches. On the other hand, if there is a sequencing error (or drop in coverage) at the end of the reads (see Fig. 6 (IV)), the graph has a "spur" – a short dead end that cannot belong to any path, but complicates the graph structure and causes ambiguities [25].

Repeated segments might cause two types of misassembly. The first type occurs when there are multiple consecutive copies of the same segment, as illustrated in Fig. 6 (II). Any assembly algorithm will have a problem in detecting the correct number of repeated copies, and, very often, fewer or more copies are included. The second type of misassembly (see Fig. 6 (I)) occurs when the repeated segments are apart from each other. There is then a possibility that the assembler will create a chimera by falsely joining two regions, and the resulting genome will be rearranged in comparison with the reference. Assembly errors that appear because of repeats are a very serious problem because sometimes the number of repeated copies is closely related to some phenotypic characteristic of an organism, and a shuffle of regions might be mistaken for a DNA rearrangement event [29].

Different strategies have been developed for different assemblers to resolve the problems caused by repeats. One of the most commonly used solutions is the use of mate pairs or paired ends. Mate pairs and paired ends are reads (of size 150–500 bp) generated in pairs from opposite ends of a longer DNA sequence. There is no significant difference between paired ends and mate pairs in terms of assembly, even though the laboratory techniques used to generate them are very different (see [23]). The pairs span larger regions, ranging from 200 to 20,000 bp. If a pair contains a repeat, it might provide enough information about the correct context of the repeated segments. For instance, in the situation depicted in Fig. 6 (I), the assembler would be able to identify the correct assembly if a mate pair which spans $M-N$ (or the first repeat r) and a mate pair that spans $P-Q$ (or the second repeat r) were available.

Another widely used approach for resolving repeats is a comparison of the depth of coverage for each contig (branch in the graph) that is in question. This method is based on the assumption that the set of reads is uniformly distributed throughout the genome (which is not generally true) and helps in estimating the number of repeated copies. This approach might be particularly helpful when one is "popping" bubbles or counting the number of repeats, as illustrated in Fig. 6. For instance, in Fig. 6 (III), if there were proportionally more reads (edges) matching the segment XY than matching the segment PN, the assembler would form a contig $MXYQ$. Also, in Fig. 6 (III), if the number of reads covering the repeated segment r was m-fold in comparison with the average depth of coverage, than m copies of r would be included in the contig [36].

Despite the attempts to resolve these problems, none of the current assemblers are applicable to all kinds of data sets while simultaneously exhibiting high accuracy and low computational complexity.

4 Theoretical Results

The results presented in this section are mainly combinatorial results on words over a given alphabet. Combinatorics on words is an area of discrete mathematics that initially dealt with problems in theoretical computer science, including formal languages, automata theory, coding theory, and the theory of computation. In mathematics, there are a number of problems that deal with reconstruction, for instance reconstruction of a function from its values at some points, of a group from its subgroups, of a graph from a subset of its subgraphs, and of an image from a sample point set, to name just a few. Combinatorics on words is an area that, among other problems, deals with the sequence reconstruction problem, and addresses the following aspects:

- Uniqueness of sequences reconstructed from the same set of subsequences;
- The existence of a reconstruction based on the size and the number of available subsequences; and
- Unambiguous reconstruction with respect to the structure and origin of the subsequences.

Some of these theoretical results can be applied directly to sequence assembly, and answer different questions related to DNA sequencing. Therefore, we have included a few of the results that we believe are the most relevant to assembly problems.

4.1 Definitions and Notation

Let Σ be a finite alphabet of symbols. The elements of Σ are called letters. The set of all sequences over Σ is denoted by Σ^*, and the elements of Σ^* are called *words* (or *sequences*). The *empty word* is denoted by λ. If w is a word over Σ, then we write $|w|$ to denote the length (measured as the number of letters) of w. We denote by Σ_n^* the set of all sequences over Σ of length n.

We often refer to the nth letter in w as the nth position. A sequence v over Σ is called a (scattered) *subsequence* of the word $w = w_1 w_2 \cdots w_n$ if $v = v_{i_1} v_{i_2} \cdots v_{i_s}$, for some $1 \leq i_1 < i_2 < \cdots < i_s \leq n$. In other words, we say that w is a *supersequence* of v. The set of all subsequences of length t of a sequence w is called the t-spectrum of w, and we denote it by $S_t(w)$. A sequence v over Σ is called a *factor* of the word w if there are words x and y over Σ such that $w = xvy$. Note that x and/or y may be empty. These factors are special types of subsequences. Given a word w such that $w = xy$, we define $w - x = y$. A factor v is said to be a *prefix* of the word $w = xvy$ if $x = \lambda$ and, similarly, v is a *suffix* of w if $y = \lambda$. The *overlap* of two words w and w', denoted by $o(w, w')$, is defined as the maximal factor v which is a suffix of w and a prefix of w'. It may also be that $v = \lambda$.

For two words w and w', we define an operation "$\circ$", called *composition*, by which $w \circ w' = w(w' - o(w, w'))$. In other words, we concatenate the two words, though a single copy of their largest overlap.

Example 3. Let $\Sigma = \{a, b, c\}$ and $w = aaaacccbabacc$, $w' = bacccbba$. The word $v = bacc$ is the maximal factor – a suffix of w and prefix of w'. Therefore, $v = o(w, w')$. The composition of w and w' is $w \circ w' = aaaacccbabacccbba$.

4.2 Sequence Reconstruction from Subsequences

The Russian mathematician V. Levenshtein investigated the problem of reconstruction of a sequence from its subsequences and supersequences. In [18], Levenshtein determined the number of subsequences needed to reconstruct an unknown sequence. The result is stated below in Theorem 1. Let X be an unknown sequence of length n. For a number $t < n$ we wish to determine how many different subsequences of length t are needed to uniquely reconstruct X. This answers the following question: What is the minimum number of reads of length t that one needs to reconstruct a genome? We note here that considering subsequences instead of factors might seem far from reality in terms of genome assembly, but in fact the mate pairs are special types of subsequences and, furthermore, each read can be viewed as a subsequence when we take sequencing errors into account.

All of the sequences in Σ_n^* are considered first, and they are compared pairwise to count the subsequences of length t that they have in common, i.e., we find the cardinality of the sets $S_t(w) \cap S_t(v)$ for every $w, v \in \Sigma_n^*$. Let $N(n, t)$ denote the maximum size of the set of subsequences of length t shared between two sequences of length n, i.e.,

$$N(n, t) = \max\{|S_t(w) \cap S_t(v)|, w, v \in \Sigma_n^*\}.$$

Theorem 1. *The minimum number of subsequences of length t that are needed to reconstruct an unknown sequence X of size $|X| = n$ equals $N(n, t) + 1$.*

Levenshtein also provided an efficient algorithm in [18] for performing the reconstruction. The number $S_t(w)$ depends on the sequence w (see [17]) and, therefore, there is no exact formula for calculating $N(n, t)$. This number can be calculated recursively as follows:

$$N(n, t) = S(n, t) - S(n - 1, t) + S(n - 2, t - 1),$$

where $S(n, t) = \max\{|S_t(w)|, w \in \Sigma_n^*\}$.

The problem with applying these results to genome assembly is the fact that there are sequences for which the existence of $N(n, t) + 1$ different subsequences is not guaranteed. For more details, we refer the reader to Example 4 below. In some cases, more than half of the sequences of a given length do not have enough different subsequences from which they can be reconstructed.

Example 4. Let $\Sigma = \{A, G, T, C\}$ and let $X \in \Sigma^*$ be such that $|X| = 5$. Let $t = 3$. According to Theorem 1, one needs $N(5, 3) + 1$ different subsequences of length 3 to reconstruct X. By examining the subsequences of length 3, one can easily conclude that $|S_3(w) \cap S_3(v)| = 7$, where $w = AGTCG$ and $v = ATGCG$. Therefore, $N(5, 3) + 1 \geq 8$, and one needs eight or more different subsequences of length 3 to uniquely construct X. Note that no sequence composed of fewer than three different symbols from Σ has eight different subsequences. This implies that those sequences composed of a single symbol (in total, four such sequences) and those composed of two different symbols (in total, $2^5 * C(4, 2) = 192$ sequences) do not have eight different subsequences. Therefore, at least 196 sequences of length 5, out of 1,024 in total, cannot be uniquely reconstructed. In other words, there are high odds, greater than $\frac{1}{6}$, that a given sequence X cannot be reconstructed.

Besides the structure of X, the reconstruction depends on the size of the available subsequences (reads). By changing the number t, one might be able to find enough subsequences to reconstruct X, as discussed in the next subsection.

4.3 Reconstruction Based on the Size of the Subsequences

There are two types of questions that can be asked about reconstruction with respect to the size of the given subsequences:

- What is the smallest k such that one can reconstruct any word of length n from the multiset of its subsequences of length k?
- What is the smallest k such that one can reconstruct a word from the set of its different subsequences of length k?

Note that a *multiset* is a set of elements such that multiple occurrences of an element are allowed and their number is known. Therefore, the first problem is about reconstruction of a sequence given all of the repeats and the number of copies of each repeat, which is a far from realistic data requirement. The second problem is more relevant to genome assembly, since it does not require input information about the repeated reads.

Manvel et al. [21] showed that the reconstruction of a sequence of length n is unique if all subsequences of size k are given, where $k \geq n/2$. In addition, it was proven that a unique reconstruction is not possible for $k < \log_2 n$. This implies that two words of length n that have identical sets of subsequences up to $n/2$ might not be identical. The lower bound on k has been improved several times. In [15], the bound $k \geq 5 + \frac{16}{7}\sqrt{n}$ was given, and Dudik et al. [6] showed that

$$k \geq 3^{(\sqrt{2/3}-o(1))\log_3^{1/2} n}.$$

These results thus provide estimates of k as an answer to the first problem above.

In any case, the condition $k \geq n/2$, which implies that to uniquely reconstruct one particular small genome namely that of a virus found in *Escherichia coli* with $n = 5,386$, the read sizes must be greater than 2,193 bp. This is when the repeats and their multiplicities are known and also all subsequences are available, conditions that are far from realistic.

The result most relevant to genome assembly is one presented in [7]. The authors of that paper examined finite words over an alphabet $\Gamma = \{a, \overline{a}, b, \overline{b}\}$ of pairs of letters, where each word $w_1 w_2 \ldots w_t$ is identified with its reverse complement $\overline{w_t} \, \overline{w_{t-1}} \ldots \overline{w_1}$. This approach takes account of the reads and their reverse complements, as is actually done in the genome assembly process. It was shown that the smallest k for which every word of length n over Γ is uniquely determined by the set of its subwords of length up to k is given by $k \approx 2n/3$ [7]. To put this result into perspective in relation to genome assembly, let us consider one of the smallest genomes, of size $n = 5,386$. Then $k \approx 2 * 5,386/3 = 3,590.6$, which implies that one needs all possible reads of length 3,590 to uniquely assemble a genome of size 5,386. This number becomes significantly larger for larger genomes. For instance, the size of the human genome is estimated at 3.2 billion bp, and thus one needs all of the $k \approx 2.13$ billion bp long subsequences to uniquely reconstruct the genome.

The following result shows that no sequence can be uniquely determined by a k-spectrum composed of factors only. Namely, the maximum length n such that every word of this length can be uniquely determined by its factors of size k, for some $k \leq n$, is k. This implies that there are words of length greater than k which cannot be uniquely reconstructed from their set of factors of length k (see [20]).

Example 5. Let $w = ababab \ldots ab$ and $v = bababa \ldots ba$ be two sequences of size n. We consider the sets of factors of length k, $F_k(w)$ for w and $F_k(v)$ for v, where $k = n-1$. We have $F_k(w) = F_k(v) = \{abab \ldots a, baba \ldots b\}$, and therefore w cannot be uniquely reconstructed from $F_k(w)$.

4.4 Reconstruction Based on the Structure and Origin of the Factors

In this section, we consider some results due to Carpi and De Luca [3] and Carpi et al. [4]. Unlike Levenshtein, whose results do not guarantee reconstruction of every possible sequence, these authors analyzed the reconstruction of sequences from sets of factors which permit reconstruction of the entire word. This analysis is based on the notion of boxes. Before we state the main result, we need some preliminary definitions.

Definition 1. The *initial box* of a word w is the shortest unrepeated prefix of w. The *terminal box* of a word w is the shortest unrepeated suffix of w.

Definition 2. *Superbox* is a factor of w that can be written as asb, where $a, b \in \Sigma$, s is repeated factor and as, sb are un-repeated factors of w.

The main result can be summarized in the following theorem.

Theorem 2. *Any finite word w is uniquely determined by the initial box, the terminal box, and the set of superboxes.*

Example 6. Consider the small genome *ATCCTATCAT*. The initial box is *ATCC*, the terminal box is *CAT*, and the set of superboxes is *CCT*, *CTA*, *TATCA*.

An efficient algorithm for finding the initial box, terminal box, and superboxes is given in [4]. In relation to the genome assembly problem, the theorem implies that one can uniquely assemble a genome if special reads are known (they all might be of different lengths). Those special reads are the maximal unrepeated subsequences of the genome, i.e., the unrepeated sequences such that each of their proper subsequences is repeated in the genome.

Long interspersed nuclear elements (LINEs) are repeated in the human genome and can be as long as a million base pairs. Based on Theorem 2, one would require a read of length approximately one million base pairs to uniquely assemble the human genome.

5 An Optimization-Based Approach

In this section, we present some preliminary results obtained from our optimization-based approach to solving the sequence reconstruction problem, by providing a set of mathematical formulations as programming problems which deal with different aspects of the assembly problem.

Let $\mathfrak{S}$ be a sequence over Σ such that $|\mathfrak{S}| = n$. Let W be a collection of (all) factors of $\mathfrak{S}$ with length k. Our goal is to reconstruct $\mathfrak{S}$ by appropriately ordering the elements of W. To this end, we determine the optimal solution with the minimum number of mismatches by encoding the problem into a quadratic integer programming problem.

Let $W = \{w_1, w_2, \ldots, w_N\}$ be a set of factors of $\mathfrak{S}$ and let $|w_i| = l$, where $l \geq 2$, for every $i \in \{1, 2, \cdots, N\}$. If there is a subset $W' = \{w'_1, \ldots, w'_{N'}\} \subset W$ and a permutation $P = (p_1, p_2, \ldots p_{N'})$ such that $\mathfrak{S} = w'_{p_1} \circ w'_{p_2} \circ \cdots \circ w'_{p_{N'}}$, then we say that $\mathfrak{S}$ is covered by W, and the pair (W', P) is called a perfect coverage of $\mathfrak{S}$. There may be no perfect coverage or there may exist multiple perfect coverage pairs for a given $\mathfrak{S}$ and W. Reconstructing $\mathfrak{S}$ from a set of factors $W = \{w_1, w_2, \ldots, w_N\}$ means finding a perfect coverage (W', P), if it exists. Let (W', P) be a pair made up of $W' \subset W$ and a permutation $P = (p_1, p_2, \ldots p_{N'})$ of the elements of W', such that $|\mathfrak{S}| = |w'_{p_1} \circ w'_{p_2} \circ \cdots \circ w'_{p_{N'}}|$. We call (W', P) a coverage of $\mathfrak{S}$. If the letters at the ith positions of $\mathfrak{S}$ and $w'_{p_1} \circ w'_{p_2} \circ \cdots \circ w'_{p_{N'}}$ differ, then we say that these two words have a *mismatch* at the ith position. In the case where a perfect coverage does not exist for $\mathfrak{G}$, it is interesting to consider a coverage with the minimum number of mismatches.

The sequence reconstruction problem described above can be translated directly into the problem of genome assembly from a large number of reads (factors). Namely, one can look at $\mathfrak{G}$ as a genome that has to be reconstructed from a set of reads $W = \{w_1, w_2, \ldots, w_N\}$. Owing to mutations and errors in sequencing, the sequence that is reconstructed from W may contain mismatches with respect to the actual genome. Therefore, finding a coverage of $\mathfrak{G}$ with the minimum number of mismatches would optimize the problem of genome assembly.

We define $x_{ijk} \in \{0, 1\}$ for $i \in \{1, 2, \ldots, N\}, j \in \{1, 2, \ldots, l\}, k \in \{1, 2, \ldots, n\}$ by

$$
x_{ik} = \begin{cases} 1 & \text{if the } j\text{th position of } w_i \text{ is assigned to the } k\text{th position in } \mathfrak{G}, \\ 0 & \text{otherwise.} \end{cases}
$$

Since $|w_i| = l$ for every i, the first constraint is given by

$$
\forall i, \ \sum_{j=1}^{l} \sum_{k=1}^{n} x_{ijk} = l. \tag{1}
$$

To satisfy the condition that each position in $\mathfrak{G}$ is covered by at least one word w_i (i.e., there are no gaps in the sequence that we reconstruct), the following constraint is added:

$$
\forall k, \ \sum_{j=1}^{l} \sum_{i=1}^{N} x_{ijk} \geq 1. \tag{2}
$$

In addition, we have to ensure the overlaps among the words are along the length l of each word:

$$
\forall i \ \forall k, \ l \leq k \leq n, \ \sum_{j=1}^{l} \sum_{m=k}^{k+l-1} x_{ijm} \leq l. \tag{3}
$$

We need three more constraints to obtain the intended result. The first of these guarantees that every symbol of w_i is matched to exactly one position in $\mathfrak{G}$. The second prevents the inclusion of multiple copies of a word, and the third does not allow stacking of words. These constraints are as follows:

$$
\forall i \ \forall j, \ \sum_{k=1}^{n} x_{ijk} = 1, \tag{4}
$$

$$\forall i \ \forall k, \ \sum_{j=1}^{l} x_{ijk} \leq 1, \tag{5}$$

$$\forall j \ \forall k, \ \sum_{i=1}^{N} x_{ijk} \leq 1. \tag{6}$$

We introduce two new binary variables z_{ik} and y_{ik}, such that $z_{ik}, y_{ik} \in \{0, 1\}$ such that

$$z_{ik} = 1, \ y_{ik} = 1 \ \text{if} \ \sum_{j=1}^{l} \sum_{m=k}^{k+l-1} x_{ijm} = l, \tag{7}$$

$$z_{ik} = 0, \ y_{ik} = 1 \ \text{if} \ l < \sum_{j=1}^{l} \sum_{m=k}^{k+l-1} x_{ijm} < 2l, \tag{8}$$

$$z_{ik} = 1, \ y_{ik} = 0 \ \text{if} \ 0 \leq \sum_{j=1}^{l} \sum_{m=k}^{k+l-1} x_{ijm} < l. \tag{9}$$

The constraints in Eqs. (10)–(14) below ensure that the new variables are properly defined (as described in Eqs. (7)–(9)):

$$\forall i \ \forall k, \ l \leq k \leq n \ \sum_{j=1}^{l} \sum_{m=k}^{k+l-1} x_{ijm} - l \leq (1 - z_{ik})l, \tag{10}$$

$$\forall i, \ \forall k, \ l \leq k \leq n \ \sum_{j=1}^{l} \sum_{m=k}^{k+l-1} x_{ijm} - l \geq -z_{ik}l, \tag{11}$$

$$\forall i \ \forall k, \ l \leq k \leq n \ \sum_{j=1}^{l} \sum_{m=k}^{k+l-1} x_{ijm} - l \geq (1 - y_{ik})(-l), \tag{12}$$

$$\forall i \ \forall k, \ l \leq k \leq n \ \sum_{j=1}^{l} \sum_{m=k}^{k+l-1} x_{ijm} - l \leq y_{ik}l, \tag{13}$$

$$\forall i \ \forall k, \ 1 \leq z_{ik} + y_{ik} \leq 2. \tag{14}$$

To ensure that a factor w_i fits at only one position, the following constraint is added:

$$\forall i \ \sum_{k=1}^{n} (z_{ik} + y_{ik} - 1) = 1. \tag{15}$$

Moreover, owing to the consecutive property, we set

$$\forall i \ \forall k, \ 1 \le k \le (l-1), \ \forall j, \ (k+1) \le j \le l, \ x_{ijk} = 0, \tag{16}$$

$$\forall i \ \forall k, \ (n-l+2) \le k \le n, \ \forall j, \ 1 \le j \le ((l-1)-(n-k)), \ x_{ijk} = 0. \tag{17}$$

The optimal solution for the number of mismatches can then be found by using the following objective function, which is to be minimized:

$$\min \sum_{k=1}^{n} \sum_{i,j,i',j'} (w_i^j \ne w_{i'}^{j'}) x_{ijk} x_{i'j'k} \le y_{ik} l \ \text{for all } i, i \ne i'. \tag{18}$$

Note that this objective function is quadratic and that the number of variables that we introduce depends on the number of reads. Furthermore, the constraints imposed are linear in the integer variables, which render this formulation a quadratic integer programming problem.

In addition to the general case of the quadratic programming problem described above, one can consider three different variants of the problem, where (1) we use a subset of the factors (instead of all possible factors), (2) we use a subset of paired factors, and (3) we use a subset of the factors including their inverses. Each of these variants is relevant to the genome assembly problem and can be matched to some of the NGS technologies that are being employed. However, because of the nature of this chapter, we shall not give details of the formulation of the optimization problem for these cases.

Finally, we should point out that solving an integer programming problem is, in general, an NP-hard problem. The computational challenges are reinforced by the large number of variables that would arise in any realistic application. Although there are efficient techniques for relaxing the integer constraints, it remains to be investigated how well this optimization-based formulation and its relaxations can solve the problem of genome assembly, even in the case of small, contrived examples.

6 Conclusion

Efficient and accurate solution of the genome assembly problem offers the possibility of improving not only our understanding of the diversity of nature but also the state of the art in medical research. Owing to its numerous applications, this problem has gained considerable attention in the bioinformatics community.

The principal aim of this chapter was to present a comprehensive overview of overlooked mathematical results which may prove relevant to improving the existing heuristic solutions to the genome assembly problem. This necessitated the inclusion of a high-level description of the existing NGS technologies, so that researchers in applied mathematics might formulate the appropriate theoretical settings. By illustrating the implications of language-theoretic results in realistic scenarios, we may have given the impression that the current technologies do not guarantee the uniqueness of the resulting genome assembly (regardless of the algorithmic approach used). However, the existing theoretical results pertain to special alphabets, use some special subsequences, or discard information about the existence of particular subsequences (which could be empirically verified). Therefore, we believe that a critical comparative view of these results could propel the development of language-theoretical tools that are closer and thus more relevant to realistic genome assembly scenarios.

To this end, we have also presented a preliminary optimization-based formulation of the genome assembly problem as a quadratic integer programming problem, whose performance, with appropriate relaxation, will be addressed in future work. The merit of the integer programming formulation is that it provides a new and potentially useful formulation of the genome assembly problem. We would like to point out that the optimization-based formulation requires information about the estimated size of the genome (here, N). Interestingly, none of the graph-based assembly methods have this as a requirement, rendering the comparison of the resulting genomes (which are likely to be of different lengths and coverage) a nontrivial task. Currently, however, we fail to see how the optimization-based formulation may be used to address the comparison of assemblies from different assemblers.

References

1. J. Adams, DNA sequencing technologies. Nat. Educ. **1**(1) (2008)
2. J. Butler, I. MacCallum, M. Kleber, I.A. Shlyakhter, M.K. Belmonte, E.S. Lander, C. Nusbaum, D.B. Jaffe, ALLPATHS, de novo assembly of whole-genome shotgun microreads. Genome Res. **18**, 810–820 (2008)
3. A. Carpi, A. De Luca, Words and special factors. Theor. Comput. Sci. **259**(1–2), 145–182 (2001)
4. A. Carpi, A. De Luca, S. Varricchio, Words, univalent factors, and boxes. Acta Inform. **38**, 409–436 (2002)
5. J.C. Dohm, C. Lottaz, T. Borodina, H. Himmelbauer, SHARCGS, a fast and highly accurate short read assembly algorithm for de nove genomic sequencing. Genome Res. **17**, 1697–1706 (2007)
6. M. Dudik, L.J. Schulman, Reconstruction from subsequences. J. Comb. Theory **A 103**, 337–348 (2003)
7. P.L. Erdos, P. Ligeti, P. Sziklai, D.C. Torney, Subwords in reverse-complement order. Ann. Comb. **10**, 415–430 (2006)
8. R.D. Fleischmann, M.D. Adams, O. White, R.A. Clayton, E.F. Kirkness, A.R. Kerlavage, C.J. Bult, J.F. Tomb, B.A. Doughherty, J.M. Merrick, K. McKenney, G. Sutton, W. FitzHugh, C. Fields, J.D. Gocyne, J. Scott, R. Shirley, L. Liu, A. Glodek, J.M. Kelley, J.F. Weidman, C.A. Phillips, T. Spriggs, E. Hedblom, M.D. Cotton, T.R. Utterback, M.C. Hanna, D.T. Nguyen, D.M. Saudek, R.C. Brandon, L.D. Fine, J.L. Fritchman, J.L. Fuhrmann, N.S.M. Geoghagen, C.L. Gnehm, L.A. McDonald, K.V. Small, C.M. Fraser,

H.O. Smith, J.C. Venter, Whole-genome random sequencing and assembly of Haemophilus influenzae Rd. Science **269**(5223), 496–512 (1995)

9. http://www.lifetechnologies.com/content/lifetech/us/en/home/about-us/news-gallery/press-releases/2012/life-techologies-itroduces-the-bechtop-io-proto.html.html. Accessed Mar 2013

10. X. Huang, A. Madan, CAP3: a DNA sequence assembly program. Genome Res. **9**, 868–877 (1999)

11. Human Genome Project Information, Genomic science program. http://www.genomics.energy.gov. Accessed Oct 2012

12. R.M. Idury, M.S. Waterman, A new algorithm for DNA sequence assembly. J. Comput. Biol. **2**(2), 291–306 (1995)

13. W.R. Jeck, J.A. Reinhardt, D.A. Baltrus, M.T. Hickenbotham, V. Magrini, E.R. Mardis, J.L. Dangl, C.D. Jones, Extending assembly of short DNA sequences to handle error. Bioinformatics **23**, 2942–2944 (2007)

14. S. Koren, M.C. Schatz, B.P. Walenz, J. Martin, J.T. Howard, G. Ganapathy, Z. Wang, D.A. Rasko, W.R. McCombie, E.D. Jarvis, A.M. Phillippy, Hybrid error correction and de novo assembly of single-molecule sequencing reads. Nat. Biotechnol. **30**, 693–700 (2012)

15. I. Krasikov, Y. Roditty, On a reconstruction problem of sequences. J. Comb. Theory **A77**, 344–348 (1997)

16. H. Lee, H. Tang, Next-generation sequencing technologies and fragment assembly algorithms. Methods Mol. Biol. **855**(2), 155–174 (2012)

17. V. Levenshtein, Reconstruction of objects from a minimum number of distorted patterns. Dokl. Math. **55**, 417–420 (1997)

18. V. Levenshtein, Efficient reconstruction of sequences from their subsequences or supersequences. J. Comb. Theory A **93**, 310–332 (2001)

19. L. Liu, Y. Li, S. Li, N. Hu, Y. He, R. Pong, D. Lin, L. Lu, M. Law, Comparison of next-generation sequencing systems. J. Biomed. Biotechnol. **2012**, 1–11 (2012)

20. J. Manuch, Characterization of a word by its subwords, in *Developments in Language Theory – Foundations, Applications, and Perspectives, Proc. DLT 2000*, ed. by G. Rozenberg, W. Thomas, pp. 210–219

21. B. Manvel, A. Meyerowitz, A. Schwenk, K. Smith, P. Stockmeyer, Reconstruction of sequences. Discret. Math. **94**, 209–219 (1991)

22. M. Margulies, M. Egholm, W.E. Altman, S. Attiya, J.S. Bader, L.A. Bemben, J. Berka, M.S. Braverman, Y. Chen, Z. Chen, S.B. Dewell, A. de Winter, J. Drake, L. Du, J.M. Fierro, R. Forte, X.V. Gomes, B.C. Godwin, W. He, S. Helgesen, C.H. Ho, S.K. Hutchison, G. Irzyk, S.C. Jando, M.L.I. Alenquer, T.P. Jarvie, K.B. Jirage, J. Kim, J.R. Knight, J.R. Lanza, J.H. Leamon, W.L. Lee, S.M. Lefkowitz, M. Lei, J. Li, K.L. Lohman, H. Lu, V.B. Makhijani, K.E. McDade, M.P. McKenna, E.W. Myers, E. Nickerson, J.R. Nobile, R. Plant, B.P. Puc, M. Reifler, M.T. Ronan, G.T. Roth, G.J. Sarkis, J.F. Simons, J.W. Simpson, M. Srinivasan, K.R. Tartaro, A. Tomasz, K.A. Vogt, G.A. Volkmer, S.H. Wang, Y. Wang, M.P. Weiner, D.A. Willoughby, P. Yu, R.F. Begley, J.M. Rothberg, Genome sequencing in microfabricated high-density picolitre reactors. Nature **437**, 376–380 (2005)

23. P. Medvedev, M. Stanciu, M. Brudno, Computational methods for discovering structural variation with next-generation sequencing. Nat. Methods **6**, S13–S20 (2009)

24. M. Metzker, Sequencing technologies – the next generation. Nat. Genet. **11**, 31–46 (2010)

25. J.R. Miller, S. Koren, G. Sutton, Assembly algorithms for next-generation sequencing data. Genomics **95**(6), 315–327 (2010)

26. E.W. Myers, G.G. Sutton, A.L. Delcher, I.M. Dew, D.P. Fasulo, M.J. Flanigan, S.A. Kravitz, C.M. Mobarry, K.H. Reinert, K.A. Remington, E.L. Anson, R.A. Bolanos, H. Chou, C.M. Jordan, A.L. Halpern, S. Lonardi, E.M. Beasley, R.C. Brandon, L. Chen, P.J. Dunn, Z. Lai, Y. Liang, D.R. Nusskern, M. Zhan, Q. Zhang, X. Zheng, G.M. Rubin, M.D. Adams, J.C. Venter, A whole genome assembly of Drosophilia. Science **287**, 2196–2204 (2000)

27. P.C. Ng, E.F. Kirkness, Whole genome sequencing. Methods Mol. Biol. **628**, 215–226 (2010)

28. A.P. Pevzner, T. Haixu, S.M. Waterman, An Eulerian path approach to DNA fragment assembly. PNAS **98**(17), 9748–9753 (2001)

29. A.M. Phillippy, M.C. Schatz, M. Pop, Genome assembly forensics: finding the elusive mis-assembly. Genome Biol. (2008). doi:10.1186/gb-2008-9-3-r55
30. M. Pop, Genome assembly reborn: recent computational challenges. Brief Bioinform. **10**(4), 354–366 (2009)
31. M. Quail, M.E. Smith, P. Coupland, T.D. Otto, S.R. Harris, T.R. Connor, A. Bertoni, H.P. Swerdlow, Y. Gu, A tale of three next generation sequencing platforms: comparison of Ion Torrent, Pacific Biosciences and Illumina MiSeq sequencers. BMC Genomics **13**(1), 341 (2012). doi:10.1186/1471-2164-13-341
32. F. Sanger, A.R. Coulson, A rapid method for determining sequences in DNA by primed synthesis with DNA polymerase. J. Mol. Biol. **94**, 441–448 (1975)
33. M.C. Schatz, A.L. Delcher, S.L. Salzberg, Assembly of large genomes using second-generation sequencing. Genome Res. **20**(9), 1165–1173 (2010)
34. J.T. Simpson, K. Wong, S.D. Jackman, J.E. Schein, S.J. Jones, I. Byrol, ABySS, a parralel asembler for short read sequence data. Genome Res. **19**, 1117–1123 (2009)
35. G.G. Sutton, O. White, M.D. Adams, A.R. Kerlavage, TIGR assembler: a new tool for assembling large shotgun sequencing projects. Genome Sci. Technol. **1**, 9–19 (1995)
36. T.J. Treangen, S.L. Salzberg, Repetitive DNA and next-generation sequencing: computational challenges and solutions. Nat. Rev. Genet. **13**(2), 36–46 (2012)
37. R.L. Warren, G.G. Sutton, S.J. Jones, R.A. Holt, Assembling millions of short DNA sequences using SSAKE. Bioinformatics **23**, 500–501 (2007)
38. K.A. Wetterstrand, DNA sequencing costs: data from the NHGRI large-scale genome sequencing program. http://www.genome.gov/sequencingcosts. Accessed Oct 2012
39. D.R. Zerbino, E. Birney, Velvet, algorithms for de novo short read assembly using de Bruijn graphs. Genome Res. **18**, 821–829 (2008)

Extracting Coevolving Characters from a Tree of Species

Alessandra Carbone

Abstract Phylogenetic proximity has guided our understanding of the evolution of species for decades. It is clear nowadays that the paradigm "phylogenetically close species should share similar characters" is just one facet of the complex process of evolution inherent in development and species differentiation. Today, there is a need for novel mathematical approaches to cluster together symbolic information organized into trees of characters that could highlight the evolutionary relations between characters and the processes of coevolution of characters. We propose a combinatorial method to do so and to derive groups of characters which appear to be correlated through their evolutionary history. This approach was first developed for protein sequences, but it is revealed to be general and applicable to any list of characters describing species. In particular, one does not need to know all characters for all species to perform coevolution analysis.

1 Introduction

Biological information is usually organized into a tree-like structure owing to the underlying evolutionary process that traces the history of the evolution of the data. Evolution is regarded as a branching process, whereby populations are altered over time and may split into separate branches, hybridize together, or terminate by extinction. The resulting tree, called a *phylogenetic tree*, represents a hypothesis about the order in which evolutionary events occurred. Unraveling a proper biological interpretation of this tree-like structure is a far from simple task,

A. Carbone (✉)
Laboratoire Génomique des Microorganismes, UMR7238, Université Pierre et Marie Curie et CNRS, Paris, France

Department of Computer Science, Université Pierre et Marie Curie, Paris, France
e-mail: Alessandra.Carbone@lip6.fr

N. Jonoska and M. Saito (eds.), *Discrete and Topological Models in Molecular Biology*,   45
Natural Computing Series, DOI 10.1007/978-3-642-40193-0_3,
© Springer-Verlag Berlin Heidelberg 2014

since the information observed is a very small part of what the evolution process could have given. Most species have been lost, for instance, and the reasons are not always a lack of fitness, that is, the ability to both survive and reproduce, nor a lack of advantage, that is, the benefit gained over competitors. Random events that cause a species to disappear or to evolve in some feasible (even though unlikely) manner might also happen.

The basic mathematical structure underlying a group of species is the phylogenetic tree associated with their evolution, where the leaves of the tree represent the species in the group and the internal nodes represent their ancestors. Based on this tree, concepts such as "fitness" and "advantage" have been promoted to a great extent, but not always accompanied by convincing explanations. In fact, quite often, those species that are located close together in the tree happen to share the same biological behavior, and hypotheses about the functional reasons for this behavior to take place have been presented for various groups of organisms. When the branches of the tree become long, though (i.e., the underlying pathways correspond to a long evolutionary history), the paradigm "phylogenetic proximity" = "biological similarity" is no longer true, and phylogenetically close species might start to display different characteristics. An interesting example is illustrated by codon bias in bacterial species. By studying this, it has been discovered that bacteria with similar codon bias are likely also to share their environment and habitat [8, 40] (see also [23, 28, 36]), but that species within a phylum do not necessarily share codon bias. The phylum of proteobacteria, for instance, spans the whole range of codon biases even though it forms a clade in the phylogenetic tree. In fact, only subsets of the proteobacteria, that is, all known "proteobacteria classes", share habitat, codon bias, and phylogenetic proximity. Another example is provided by protein domain evolution within species. For conserved protein domains, it has been found that phylogenetically close species share conserved patterns in homologous sequences. For proteins that have diverged by over 50 % sequence identity, we can still find homologous sequences that display specific conserved patterns, but this is more likely in distant species than in phylogenetically close ones [4].

Hence, to learn about the evolutionary process underlying the species that exist today, it becomes particularly important to develop methods that take into account the topology of the phylogenetic tree (and possibly the metric of its branches) to identify where conservation takes place along the tree and whether characters are correlated within conserved subtrees. The hope is that a careful handling of the signals of evolution will lead to accurate interpretations of the coevolution of characters.

The theory that we shall develop here concerns the identification of the correlated evolutionary changes of characters in species. It is a generalization of the work presented in [3], where the main concepts and the approach used were introduced for biological sequences and the detection of coevolution of protein residues. These residues are important in protein folding and allosteric movements, and our approach allows the identification of networks of correlated residues [2, 3, 13, 15, 17, 19, 20, 27, 31, 33, 35, 39, 41] (see also [11, 12, 30]). We shall demonstrate here that an approach based on protein residues can be generalized to a collection of characters describing a set of species.

The applications may concern protein sequences, but also other biological data where trees describe relations between species. Trees might have a phylogenetic origin (where intermediate nodes v represent potential ancestors of the species labeling the leaves of the subtree rooted at v); they might be trees capturing environmental similarities and organismal differences; they might be trees tracing the evolution of a population; or they might be distance trees arising from clustering of characters (e.g., sequence alignment). The important point is that the trees may be constructed from data that either are independent of the characters to be analyzed (for instance, think of a tree of bacterial species constructed using codon bias information, and a correlation analysis of the environmental properties of the species in the tree) or they use the full set of characters associated with the species. In the second approach, the tree is generated by taking into consideration the whole set of characters (possibly many characters compared with those that will be analyzed) and by solving a global optimization problem where the number of character transformations in the tree is minimized. Coevolution of character pairs (or tuples) is then studied, with local optimization analysis.

1.1 Evolution and Characters

In biology, coevolution of characters refers to a reciprocal genetic change in two or more species [37]. Common examples are (1) a predator and its prey, where the predator kills and eats the prey; (2) a herbivore and a plant, where the herbivore eats the plant; (3) a host and a parasite, where the host is adversely affected and the parasite benefits; (4) mutualism, where both species benefit from the change; and (5) competition between two or more species. Given a number of characters describing each species and a tree of species, one wants to identify correlations among a subset of these characters. Ideally, one would like to identify several subsets of characters that coevolve, and interpret the results within the categories above.

1.2 Evolution of Microbial Populations

The exponential growth of cellular cultures and the tracking of genetic differences such as chromosomal rearrangements within populations provide a setting where our theory might apply. Populations are intended here to be sets of individuals of the same species for which an underlying tree organization describes their evolutionary relations. A description of differences among individuals is usually less precise for populations than for ensembles of species, owing to the strong similarity between individuals. In practical terms, this means that events concerning populations are studied in an isolated manner, one kind at a time, and that the analysis targets the reconstruction of the distribution of events within the population.

Unfortunately, population trees are often missing. Without knowing the tree, the effects detectable from a population study are those that occur in the very early stages of the evolutionary process generating the population. This is because early events are sufficiently represented in later generations, while events occurring later on the process are not able to emerge quantitatively from the observed population, since the exponential process underestimates their number [18, 21]. Having a tree is therefore key to a thorough understanding of the evolution of a population. Ideally, experiments on a population should be performed to track several characters at once; in this case, a tree could be constructed based on all observations (it will not be a tree arising from ancestral relations but a tree constructed from the events observed in the experiment, all taken together) and screened for correlation discovery. Our theory would help to find correlations among events in the tree.

1.3   Evolution and Single-Nucleotide Polymorphism (SNP)

One might envisage using the theory to study correlations among SNPs, that is, DNA sequence variations that occur when a single nucleotide (A, T, C, or G) in the genome sequence (or some other shared sequence) differs between members of a species. Each individual has many single-nucleotide polymorphisms that together create a unique DNA pattern for that individual. Our theory might help to define correlations between SNPs, for both prokaryotic and eukaryotic populations. In fact, specific phenotypes due to sequence variations might be induced by a single SNP or by multiple SNPs. We expect that a functional relation among several SNPs could be detectable through coevolution analysis. Trees of individuals could be constructed from sequence similarity, with a sequence containing the SNPs under investigation for each individual labeling the leaves.

1.4   Gene Responses in Microarray Data

Matrices of microarray data might be another interesting biological set to be considered for analysis with our approach, where genes play the role of characters and patients or organisms play the role of species. After clustering the data, one is usually interested in determining the correlations between different genes, and our approach might provide a way to do this, based on the tree obtained after clustering.

2   Why Look at Trees

Conservation and coevolution of characters are intimately linked concepts. Conservation refers to the persistence of a character during species variation, while coevolution refers to the correlated changes of character values in two or more species.

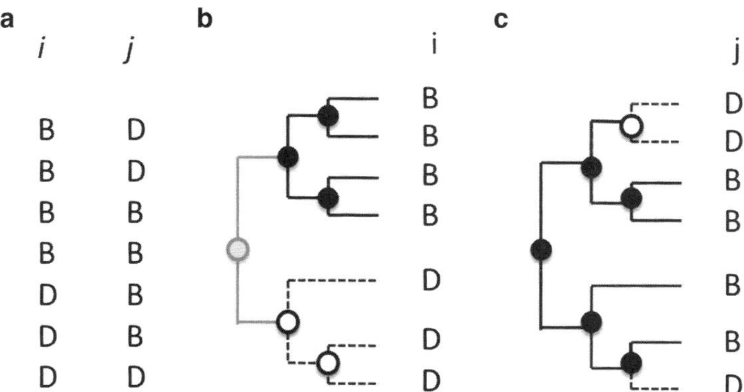

Fig. 1 (**a**) Two lists of character values for characters i and j, displaying the same information content. (**b**) Tree associated with the distribution of character i in (**a**), whose leaves are labeled by character values. Two subtrees organize the character values B (*solid black line*) and D (*dashed line*). The *black dots* at the roots of the subtrees represent preservation of character B for the ancestors at the roots. Similarly, *white dots* represent character D at the roots. The *gray dot* highlights the fact that neither B nor D can be inferred to hold (by the parsimony principle that minimizes the number of changes in evolution). (**c**) Tree associated with the distribution of character j in (**a**). The character values B and D are not organized in subtrees. *Black* and *white dots* as in (**b**)

There are several ways to evaluate conservation. If the values of character lie in a discrete interval $[1, \ldots, N]$, the "information content" of a character, that is, its level of conservation, can be computed as the Kullback–Leibler relative entropy [10, 24], for instance; that is, $IC(s) = -\sum_{i \leq N} p_s(v_i)\log_N(p_s(v_i)/q_s(v_i))$, where $p_s(v_i)$ is the observed frequency of value v_i for character s, q_s is a background frequency distribution of character values, and the logarithm is to base N, where N is the number of possible values that s can take [5, 34].

The notion of information content does not take into account the extra information associated with biological data, such as the phylogenetic organization of species. In this respect, we wish to distinguish between distributions of character values in a tree of species that provide the same information content (Fig. 1a) but display a different tree-like organization (see Fig. 1b, c). We make the hypothesis that a tree-like structure governs the different species considered and that the way characters change on the tree can be important for properly evaluating conservation of characters. The tree in Fig. 1b presents the character value B as more conserved than it is in the tree of Fig. 1c, where B occurs in different subtrees representing species that display the character value D also. In Sect. 3, we shall develop such a notion.

2.1 Conserved Residues and Prediction of Interaction Sites in Proteins

An important example of the effect of a proper detection of conservation signals is given by families of homologous proteins. These families are characterized by alignments, with conserved positions corresponding to residues sitting either in the protein core or on the surface [38]. For the latter, it has been noticed that they form, in three dimensions, patches of residues corresponding to interaction sites [25, 26]. The level of conservation of the residues could have been detected by using the notion of information content (see below), but it has been observed that a careful analysis of the topology of the distance tree of sequences provides more refined predictions [1, 9, 16, 22, 26, 29, 32]. Automatic detection of interaction sites, combining conservation signals with preservation of physicochemical properties, has become a feasible task. A tool for large-scale protein analysis was proposed in [16]. In [7], we showed that the usage of trees for capturing conservation ameliorates the problem of detection of protein binding sites compared with the use of *IC*.

Since conservation is detected better with the tree topology, we have extended the use of the tree topology to the study of coevolution of characters. The idea is to analyze the coexistence of pairs of character values in a tree of species by checking that character values are conserved in the same "regions" of the tree, where a region in the tree refers to branches close together in the tree. These might be subtrees, but not only those captured by the notion of an "inner tree" presented in Sect. 3. For a presentation of previously introduced methods for the study of coevolution in biological sequences, see [6].

3 Coevolution of Trees and Characters

Let T be a tree of species, and suppose that each species labeling a leaf in T is characterized by a set of characters. These characters are mostly shared by all species, but some species might be missing some. In our analysis, each character may or may not have the same alphabet of values, and the number of characters is arbitrary. No order is imposed on the characters. They will be considered as a set.

Given T and a character s, we identify the maximal subtrees (MSTs) for s as the largest subtrees of T that conserve a value of s. Formally speaking, let T be a tree associated to some set of characters, let $N(T)$ be its nodes, let $L(T)$ be its leaves, each labeled with values of characters, let $T(x)$ be the subtree of T rooted at $x \in N(T)$, and let $f(x)$ be the father node of $x \in N(T)$, if it exists. We distinguish S different characters and we allow certain species not to have a character. Let $R(s)$ be the set of values for a character $s \in [1 \ldots S]$, and let

$$R(S) = \bigcup_{s \in [1 \ldots S]} R(s).$$

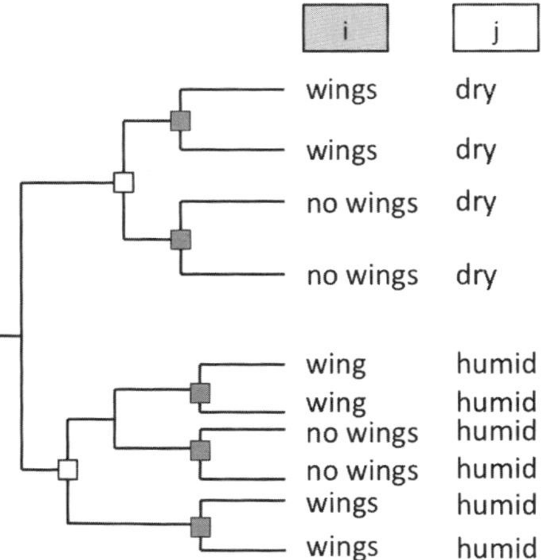

Fig. 2 Maximal subtrees (MSTs) of a tree T, **and ranks**. The tree T is made up of seven species characterized by the two characters i and j, where i indicates species with or without wings, and j indicates species living either in a dry or in a humid environment. Maximal subtrees in the tree T defined with respect to characters i and j are highlighted by their roots: *gray squares* for MSTs of i and *white squares* for MSTs of j. Character i has rank 5, since there are five distinguished MSTs associated to it, and j has rank 2

The function *charvalue* $: L(T) \times [1 \ldots S] \rightarrow R(S)$ associates to a leaf l of T and a character s the value r corresponding to the species labeling the leaf l, with $r \in R(s) \subseteq R(S)$.

A subtree $T(x)$ is *conserved for character s* if all leaves in T have the same value of s, that is, $\forall l_1, l_2 \in L(T(x))$, *charvalue*$(l_1, s) =$ *charvalue*(l_2, s). By convention, species are allowed not to have a character, and if at least one of the leaves l_1, l_2 has an undefined value, then *charvalue*$(l_1, s) \neq$ *charvalue*(l_2, s). A subtree $T(x)$ is *maximal for character s* if $T(x)$ is conserved for character s and, if $f(x)$ exists, then $T(f(x))$ is not fully conserved for s.

3.1 Importance of a Character Within a Set of Characters

The importance of a character for a set of species is analyzed using the notion of *rank*. The *rank of a character s in T* is defined as

$$R(T, s) = |\{x \in N(T) \mid T(x) \text{ is maximal for } s\}|,$$

with $1 \leq R(T, s) \leq |L(T)|$ (see Fig. 2). Namely, the rank of a character s in a tree T is the number of MSTs that decompose T with respect to the character s, that is, the number of largest MSTs conserving the same value for s. Figure 2 illustrates the concept and shows that the rank of a character is a piece of information extracted purely from the tree. This definition of the rank corresponds to the one used in [3]. A rank $R(T, s) = 1$ means that T is maximal for character s, that is, the value of s is conserved in all species, and a rank $R(T, s) = |L(T)|$ means that each leaf in T for s is an MST, that is, each pair of neighboring leaves in the tree is associated to different values of s. Intuitively, characters with small and large rank have undergone strong and weak evolutionary pressure, respectively.

3.2 Combinatorics of MSTs and Correspondence Scores Between Pairs of Values for the Same Character

To evaluate the coevolution of a pair of characters, we proceed in two steps. First, we analyze the combinatorics of the MSTs associated to a pair of values for these characters and construct a correspondence matrix summarizing the degree of coevolution between all pairs of character values. In the second step, coevolution scores for pairs of characters are inferred from the correspondence matrix. These represent how well the MSTs associated to a character mirror the MSTs associated to another character compared with what would be expected for ideally coevolved characters (see "perfect inclusion" in Figs. 4 and 5a).

3.2.1 Correspondence Matrix Construction

Let A_i be a character value for a character i. For each pair of values A_i, A_j of characters i, j, we consider the "inner" tree $T(A_i, A_j)$ of T, for which only the leaves of T which are labeled by the value A_i for character i or by A_j for character j are considered (see the examples in Fig. 3). The inner tree is used to evaluate the overlap of the MSTs associated to A_i and A_j. We denote by $MST(A_i)$ the set of all MSTs associated to a value A_i of character i.

A correspondence score $C(A_i, A_j)$ is assigned to each pair of values A_i, A_j of characters i, j:

$$C(A_i, A_j) = \frac{N^{A_i + A_j}}{N^{A_i + A_j} + N^{A_i - A_j} + N^{A_j - A_i}},$$

where $N^{A_i + A_j}$ is the number of nodes (leaves excluded) that are common to $MST(A_i)$ and $MST(A_j)$, and $N^{A_i - A_j}$ and $N^{A_j - A_i}$ are the numbers of nodes (leaves excluded) of $MST(A_i)$ and $MST(A_j)$ that do not belong to $MST(A_j)$ and $MST(A_i)$, respectively. Correspondence scores vary between $0 \leq C(A_i, A_j) \leq 1$ with $C(A_i, A_j) = 0$ in the case of a perfect disjunction of $MST(A_i)$ and $MST(A_j)$, and $C(A_i, A_j) = 1$ in the case of a perfect inclusion of $MST(A_i)$ and $MST(A_j)$ (Fig. 4).

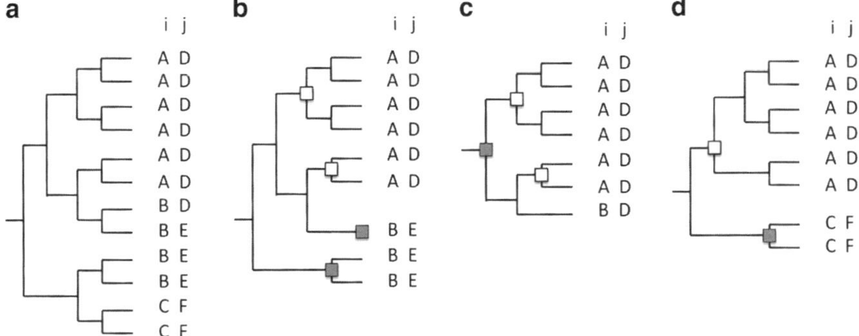

Fig. 3 Inner trees. A tree T (**a**) and "inner" trees: $T(A, E)$, specific to the values A and E of characters i and j, respectively (**b**); $T(A, D)$ (**c**); and $T(A, F)$ (**d**). Only values of characters i and j are taken into consideration. The branches of T labeled with A for character i and with E for character j are shown by *white* and *gray squares*, respectively, and determine the inner tree $T(A, E)$ (**b**). The inner trees $T(A, D)$ (**c**) and $T(A, F)$ (**d**) are determined in a similar way. The *white squares* in $T(A, E)$, $T(A, D)$, and $T(A, F)$ identify the roots of MSTs associated to the value A for character i, and the *gray squares* identify roots of MSTs associated to values E, D, and F of character j (Figure reproduced from [3] under Creative Common Attribution Licence (CCAL))

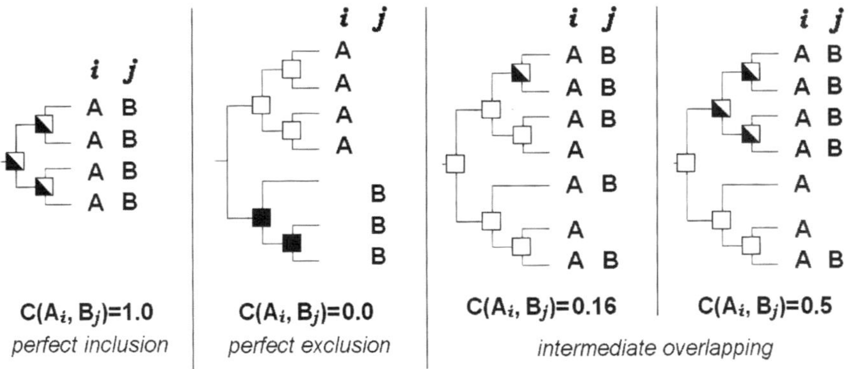

Fig. 4 Overlap of MSTs, and correspondence scores. Different inner trees specific to values A and B of characters i and j and their corresponding correspondence scores. *White squares* identify nodes of $MST(A)$ (leaves excluded), and *black squares* identify nodes of $MST(B)$. The *white* and *black squares* identify common nodes $MST(A)$ and $MST(B)$. The first two trees illustrate perfect inclusion and exclusion. The last two trees illustrate intermediate cases where the numbers of species with values A and B are equal but the correspondence scores are different owing to different distributions of species in the tree (Figure reproduced from [3] under CCAL)

Correspondence scores are calculated for each pair of values A_i, A_j for characters i, j and are organized into a correspondence matrix $C_{i,j}$, indexed by values from the most to the least frequent (an arbitrary order is followed for equal frequencies). A row or column indexed by A_i or A_j contains all correspondence

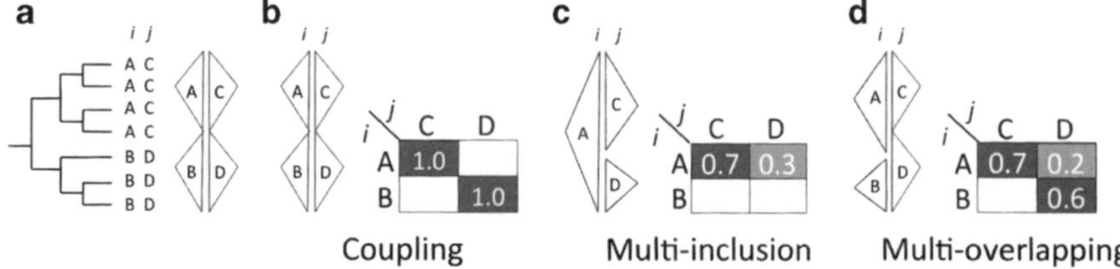

Fig. 5   Correspondence matrices and matrix patterns. (**a**) Tree composed of two MSTs associated to character values A, C and B, D; characters i, j take values A, B and C, D, respectively (*left*). The *right* part shows a representation of the tree on the *left* where the MSTs associated to character values A, B, C, D are distinguished and denoted by *triangles*, pointing to the *left* for character i and to the *right* for character j; the MSTs for these values are represented by *triangles* with the associated character indicated in the *center*. (**b**) Coupling pattern with an identity correspondence matrix. (**c**) Multi-inclusion pattern, where a single value of character i is associated to several values of character j. (**d**) Multi-overlapping pattern, where several values of character i are associated to several values of character j ((**b**), (**c**), (**d**) are reproduced from [3] under CCAL)

scores obtained for A_i or A_j with values of character j or i, respectively. The sum of the correspondence scores in each line and in each column of the matrix $C_{i,j}$ is at most 1.

3.2.2   Patterns in a Correspondence Matrix

Specific patterns may appear in the correspondence matrix according to the combinatorics of the MSTs associated to pairs of character values. The evolution of a character i with itself, for instance, corresponds to the ideal case of coevolution and is characterized by a *perfect inclusion* of the MSTs associated to the same character value ($C_{i,i}(A_i, A_i) = 1$) and by a perfect disjunction of the MSTs associated to all other character values ($C_{i,i}(A_i, B_i) = 0$). This "perfect" configuration corresponds to an identity matrix. In the case of a pair of independent characters i, j, a random overlap of the MSTs $MST(A_i)$ and $MST(A_j)$ is expected instead.

The patterns in these matrices capture three kinds of relations between MSTs associated to pairs of characters:

1. *Coupling.* The MSTs associated to values of character i mirror the MSTs associated to values of character j. This correspondence is represented by an identity correspondence matrix (Fig. 5a).
2. *Multi-inclusion.* The MST associated to a value of character i or j includes several MSTs associated to different values of character j or i, respectively. In Fig. 5b, value A obtains its best correspondence score with value C (since it overlaps mostly with C), but it *lacks specificity* for C since $MST(A)$ also includes $MST(D)$. Values C and D are A-specific, since they do not overlap with any other MST for character i.

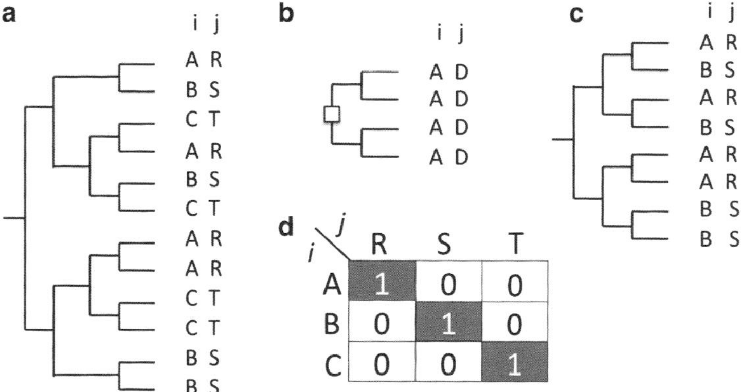

Fig. 6 **Lack of conservation, and perfect coevolution.** (**a**) Tree of characters where the pairs of values $(A, R), (C, T), (B, S)$ for characters (i, j) occur in distinguished leaves. Characters are not conserved (three distinct character values occur with the same frequency), but they coevolve perfectly. The tree is composed of two main subtrees with six leaves each. One of the subtrees shows no MST associated to character values. The second subtree has three MSTs with two leaves each. (**b**) The inner tree $T(A, R)$ of the tree in (**a**) shows perfect inclusion. The inner trees $T(B, S)$ and $T(C, T)$ have the same topology. Their correspondence score is 1. (**c**) The inner tree $T(A, S)$ of the tree in (**a**) shows perfect exclusion. The inner trees $T(A, T)$, $T(B, R)$, $T(B, T)$, $T(C, R)$, and $T(C, R)$ have the same topology. Their correspondence score is 0. (**d**) Correspondence matrix for the tree in (**a**), showing coupling for positions i and j

3. *Multi-overlapping.* The MSTs associated to different values of character i overlap with the MSTs associated to several values of character j. In Fig. 5c, value A shares values D with B. The *interference* of $MST(D)$ with $MST(A)$ neither *excludes* $MST(D)$ from $MST(A)$ nor *includes* it in $MST(A)$.

Coupling describes perfect coevolution between two characters. Since it is unlikely to be observed in real data, the evaluation of coevolution between pairs of characters cannot be reduced to a simple assessment of the presence or absence of a perfect identity matrix. In particular, even for a pair of characters with a good overlap of the MSTs, noise in the data caused by a single value that disrupts the maximality of the tree can lead to a diagonal matrix which is not an identity matrix. Thus, we define a coevolution score between two characters by evaluating the "distance" between an ideal identity matrix (coupling) and the actual correspondence matrix, which displays less regularity (because of a possible combination of multi-overlapping and multi-inclusion), for all values associated to the character.

Conservation of characters is not a necessary requirement for coupling. The tree of characters in Fig. 6 illustrates the idea. Namely, characters i and j in the tree do not display conservation, and the MSTs that preserve a character value at i or j have at most two leaves. Nevertheless, notice that the associated correspondence matrix highlights coupling, since all inner trees for pairs of character values display either a perfect inclusion or a perfect exclusion of the values.

3.3 Coevolution Score for Pairs of Characters

The *coevolution score of two characters* i, j is the sum of two subscores, one evaluated for each value of character i according to all values taken by character j, and the other evaluated for each value of j according to all values taken by i. For each character value, three multiplicative factors are computed. Intuitively, they numerically describe the divergence of the correspondence matrix from the identity matrix, which would be expected in the ideal case. In case of perfect coevolution, the three factors will provide no penalties; they equal 1 for all pairs of values of i, j, and make the two subscores equal to 1. The more the correspondence matrix diverges from the identity matrix, the more these factors tend to 0 and penalize the coevolution score.

3.3.1 Maximal Correspondence Factor

This is defined as

$$S^j_{\max}(A_i) = \max_{X \in R_j} \{C_{i,j}(A_i, X)\},$$

where R_j is the set of all values of character j; this corresponds to the highest correspondence score obtained for A_i when the scores for all values of character j are compared. Note that $0 \leq S^j_{\max}(A_i) \leq 1$. We denote

$$R^j_{\max}(A_i) = \arg \max_{X \in R_j} \{C_{i,j}(A_i, X)\},$$

where, by convention, if the maximum of the function is reached on several values, then $R^j_{\max}(A_i)$ is the most frequent value among them for character j. The maximal correspondence factor penalizes the lack of perfect inclusion among MSTs, which can be due to noise in the data, multi-inclusion, or multi-overlapping.

3.3.2 Specificity Factor

This is defined as

$$S^j_{\text{spec}}(A_i) = \frac{S^j_{\max}(A_i)}{\displaystyle\sum_{X \in R_j} C_{i,j}(A_i, X)},$$

where R_j is the set of values of character j; this evaluates the specificity of A_i for the value $R^j_{\max}(A_i)$. Note that $0 \leq S^j_{\text{spec}}(A_i) \leq 1$. This factor penalizes the lack of specificity which is observed in the case of multi-inclusion and multi-overlapping.

3.3.3 Interference Factor

This is defined as

$$S_{\text{inter}}^{j}(A_i) = 1 - \sum_{X \in R_j \setminus R_{\text{max}}^{j}(A_i)} \left(\frac{C_{i,j}(A_i, X)}{\sum\limits_{Y \in R_i} C_{i,j}(Y, X)} \times \omega_j(X) \right),$$

where R_i is the set of values of character i, R_j is the set of values of character j, and $\omega_j(X)$ is the frequency of value X for j. This evaluates the overlap between $MST(A_i)$ and $MST(X_j)$, with $X_j \neq R_{\text{max}}^{j}(A_i)$. Note that $0 \leq S_{\text{inter}}^{j}(A_i) \leq 1$. This factor penalizes interference of MSTs at j, associated to X_j's that are not $R_{\text{max}}^{j}(A_i)$, with $MST(A_i)$. Interference is observed in the case of multi-inclusion and multi-overlapping.

Toy examples of 2×2 correspondence matrices are presented in Fig. 5. For coupling (Fig. 5a), the factors for value A of character i are $S_{\text{max}}^{j}(A) = 1$, $S_{\text{spec}}^{j}(A) = 1/1 = 1$, and $S_{\text{inter}}^{j}(A) = 1 - (0/1) = 1$, which give $S_{\text{max}}^{j}(A) \times S_{\text{spec}}^{j}(A) \times S_{\text{inter}}^{j}(A) = 1$. The perfect mirroring of the inner trees ensures that the correspondence matrix is the identity matrix.

For multi-inclusion (Fig. 5b), the factors for value A of character i are $S_{\text{max}}^{j}(A) = 0.7$, $S_{\text{spec}}^{j}(A) = 0.7/(0.7 + 0.3) = 0.7$, and $S_{\text{inter}}^{j}(A) = 1 - (0.3/0.3) = 0$, which give $S_{\text{max}}^{j}(A) \times S_{\text{spec}}^{j}(A) \times S_{\text{inter}}^{j}(A) = 0$. No correlation is observed between characters i and j, since a value of j is associated to two values of i, leading to a correspondence matrix far away from the identity matrix. The product of the subscores equals 0 and penalizes the configuration. In the more general case of a combination of several values of i and j displaying overall a good overlap of their MSTs, local multi-inclusion between pairs of values might induce a weak penalizing effect on the final score.

For multi-overlapping (Fig. 5c), the factors for value A of character i are $S_{\text{max}}^{j}(A) = 0.7$, $S_{\text{spec}}^{j}(A) = 0.7/(0.7 + 0.2) = 0.77$, and $S_{\text{inter}}^{j}(A) = 1 - (0.2/(0.2 + 0.6)) = 0.75$, which give $S_{\text{max}}^{j}(A) \times S_{\text{spec}}^{j}(A) \times S_{\text{inter}}^{j}(A) = 0.4$. Here the correspondence matrix is closer to the identity matrix, and the score is penalized less than in the previous case. However, the important multi-overlaps of $MST(D)$ with $MST(A)$, and of $MST(D)$ and $MST(A)$ with $MST(C)$ and $MST(D)$ lead to a rather low product of the subscores, namely 0.4.

3.3.4 Coevolution Score

The coevolution score $CoE(i, j)$ sums the products of the three factors calculated for each value of the pair of characters i, j and weights each product according to the frequency of the value of a given character. We define

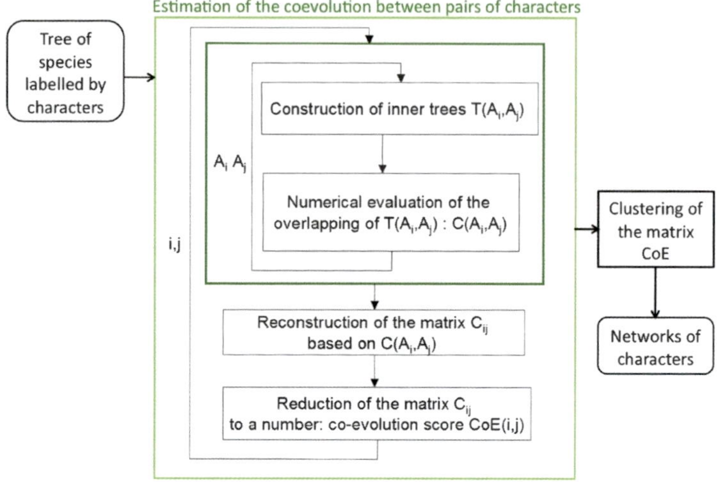

Fig. 7 Flowchart of the analysis. The main algorithmic steps of the analysis are represented by the two main boxes. *Inner box*: estimation of coevolution between pairs of characters; the indices i, j run over characters only. *Outer box*: construction of the coevolution matrix for all characters (Figure modified from [3] under CCAL)

$$CoE^j(i) = \sum_{X \in R_i} S^j_{\max}(X) \times S^j_{\mathrm{spec}}(X) \times S^j_{\mathrm{inter}}(X) \times \omega_i(X),$$

$$CoE^i(j) = \sum_{Y \in R_j} S^i_{\max}(Y) \times S^i_{\mathrm{spec}}(Y) \times S^i_{\mathrm{inter}}(Y) \times \omega_j(Y),$$

and

$$CoE(i, j) = CoE^j(i) + CoE^i(j),$$

where R_j is the set of values of character j, R_i is the set of values of character i, and $\omega_i(X)$ and $\omega_j(Y)$ are the frequencies of values X and Y for characters i and j, respectively. Note that $0 \leq CoE(i, j) \leq 2$ and that $CoE(i, j) = CoE(j, i)$.

Notice that pairs of very conserved characters will show a high overlap of their MSTs and obtain high coevolution scores. In the extreme case of two completely conserved characters, the unique MSTs associated to the two characters will perfectly mirror each other and lead to a maximal coevolution score of 2.

3.4 Networks Reconstruction

The matrix of coevolution CoE provides all of the information needed to extract a network of coevolving characters for the species under consideration (see Fig. 7).

There are clustering algorithms that can extract such a network of characters; see [14] for a method for automatic selection of networks from a matrix and [3] for an automatic method to cluster the matrix followed by manual extraction of networks from the clustered matrix. Such networks are helpful for data interpretation.

3.5 Computational Complexity and Selection of Character Seeds

When there are many characters and the combinatorics of the characters (that is, the number of pairs of characters) to be analyzed results in a significant computational time [3], it may be useful to restrict the analysis to characters that are "sufficiently conserved" and to identify coevolution of them. We call these characters *seeds* and select them by reading the persistency of the conservation signal along the subtrees of the tree.

3.5.1 Conserved Characters and Stability of a Character Along the Tree

We assume that the absence of a character for a species is represented by a special value. This implies that whenever a character is missing in several species then it will be ranked high. We could have chosen to consider the absence of a character as a specific value, and in this case a character missing in several species would imply a low rank. The rank distribution and the mean rank calculated over all characters turn out to be strictly dependent on the definition one chooses.

Let $R_D(T, s)(= R(T, s))$ be the rank of the character s in T and let $\overline{R_D(T)}$ be the mean rank calculated over all characters in T, when missing values of a character are considered as *different* (D). $R_I(T, s)$ and $\overline{R_I(T)}$ denote the rank of character s in the tree T and the mean rank calculated over all characters in T, respectively, when missing values of a character are considered as an *identical* (I) character. A *stable* character s in T is such that $R_D(T, s) - R_I(T, s) < \overline{R_D(T)} - \overline{R_I(T)}$, that is, a character whose rank is not affected much by missing values.

Let $\overline{R(T)}$ be the mean rank calculated over all stable characters in T. A character s in T is a *character seed* if $R(T, s) < \overline{R(T)}$. The intuition here is that we identify (and select for coevolution analysis) as seeds those characters which exhibit a stronger signal of conservation than the average.

Since simple variations in values of a character can lead to different tree decompositions of T into MSTs, and different character ranking and mean ranks, we check the robustness of the conservation of a character over a number of landmark points on T, called *checkpoint nodes*. Below, we formally describe how to select checkpoint nodes in T, and a detailed parameterization of the method for dealing with protein sequences, handling sequence divergence, and evaluating *persistency* of conservation of a character in all subtrees of T rooted at checkpoint nodes is described.

3.5.2 Checkpoint Nodes

Checkpoint nodes are selected in T going from the leaves of the tree up to the root. The first checkpoint nodes are the roots of the smallest subtrees of T whose corresponding species show enough diversity. We require that more than 50 % of the characters have values that are not fully conserved across all species; that is, more than 50 % of the characters will have rank greater than 1. This ensures that species are sufficiently well represented by the pool of characters and are not too similar. The intuition is that conserved characters detected using this threshold are supposed to undergo strong evolutionary pressure and be relevant.

Checkpoint nodes with higher character divergence are defined inductively to be nodes x in T that show at least 10 % of mutated characters more than in the checkpoint node y with highest divergence lying below x. A minimum increase of 10 % in character divergence in x is required between successive checkpoint nodes in order to favor diversity of the subtrees in which positional conservation is evaluated. Jumps of 10 %, the number of mutated characters provide a way to discretize the tree by avoiding an evaluation on all its nodes, which could be affected by phylogenetic effects (certain branches could be populated more heavily with very similar sequences), leading to an overestimation of conservation signals.

Finally, a node in the tree that has reached 90 % of mutated characters is considered to be a checkpoint node, as are its immediate children.

3.5.3 Persistent Conservation of a Character

At each checkpoint node x, the mean rank $\overline{R(T(x))}$ calculated over all stable characters in $T(x)$ is compared with the rank $R(T(x), s)$ for all characters s. The *persistency of conservation of a character s* is measured at each checkpoint node in the tree according to the conserved status of the character s within the subtree rooted at that node. We define a function P_s for this as follows. If a character s is conserved at checkpoint node x (i.e., $R(T(x), s) < \overline{R(T(x))}$), P_s is incremented by a weight $i(T(x), s)$ corresponding to the maximum number of consecutive checkpoint nodes encountered on a path of the tree $T(x)$ from x down to some leaf. If a character s is not conserved (i.e., $R(T(x), s) \geq \overline{R(T(x))}$), P_s is decremented by a weight $d(T(x), s)$ corresponding to the maximum number of consecutive checkpoint nodes where s is not conserved encountered on a path of the tree $T(x)$ from x (including that node) down to some leaf.

At the root of T, P_s measures the stability of conservation for the character s in T. Characters that are conserved in all subtrees rooted at checkpoint nodes have a positive persistency score $P_s \gg 0$, and characters conserved in none of the subtrees rooted at checkpoint nodes have a negative persistency score $P_s \ll 0$. The persistency score of other characters might take a positive or negative value according to the global conservation evaluated at different checkpoint nodes. Characters with a positive persistency score $P_s > 0$ at the root of T are considered as *persistently conserved* and are selected as seed characters for the analysis of coevolving characters.

Note that not all seed characters are guaranteed to belong to some coevolving network at the end of the analysis. Seeds display some evolutionary pressure and consistent behavior along the tree, and in this respect they form a set of potential coevolving characters to which one can restrict the analysis so as to reduce the overall computational time arising from the large number of pairs of combinatorial characters. Note also that the thresholds defining the checkpoint nodes along the distance tree provide a computationally fast way to avoid phylogenetic effects that might contribute negatively to persistency conservation.

4 Discussion

The method introduced in this chapter provides a mathematical framework where the concept of coevolution of characters can be defined combinatorially. The advantage of developing combinatorial approaches compared with more implicit statistical approaches is that combinatorial methods are based on a direct understanding of the building blocks involved in a structure. In this specific case, the definition of coevolution helps to describe the interaction of coevolving information within inner subtrees of a tree of species. In contrast, implicit approaches provide little intuition about these building blocks.

One of the main questions for future investigation is to find mathematical properties that will highlight the interaction of networks of coevolving characters. Some of these properties might correspond to nonobvious overlapping of inner subtrees. This kind of question has never been addressed by the available approaches, and we expect combinatorics to help provide new insights into evolutionary signals in species. In the case of protein sequences, an attempt to unravel interactions of networks of coevolving residues (providing insights to protein folding) can be found in [13].

4.1 Coevolution and Number of Species

Coevolution analysis methods usually require sets of data to be of a sufficiently large size. This is especially true for statistical approaches to coevolution analysis. The combinatorial method presented here makes the analysis less subject to sample size effects due to a small number of species. Novel approaches that allow the extraction of coevolving characters for a small number of species (say, less than 50) need to be developed. The answer provided in [13] for coevolution analysis of protein sequences might plausibly be generalizable to characters. Other propositions are envisageable and the question is still open, even though it is extremely important in biology, where the space of characters is typically very large and the number of species small.

4.2   Conservation and Coevolution

At first, the notions of conservation and coevolution might appear to be distinct concepts, but our combinatorial approach exploits the idea that during evolution, conservation comes before coevolution in time (this is inherent in the counting of the number of inner subtrees $N^{A_i+A_j}$ in the correspondence score $C(A_i, A_j)$) and that conservation occupies a specific position in the continuous spectrum on which we measure different degrees of coevolution. The model that we intend to use identifies a set of species as an ensemble that evolves through mutational changes driven by the functional roles of the characters. If two or more characters cooperate in serving a function, they will coevolve together. Depending on the evolutionary constraints due to the degree of specificity of the interaction between species, the signals of coevolution will be stronger or weaker. Note that two characters which are fully conserved are treated by the method as being "perfectly coevolving" and that, in this sense, conservation can be treated mathematically as an extreme case of coevolution.

4.3   Random Processes and Constraints

Protein evolution occurs by random processes of mutation subject to a number of functional and structural constraints governing the interactions of the molecule. A way to think of this is to take an ancestor and imagine that within a set of constraints on the ancestor, the process bifurcates by generating new evolutionary solutions and develops until it produces the molecules observed today. We have shown [4] that within the same protein, different domains follow evolutionary pathways that we observe in very different branches of the tree of species. This observation highlights the fact that the evolutionary process of proteins is in the first place a random process, and that functional and structural constraints on a molecule cannot be ignored if we wish to understand the process, since these constraints can give rise to different functional solutions. Studies of the process of evolution on a large scale do not take account of these constraints, which are molecule-specific. Because of this, the computational exploration of divergent homologous proteins has reached its limits and new insights are needed to overcome the difficulties. Methods such as the one we have described, highlight new signals that can be taken into consideration while searching for homologous proteins. In particular, the paradigm that we have described for protein evolution might be reconsidered for evolutionary processes in general.

4.4 Application of the Theory to Protein Sequence Alignments

The theory of coevolving characters has been validated on families of protein sequences, where residue positions in multiple alignments are considered as characters, and the residues are their values. Based on this theory, networks of coevolving positions within the alignments have been identified [3]. These networks are clusters of residues, often entering into physical contact with one another, and they relate residues which are located far apart in the three-dimensional structure. Coevolved residues often play a major biological role in proteins, and the nature of their interactions might be multiple, spanning binding specificity, allosteric regulation, and conformational changes of the protein. By carefully tracing the way residues evolved within the phylogenetic tree of sequences of a protein family, the maximal-subtree method captures the transition from a conserved position to a coevolved position during evolution, and provides a numerical evaluation of the degree of coevolution of pairs of coevolved residues in a protein. This combinatorial approach drops the constraints on high sequence divergence that limit the range of applicability of the statistical approaches previously proposed, and it can be applied with high accuracy to families of protein sequences with variable divergence. Relative coevolution score matrices have been successfully computed for four protein families: the hemoglobin, serine protease, leucine dehydrogenase, and PDZ domain families, for which associated networks have been identified. The program for the coevolution analysis and the clusterization procedure is publicly available [3]. It can be used as a basis to develop new tools for coevolution of characters. Further evaluations of the performance of this system compared with others presented in [13, 19, 35, 41] were reported in [13].

References

1. A. Armon, D. Graur, N. Ben-Tal, ConSurf: an algorithmic tool for the identification of functional regions in proteins by surface mapping of phylogenetic information. J. Mol. Biol. **307**, 447–463 (2001)
2. W.R. Atchley, K.R. Wollenberg, W.M. Fitch, W. Terhalle, A.W. Dress, Correlation among amino acid sites in bHLH protein domains: an information theoretic analysis. Mol. Biol. Evol. **17**, 164–178 (2000)
3. J. Baussand, A. Carbone, A combinatorial approach to detect co-evolved amino-acid networks in protein families with variable divergence. PLoS Comput. Biol. **5**(9), e1000488 (2009)
4. J. Bernardes, G. Zaverucha, C. Vaquero, A. Carbone, High performance domain identification in proteins explores a multitude of diversified profiles with grid computing (2012). Manuscript submitted
5. L. Brillouin, *Science and Information Theory* (Dover Publications, Mineola, 2004), p. 293
6. A. Carbone, L. Dib, Co-evolution and information signals in biological sequences. Theor. Comput. Sci. (2010). doi:10.1016/j.tcs.2010.10.040
7. A. Carbone, S. Engelen, Information content of sets of biological sequences revisited, in *Algorithmic Bioprocesses*, ed. by A. Condon, D. Harel, J.N. Kok, A. Salomaa, E. Winfree. Natural Computing Series (Springer, Berlin/Heidelberg, 2008)

8. A. Carbone, F. Képès, A. Zinovyev, Codon bias signatures, organisation of microorganisms in codon space and lifestyle. Mol. Biol. Evol. **22**(3), 547–561 (2004)
9. G. Cheng, B. Qian, R. Samudrala, D. Baker, Improvement in protein functional site prediction by distinguishing structural and functional constraints on protein family evolution using computational design. Nucleic Acids Res. **33**, 5861–5867 (2005)
10. T. Cover, J. Thomas, *Elements of Information Theory* (Wiley, New York, 1991)
11. A. Del Sol, M.J. Arauzo-Bravo, D. Amoros, R. Nussinov, Modular architecture of protein structures and allosteric communications: potential implications for signaling proteins and regulatory linkages. Genome Biol. **8**, R92 (2006)
12. A. Del Sol, H. Fujihashi, D. Amoros, R. Nussinov, Residues crucial for maintaining short paths in network communication mediate signaling in proteins. Mol. Syst. Biol. **2**, 2006.0019 (2006)
13. L. Dib, A. Carbone, Protein fragments: functional and structural roles of their coevolution networks. PLoS ONE **13**, 194 (2012)
14. L. Dib, A. Carbone, CLAG: an unsupervised non hierarchical clustering algorithm handling biological data. BMC Bioinform. **13**, 194 (2012)
15. R.I. Dima, D. Thirumalai, Determination of networks of residues that regulate allostery in protein families using sequence analysis. Protein Sci. **15**, 258–268 (2006)
16. S. Engelen, L. Trojan, S. Sacquin-Mora, R. Lavery, A. Carbone, Joint evolutionary trees: detection and analysis of protein interfaces. PLoS Comput. Biol. **5**(1), e1000267, 1–17 (2009)
17. M. Fares, S.A.A. Travers, A novel method for detecting intramolecular coevolution: adding a further dimension to select constraints analyses. Genetics **173**, 9–13 (2006)
18. J.H. Gillespie, *Population Genetics: A Concise Guide* (Johns Hopkins Press, Baltimore, 1998)
19. G.B. Gloor, L.C. Martin, L.N. Wahl, S.D. Dunn, Mutual information in protein multiple sequence alignments reveals two classes of coevolving positions. Biochemistry **44**, 7156–7165 (2005)
20. C.C. Goh, A.A. Bogan, M. Joachmiak, D. Walther, F.E. Cohen, Coevolution of proteins with their interaction partners. J. Mol. Biol. **299**, 283–293 (2000)
21. D. Hartl, *Principles of Population Genetics* (Sinauer Associates Publisher, Sunderland, 2007)
22. C.A. Innis, siteFiNDER-3D: a web-based tool for predicting the location of functional sites in proteins. Nucleic Acids Res. **35**(Web-Server-Issue), 489–494 (2007)
23. P.D. Kreil, C.A. Ouzounis, Identification of thermophilic species by the amino-acids composition deduced from their genomes. Nucleic Acids Res. **29**, 1608–1615 (2001)
24. S. Kullback, R.A. Leibler, On information and sufficiency. Ann. Math. Stat. **22**(1), 79–86 (1951)
25. O. Lichtarge, M.E. Sowa, Evolutionary predictions of binding surfaces and interactions. Curr. Opin. Struct. Biol. **12**, 21–27 (2002)
26. O. Lichtarge, H.R. Bourne, F.E. Cohen, An evolutionary trace method define binding surface common to protein families. J. Mol. Biol. **257**, 342–358 (1996)
27. S.W. Lockless, R. Ranganathan, Evolutionary conserved pathways of energetic connectivity in protein families. Science **286**, 295–299 (1999)
28. D.J. Lynn, G.A. Singer, D.A. Hickey, Synonymous codon usage is subject to selection in thermophilic bacteria. Nucleic Acids Res. **30**, 4272–4277 (2002)
29. I. Mihalek, I. Res, O. Lichtarge, A family of evolution-entropy hybrid methods for ranking protein residues by importance. J. Mol. Biol. **336**, 1265–1282 (2004)
30. N. Ota, D.A. Agard, Intramolecular signaling pathways revealed by modeling anisotropic thermal diffusion. Eur. J. Mol. Biol. **351**, 345–354 (2005)
31. D.D. Pollock, W.R. Taylor, Effectiveness of correlation analysis in identifying protein residues undergoing correlated evolution. Protein Eng. **10**, 647–657 (1997)
32. T. Pupko, R.E. Bell, I. Mayrose, F. Glaser, N. Ben-Tal, Rate4Site: an algorithmic tool for the identification of functional regions in proteins by surface mapping of evolutionary determinants within their homologues. Bioinformatics **18**, S71–S77 (2002)
33. A.K. Ramani, E.M. Marcotte, Exploiting the coevolution of interacting proteins to discover interaction specificity. J. Mol. Biol. **327**, 273–284 (2003)

34. C.E. Shannon, A mathematical theory of communication. Bell Syst. Tech. J. **27**(3), 379–423 (1948)
35. G.M. Suel, S.W. Lockless, M.A. Wall, R. Ranganathan, Evolutionary conserved networks of residues mediate allosteric communication in proteins. Nat. Struct. Biol. **23**, 59–69 (2003)
36. F. Tekaia, E. Yeramian, B. Dujon, Amino acid composition of genomes, lifestyles of organisms, and evolutionary trends: a global picture with correspondence analysis. Gene **297**, 51–60 (2002)
37. J.N. Thompson, *The Geographic Mosaic of Coevolution* (University of Chicago Press, Chicago, 2005)
38. J.D. Watson, R.A. Laskowski, J.M. Thornton, Predicting protein function from sequence and structural data. Curr. Opin. Struct. Biol. **15**, 275–284 (2005)
39. M. Weigt, R.A. White, H. Szurmant, J.A. Hoch, T. Hwa, Identification of direct residue contacts in protein-protein interaction by message passing. Proc. Natl. Acad. Sci. U.S.A. **106**, 67–72 (2009)
40. H. Willenbrock, C. Friis, A.S. Juncker, D.W. Ussery, An environmental signature for 323 microbial genomes based on codon adaptation indices. Genome Biol. **7**(12), R114 (2006)
41. C.-H. Yeang, D. Haussler, Detecting coevolution in and among proteins domains. PLoS Comput. Biol. **3**, 2122–2134 (2007)

When and How the Perfect Phylogeny Model Explains Evolution

Paola Bonizzoni, Anna Paola Carrieri, Gianluca Della Vedova,
Riccardo Dondi, and Teresa M. Przytycka

Abstract Character-based parsimony models have been among the most studied
notions in computational evolution, but research in the field stagnated until some
important, recent applications, such as the analysis of data from protein domains,
protein networks, and genetic markers, as well as haplotyping, brought new life
into this sector. The focus of this survey is to present the perfect phylogeny model
and some of its generalizations. In particular, we develop the use of persistency
in the perfect phylogeny model as a new promising computational approach to
analyzing and reconstructing evolution. We show that, in this setting, some graph-
theoretical notions can provide a characterization of the relationships between
characters (or attributes), playing a crucial role in developing algorithmic solutions
to the problem of reconstructing a maximum parsimony tree.

1 Introduction

Evolution is a lens that allows us to study and understand a lot of phenomena in
molecular biology [8]. The prototypical representation of any evolutionary history
is a phylogeny, that is, a labeled tree whose leaves are extant species, or individuals,

P. Bonizzoni (✉) · A.P. Carrieri · G. Della Vedova
DISCo, Università degli Studi di Milano–Bicocca, Milan, Italy
e-mail: bonizzoni@disco.unimib.it; annapaola.carrieri@disco.unimib.it;
gianluca.dellavedova@unimib.it

R. Dondi
Dipartimento di Scienze Umane e Sociali, Università degli Studi di Bergamo, Bergamo, Italy
e-mail: riccardo.dondi@unibg.it

T.M. Przytycka
National Center for Biotechnology Information, National Library of Medicine, National Institutes
of Health, Bethesda, MD, USA
e-mail: przytyck@ncbi.nlm.nih.gov

N. Jonoska and M. Saito (eds.), *Discrete and Topological Models in Molecular Biology*,
Natural Computing Series, DOI 10.1007/978-3-642-40193-0_4,
© Springer-Verlag Berlin Heidelberg 2014

or simply data that we are currently able to analyze [11]. Phylogenetics is the research area of computational biology devoted to computing phylogenies. In this field, the focus has shifted over the years. The initial developments date back to the pioneering work of Cavalli-Sforza and Edwards [6, 9] in the 1960s, where some fundamental ideas in the study of phylogenies were introduced, namely the fact that evolution is a branching process where characters change, that an intuitive approach is to find the minimum total number of evolutionary events compatible with the available data, and the idea of maximizing the likelihood of the proposed interpretation.

The limited computational resources at the time, together with the kind of data available (phenotypical data were much more frequent than genomic data), initially put the emphasis on maximum parsimony character-based approaches. Subsequent advances, including some in the statistical modeling of evolution [11], made approaches based on inferring maximum likelihood phylogenies more attractive.

More recently, the pendulum has swung again, as parsimony methods have found new relevance, mostly as a result of new applications. The perfect phylogeny model, which is conceptually the simplest, is based on the infinite-sites assumption; that is, no character can mutate more than once in the whole tree. Although this assumtion is quite restrictive, bordering on plainly wrong in some cases, the perfect phylogeny model has turned out to be splendidly coherent in the context of the haplotyping problem [3, 18], where we want to distinguish between the two haplotypes present in each individual when given only genotype data. More precisely, the interest here is in computing a set of haplotypes and a perfect phylogeny such that the haplotypes (i) label the vertices of the perfect phylogeny and (ii) explain the input set of genotypes. This context has been studied deeply in the last decade, giving rise to a number of beautiful algorithms [2, 7]. Those algorithms (and others on the same topics) exploit a number of nice combinatorial properties of perfect phylogenies and graphs. In [2], a graph-theoretical characterization of genotype matrices admitting a tree representation was given by using properties of partial orders and Dilworth's Theorem [12]. In its original formulation, the haplotyping problem, under the perfect phylogeny model [7, 18], has revealed an interesting connection with the graph realization problem [29], a well-known graph problem used to decide whether a matroid is graphic.

Still, the perfect phylogeny model and the assumptions that have been central in previous decades cannot be employed without adaptations or improvements. One of the main open problems regarding the model is finding generalizations that retain the computational tractability of the original model but are more flexible in modeling biological data. Following this research direction, we explore here some extensions of the perfect phylogeny model that are capable of modeling some processes whose study has been motivated by some recent applications.

In particular, we present two recent applications that can find only a partial solution in perfect phylogenies. The first application is to carcinogenesis, i.e., the factors and mechanisms that cause the onset of cancer in cells. Carcinogenesis can result from many combinations of mutations, but only a few sequences of mutations, called *progression pathways*, seem to account for most human tumors [28]. The main issue here is characterizing the common progression pathways as a first step towards identifying therapeutic targets and reliable diagnostic tests. The natural observation that tumors are evolving cell populations leads to phylogeny-based studies. At the same time, the intrinsic nature of cancer cells, that is, cells that proliferate quickly and in a degenerate way, results in a relatively high number of sites with multiple mutations (in violation of the infinite-sites assumption).

The second application concerns the study of protein domains. A protein domain is a part of the sequence and structure of a protein that can evolve, function, and exist independently of the rest of the protein chain. Many proteins consist of several structural domains, and a domain may appear in a variety of different proteins. In this case, it is quite frequent for a protein to acquire a domain and then to lose it (this is much more frequent than acquiring and then losing a whole gene). Again, the infinite-sites assumption can be violated.

In this survey, we pay special attention to an approach proposed in [23], based on the notion of persistent characters in the perfect phylogeny model and on its use to exclude some characters from the construction of the phylogeny. The general focus will be on computational issues, such as efficient algorithms.

2 Maximum Parsimony and the Perfect Phylogeny

Parsimony models, just like all models, are characterized by specific constraints that are based on biological assumptions. The first basic assumption states that each species or taxon is described by a set of attributes, called characters, where each character is inherited independently, and each character can assume one of a finite set of values, called *states*. Alternatively, the input is a matrix whose rows are the taxa and the columns are the characters. Another basic assumption about the evolution of characters, called *homology*, assumes that characters that are present in more than one species must be inherited from a common ancestor.

The natural computational problem has, as input, a matrix M with n rows and m columns, where each row can be viewed as an m-vector over the set of states of characters. The matrix describes a set of n taxa (species or individuals), corresponding to the rows of M, and a set of m characters, corresponding to the columns of M, and we seek a minimum-cost tree that explains the input matrix M. In a tree T *explaining* a matrix M, (i) the nodes are labeled by vectors of states, of length m (ii) each row of M labels exactly one node of T, (iii) the leaves are labeled by some rows of M, and (iv) each edge (r_1, r_2) of T is labeled by the character c of M whose state in r_1 differs from that in r_2 (see Fig. 1 for an example).

The cost of a tree is the number of mutations in the tree or, more formally, the sum over all edges of the tree of the cost of each edge, given by the number of characters with different states in the two nodes that make up the edge. In binary parsimony models – the most widely used – characters can take only the values (or

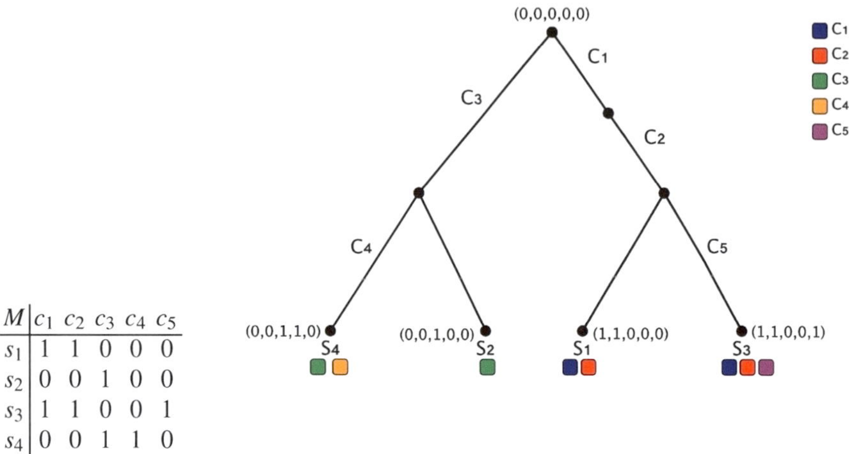

$$
\begin{array}{c|ccccc}
M & c_1 & c_2 & c_3 & c_4 & c_5 \\
\hline
s_1 & 1 & 1 & 0 & 0 & 0 \\
s_2 & 0 & 0 & 1 & 0 & 0 \\
s_3 & 1 & 1 & 0 & 0 & 1 \\
s_4 & 0 & 0 & 1 & 1 & 0
\end{array}
$$

Fig. 1 Example of perfect phylogeny over a binary matrix M of five characters and four species

states) zero and one, usually interpreted as the presence or absence of an attribute in the taxon.

We will now discuss how computing the maximum parsimony phylogeny can be framed as a Steiner tree problem, which is one of the most widely studied problems in operations research. Recall that each input taxon is viewed as a binary vector of length m. The set of all possible binary vectors of length m forms a hypercube H, whose edges are exactly the pairs of vertices (u, v) where u and v differ in exactly one position. Let S be the set of input species in the phylogeny problem, and notice that S is also a subset of the vertices of the hypercube H. Then the Steiner tree problem asks for a minimum-cost subtree T of H such that all vertices in S are also in T. The cost of the solution T is the number of edges of T. The Steiner tree problem is NP-hard [21], even in the case of a binary alphabet with the metric induced by the Hamming distance [14], which is a restriction derived from the reduction from the maximum parsimony phylogeny to the Steiner tree on a hypercube. Extensive recent work, both experimental and theoretical, has focused on the binary character set with the Hamming metric [25, 27].

We can now introduce some specific parsimony models, starting from the simplest: the perfect phylogeny. A tree is called a *perfect* phylogeny if each character i mutates exactly once (i.e., there is exactly one edge such that its vertices are labeled by vectors differing in position i). Note that a perfect phylogeny (if it exists) minimizes the overall cost, as any perfect phylogeny has cost m. We call a perfect phylogeny *directed* or *rooted* if there is a distinguished node corresponding to the vector $[0, \ldots, 0]$. It can be noticed immediately that we can transform a perfect phylogeny into a rooted perfect phylogeny by choosing an arbitrary node x and flipping (for each species) the state of each character that initially has value 1 in x (those characters are called *active* in x). There is a well-known linear-time algorithm for computing a binary perfect phylogeny [17], if it exists, and some more complicated fixed-parameter algorithms for the general perfect phylogeny problem, where the parameter is the maximum number of states for each character [1, 20]. In the following, unless specified differently, we mean by "perfect phylogeny" a rooted perfect phylogeny, that is, characters mutate only from state zero to one.

3 The Dollo Parsimony Model and Its Variants

Unfortunately, there are some evolutionary phenomena, such as *homoplasy*, that violate the fundamental assumptions of perfect phylogeny [11]. Two kinds of homoplasy are *recurrent mutations* and *back mutations*. A recent mutation occurs when a character changes state along divergent branches of the tree, and a back mutation implies that a character may go back to the ancestral state in descendant species after changing state. These two types of events justify the introduction of different models, differing mainly in the allowed homoplasies. Although the perfect phylogeny model does not allow any homoplasy, some extended models have been introduced to allow recurrent or back mutations.

One extended model is the Camin–Sokal parsimony model [5] (Fig. 2), where characters are *directed*; that is, only changes from zero to one are possible on any path from the root to a leaf. This fact means that the root is assumed to be labeled by the ancestral state with all zeros, and no back mutation is allowed, but any character can be acquired more than once, i.e., recurrent mutations are possible.

Another possible way of extending the perfect phylogeny model is the Dollo parsimony model, which allows any character to change state from zero to one only once, but puts no restriction on the number of times that it mutates from one to zero [11] (Fig. 3); that is, back mutations are allowed, but recurrent mutations are not. The definition of the Dollo parsimony model implies that characters are acquired at most once in the tree, but may be lost multiple times.

An interesting application of the Dollo parsimony model is to the analysis of dynamic protein interactions [31], which has also shown an interesting connection with graph theory. Protein networks are graphs that model protein interactions. More precisely, the nodes of the graph are the proteins studied and the edges represent the interactions. A functional module is a subset of the proteins that have a common biological function. Usually, a functional module is not a generic graph, as it is made of overlapping cliques or quasi-cliques (which are called functional groups or complexes). It is possible to represent the interactions of those functional groups and complexes by a tree, called a *tree of complexes*, whose nodes are the functional groups (to be identified as cliques or quasi-cliques of an original protein network) and are such that the set of nodes consisting of the functional groups containing any given common protein is connected.

Let us denote each complex or protein with a distinct symbol from an alphabet Σ. Then the tree-of-complexes (TC) problem, over an instance consisting of a set $A = \{a_1, a_2, \cdots, a_m\}$ of subsets of Σ, asks for a tree T, if it exists, whose nodes are the input sets and are such that for each $\sigma \in \Sigma$, the set of nodes to which σ

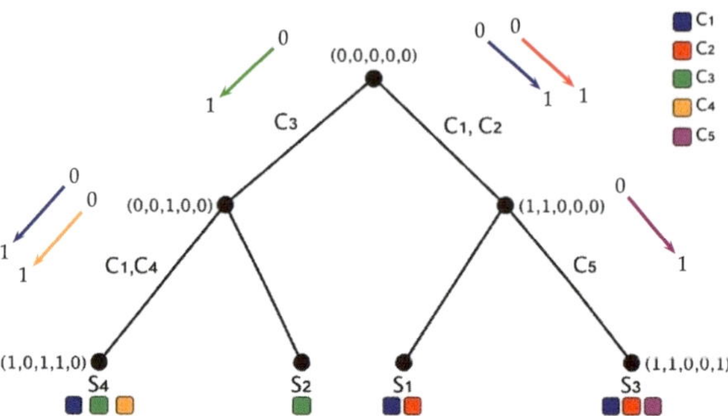

Fig. 2 Example of a Camin–Sokal parsimony model over the same set of characters as in Fig. 3. Observe that character c_1 is gained twice in the tree

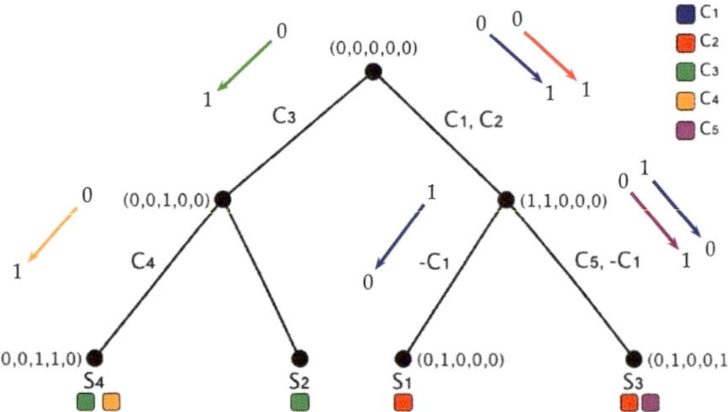

Fig. 3 Example of a Dollo parsimony model over a matrix of five characters. Observe that character c_1 is the only one that is lost in the tree

belongs is a subtree of T. Clearly, the TC problem admits several solutions that may explain a set S. The following property has never been, to the best of our knowledge, explicitly pointed out previously.

Lemma 1. *Let* $A = \{a_1, a_2, \cdots, a_m\}$ *be an instance of the TC problem admitting a tree of complexes* T. *Then* T *is compatible with the Dollo parsimony model (i.e., no two characters are acquired more than once).*

Proof. Let σ be a generic symbol, and let $N(\sigma)$ be the set of nodes of T with σ. By the definition of the tree of complexes, $N(\sigma)$ induces a connected subtree of T. It is not a limitation to assume that $|N(\sigma)| > 1$. Let x be the least common ancestor of $N(\sigma)$. We claim that the incoming arc in x is the only one where σ is acquired. By the definition of the least common ancestor in a tree, (i) only nodes that are descendants of x can have the symbol σ, and (ii) there are two nodes $v_1, v_2 \in N(\sigma)$ such that all paths in T connecting v_1 and v_2 pass through x; therefore σ is active in x, for otherwise $N(\sigma)$ would be disconnected. Consequently, σ is acquired in x. Assume now, on the contrary, that σ is also acquired in node x_1, which is a descendant of x, and let x_2 be the parent of x_1. Since σ is acquired in node x_1, σ is active in x_1 but not in x_2. Since all paths in T connecting x and x_1 must pass through x_2, $N(\sigma)$ is disconnected, contradicting the hypothesis that T is a tree of complexes.

The connection between trees of complexes and graph theory is deeper. For instance, when S is the set of cliques of a chordal graph, the tree of complexes can be obtained from the clique tree associated with the chordal graph [31]. In fact, chordal graphs are exactly those that admit a clique tree representation. Recall that a graph is chordal if the only vertex-induced subgraphs that are also cycles have exactly three vertices [16].

One of the main open questions in [31] is how to provide a characterization of the protein networks that admit a tree-of-complexes representation. Lemma 1 shows the equivalence of this open problem to the question of finding the protein networks that admit an evolutionary representation of functional groups compatible with a Dollo parsimony model.

As pointed out in the introduction, the perfect phylogeny model is too restrictive for some applications, since it cannot explain the evolution of characters in the presence of homoplasy events. On the other hand, the optimization problems associated with the Dollo and Camin–Sokal parsimony models are NP-hard [11]. Moreover, these models are too general to be useful in practical applications where interesting characters are usually affected by only a few back mutations or recurrent mutations. Therefore, research activity has focused on finding models that couple computational tractability with the capability to adequately model actual phenomena, for example in the context of proteomics when one is analyzing the properties of multidomain proteins [23, 24].

Notice that, unlike a perfect phylogeny, a Dollo phylogeny always exists. This can be seen by assuming a special internal node $[1, \ldots, 1]$ that is also the least common ancestor of all leaves. This fact implies that a rooted Dollo parsimony model always exists for any input matrix. Since there is no restriction on mutations from 1 to 0, any binary vector can be generated. Although there is no guarantee that such a tree is optimal, it suffices to prove the existence of a Dollo phylogeny. However, such a tree makes no sense from a biological point of view, because it implies the existence of an ancestral taxon that has all of the characters in the extant taxa.

We have already pointed out that the problem of constructing a maximum parsimony tree is a special case of the well-studied problem of a Steiner tree on a hypercube, but the set of allowed homoplasies can influence in a fundamental way the computational complexity of the resulting problem. An initial effort towards describing new, relevant variants of the Dollo parsimony model has been the introduction of the *conservative Dollo* and *static Dollo parsimony* models [24].

The *static Dollo parsimony* model is a Dollo parsimony model where, for each node x and for each active character c in x, there exists a leaf l that is a descendant of x and where c is active. The *conservative Dollo parsimony* model is a Dollo parsimony model where, for each node x and for each pair c_1 and c_2 of active characters, there exists a leaf l that is a descendant of x and where both c_1 and c_2 are active. Notice that both of these models forbid the presence of an ancestral active character that is not shared with some extant species. The main motivations for those models arise in the study of multidomain protein evolution in terms of domain insertions and losses. A protein domain is a part of the sequence and structure of a protein that can evolve, function, and exist independently of the rest of the protein chain; the approach followed represents the domain structure as taxa, and the domains are the characters. A character that is, active for a certain taxon represents the fact that a domain is part of a given architecture. Hence, a state change from 0 to 1 corresponds to the addition of a domain, and a change from 1 to 0 corresponds to a domain loss. A conservative Dollo parsimony model for a protein family is a history where each domain pair that is observed in an extant taxon has been generated from a single merge event. Since the simultaneous presence of two domains in one protein often enhances the functionality of that protein, the model suggests it is highly unlikely that such a pair has been separated (and its enhanced functionality has not survived) in all extant species.

Although optimization problems associated with the static and conservative Dollo parsimony model, where the number of back mutations is minimized, are both NP-hard, there are two fast algorithms for testing if such a phylogeny exists [24]. However, an experimental analysis [24] shows that a sizable minority of multidomain protein superfamilies do not admit a static Dollo parsimony model (and, a fortiori, a conservative Dollo parsimony model). Hence an even less restrictive model is necessary to successfully model those cases.

4 Persistent Phylogeny

An important ingredient that may affect the applicability and success of parsimony methods is the set of characters used to infer the phylogeny. The issue of selecting characters was addressed in [23], where the notion of a *persistent* or *stable* character was proposed. Such characters are allowed to violate the properties of a perfect phylogeny, as a persistent character is gained exactly once but can be lost at most once in the tree.

Based on this notion, a different model, which is intermediate between the perfect phylogeny and the Dollo parsimony models, called the *persistent phylogeny* has been proposed [4]. Notice that a persistent perfect phylogeny is also a Dollo phylogeny, and even a static Dollo parsimony model. In fact, a persistent phylogeny is a static Dollo parsimony model where all but at most one of the descendants of a species with any given character must retain that character. Moreover, differently from the Dollo parsimony model, some matrices may not admit a persistent perfect phylogeny. Therefore, the main computational problem that we will discuss in this section is to compute (if it exists) a persistent perfect phylogeny compatible with a given matrix M. The computational complexity of this problem is still unsettled; there exists an algorithm that is exponential in the number of characters but polynomial in the number of species [4]. This time complexity makes the algorithm of practical interest for the biological applications discussed above, as usually the number of species is large, whereas the number of characters is bounded.

The notion of an *overlap graph*, which is a graph whose nodes are the characters and where two characters are adjacent if and only if there exists a species with both characters, is useful in this context. In fact, if a matrix M admits a persistent phylogeny, then the corresponding overlap graph is chordal [23].

One of the first applications of the persistent phylogeny model was to the study of introns, which are sequences of noncoding DNA in eukaryotic genes. In fact, the Dollo parsimony model has led to an incorrect evolutionary tree for such data, whereas assuming the persistent phylogeny model has resulted in an evolutionary tree consistent with the Coelomata hypothesis, that is, that there is a clade comprising arthropods and chordates. In contrast, an analysis of more variable introns favored the Ecdysozoa topology, that is, a clade of arthropods and nematodes [30]. The controversy about the Coelomata and Ecdysozoa topologies is one of the most discussed and persistent problems in animal phylogeny.

For the sake of completeness, we recall here the definition of a persistent phylogeny given in [4]. Let M be a binary matrix of size $n \times m$. The *persistent phylogeny* for M is a rooted tree T that satisfies the following properties:

1. Each node x of T is labeled by a vector l_x of length m.
2. The root of T is labeled by a vector of all zeros, and for each node x of T the value $l_x[j] \in \{0, 1\}$ is the state of character c_j at this node.
3. For each character c_j, there are at most two edges $e = (x, y)$ and $e' = (u, v)$ such that $l_x[j] \neq l_y[j]$ and $l_u[j] \neq l_v[j]$ (representing a change in the state of c_j) and such that e, e' occur along the same path from the root of T to a leaf of T; if e is closer to the root than e', then the edge e where c_j changes from 0 to 1 is labeled c_j^+, and while edge e' is labeled c_j^-.
4. Each row of M labels exactly one leaf of T.

Thus the main problem investigated in this section, called the persistent phylogeny problem, is this: given a binary matrix M as input, find a persistent phylogeny for M if such a tree exists.

We will devote the remainder of the section to the discussion of the algorithm presented in [4] for determining whether an input matrix M admits a persistent phylogeny and, if that is the case, for computing such a phylogeny (although the solution computed might not be the most parsimonious).

First of all, we recall that there exists a very simple test to determine if M admits an unrooted perfect phylogeny. Two characters c_1 and c_2 are in *conflict* in the matrix M if and only if the two corresponding columns of M contain the four possible rows $(0, 0)$, $(0, 1)$, $(1, 1)$, $(1, 0)$, called the four gametes. A matrix M has an unrooted perfect phylogeny if and only if no two of its characters are in conflict. The test for a matrix M in the rooted case consists of verifying that M has no induced matrix consisting of the three configurations $(0, 1)$, $(1, 1)$, $(1, 0)$.

Conflicting characters in a matrix can be represented by an undirected *conflict graph* $G_c = (C, E \subseteq C \times C)$, where the nodes are the characters and two characters are adjacent if they are in conflict in M. Clearly, having an edgeless conflict graph is a necessary but not a sufficient condition for having a rooted perfect phylogeny, but it implies that in that case a persistent phylogeny exists [4]. Moreover the conflict graph is also a measure of the complexity of an instance of the reconstruction of the persistent perfect phylogeny.

4.1 A Graph Theoretical Solution of the Persistent Phylogeny Problem

We can associate an *extended matrix* M_e to the input matrix M, by replacing each column c of M by a pair of columns (c^+, c^-), where c^+ is called the *positive* character and c^- is called the *negated* character. Moreover for each row s of M, $M_e[s, c^+] = 1$ and $M_e[s, c^-] = 0$ whenever $M[s, c] = 1$, and $M_e[s, c^+] = M_e[s, c^-] = ?$ otherwise. We want to complete the extended matrix M_e, obtaining a new matrix M_f which is equal to M_e for all species s and characters c such that $M_e[s, c] = 1$, while $M_f[s, c^+] = M_f[s, c^-]$ whenever $M_e[s, c] = 0$ (in this case we can interpret $M_f[s, c^-] = 1$ as the fact that the species s does not have the character c, but some of its ancestors used to have it). The idea of completing a matrix with missing data in order to obtain a perfect phylogeny was introduced in [22], but in our case the completion has some constraints, making the algorithm of [22] inapplicable. Finding such a matrix M_f that admits a perfect phylogeny is equivalent to computing a persistent phylogeny on the original matrix M. The following theorem was proved in [4].

Theorem 1. *Let M be a binary matrix and let M_e be the extended matrix associated with M. Then M admits a persistent phylogeny if and only if there exists a completion M_f of M_e admitting a perfect phylogeny.*

Figure 4b provides an example of an extended matrix M_e, with respect to Fig. 4a, whose conflict graph is given in Fig. 5.

4.1.1 The Red–Black Graph and the Realization of a Character

In order to find a completion of the input matrix M_e, another graph representation of the input matrix, called the *red–black graph*, denoted by G_{RB}, can be used. The latter consists of the edge-colored graph (V, E), where $V = C \cup S$, with $C = \{c_1, \cdots, c_m\}$ and $S = \{s_1, \cdots, s_n\}$ being the sets of positive characters and species of the matrix M_e, and E is defined as follows: $(s, c) \in E$ is a black edge if and only if $M_e[s, c] = 1$ and $M_e[s, c^-] = 0$. The algorithm for finding a persistent phylogeny

a

M	a	b	c	d	e
s_1	0	0	0	1	0
s_2	0	0	1	1	1
s_3	0	1	1	0	0
s_4	1	1	0	0	0
s_5	1	1	1	0	1

b

M'	a^+	a^-	b^+	b^-	c^+	c^-	d^+	d^-	e^+	e^-
s_1	?	?	?	?	?	?	1	0	?	?
s_2	?	?	?	?	1	0	1	0	1	0
s_3	?	?	1	0	1	0	?	?	?	?
s_4	1	0	1	0	?	?	?	?	?	?
s_5	1	0	1	0	1	0	?	?	1	0

Fig. 4 An example of a binary matrix M which is the input of the persistent phylogeny problem, and its associated extended matrix. (**a**) Binary matrix M. (**b**) Extended matrix M'

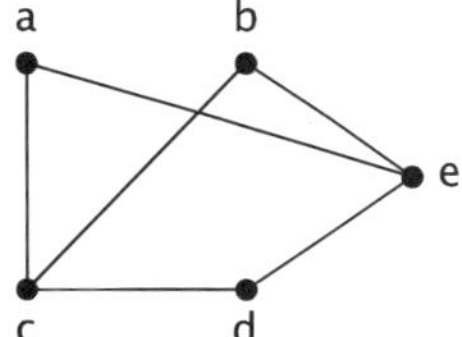

Fig. 5 The conflict graph G_c associated with the binary matrix M of Fig. 4a

basically determines a sequence of *character realizations*, which are represented as very specific operations on the red–black graph. The graph operation is called a *realization of a character* and consists of removing black edges and adding or removing red edges.

Let c be a character, and let $\mathscr{C}(c)$ be the connected component of the graph G_{RB} containing the node c. Then realizing the character c on G_{RB} consists of the following steps:

(i) Adding the red edges (c, s) for all species $s \in \mathscr{C}(c)$ such that (c, s) is not an edge of G_{RB},
(ii) Removing all black edges (c, s) (in this case c is called *active*);
(iii) If an active character c_1 is connected by red edges to all species on $\mathscr{C}(c_1)$, then all edges incident of c_1 are deleted and c is called *free*.

Realizing a character c is associated with a *canonical completion* of c in the matrix M_e by completing incomplete pairs of characters c^+, c^- as $M_f(c^+, s) = M_f(c^-, s) = 1$ for each species $s \in \mathscr{C}(c)$, while $M_f(c^+, s) = M_f(c^-, s) = 0$ for the other species – we recall that in a completion, $M_f(c^+, s) = M_e(c^+, s)$ and $M_f(c^-, s) = M_e(c^-, s)$ if $M_e(c^+, s) \neq M_e(c^-, s)$.

Consequently, any ordering $\langle c_{i_1}, \ldots, c_{i_m} \rangle$ of the character set represents a possible solution, obtained by realizing the characters according to the ordering. Not all orderings lead to an actual feasible solution, though, but only those whose resulting red–black graph is edgeless [4]. Nevertheless, the fundamental result of [4] is the following.

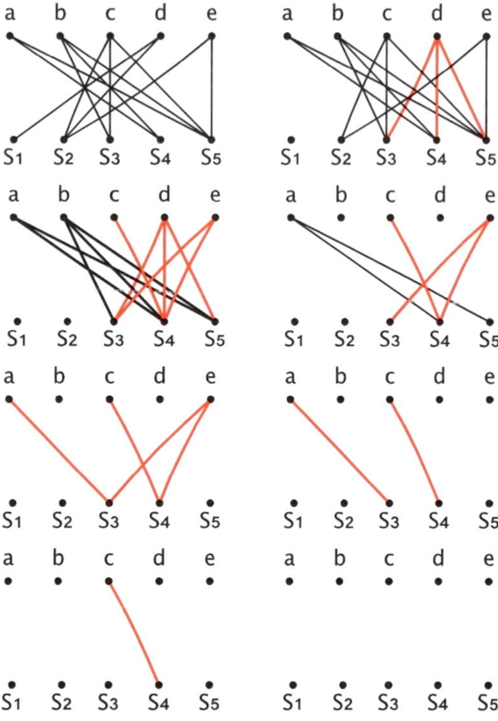

Fig. 6 The realization of $\langle d, c, e, b, a \rangle$ on the red–black graph $G_{\mathrm{R.B}}$

Theorem 2. *Let M be a binary matrix and G_{RB} be the red–black graph for the matrix M. Then M admits a persistent phylogeny if and only if there exists an ordering of the characters of M such that the realization of characters in that ordering in the graph G_{RB} results in an edgeless red–black graph.*

The main consequence of Theorem 2 is that one algorithm for finding a persistent phylogeny, if it exists, is to enumerate all possible orderings of the character set and to compute the red–black graph resulting from realizing the characters in each such order. In fact, the algorithm of [4] builds a decision tree that explores all orderings of the set C of characters. An experimental analysis of the computational performance of this algorithm in building a persistent phylogeny has been presented in [4].

Example 1. Consider the matrix M given in Fig. 4a. In Fig. 6 shows an example of a realization of characters in the red–black graph according to the ordering $\langle d, c, e, b, a \rangle$. The binary matrix M has associated with the conflict graph G_{c} represented in Fig. 5. The pairs of characters in conflict are (a, c), (b, c), (c, d), (a, e), (b, e), and (d, e). The ordering $\langle d, c, e, b, a \rangle$ leads to the canonical completion M' shown in Fig. 7. The perfect phylogeny compatible with M' is also a persistent phylogeny for the input matrix M and is represented in Fig. 8.

Fig. 7 A completion M' of the extended matrix M_e of Fig. 4b

M'	a^+	a^-	b^+	b^-	c^+	c^-	d^+	d^-	e^+	e^-
s_1	0	0	0	0	0	0	1	0	0	0
s_2	0	0	0	0	1	0	1	0	1	0
s_3	1	1	1	0	1	0	1	1	1	1
s_4	1	0	1	0	1	1	1	1	1	1
s_5	1	0	1	0	1	0	1	1	1	0

Fig. 8 Realizing the characters in the ordering $\langle d, c, e, b, a \rangle$ results in a persistent phylogeny for M

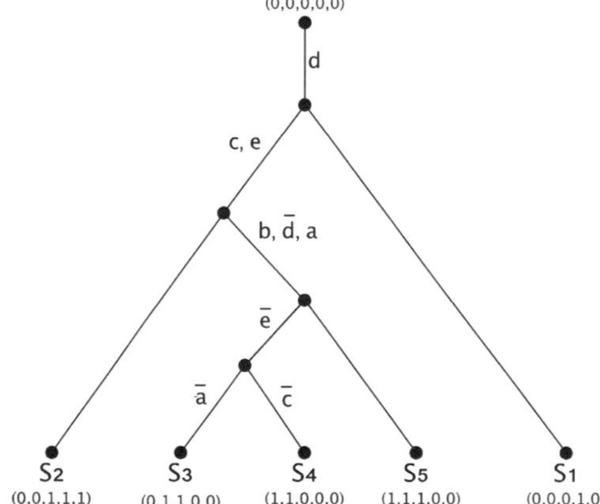

5 The Near-Perfect Phylogeny

We recall that there are instances M of the perfect phylogeny problem that cannot be solved, motivating the need for different models. Another approach to the construction of a most likely phylogeny in accordance with the input data is to move towards an optimization problem, such as identifying a largest subset of characters that admits a perfect phylogeny or, equivalently, removing the minimum number of columns from the input matrix M so that the resulting matrix has a perfect phylogeny. This problem is also called the character compatibility problem. Unfortunately, these optimization problems are intractable, as identifying the largest subset of characters admitting a perfect phylogeny is equivalent to MAX CLIQUE [15], and also shares its inapproximability [19]. Consequently, different versions must be sought.

An interesting problem stems from the observation that the perfect phylogeny is the minimum-cost Steiner tree where the set of species sharing a common state for any given character forms a connected subtree, and that the minimum-cost is exactly equal to m (i.e., the number of characters). The near-perfect phylogeny problem (NPPP) has a matrix M as input and asks for a minimum-cost Steiner tree whose leaves are taken from the species (i.e., the rows of M) and in which all species label some vertices of the tree. By the previous argument about the optimum, the cost of any solution can be expressed as $m + q$, where q (which is always positive) is called the *penalty*. Note that the penalty can be related to a back mutation or a recurring mutation, since we have no way to distinguish or prioritize these.

The first result in this setting was an $O(nm^q 2^{q^2 r^2})$ time algorithm [13] which draws upon some of the ideas of the first fixed-parameter algorithm for the perfect phylogeny problem [1] to find a solution with a penalty of at most q, if such a tree exists. Unfortunately, such a time complexity makes the algorithm impracticable; in particular, the m^q factor limits its usefulness to very small values of q. From a theoretical point of view, the main question left open in [13] was whether the NPPP admits a fixed-parameter (FPT) algorithm [10] when the parameters are q and r, r being the maximum number of states of any character.

The question was answered positively in [27] for the binary perfect phylogeny, that is, in the case $r = 2$. This case is especially important in both theory and practice. In fact, a study of the perfect phylogeny problem shows that ideas originating from the two-state case have, in time, percolated up to three-state and four-state cases and then up to the r-state case, for any fixed r. Therefore, the binary-state algorithm is a strong hint that an FPT algorithm exists for any fixed r. From a practical point of view, most of the available data are binary or can be transformed into binary characters via opportune clustering, and therefore the algorithm of [27], which has $O(72^q + 8^q nm^2)$ time complexity, can be applied.

We will briefly sketch the main ideas of the algorithm of [27], which follows a randomized divide-and-conquer approach where, at each stage, a conflicting character c is picked at random, and then c is allowed to mutate only once in the tree. Since c mutates only once, the Steiner tree instance is partitioned into two subtrees T_0 and T_1, according the state that the species assumes in c. Then two vertices r_0 and r_1 are chosen at random from T_0 and T_1, respectively (note that r_0 and r_1 might be Steiner vertices, so we cannot sample directly from the leaves or the species). A new edge (r_0, r_1) is created and labeled by the character c. Then the algorithm operators recursively on T_0 and T_1, by guessing no more than q edges overall and checking that, at the end, the conflict graph is sufficiently small to be solved via exhaustive enumeration.

The correctness of the algorithm derives mainly from the observation that at most q characters can mutate more than once. Therefore, when the conflict graph is large, the random choice of c has a high probability of being correct (i.e., there exists a solution where c mutates once), whereas if the conflict graph is small, then the optimal solution can be computed via brute force. The analysis of the time complexity is quite involved, as computing the vertices r_0, r_1 requires efficiently some combinatorial properties of Buneman graphs [26] (which are related to Steiner trees). The aforementioned $O(72^q + 8^q nm^2)$ time complexity is for a derandomized version of the algorithm. If we settle for finding the optimal solution with probability at least 8^{-q}, then the time complexity can be lowered to $O(18^q + 8nm^2)$.

A related problem studied in [25] is $H(p, q)$-NPP, where the input is a set of genotypes and we want to compute a phylogeny where the vertices are labeled with haplotypes so that (i) at most p sites can mutate, each at most q times (i.e., have at most q homoplasy events), and (ii) the set of haplotypes labeling the vertices is able to explain the input genotypes. An algorithm for $H(1, q)$-NPP, that is, when only one character is allowed to have at most q recurrent mutations, was presented in [25]. That algorithm nicely complements that of [27], where no restriction on the number of characters affected exists, and is based on an analysis of the conflict graph, the main point being the property that the character with recurrent mutations must be the only one with two adjacent characters in the conflict graph.

6 Open Problems

This chapter has presented some generalizations of the perfect phylogeny model motivated by recent biological applications in which evolution was investigated as a character-based process. The availability of a large amount of genomic and proteomic data makes the use of genetic attributes or biological markers quite appealing in evolution analysis, thus giving even more importance to applying computationally efficient parsimony models. On the other hand, there is a huge gap between tractable and NP-hard parsimony models that needs to be filled. In fact, one extreme is the perfect phylogeny model, which has a linear time solution but only a few specific biological applications. On the other hand, we have models such as the Dollo and Camin–Sokal parsimony models, which are often too generic from a biological viewpoint and computationally impracticable. A middle ground is occupied by the persistent perfect phylogeny model, for which some efficient, practical algorithms have recently been presented [4], and for which some specific applications such as the analysis of protein networks and domains have been found [23, 31]. However this research direction still needs to be explored. In particular, finding a polynomial time algorithm for the persistent phylogeny model is still an open problem, and the novelty of the algorithm of [4] hints that even more practical approaches are possible, even for some optimization versions of the problem that deserve to be investigated. It must be pointed out that the persistent phylogeny model is useful for detecting persistent characters that can be excluded from the evolutionary reconstruction process. In fact, having computational tools to detect characters that should or should not be included in a parsimony model analysis can improve the correctness of the tree that is built from such characters [23]. From a theoretical point of view, the investigation of variants of the perfect phylogeny model and restrictions of the Dollo parsimony model other than those presented here is still an important research direction. In particular, the tree-of-complexes problem discussed in this chapter reveals that there may be interesting, strong connections between graph theory and parsimony models representing the evolutionary relationships between functional modules in a protein network. To conclude, characterization of the structural properties of protein networks and of the overlap graphs of characters seems to be a promising novel direction for building parsimony models in a more efficient and biologically meaningful way.

Acknowledgements PB and APC are supported by the Fondo di Ateneo 2011 grant "Metodi algoritmici per l'analisi di strutture combinatorie in bioinformatica". GDV is supported by the Fondo di Ateneo 2011 grant "Tecniche algoritmiche avanzate in Biologia Computazionale". PB, GDV and RD are supported by the MIUR PRIN 2010–2011 grant "Automi e Linguaggi Formali: Aspetti Matematici e Applicativi", code H41J12000190001. TMP is supported by the Intramural Research Program of the National Institutes of Health, National Library of Medicine.

References

1. R. Agarwala, D. Fernandez-Baca, A polynomial-time algorithm for the perfect phylogeny problem when the number of character states is fixed. SIAM J. Comput. **23**(6), 1216–1224 (1994)
2. P. Bonizzoni, A linear time algorithm for the Perfect Phylogeny Haplotype problem. Algorithmica **48**(3), 267–285 (2007)
3. P. Bonizzoni, G. Della Vedova, R. Dondi, J. Li, The haplotyping problem: an overview of computational models and solutions. J. Comput. Sci. Technol. **18**(6), 675–688 (2003)
4. P. Bonizzoni, C. Braghin, R. Dondi, G. Trucco, The binary persistent perfect phylogeny. Theor. Comput. Sci. **454**, 51–63 (2012)
5. J. Camin, R. Sokal, A method for deducting branching sequences in phylogeny. Evolution **19**, 311–326 (1965)
6. L.L. Cavalli-Sforza, A.W.F. Edwards, Phylogenetic analysis. Models and estimation procedures. Am. J. Hum. Genet. **19**(3 Pt 1), 233 (1967)
7. Z. Ding, V. Filkov, D. Gusfield, A linear time algorithm for Perfect Phylogeny Haplotyping (pph) problem. J. Comput. Biol. **13**(2), 522–553 (2006)
8. T. Dobzhansky, Nothing in biology makes sense except in the light of evolution. Am. Biol. Teach. **35**(3), 125–129 (1973)
9. R.G. Downey, M.R. Fellows, *Parameterized Complexity*, Monographs in Computer Science, (Springer-Verlag, New York, 1999). ISBN 978-0-387-94883-6
10. A.W.F. Edwards, L.L. Cavalli-Sforza, The reconstruction of evolution. Heredity **18**, 553 (1963)
11. J. Felsenstein, *Inferring Phylogenies* (Sinauer Associates, Sunderland, 2004)
12. S. Felsner, V. Raghavan, J. Spinrad, Recognition algorithms for orders of small width and graphs of small Dilworth number. Order **20**, 351–364 (2003)
13. D. Fernandez-Baca, J. Lagergren, A polynomial-time algorithm for near-perfect phylogeny. SIAM J. Comput. **32**(5), 1115–1127 (2003)
14. L. Foulds, R. Graham, The Steiner problem in phylogeny is NP-complete. Adv. Appl. Math. **3**(1), 43–49 (1982)
15. M. Garey, D. Johnson, *Computer and Intractability: A Guide to the Theory of NP-Completeness* (W.H. Freeman, San Francisco, 1979)
16. M. Golumbic, *Algorithmic Graph Theory and Perfect Graphs* (Academic, New York, 1980)
17. D. Gusfield, *Algorithms on Strings, Trees and Sequences: Computer Science and Computational Biology* (Cambridge University Press, Cambridge, 1997)
18. D. Gusfield, Haplotyping as perfect phylogeny: conceptual framework and efficient solutions, in *Proceedings of the 6th Annual Conference on Research in Computational Molecular Biology (RECOMB)*, Washington, DC, 2002, pp. 166–175
19. J. Håstad, Clique is hard to approximate within $n^{1-\epsilon}$. Acta Math. **182**, 105–142 (1999). doi:10.1007/BF02392825
20. S. Kannan, T. Warnow, A fast algorithm for the computation and enumeration of perfect phylogenies. SIAM J. Comput. **26**(6), 1749–1763 (1997)
21. R.M. Karp, Reducibility among combinatorial problems, in *Complexity of Computer Computations*, ed. by R.E. Miller, J.W. Thatcher. The IBM Research Symposia Series (Plenum Press, New York, 1972), pp. 85–103
22. I. Peer, T. Pupko, R. Shamir, R. Sharan, Incomplete directed perfect phylogeny. SIAM J. Comput. **33**(3), 590–607 (2004)
23. T.M. Przytycka, An important connection between network motifs and parsimony models, in *Proceedings of the 10th Annual Conference on Research in Computational Molecular Biology (RECOMB)*, Venice, 2006, pp. 321–335
24. T. Przytycka, G. Davis, N. Song, D. Durand, Graph theoretical insights into Dollo parsimony and evolution of multidomain proteins. J. Comput. Biol. **13**(2), 351–363 (2006)

25. R.V. Satya, A. Mukherjee, G. Alexe, L. Parida, G. Bhanot, Constructing near-perfect phylogenies with multiple homoplasy events, in *ISMB (Supplement of Bioinformatics)*, Fortaleza, 2006, pp. 514–522
26. C. Semple, M. Steel, *Phylogenetics*. Oxford Lecture Series in Mathematics and Its Applications (Oxford University Press, Oxford, 2003)
27. S. Sridhar, K. Dhamdhere, G. Blelloch, E. Halperin, R. Ravi, R. Schwartz, Algorithms for efficient near-perfect phylogenetic tree reconstruction in theory and practice. IEEE/ACM Trans. Comput. Biol. Bioinf. **4**(4), 561–571 (2007)
28. A. Subramanian, S. Shackney, R. Schwartz, Inference of tumor phylogenies from genomic assays on heterogeneous samples. J. Biomed. Biotechnol. **2012**, 1–16 (2012)
29. W.T. Tutte, An algorithm for determining whether a given binary matroid is graphic. Proc. Am. Math. Soc. **11**(6), 905–917 (1960)
30. J. Zheng, I.B. Rogozin, E.V. Koonin, T.M. Przytycka, Support for the Coelomata clade of animals from a rigorous analysis of the pattern of intron conservation. Mol. Biol. Evol. **24**(11), 2583–2592 (2007)
31. E. Zotenko, K.S. Guimarães, R. Jothi, T.M. Przytycka, Decomposition of overlapping protein complexes: a graph theoretical method for analyzing static and dynamic protein associations. Algorithms Mol. Biol. **7**(1), 1–11 (2006)

An Invitation to the Study of Brain Networks, with Some Statistical Analysis of Thresholding Techniques

Mark Daley

Abstract We provide a brief introduction to the nascent application of network theory to mesoscale networks in the human brain. Following an overview of the typical data-gathering, processing, and analysis methods employed in this field, we describe the process for inferring a graph from neural time series. A crucial step in the construction of a graph from time series is the thresholding of graph edges to ensure that the graphs represent physiological relationships rather than artifactual noise. We discuss the most popular currently employed methodologies and then introduce one of our own, based on the theory of random matrices. Finally, we provide a comparison of our random-matrix-theory thresholding approach with two dominant approaches on a data set of 1,000 real resting-state functional magnetic resonance imaging scans.

1 Introduction

In an age where humans are able, for the first time, to collect huge quantities of data on the relationships between objects, network/graph theory has become a rising star as a tool to help make sense of raw relational data. Nowhere is this more apparent than in the field of neuroimaging, which has begun to embrace graph theory as a tool for understanding relationships between brain areas at the mesoscale. In this field, one does not study individual neurons and traditional "neural networks", as the imaging modalities available to us have a resolution which is far too coarse for this application; instead, one studies relationships between medium-scale (mesoscale) assemblies of tens of thousands, or more, neurons.

M. Daley (✉)
The University of Western Ontario, London, ON, Canada
e-mail: mdaley2@uwo.ca

N. Jonoska and M. Saito (eds.), *Discrete and Topological Models in Molecular Biology*, 85
Natural Computing Series, DOI 10.1007/978-3-642-40193-0_5,

Viewing this work from the standpoint of neuroscience, the motivations are immediate: if large, four-dimensional, data sets can be reduced to single static graphs – the topology of which varies according to the properties of interest – then the complexity of the analysis can be greatly reduced. For example, a developmental neuroscientist may hypothesize that global connectivity varies in the brain as we age from children to adults. Given a method for converting functional neuroimaging data into graphs, one now has an immediate way to test this hypothesis: scan the brains of children and adults, convert the scans to graphs, and compute a topological metric for each graph that captures the intuition of "global connectivity"; indeed, this precise study has been performed [9]. Likewise, one can imagine a clinician interested in whether or not there are metrics which can be applied to such graphs that would distinguish clinical conditions such as Alzheimer's disease [7] or schizophrenia [17]. A broad introduction to the neuroscientific implications of this line of research can be found in [21].

Viewing this work as a mathematician interested in graph theory, one sees the opportunity to study very special classes of graphs and to investigate, for example, new metrics better suited to quantifying the topological differences between graphs in this class. As a computability theorist, one sees the opportunity to study snapshots of an intriguing computing machine (the brain) within the familiar framework of graph theory.

This chapter is meant to serve as a very brief introduction to the increasingly popular application of graph-theoretic techniques to data derived from neuroimaging experiments. In the background section, we will quickly review the fundamental concepts and procedures employed to move from neural activity in the brain, through several processing steps, to a static graph representation of the same activity. We note that the ability to threshold signal from noise in our final graph representation is a critical step in building physiologically credible graphs and, after reviewing the currently popular techniques for thresholding, we suggest a new approach based on the theory of random matrices. Finally, we conclude with a summary and discussion of the many classes of open problem in this field, both pragmatic and abstract.

2   Background

We now bring together the necessary background to understand how to map mesoscale neural time series, derived from neuroimaging, to static graphs. We begin with a short introduction to magnetic resonance imaging (MRI) and functional magnetic resonance imaging (fMRI) and clarify our use of graph-theoretic concepts and terminology. We discuss the specifics of the somewhat involved process of preprocessing such data to remove artifacts and make it suitable for analysis. With suitably preprocessed data, the next step towards constructing a graph involves the pairwise comparison of time series. We review three dominant approaches in current use: one linear time-domain measure (correlation), one linear spectral measure (coherence), and one nonlinear measure (mutual information). Finally, we discuss the construction and analysis of a graph-theoretic representation of neuroimaging data.

2.1   MRI, fMRI, the Resting State, and Functional Connectivity

The data with which we will build our graphs is most commonly derived from resting-state functional magnetic resonance imaging (though similar approaches exist for other neuroimaging modalities). Magnetic resonance imaging leverages nuclear magnetic resonance in a carefully controlled fashion to generate volumes containing information about spin-relaxation properties in particular voxels.[1]

We now give a very brief characterization of imaging via magnetic resonance. A subject is placed in a very strong magnetic field (typically 1.5–7 tesla), which causes the spin states of individual protons to become increasingly uniform, with each proton precessing around the direction of the magnetic field.[2] Electromagnetic radiation at the resonant frequency of the protons is then introduced, causing the protons to absorb the emitted photons and flip spin state; at the same time, this process causes a synchronization of the precession of individual protons. When the influx of photons from the electromagnetic field is removed, the spin states of the protons flip back to the lower-energy state, while the individual precessions drift increasingly out of synchronization. By measuring the photons emitted during this process, the magnetic resonance scanner can determine the length of time it takes for the spin vector of the bulk system to return to being parallel with the static magnetic field and the length of time for the individual precessions to become completely decoherent.

By superimposing a variable magnetic gradient on top of the static magnetic field, one can spatially localize this process and extract resonance contrast for a given spatial frequency in a particular location. After this k-space has been sampled densely, a simple Fourier transform yields a volume in Euclidean 3-space. By carefully choosing which properties (e.g., relaxation time, decoherence time, proton density, flow, and spectral shift) one records, and developing different pulse sequences,[3] one can optimize the contrast for different tissue types. For a detailed

---

[1]A *voxel* is simply the three-dimensional analog of a pixel: a (usually cube-shaped) volume assigned a homogenous scalar or vector value. In the context of fMRI, a typical voxel will represent a volume of size roughly 1–5 mm^3, depending on several technical details of the hardware and software used for the imaging.

[2]Spin is a quantum property of elementary particles and, as such, may take on only a discrete number of states (or some quantum superposition thereof). In the case of fermions such as the proton, the two possible spin states are $\{\frac{1}{2}, -\frac{1}{2}\}$.

[3]Pulse sequences contain integrated information about the application of magnetic gradients, electromagnetic pulses, and the recording of electromagnetic radiation emitted from the subject. For more detail, see [4].

exposition of the physics of nuclear magnetic resonance imaging, we direct the reader to [11].

The contrast of interest for our purposes is the T2* contrast, which allows imaging of the BOLD (blood-oxygen-level-dependent) response, which, roughly, forms contrast from the oxygenation level of blood in a volume of tissue. Using earlier magnetic resonance techniques, it is difficult to visualize oxygenated blood versus deoxygenated blood without the use of an injectable contrast agent. The ability to image the BOLD signal without the use of exogenous contrast agents was first reported in [14, 18], with the suggestion that variation in blood flow to a particular brain region was correlated with activity in that region. Follow-up work, notably that in [16], demonstrated that this modulation of blood flow is strongly linked with underlying electrical activity, although the exact nature of the BOLD signal is still under active investigation. Interestingly, it has recently been linked to astrocytes [19], suggesting a greater than previously acknowledged functional role for glial cells. The BOLD signal is very slow, lagging behind electrical recordings of neural activation by several seconds and effectively acting as a temporal low-pass filter for the underlying neural activity. It is also spatially diffuse, as it is intrinsically linked to the vasculature surrounding the recently activated area rather than the area itself – one might view this as a spatial form of low-pass filtering. A thorough introduction to the methodologies of fMRI can be found in [12].

Traditional fMRI studies involve measuring the BOLD response in block-designed or event-related experiments and contrasting the BOLD signal between different experimental conditions. Beginning with [5], and to an increasingly extent recently, a different question has been addressed: "what can we learn about the brain by observing basal, 'spontaneous', fluctuations in the BOLD signal during rest?". In this chapter, we shall describe work on resting-state data in which we record the BOLD signal from subjects lying in the scanner and given no particular task other than the instruction to relax and not think about anything in particular.

Resting-state fMRI data is typically used to build models of *functional connectivity*, in which two brain regions are said to be functionally connected if their BOLD time courses are highly similar according to some metric such as correlation. It is important to note that despite the use of the term "connectivity", the information obtained in this type of analysis is purely correlational and cannot give any definitive information about actual anatomical connectivity. When we say that two brain regions are functionally connected, what we mean is simply that their BOLD signals "looked similar" during the period of time in which we recorded them. The study of functional connectivity inferred from resting-state data is referred to as resting-state functional-connectivity MRI (rs-fcMRI).

2.2 Graph Theory

Graph theory, at its simplest, is the study of objects which encode relationships. The two fundamental building blocks of graph theory are a *vertex* (also, equivalently, called a *node*) and an *edge*. The common interpretation of a graph is very straightforward: the vertices represent entities of some sort and the edges represent relationships between those entities.

Here, we will limit ourselves to studying undirected graphs; while it is clear that on the lowest physiological level, the connectivity between individual neurons is directional, the issue becomes more complex at the macroscopic scale we wish to study. Connections between gross anatomical regions are complex and, in many cases, bidirectional, though both directions are likely not to be equally strong. We choose to study undirected graphs not because we believe that the true underlying system is undirected, but rather because of the limitations of our methods. The time series analysis methods used here are inherently symmetric. And, worse, even when one explicitly attempts to infer directionality, the best techniques are no more than about 60 % accurate,[4] and some common techniques, such as Granger causality, perform no better than chance [20].

Graphs are often classified into broad categories according to their topological properties. For example, a graph where every vertex has the same number of edges is referred to as a regular graph. Clearly, regular graphs are very structured entities, and if spatial constraints regarding how the edges are connected are added (viz., an edge vertex is connected to its closest neighbors only), we reach the extreme of structure in a graph: a regular lattice. Imagine extending a small 4-regular graph to several hundred nodes, but keeping each vertex of degree 4. This graph would have high transitivity but a very long average path length.

At the far opposite extreme of structure, we have the Erdös–Renyi random graph, which one can imagine being generated thus: after picking a fixed set of vertices, we flip a coin for each pair and add an edge only if the coin comes up "heads". With edges chosen randomly, this graph will, with high probability, have very low transitivity; however, because some "long-distance" edges will be chosen purely by luck, the average path length in the graph will typically be very small – the long-distance edges act as a shortcut through the graph.

Between these two extremes exists the class of small-world graphs. Small-world graphs are characterized by having both relatively high transitivity and relatively short path lengths; this is typically achieved with a structure consisting of densely connected local clusters (thus achieving high transitivity) with the occasional long-distance edge (thus shortening significantly the average path length). Small-world networks appear to be common in both natural systems (e.g., cellular metabolic networks, genetic transcription networks, food chains, and social networks) and artificial systems (e.g., road networks and power grids). Most significantly for the present chapter, the small-world structure has been proposed as a dominant feature of brain networks (see, e.g. [3] for a review).

---

[4]This figure comes from an analysis in [20] in which known "ground truth" networks we used to construct simulated fMRI data, to which various network inference techniques we then applied. The accuracy reflects how much of the topology of the ground truth network the inference method managed to capture.

2.3 Preprocessing

Raw fMRI data is noisy and contaminated by a host of artifacts, necessitating fairly aggressive preprocessing. Although the specific details of preprocessing pipelines vary somewhat between studies, we present here a "consensus" preprocessing pipeline based on the protocol optimization results in [23]. This is precisely the pipeline that we use for our own work on thresholding. For each subject, we acquire both a single high-resolution "anatomical" volume (3D) and a series of lower-resolution "functional" volumes (a time series of lower-spatial-resolution volumes that have been acquired sequentially).

The initial preprocessing steps consist of the following steps, applied to functional data:

1. Deletion of first four low-resolution volumes acquired to allow for stabilization of the T1 signal.
2. Brain extraction/skull stripping.
3. Slice-time correction – each "slice" in an fMRI volume is acquired at a slightly later time than the previous slice. For our time series to be comparable along the z-axis, it is necessary to compensate for this.
4. Motion correction – even when the subject is instructed to lie still, subject motion is unavoidable. Images can be corrected to a single spatial baseline using standard image registration algorithms.
5. Spatial smoothing (5 mm full width at half maximum Gaussian).
6. Prewhitening (removal of spurious/unwanted autocorrelations).
7. Removal of residual motion by linear regression.

This is followed by anatomically driven preprocessing on both the high-resolution anatomical data and the functional data:

1. Brain extraction/skull stripping of the anatomical image.
2. Tissue-type segmentation into gray matter, white matter, and cerebrospinal fluid. (white matter, and cerebrospinal fluid do not contain neuron soma, and any signal detected in these areas is noise, rather than functionally induced).
3. Transformation of tissue masks generated in higher-spatial-resolution imagery (viz., cerebrospinal fluid, white matter, or gray matter masks) into the corresponding lower-resolution coordinate space of the functional imagery.
4. Extraction of mean time series for the cerebrospinal fluid and white matter to use as regressors in the final processing step.
5. Registration of functional images to anatomical images according to the Montreal Neurological Institute's MNI 152 T1 2 mm standard space. This is a purely linear, rigid, transformation.

In the penultimate step, we regress out the signals of the cerebrospinal fluid and white matter from our functional data set. We then perform temporal band-pass filtering ($0.009\,\mathrm{Hz} < f < 0.08\,\mathrm{Hz}$) and mask the output so that it includes only time series derived from voxels containing gray matter. Cortical gray matter is distributed as a sheet on the surface of the brain which is approximately 2–4 mm thick; the gray matter contains the neuron cell bodies of the cortex and, consequently, is where we expect to find the BOLD signal when imaging.

2.4   Time Series Analysis

The first step towards building a graph of functional connectivity involves analyzing which voxels in the data under study have related time courses. We generate an $n \times n$ (where n is the number of gray-matter voxels) similarity matrix by analyzing every pair of gray-matter voxel time series. In principle, one may use any time series comparison method to evaluate the similarity of the time series obtained from a pair of voxels. A full description of every metric currently in use is beyond the scope of this review, so instead we explain here three commonly used, but significantly different, similarity measures: the Pearson correlation, which is a linear, time-domain measure; the band-averaged coherence, a linear frequency-domain (spectral) measure; and the mutual information, a nonlinear measure.

Given the time series of two voxels, x and y, from an fMRI data set preprocessed as described above (and, thus, hopefully consisting primarily of the low-frequency BOLD signal), we will now define our similarity measures.

Formally, we define the population Pearson correlation coefficient in the usual way:

$$\frac{\sum_{i=1}^{m}(x[i] - \bar{x})(y[i] - \bar{y})}{\sqrt{\sum_{i=1}^{m}(x[i] - \bar{x})^2}\sqrt{\sum_{i=1}^{m}(y[i] - \bar{y})^2}},$$

where $x[i]$ indicates the ith element in the time series x, and $\bar{x}$ is the mean of that time series. Intuitively, one can think of the correlation coefficient in a very simple way: step through the two time series under consideration, in a parallel fashion, and construct an ordered pair (x, y) from the values of the two series at each point in time. Plot these pairs and fit a straight a line to the plot. The closer the plotted points are to the best-fit line, the higher the correlation of the two series. Note carefully here the importance of the line. If two series are linearly correlated, then they will surely have a high correlation coefficient; imagine, though, that you were to perform such a plot and find a perfect monochrome rendering of the Sierra Nevada mountains. The Pearson correlation between these two series would be quite low (since there is no single line fitting a photograph of the Sierra Nevada mountains), but you might feel justified in supposing that, in truth, these two series do indeed have a very special relationship. In principle, this is an inherent weakness of linear methods – they can find only lines. In the case of low-frequency BOLD signals, however, this may pose less of a problem than it seems as the bulk of the informative relationships do indeed appear to be linear [20].[5]

It is clear from the description of the correlation that it inherently provides a time-domain method answering the question "how similar is the variation in amplitude of these series over time?" An alternative question one might wish to ask is "how similar are the power spectra of these two time series?" That is to say, "how similar are the *frequency* components in each time series?" Instead of comparing the time series directly at the temporal level, we are suggesting transforming them into the frequency domain and comparing their spectra. We desire a metric that will be high if two series are composed of components having very similar frequencies and amplitudes, and low otherwise. The band-averaged spectral coherence is just such a metric.

In its most general form, the coherence C_{xy} between x and y can be computed as

$$C_{xy} = \frac{|G_{xy}|^2}{G_{xx}G_{yy}},$$

where G_{xx} is the autospectral density of x, and G_{xy} is the cross-spectral density between x and y. Note that C_{xy} is a function of frequency, and thus, to obtain a scalar value, we average C_{xy} over the frequency band of interest. In a typical implementation, one estimates power and cross spectra using Welch's modified periodogram averaging methods, with a window length of 50 and an overlap of 25. Each windowed segment is then normalized and weighted by a Hanning window (also of length 50). The power spectral density is estimated as

$$G_{xx}(\lambda) = \frac{1}{N} \sum_{n=1}^{N} |X_n(\lambda)|^2,$$

where X_n is the discrete Fourier transform of the nth windowed segment of x, and λ is a variable depending on frequency. The cross-spectral density is estimated as

$$G_{xy}(\lambda) = \frac{1}{N} \sum_{n=1}^{N} X_n(\lambda)Y_n^*(\lambda).$$

Finally, we estimate the coherence, averaged over the band of interest ($\bar{\lambda}$), as

$$C_{xy}(\bar{\lambda}) = \frac{|\sum_\lambda G_{xy}(\bar{\lambda})|^2}{\sum_\lambda G_{xx}(\bar{\lambda}) \sum_\lambda G_{yy}(\bar{\lambda})}$$

---

[5]The careful reader may find, as does this author, this conclusion somewhat disturbing. The brain is a Turing-complete computational system, and accurate measures of its state should show useful statistics far transcending the first moment; yet, the smoothed, preprocessed BOLD signal does not, it would appear.

Although the coherence moves us from the time domain to the frequency domain – which enables the detection of temporal relationships regardless of phase shifts – it is still a linear measure, as it is only capable of detecting linear spectral relationships. To search for nonlinear relationships in time series, we turn to the mutual information.

Where as the correlation and coherence operate directly on the time series (or the Fourier transforms thereof) under investigation, the mutual information operates instead on the probability distributions of the hidden sources generating the series. Intuitively, if I am gathering time series data on two possibly related processes called A and B, the mutual information answers the following question: "How much does knowing the statistics of A allow me to infer about the statistics of B?"

More formally; let A and B be the random variables modeling the process generating x and y, respectively. We compute the mutual information $I(A; B)$ between A and B as

$$I(A; B) = \sum_{b \in B} \sum_{a \in A} p(a, b) \log \left(\frac{p(a, b)}{p(a) p(b)} \right),$$

where $p(a, b)$ is the joint probability distribution function of A and B, and $p(a)$ and $p(b)$ are the marginal probability distribution functions of A and B, respectively. The quantity $I(A; B)$ measures how much knowing the distribution of A tells us about B, and vice versa. For example, if A and B are independent, then clearly $I(A; B) = 0$. If there is some statistical dependency between A and B, then $I(A; B)$ can quantify this dependency in units of bits.

In reality, of course, we do not have closed, analytical, expressions for A and B and must instead estimate them, ad hoc, from our data for example, with a k-nearest-neighbor estimation technique such as that of [13].

Even for relatively large data sets, the computational burden of these approaches is reasonably small. A Pearson correlation matrix for a data set with 15,000 gray-matter voxels can be computed in under 3 min on a typical quality laptop available in 2012 (viz., a hyperthreaded two-core (four effective cores), 2.2 GHz Intel i7-based 2011 MacBook Pro), and the calculation has only very modest memory requirements, as the results are streamed to disk. The band-averaged spectral coherence for the same data set, on the same machine, can be computed in approximately 1 h. Estimation of a mutual information matrix, however, requires more significant computational resources and would require approximately 1 day to complete.

2.5 Graph-Theoretical Analysis

Once a correlation (or coherence or mutual information) matrix has been generated, we may reinterpret it as the adjacency matrix of a graph thus:

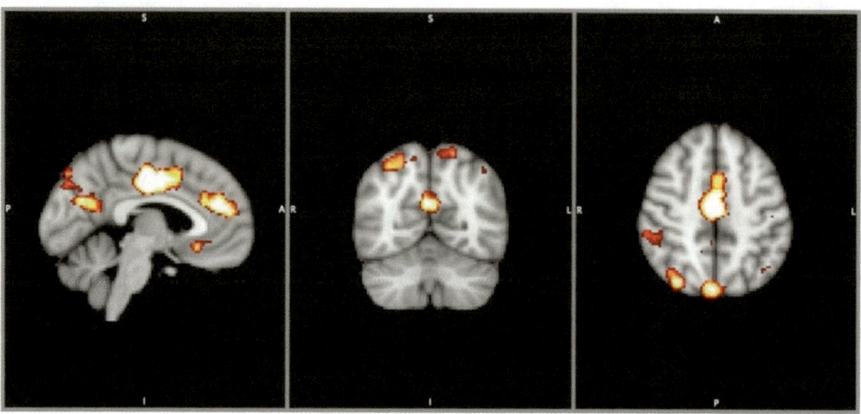

Fig. 1 Node degree projected back into voxel space and superimposed on an anatomical image. The degree is heatmapped so that lighter colors indicate higher degrees. The views from *left* to *right* are sagittal, coronal, and axial

- Each voxel is treated as a node in the graph.
- For every entry (i, j) in the correlation matrix:

  - If the entry falls below a user-specified threshold, do nothing.
  - If the entry exceeds the threshold, add an edge between nodes i and j, having a weight specified by this entry.

With the graph constructed, we can now begin to investigate its topological properties. A currently popular approach is to characterize such graphs in terms of both per-vertex and whole-graph metrics; for example, one might calculate the degree of each vertex in the hope of identifying regions of the brain which are "more connected" than others. In Fig. 1, we show the result of computing the degree of each vertex for a graph derived from a resting-state fMRI scan for a single subject. Once we have computed the degrees on the graph, we invoke our trivial mapping from graph vertices to voxels to visualize the degree in "brain space" rather than "graph space"; here, the degree is heatmapped so that brighter colors represent vertices (voxels) with higher degrees. The degree is a very coarse metric, and putatively neuroscientifically interesting results have been obtained using more sophisticated combinatorial and spectral methods such as those based on vertex betweenness, which can be informally characterized as measuring the fraction of shortest paths which travel through a given vertex.

An interesting consequence of using more sophisticated metrics such as betweenness is that although the topological interpretation of these metrics on a graph is as straightforward as for the degree, the same cannot be said of the neuroscientific interpretation of the results in "brain space". A vertex with high degree represents a "more connected" brain region, but what does a vertex with high "betweenness" represent? Recall that the relationships in our graph are *not* defined by anatomical connectivity, but by *functional* connectivity, which is simply to say that two regions

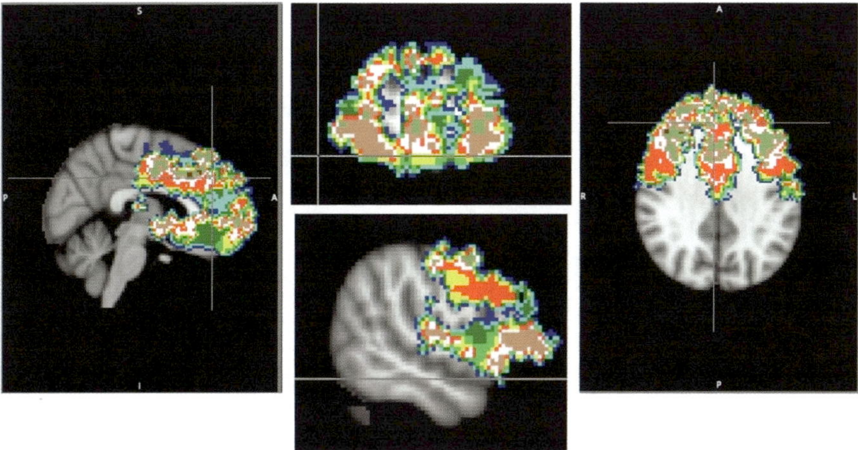

Fig. 2 Node modularity (computed using Markov clustering) projected into voxel space and superimposed on an anatomical image. Nodes with the same color belong to the same module

are considered related if their patterns of activity are similar enough. In such a framework the intuitive definition of betweenness, and very many other more complex graph metrics, breaks down; however, even without clear physiological semantics, such metrics have already proven useful in characterizing, for example, developmental trajectories in the brain [9].

Work in this area is still very exploratory in nature. It is clear that different metrics can yield different insights, although the correct physiological interpretation of these results is often less clear. It is equally apparent that many metrics recapitulate the same results; sometimes this is a consequence of a rigorously provable relationship between metrics, and other times it is a consequence of the particular topologies of the graphs themselves. We have recently provided a comparison of the repeatable intra-metric similarity (and robustness to parameters) for several commonly used metrics [8], but it seems likely that a more interesting direction for future work will be to develop metrics which are specialized for the restricted class of graphs derived from neuroimaging data.

An alternative approach to studying the topology of neuroimaging-derived graphs is to attempt a modular decomposition. In Fig. 2 we show an example of the result of applying Markov clustering [24] to the same resting-state scan used for Fig. 1, but restricted here to the prefrontal cortex. Following convergence of the clustering algorithm, the vertices were again mapped back into brain space and colored according to module membership. The modularization here yields strong support for a dorsal/ventral[6] distinction in the prefrontal cortex. Support for a

---

[6]"Dorsal" refers here to the top of the brain – the part above the eyeline in a human standing upright – and "ventral" refers to the bottom part.

rostro-caudal[7] gradient is less clear; instead, we find a complex pattern of inter-acting, nesting, and embedded networks along the rostro-caudal axis. Intriguingly, this challenges the model of two parallel strict hierarchies in dorsolateral prefrontal cortex and dorsomedial prefrontal cortex recently suggested in [22].

3   On Thresholding Graphs

Determining a reasonable threshold to use in the generation of the connectivity graph is a crucial issue, and we must be aware that, by design, any similarity matrix is composed of multiple comparisons (quadratic in the number of time series). We begin this section with a brief theoretical discussion of an overly conservative thresholding approach that applies only to correlation matrices. This is followed by a description of two thresholding procedures which are dominant in the current literature: thresholding by fixing topological expectations (the so-called S-value method) and thresholding by using the false discovery rate (FDR) to control the error. Noting the limitations of these techniques, we extend them with a new approach which computes thresholds based on the expectation of modular structure in the underlying graph; we implement this approach using the tools of random matrix theory (RMT).

3.1   Current Approaches

In the case of the Pearson correlation, we can rely upon theory to guide our thresholding. Given a desired global p-value, we apply a Bonferroni correction to obtain a corrected p-value, which we may then convert into a minimum r-score for the observed correlation.

For example, our preprocessing pipeline typically produces approximately 15,000 gray matter voxel time series, which results in

$$\frac{1}{2} \cdot (15{,}000^2 + 15{,}000) = 112{,}507{,}500$$

comparisons during correlation analysis. Assuming that our acceptable global threshold for significance is $p = 0.05$, we can correct to $p \approx 4.4 \times 10^{-10}$.

---

[7]The rostro-caudal axis follows a curved path through the head, beginning roughly in the region of the nose, and proceeding straight back towards the midbrain, at which point it bends 90° downwards to follow the spine.

Although this looks bleak, for time series of length 300, this yields a threshold of $r \approx 0.34$.[8]

This is, of course, overly conservative. Following this train of thought, a more statistically appropriate approach might be bootstrapping the correlation matrix. Unfortunately, the computational demands of computing the full correlation matrix make not only permutation analysis, but also a Monte Carlo approach, relatively infeasible. Worse still, a resampling approach to the band-averaged coherence and mutual information may be out of reach owing to significant computational requirements. Fortunately, in the discussion below, we find that these problems may be irrelevant.

Although the Bonferroni-corrected approach *should be* overly conservative, when we apply it to real data we find that it is, in fact, *overly liberal*. The Bonferroni-corrected r-value introduces edges in the graph which are well below the physiological noise floor in our data (viz., they are the product of spurious correlations). This is a consequence of the fact that our simple analysis treats the time series of each voxel as if it were generated by a unique, statistically independent source. The reality of the situation could scarcely be more different: voxels are highly temporally and spatially correlated as a consequence of both the underlying physiological processes being measured and the physical properties of the machines performing the measurement. Without appealing to detailed statistical models of these processes (in both spatial and temporal dimensions), a completely analytical approach to this problem is untenable.

An alternative approach is to define the cutoff threshold post hoc, attempting to match the properties of the induced graph to those expected from other analyses presented in the literature. A straightforward approach is to demand a constant relationship, S, between the number of nodes in the graph and the average node degree. The relationship $S = \log |V| / \log K$ (where $|V|$ is the number of vertices in the graph and K is the average degree) has been suggested in the literature. If one chooses a fixed S value at which to compare graphs, the only value that then needs to be computed is K, since the number of voxels, $|V|$, in each scan is fixed. With K computed from the formula, finding a threshold value for each particular graph is simply a matter of determining what threshold retains E edges in the graph, where $K = E/|V|$.

Note also that our definition of S is an approximation to the average path length in a large Erdös–Rényi graph (see the scaling relation of [1]), so we are, in some sense, fixing a target minimum path length. Of course, the graphs we are dealing with in practice are most certainly not random graphs (in the sense of an Erdös–Rényi graph), so this characterization of the metric S is at best heuristic, and at worst completely misleading.

That said, this approach has one overwhelmingly positive attribute: it facilitates easier cross-modality, cross-analysis, cross-subject comparison. By setting the

---

[8]We derived is r-value threshold by converting the corrected p threshold to a t-score with 298 degrees of freedom, giving $t = -6.33$, and computing $6.33/\sqrt{298 + 6.33^2}$.

threshold based on structural properties of the graph, rather than statistical properties of the underlying analysis (such as r-values), we are more likely to end up "comparing apples with apples". At the same time, we also run the risk of missing significant structural differences by forcing the structures to be similar.

Recently, a thresholding approach based upon controlling the false discovery rate (FDR) has become popular. When performing multiple comparisons, we refer to the rate of *false positives* (for us this would be ascribing a functional relationship to a pair of time series which are, in fact, unrelated) as the *false discovery rate*. We control the FDR by specifying a rate of false discovery, between 0 and 1, which we are willing to accept, and applying a procedure to ensure that, on average, this rate is maintained. Specifically, let E be the set of measurements for potential edges in our graph (equivalently, the set of entries in our similarity matrix). We then do the following:

1. Select an acceptable FDR bound α (a common choice is $\alpha = 0.05$).
2. Sort E into an ordered set, from smallest to largest: $E' = \{e_0, e_1, \ldots, e_n\}, e_0 \leq e_1 \leq \ldots \leq e_n$.
3. Find the largest $i \in \mathbb{N}$ such that $e_i < \alpha \cdot i / |E|$.
4. Select e_i as the threshold.

One potential drawback to the FDR approach is that it necessarily treats all values in E as exchangeable, in the sense that one can reasonably expect the semantics of comparing a pair of values to be constant across any pair in E. This conflicts with our understanding of real neuroimaging results, in that some brain areas (e.g., the primary visual cortex) are dominated by extremely strong correlations in activity while other areas (e.g., the parietal cortex) show correlations which are still very much physiologically "real" and scientifically interesting, but up to an order of magnitude weaker. A possible solution is to consider an approach in which one allows *local structure* to influence the threshold, rather than a simple globally compared magnitude.

3.2 Thresholds from Random Matrix Theory

We suggest here a mathematically elegant approach to choosing graph thresholds starting from a statistical basis rather than a post hoc heuristic. In particular, we note that a failing of the currently popular thresholding techniques is that they are conceived at the level of abstraction of the *graph* rather than that of the *correlation matrix*. Although a correlation matrix can certainly be profitably viewed as the adjacency matrix of a graph, it is an error to jump to this view prior to examining the statistics of the matrix. While all correlation matrices can be viewed as graphs, *most* graphs do not have adjacency matrices that are correlation matrices, and by analyzing thresholds at the graph level, we are thus throwing away our knowledge of the special structure of correlation matrices.

The theory of random matrices was proposed originally by Wigner and Dyson, in the context of studying the spectra of complex nuclei (see, e.g., [25]), and is specifically suited to studying phase transitions between disordered and ordered, modular, systems defined by correlation matrices. Consider the matrices studied here of correlations between the (filtered) time series of gray-matter voxels. A simple model for the value of the correlation between voxels i and j might look like this:

$$r_{i,j} = r_{i,j}^{\dagger} + \varepsilon,$$

where $r_{i,j}^{\dagger}$ represents the *true correlation* between the physiological processes underlying the signals observed at voxels i and j and ε represents the sum of the many noise sources (e.g., physiological noise and scanner noise). For a single correlation, we have no way of separating $r^{\dagger}$ from ε, but if we have a priori a model for the global structure of the matrix of true correlations $r^{\dagger}$, then we can attempt to extract a reasonable estimate of $r^{\dagger}$ from $r_{i,j}^{\dagger} + \varepsilon$.

In essence, this is exactly what the currently popular thresholding methods attempt to do, albeit in a nonrigorous way: remove ε from the matrix by setting a threshold that yields a matrix corresponding to a graph with an expected structure.

In random matrix theory, one studies – amongst many other things – the spectra of real, symmetric matrices representing systems composed of a sum of signal and noise. In particular, we are interested here in looking at the statistical properties of the eigenvalue spacing of our correlation matrices. RMT tells us that when we observe the spacing of eigenvalues of a correlation matrix, we should expect to find a distribution of spacings conforming to one of two possibilities: the Gaussian orthogonal ensemble (GOE), where there are strong correlations everywhere, and, at the other extreme, Poisson statistics, where there exist strong correlations only along the (block) diagonal of the matrix.

In the context of our correlation matrices, the former case – in which the eigenvalue spacings follow a GOE distribution – is indicative of a matrix which is dominated by noise and spurious correlations; the latter case, where the eigenvalues follow Poisson statistics, is indicative of a matrix describing a highly modular system.

If one is prepared to accept the hypothesis that the physiological networks generating the observed BOLD signal in our neuroimaging data sets are indeed modular, then we now have exactly the statistical tools we need to separate these modular networks from noise. We must simply find the threshold at which the distribution of eigenvalue spacings for our matrix completes the transition from GOE to Poisson statistics. Once we have identified a threshold which generates Poisson eigenvalue spacings, we can be mathematically sure that we have extracted, to the best of our ability given the signal-to-noise ratio of the data, the modular true correlations present in the data while having sacrificed the minimum number of true correlations during the elimination of spurious, noisy correlations.

More formally, for a correlation matrix of order n, let E_i for $i \in \{1, \ldots, n\}$ denote the magnitude-ordered list of eigenvalues of the matrix. We perform a spectral unfolding procedure to obtain a distribution with the eigenvalue spacing represented in units of the local mean eigenvalue spacing. In particular, we unfold our eigenspectrum by estimating the integrated density of the spectrum, which we then fit to a cubic spline.[9] Individual eigenvalues are then projected into the unfolded representation by evaluation on the spline; we denote these transformed eigenvalues by e_i. We then simply compute the pairwise difference between adjacent transformed eigenvalues (i.e., $d = e_{i+1} - e_i$), and from this generate the probability density $P(d)$ of the unfolded eigenvalue spacing. Formalizing the relationships noted above, we consider two extreme distributions for $P(d)$: the Wigner–Dyson distribution (for the GOE case), where

$$P(d) \approx \frac{1}{2}\pi d e^{-\pi d^2/4},$$

and the Poisson distribution

$$P(d) \approx e^{-d}.$$

Indeed, we can see in Fig. 3 that the nearest-neighbor spacing distribution (NNSD) for the eigenvalues of a raw, unthresholded, fMRI-derived correlation matrix pictured in the upper left very closely follows GOE statistics. If the same matrix is subjected to strict thresholding, the empirical NNSD looks a great deal "more Poisson".

For a candidate threshold value, we can do the following: threshold the matrix at the candidate value, unfold the eigenspectrum and compute $P(d)$, and compare $P(d)$ with the GOE and Poisson distributions. If $P(d)$ follows the GOE, it is dominated by noise and our threshold is too low. If $P(d)$ is a Poisson distribution, it represents a modular network, and we assume it is thus dominated by signal rather than noise. Maximizing the signal-to-noise ratio then simply becomes a game of identifying the threshold at which the statistics of the unfolded eigenvalue difference (UED) change from Wigner–Dyson to Poisson statistics. The transition between GOE and Poisson distributions cannot occur instantaneously between thresholds, but rather resembles a phase transition, with intermediate threshold values having some degree of "Poissonness" and some degree of "Wignerness".

A very liberal estimate of the threshold, certain to still include a great deal of noise, can be obtained by finding the first point at which the UED distribution begins to differ significantly from the Wigner–Dyson distribution. Likewise, a conservative

---

[9]One may, of course, fit the curve to an arbitrarily sophisticated function; we chose cubic splines here, as they have been demonstrated to work well in applications ranging from neutron scattering to quantitative finance.

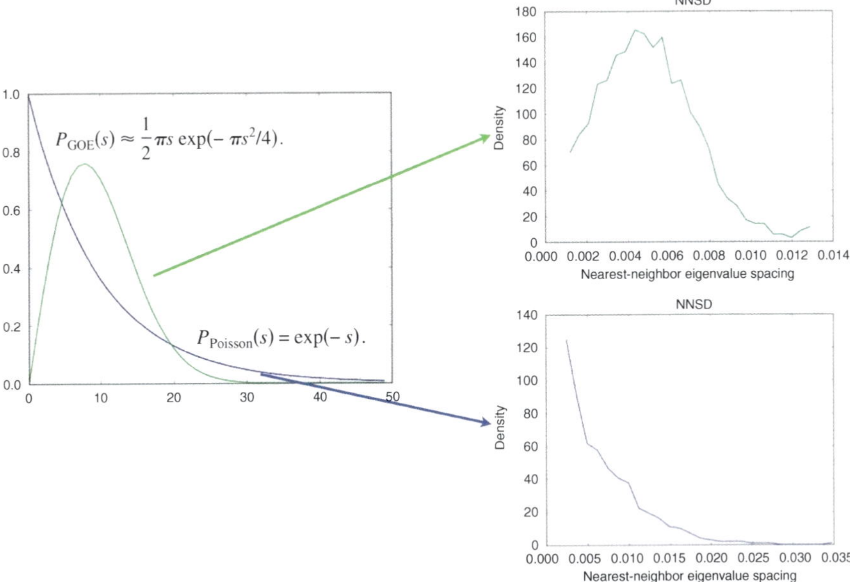

$$P_{\mathrm{GOE}}(s) \approx \frac{1}{2}\pi s \exp(-\pi s^2/4).$$

$$P_{\mathrm{Poisson}}(s) = \exp(-s).$$

Fig. 3 *Left*: a plot of the GOE (*green*) and Poisson (*blue*) distributions. *Right*: above, the NNSD of eigenvalues from an unthresholded fMRI correlation matrix; below, the NNSD for the same matrix following strict thresholding

estimate may be obtained by identifying the threshold at which the UED distribution becomes significantly Poisson.

We now propose a very simple algorithm for finding such a threshold for a given correlation matrix M. Begin with a very low threshold $t \approx 0$. Compute $M' \leftarrow M$, where all matrix entries $m_{i,j} < t$ are set to zero, followed by $E \leftarrow$ eigenvalues of M' and $e \leftarrow U(E)$, where $U(E)$ is the smoothed, integrated eigenvalue density. Then, compute the distribution P of nearest-neighbor spacings in $U(E)$. If an Anderson–Darling goodness-of-fit test of P against a *Poisson* distribution yields $p < \alpha$ (for, say, $\alpha = 0.05$), terminate and report the threshold t; otherwise, increase t by some small increment δ and repeat until this process terminates. The choice of value for δ controls a trade-off between computation time and threshold precision.

The result of applying this approach to matrices derived from ten resting-state fMRI scans can be seen in Fig. 4. The unthresholded matrices have very high Anderson–Darling scores (when tested against an exponential distribution), but these scores rapidly drop as we increase the threshold to an r-value of approximately 0.15. Beyond this value, the scores remain relatively stable, modulo some apparently stochastic factor. Thus, for the matrices considered in this diagram, thresholding in the neighborhood of $r > 0.15$ would be appropriate.

The computational complexity of this approach is tractable, even for large matrices, in a high-performance computing environment given the wide availability

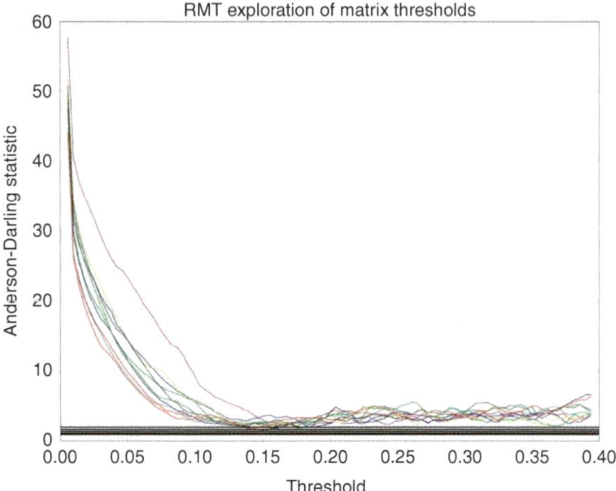

Fig. 4 Plots of Anderson–Darling test statistic, testing the fit of the empirical NNSD to an exponential distribution, for increasing thresholds (values below the threshold in the correlation matrix are set to zero) for ten resting-state fMRI scans. *Horizontal black lines* denote critical values for the Anderson–Darling score

of excellent parallel eigenvalue finding libraries[10] (see, e.g., LAPACK [2]); finding the eigenvalues is the dominant computational step in our algorithm.

3.3 Evaluation of RMT-Based Thresholding

We now compare the real-world effectiveness of our RMT-derived thresholding approach with two approaches commonly used in the neuroimaging literature, based on the S-value and the false discovery rate. Our test data set consists of 1,000 resting-state fMRI scans from the 1,000 Functional Connectomes Project [6]. Each scan was preprocessed and then converted into a correlation matrix according the protocol we outlined above. We focus solely on correlation matrices here, as these are presently by far the mostly commonly used basis for building graphs in the neuroimaging literature.

For each matrix, we computed an optimal threshold (below which all entries in the matrix should be set to zero) using the S-value approach (with $S = 2.0$), The FDR, and our RMT approach. The resulting thresholds are plotted in Fig. 5. Three trends are immediately visible:

---

[10]It is also worth noting that significant progress has been made in spectral decomposition on commodity GPU hardware; see, e.g., [15].

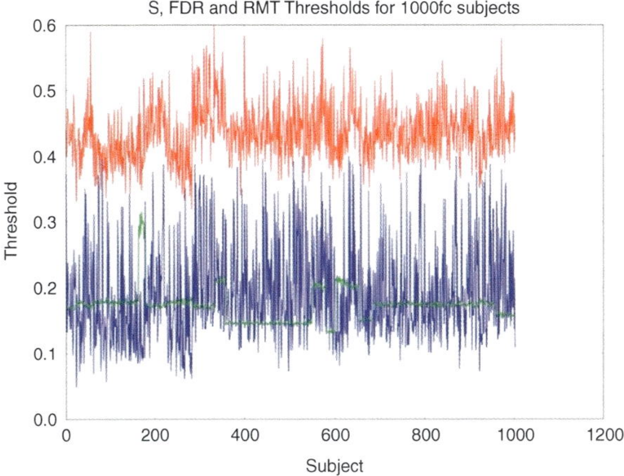

Fig. 5 Optimal thresholds for 1,000 resting-state fMRI scans as determined by the S-value approach (*red*), the FDR (*green*), and the proposed RMT approach (*blue*)

1. The variance in thresholds for the FDR approach is significantly lower than for either the S-value or the RMT-based approach.
2. The RMT and FDR thresholds are almost always lower than the S-value-based thresholds.
3. There is no consistent greater/less than relationship between the FDR- and RMT-based approaches, though it appears that were the variance of the RMT thresholds smaller, the two values would track each other relatively well.

It is not possible to draw a clear conclusion about which method is "best" from the data at hand; we can, however, offer some observations. If one expects thresholds to vary very little between subjects, then the significantly lower variance in the FDR scores would make this approach quite attractive. Conversely, if one expects some nontrivial physiological differences between subjects, then one might interpret the higher variance in the results from the RMT and S-value approaches as higher intrasubject sensitivity. We can investigate this hypothesis by looking at the distributions of thresholds while limiting ourselves to considering subjects from a single site (to avoid possible confounds from the use of different equipment at different sites contributing to the 1000 Functional Connectomes Project (http://fcon_1000.projects.nitrc.org/) data set).

In Fig. 6 we plot the distribution of thresholds for the S-value, FDR, and RMT approach for 198 scans from the Harvard site only. As expected, the width of the histogram is significantly lower for the FDR approach, but it is the shape of the histograms which is more interesting. While the FDR and S-value histograms have a roughly Gaussian-appearing shape, the RMT histogram appears to have a long right tail. It is, of course, impossible to make any definitive inferences from such a crude analysis, but this does suggest that, at the very least, the source of variance in the RMT thresholds is different from that in the FDR and S-value thresholds.

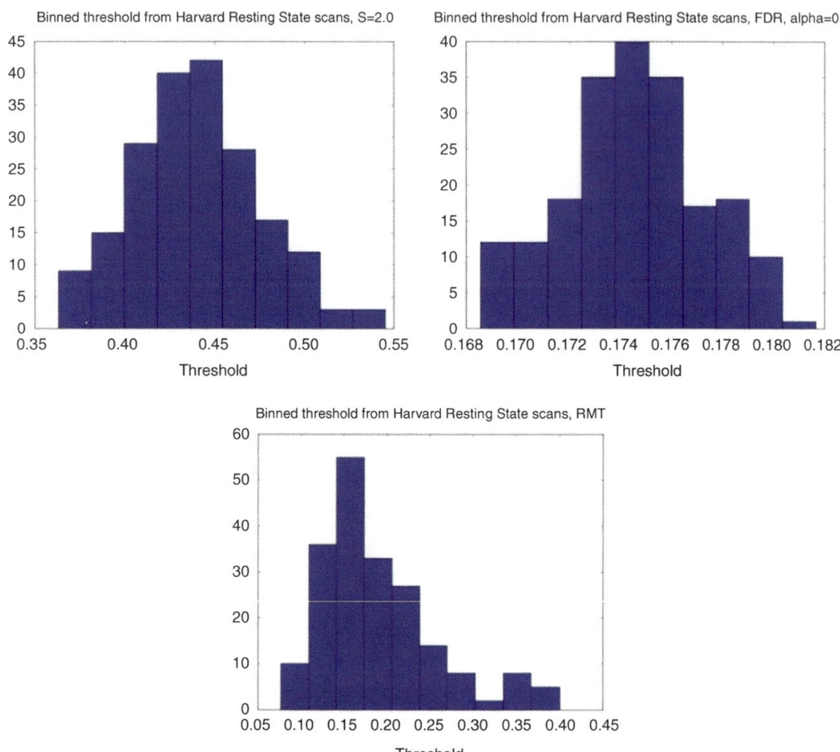

Fig. 6 Histograms of threshold values obtained for 198 resting-state fMRI scans from a single site for the S-value, FDR, and RMT approaches

Although providing food for thought, these results cannot conclusively identify a "best" approach to thresholding. Future work in which physiologically and physically accurate models generate simulated fMRI data based on precisely specified networks will allow qualitative comparison of thresholding techniques. If we have a contrived "ground truth" mesoscale brain network to hand, and a reasonable model for the physiological and physical processes that would produce a BOLD signal for this network (e.g., inverse dynamic casual modeling [10]), we can directly test the results of our thresholding against a 100 % known ground truth – a concept not available to us in real data sets.

4   Closing Thoughts

We have provided a brief overview of the growing interest in applying tools from graph theory to the analysis of graphs derived from neuroimaging data. We introduced fMRI – currently the most popular functional neuroimaging technique – and touched on the details of what is required to preprocess fMRI data. Moving from preprocessed neuroimaging-derived time series, we discussed methods for comparing time series and thus inferring a static graph describing relationships between these time series. We then focused in on the critical issue of choosing a threshold to separate "real" physiologically relevant relationships from noise in our graphs, including the description of a new thresholding technique based on random matrix theory.

The bulk of the work done in this area to date has been very pragmatic and applied in nature; the field is driven by neuroscientists who seek tools which can help them further understand the brain, with relatively less interest in the nature of the tools themselves. Several clear, pragmatic problems remain open, including two quite significant ones. First, the correct interpretation of complex graph metrics (e.g., betweenness or PageRank) on functionally derived graphs is still unclear. Second, the metrics that are typically employed are classical metrics optimized for studying abstract graph topologies, and networks which arise in physics and the social sciences. It would be surprising if there did not exist better metrics, tailored specifically to the questions most interesting to neuroscientists.

Moving to a more abstract, theoretical level, there has been little investigation of the deeper topological properties of neuroimaging-derived graphs. Is there a theory of descriptional complexity for these graphs? Can we define hierarchies of structure which correspond to, for example, clinical metrics of impaired neurological function? This has been done, for example, for simple measures such as the clustering coefficient, but is there a deeper unifying theory underlying these observations? Questions of computability arise naturally as well: if we imagine these graphs to represent a snapshot of synchronizations between the nodes of a computing machine, what can we say about the nature of the machine from such a snapshot? Is it even possible to reason about questions of computability and complexity given such a coarse projection of the dynamics of a computing system?

For one who studies the theory of computation and its application to natural systems, it is difficult to ignore the attraction of the dynamic computational system that is the human brain. With the recent, and rising, interest in graph theory from neuroscientists, it is now possible to find collaborators in the mathematical sciences and neurosciences who speak a common language and have a common interest in the dynamics of complex computational systems. Many problems – both straightforward and pragmatic and deeper, more philosophical – remain open and ready for investigation; and, as always, with each solution comes even more open problems.

References

1. R. Albert, A.-L. Barabási, Statistical mechanics of complex networks. Rev. Modern Phys. **74**(1), 47–97 (2002)
2. E. Anderson, Z. Bai, C. Bischof, S. Blackford, J. Demmel, J. Dongarra, J. Du Croz, A. Greenbaum, S. Hammarling, A. McKenney, D. Sorensen, *LAPACK Users' Guide*, 3rd edn. (Society for Industrial and Applied Mathematics, Philadelphia, 1999)

3. D.S. Bassett, E. Bullmore, Small-world brain networks. Neuroscientist **12**(6), 512–523 (2006)
4. M. Bernstein, K. King, X. Zhou, *Handbook of MRI Pulse Sequences* (Elsevier/Academic Press, Burlington, 2004)
5. B. Biswal, F.Z. Yetkin, V.M. Haughton, J.S. Hyde, Functional connectivity in the motor cortex of resting human brain using echo-planar MRI. Magn. Reson. Med. **34**(4), 537–541 (1995)
6. B.B. Biswal, M. Mennes, X.N. Zuo, S. Gohel, C. Kelly, S.M. Smith, C.F. Beckmann, J.S. Adelstein, R.L. Buckner, S. Colcombe et al., Toward discovery science of human brain function. Proc. Natl. Acad. Sci. **107**(10), 4734–4739 (2010)
7. R.L. Buckner, J. Sepulcre, T. Talukdar, F.M. Krienen, H. Liu, T. Hedden, J.R. Andrews-Hanna, R.A. Sperling, K.A. Johnson, Cortical hubs revealed by intrinsic functional connectivity: mapping, assessment of stability, and relation to Alzheimer's disease. J. Neurosci. **29**(6), 1860–1873 (2009)
8. M. Daley, Optimizing voxel scale graph theoretical analysis of fMRI-derived resting state functional connectivity. Master's thesis, The University of Western Ontario, 2012
9. N.U.F. Dosenbach, B. Nardos, A.L. Cohen, D.A. Fair, J.D. Power, J.A. Church, S.M. Nelson, G.S. Wig, A.C. Vogel, C.N. Lessov-Schlaggar, K.A. Barnes, J.W. Dubis, E. Feczko, R.S. Coalson, J.R. Pruett, D.M. Barch, S.E. Petersen, B.L. Schlaggar, Prediction of individual brain maturity using fMRI. Science **329**(5997), 1358–1361 (2010)
10. K.J. Friston, L. Harrison, W. Penny, Dynamic causal modelling. Neuroimage **19**(4), 1273–1302 (2003)
11. E.M. Haacke, R.W. Brown, M.R. Thompson, R. Venkatesan, *Magnetic Resonance Imaging: Physical Principles and Sequence Design* (Wiley, New York, 1999)
12. S.A. Huettel, A.W. Song, G. McCarthy, *Functional Magnetic Resonance Imaging* (Sinauer Associates, Sunderland, 2004)
13. A. Kraskov, H. Stögbauer, P. Grassberger, Estimating mutual information. Phys. Rev. E **69**(6), 066138 (2004)
14. K.K. Kwong, J.W. Belliveau, D.A. Chesler, I.E. Goldberg, R.M. Weisskoff, B.P. Poncelet, D.N. Kennedy, B.E. Hoppel, M.S. Cohen, R. Turner, Dynamic magnetic resonance imaging of human brain activity during primary sensory stimulation. Proc. Natl. Acad. Sci. **89**(12), 5675–5679 (1992)
15. S. Lahabar, P.J. Narayanan, Singular value decomposition on GPU using CUDA, in *IEEE International Symposium on Parallel & Distributed Processing, 2009. IPDPS 2009*. IEEE (2009)
16. N.K. Logothetis, J. Pauls, M. Augath, T. Trinath, A. Oeltermann, Neurophysiological investigation of the basis of the fMRI signal. Nature **412**(6843), 150–157 (2001)
17. M.E. Lynall, D.S. Bassett, R. Kerwin, P.J. McKenna, M. Kitzbichler, U. Muller, E. Bullmore, Functional connectivity and brain networks in schizophrenia. J. Neurosci. **30**(28), 9477–9487 (2010)
18. S. Ogawa, D.W. Tank, R. Menon, J.M. Ellermann, S.G. Kim, H. Merkle, K. Ugurbil, Intrinsic signal changes accompanying sensory stimulation: functional brain mapping with magnetic resonance imaging. Proc. Natl. Acad. Sci. **89**(13), 5951–5955 (1992)
19. J. Schummers, H. Yu, M. Sur, Tuned responses of astrocytes and their influence on hemodynamic signals in the visual cortex. Science **320**(5883), 1638–1643 (2008)
20. S.M. Smith, K.L. Miller, G. Salimi-Khorshidi, M. Webster, C.F. Beckmann, T.E. Nichols, J.D. Ramsey, M.W. Woolrich, Network modelling methods for fMRI. NeuroImage **54**(2), 875–891 (2011)
21. O. Sporns, *Networks of the Brain* (The MIT Press, Cambridge, MA 2010)
22. A.A. Taren, V. Venkatraman, S.A. Huettel, A parallel functional topography between medial and lateral prefrontal cortex: evidence and implications for cognitive control. J. Neurosci. **31**(13), 5026–5031 (2011)
23. K.R.A. Van Dijk, T. Hedden, A. Venkataraman, K.C. Evans, S.W. Lazar, R.L. Buckner, Intrinsic functional connectivity as a tool for human connectomics: theory, properties, and optimization. J. Neurophys. **103**(1), 297–321 (2010)

24. S. Van Dongen, Graph clustering via a discrete uncoupling process. SIAM J. Matrix Anal. Appl. **30**(1), 121–141 (2008)
25. E.P. Wigner, Random matrices in physics. SIAM Rev. **9**(1), 1–23 (1967)

Simplicial Models and Topological Inference in Biological Systems

Vidit Nanda and Radmila Sazdanović

Abstract This article is a user's guide to algebraic topological methods for data analysis with a particular focus on applications to datasets arising in experimental biology. We begin with the combinatorics and geometry of simplicial complexes and outline the standard techniques for imposing filtered simplicial structures on a general class of datasets. From these structures, one computes topological statistics of the original data via the algebraic theory of (persistent) homology. These statistics are shown to be computable and robust measures of the shape underlying a dataset. Finally, we showcase some appealing instances of topology-driven inference in biological settings, from the detection of a new type of breast cancer to the analysis of various neural structures.

1 Introduction

Recent advances in genomics [40] have made it possible to sequence the entire DNA of an individual from a very small amount of that person's genetic material, say in the form of a saliva sample or a hair follicle. For each individual, one obtains as the raw output of this full sequencing process an ordered list of roughly 15 billion letters, representing the base pairs which comprise that person's DNA. This technological achievement is absolutely amazing in itself, but in all probability the bulk of its benefits will materialize over time as scientists analyze the structure of such sequences in detail. Now consider another marvel of modern engineering: the

V. Nanda
The University of Pennsylvania, Philadelphia, PA 19104, USA
e-mail: vnanda@sas.upenn.edu

R. Sazdanović (✉)
North Carolina State University, Raleigh, NC 27695, USA
e-mail: rsazdanovic@math.ncsu.edu

N. Jonoska and M. Saito (eds.), *Discrete and Topological Models in Molecular Biology*,
Natural Computing Series, DOI 10.1007/978-3-642-40193-0_6,
© Springer-Verlag Berlin Heidelberg 2014

Fig. 1 A dataset consisting of points sampled from a *circle*. Although traditional line-fitting methods are likely to be uninsightful for such datasets, the methods of persistent homology can extract knowledge about the underlying shape from the point samples alone

Protein Data Bank [39] contains a wealth of structural information about protein molecules, down to the location of individual atom centers. Again, the fact that such data can now be effectively measured and collected is fascinating, but ideally one desires the ability to understand how the physical structure of a protein relates to its role in the body.

In both cases, one is confronted with enormous quantities of high-dimensional *data* prone to the usual amounts of noise or errors. From such data, one would like to extract *robust, qualitative information* and gain insight into the processes which generated the data in the first place. The standard toolkit for such inference is *statistical* at its core, and it provides computable, noise-tolerant answers to questions such as "what does the average data point look like?" or "what is the line or plane of best fit through the data?" These statistical tools are well understood, accessible to the experimentalist with a rudimentary mathematical background, and efficiently implemented in various standard software packages.

However, when the experimental data in question is produced by an essentially *nonlinear* process, the utility of our ordinary statistical tools is somewhat diminished even in the simplest of cases. Consider the dataset of Fig. 1, consisting of points sampled uniformly from a large circular figure sitting in the plane. With high probability, the average point lies near the center (but far away from the actual circle), and there is no reasonable line of best fit. It is not clear how to recover knowledge about the circle from statistics alone. Perhaps one might get lucky by noticing that the mean is roughly equidistant from all the data points, but it is easy to create slightly more complicated examples where recovering the underlying objects with any reasonable degree of accuracy from the standard statistical tools becomes hopeless. Thus, one might ask, *is there a complementary set of tools which detects the shape of the object underlying a dataset?*

A partial answer to this question comes from a previously esoteric branch of mathematics called *algebraic topology*. In particular, the theory of *persistent homology* has witnessed some success in the context of analyzing large-scale nonlinear data [16]. The basic idea behind this theory is to build an increasing family of *simplicial complexes* (indexed by a scale parameter) around the data points while carefully keeping track of the appearance and disappearance of topological features – connected components, tunnels, cavities, and their higher-dimensional cousins – as the scale parameter is increased. Numerous applications of persistent homology to various problems in the experimental sciences have been thoroughly documented elsewhere [4, 13, 18].

Although one can easily find efficient software [24, 29] for computing the persistent homology of filtered simplicial complexes, two key obstacles undermine the effective use of persistent homology to analyze experimental data. The first obstacle is an issue of *input*: how should one build a simplicial complex that captures the interesting aspects of one's data? The second issue involves the *output*: how does one make inferences about the data from the persistent homology of the input complex? With these issues in mind, the purpose of our work is threefold.

1. We provide a gentle and example-filled introduction to the mathematical *theory* which underlies (filtered) simplicial complexes. Starting with elementary combinatorial properties, we describe the connection between simplicial complexes and piecewise-linear geometry. We also discuss those algebraic objects which generate persistent homology, how they relate to simplicial geometry, and how one computes them in practice.
2. We highlight the standard *methods* of constructing filtered simplicial complexes around point cloud data via the Vietoris–Rips and Čech filtrations. We mention the relative advantages and disadvantages of these filtrations.
3. We showcase some *examples* of persistent homology in action on biological data. Recent applications have involved detection of a certain subtype of breast cancer [31] and yielded insight into the nature of neural activity – of crickets [3], monkeys, and rats [35]!

The outline of this chapter is as follows. The fundamentals of simplicial complexes and their filtrations are described in Sect. 2. Section 3 contains the core ideas needed for establishing connections between experimental data and filtrations. Section 4 describes the linear algebra of (persistent) homology and formally defines the topological features which can be detected by the theory. Finally, in Sect. 5, we survey several biological applications of persistent homology and closely related topological methods.

2   The Yoga of Simplicial Complexes

Our main goal throughout this Sect. 2 is to understand *simplicial complexes* and various related constructions. These combinatorial objects serve as a bridge between the discrete, computable world of data on one side and the continuous realm of geometric or topological spaces on the other. Our presentation here is far from complete, so we invite the interested reader to consult the wonderful texts of Munkres [28, Chaps. 1 and 2] and Spanier [36, Chap. 3] for the many gory details which we have omitted.

2.1 Simplicial Complexes

We start with a finite set V, whose elements we call *vertices*. A simplicial complex with vertex set V is a collection K of subsets of V which is closed under inclusion. More precisely, we require that the following two conditions hold:

- For each vertex v in V, the one-element set $\{v\}$ lies in K, and
- If τ is in K and $\sigma \subset \tau$ is a subset, then σ is also in K.

Each element τ of K is called a *simplex*, and its *dimension* (written dim τ) is defined to be $\#(\tau) - 1$, where $\#$ denotes the cardinality (i.e., it counts the number of vertices of τ). Any subset σ of τ is called a *face* of τ, and this relationship is denoted by $\sigma \preceq \tau$. We write K_d to indicate the collection of d-dimensional simplices in K for each $d \geq 0$. It is clear from the first property of simplicial complexes that the elements of V correspond in a one-to-one manner with those of K_0, and it is therefore customary to speak of the two sets interchangeably. Consequently, one often encounters phrases resembling "let K be a simplicial complex" with no explicit mention of the underlying vertex set. Before proceeding any further, we will examine a small simplicial complex in some detail.

Example 1. Given a vertex set $V = \{a, b, \ldots, f, g\}$, we may construct a simplicial complex K in layers, one dimension at a time. We denote subsets of V by their elements in alphabetical order, so that $\{a, b, c\}$ is simply written abc. As we have already seen, K_0 is completely determined by V. Next, K_1 can contain any pair of distinct vertices in V and there is some freedom to choose such pairs. For instance, we can select

$$K_1 = \{ab, ac, ae, bc, bd, be, bg, cd, cg, dg, ef\}.$$

Fixing K_1 immediately constrains which simplices can lie in K_2. For instance, abe is allowed in K_2, since all of its one-dimensional faces ab, ae, be are in K_1. However, acd is banned because $ad \prec acd$ but ad is not present in K_1. We add the following (legal!) two-dimensional simplices to K:

$$K_2 = \{abe, bcg, bcd, bdg, cdg\},$$

and note that the only three-dimensional simplex whose faces all exist in K_2 is $bcdg$. Let us include that simplex as well, and we obtain

$$K_3 = \{bcdg\}.$$

No four-dimensional simplices are allowed, since the presence of a single such simplex would require K_3 to have at least four elements, so our K is just the union of K_d for d in $\{0, 1, 2, 3\}$. This complex K is reasonably small and low-dimensional; it is often useful to visualize such simplicial complexes as embedded in Euclidean space (see Fig. 2).

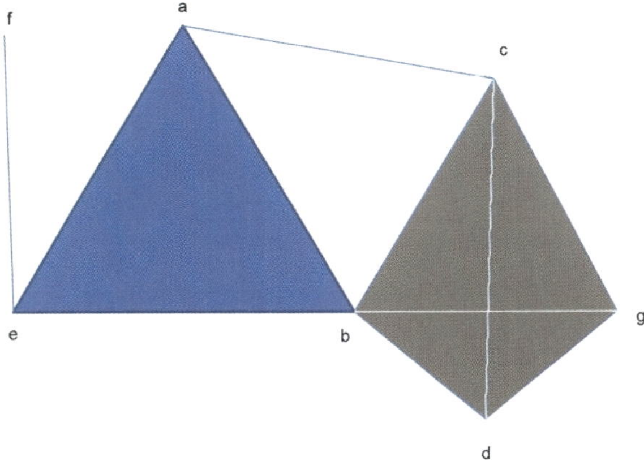

Fig. 2 A pictorial representation of the simplicial complex K of Example 1 with points representing vertices. The *lines*, *triangles*, and tetrahedra stand in for one, two, and three-dimensional simplices, respectively. We note that K consists of a single connected component and that the 1-simplices ab, ac, bc form a loop

2.2 Subcomplexes, Filtrations, and Sublevelsets

Let K be any simplicial complex. A subcollection L of simplices from K which forms a simplicial complex in its own right is called a *subcomplex* of L, written $L \hookrightarrow K$. In other words, *if a simplex τ lies in L, then all of its faces in K are also present in L*. In general, the vertex set of L may be strictly smaller than that of K, with equality only occurring when $L_0 = K_0$. The reader may enjoy proving the following result, but we have our doubts.

Proposition 1. *If simplicial complexes K, L, and M satisfy $L \hookrightarrow K$ and $K \hookrightarrow M$, then we also have $L \hookrightarrow M$.*

Let $N \geq 1$ be a natural number and K a simplicial complex. A *filtration* $\mathscr{F}$ of the simplicial complex K is a nested collection of subcomplexes $\mathscr{F}_n K \hookrightarrow K$ for n in $\{0, \ldots, N\}$ which ascends from the empty set $\emptyset$ all the way up to K like this:

$$\emptyset = \mathscr{F}_0 K \hookrightarrow \mathscr{F}_1 K \hookrightarrow \mathscr{F}_2 K \hookrightarrow \cdots \hookrightarrow \mathscr{F}_{N-1} K \hookrightarrow \mathscr{F}_N K = K.$$

Here, N is called the *length* of $\mathscr{F}$. The simplicial complex K trivially forms a length-1 filtration, since we have $\emptyset \subset K$. A slightly less obvious filtration could be constructed by dimension: let $\mathscr{F}_n K$ be the collection of all simplices of dimension at most n. But we will consider a more interesting example. In particular, we would like to illustrate the fact that the process of building K from subcomplexes along $\mathscr{F}$ causes various interesting intermediate features to appear and disappear.

Example 2. Let K be the simplicial complex of Example 1. We will define a filtration $\mathscr{F}$ of K which has length 4 by describing each subcomplex $\mathscr{F}_n K$

individually. Since $\mathscr{F}_0 K$ is empty, we ignore it and move on to the first subcomplex,

$$\mathscr{F}_1 K = \{a, b, c, d, f, ac, cd, eb\}.$$

There are three pieces in this subcomplex, as the first quarter of Fig. 3 reveals. Next, we consider

$$\mathscr{F}_2 K = \mathscr{F}_1 K \cup \{g, ab, ae, bc, bd, ef, bcd\},$$

where $\cup$ indicates a union of sets. The addition of the vertex g adds yet another piece to the three already present in $\mathscr{F}_1 K$, but ab and ef join three of those pieces into a single large component. The sequences (ab, ae, be) and (ab, ac, bc) of one-dimensional simplices form two *loops*. A similar loop formed by (bc, bd, cd) is immediately filled by the two-dimensional simplex bcd. Moving on, we define

$$\mathscr{F}_3 K = \mathscr{F}_2 K \cup \{abe, bcg, bdg, cdg\}.$$

The simplex abe fills up the loop (ab, ae, be) consisting of its faces. The addition of the other simplices reveals a new feature: a void, or *cavity*, formed by (bcd, bcg, bdg, cdg). This cavity is very different from the loops that we have encountered before, in the sense that – at least as pictured in Fig. 3 – it encloses a three-dimensional region rather than a planar one. Finally, we add

$$\mathscr{F}_4 K = \mathscr{F}_3 K \cup \{bcdg\},$$

and this last simplex fills the cavity obtained from $\mathscr{F}_3 K$.

Let K be any simplicial complex, and let $\mathbf{N}$ denote the natural numbers. Consider a function $g : K \rightarrow \mathbf{N}$ which assigns to each simplex σ a natural number $g(\sigma)$. Then, the *sublevelset* of g at the natural number n is defined by $S_n(g) = \{\sigma \in K \mid g(\sigma) \leq n\}$. Clearly, we have $S_n(g) \subset S_{n+1}(g)$ as sets. Unfortunately, $S_n(g)$ is not always a subcomplex of K for arbitrary functions g: if $g(\sigma) > n \geq g(\tau)$ with $\sigma \prec \tau$, then $S_n(g)$ contains τ but not its face σ. It turns out that this is the only obstruction to having a filtration by sublevelsets, so we will restrict our choice of g to functions which avoid this behavior.

We call $g : K \rightarrow \mathbf{N}$ *monotone* if it increases with dimension along faces. Thus, g is monotone (or *order-preserving*) if $g(\sigma) \leq g(\tau)$ whenever $\sigma \preceq \tau$. In this case, it is easy enough to check that setting

$$\mathscr{F}_n K = S_n(g) = \{\tau \in K \mid g(\tau) \leq n\}$$

yields a filtration of K whose length equals the maximum number of distinct values attained by g on K. One could also consider monotone maps $g : K \rightarrow \mathbf{R}$ to the real numbers, but this would be largely for convenience. Since there are only finitely many simplices in K, the image of g may assume only finitely many distinct real

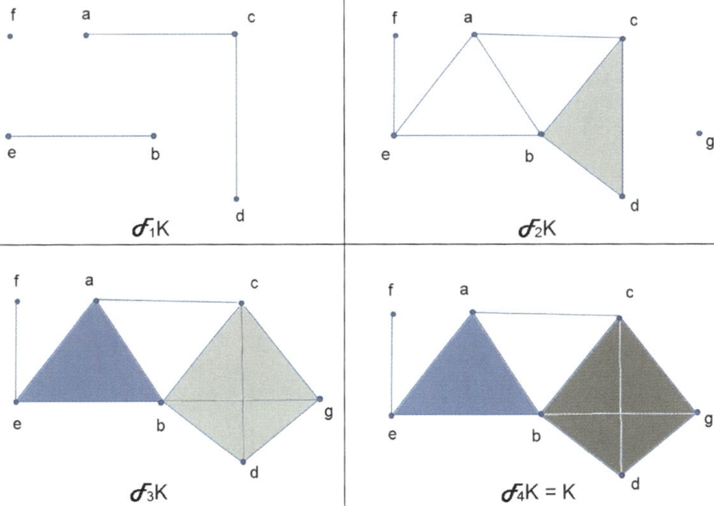

Fig. 3 An illustrated view of the filtration $\mathscr{F}$ defined in Example 2. Note that the intermediate stages of the filtration look very different from K! A systematic study of the appearance and disappearance of features such as connected components, loops, and cavities provides a coarse-grained view of how K is incrementally built along its subcomplexes in $\mathscr{F}$. See Example 2 for details

values. Indexing these values $\{c_1, \ldots, c_N\}$ in ascending order yields a one-to-one monotone correspondence with a subset of $\mathbf{N}$: just send c_n to n. Thus, an $\mathbf{R}$-valued g can easily be replaced by an $\mathbf{N}$-valued cousin with no essential change in the structure of the sublevelset filtration. Sublevelset filtrations are ubiquitous for the following simple reason.

Proposition 2. *For any filtration $\mathscr{F}$ of a simplicial complex K, there is a unique monotone function $g : K \to \mathbf{N}$ such that $\mathscr{F}$ is the sublevelset filtration of g.*

The proof is easy: let g be the function that sends each simplex σ in K to the smallest n such that σ is a simplex in $\mathscr{F}_n K$. The reader may wish to check, for example, that setting $g(\sigma) = \dim(\sigma)$ retrieves the filtration by dimensions mentioned before Example 2.

2.3   The Geometry of Simplices: Realizations and Simplicial Maps

As we have remarked before, a primary advantage of simplicial complexes is their ability to interface between discrete and continuous spaces. When we visualize simplicial complexes (see Fig. 2, for example), we use nondiscrete geometric objects such as lines, triangles, and tetrahedra. The reader may have noticed that much of the terminology for simplicial complexes (for instance "dimension" and "vertex") appears to have been borrowed from corresponding notions for these familiar and concrete geometric objects. There is a standard protocol underlying this dictionary between simplices and these objects, which we will now describe. All that is assumed of the reader is a basic understanding of d-dimensional Euclidean real space $\mathbf{R}^d$, each point of which consists of an ordered sequence of d real numbers. For each j between 1 and d, the j-th *basis vector* $\mathbf{e}_j$ of $\mathbf{R}^d$ is identified with the point which contains a 1 in the j-th component and 0's everywhere else.

We fix a dimension d, and let $\mathbf{u} = \{\mathbf{u}_1, \ldots, \mathbf{u}_M\}$ be a collection of $M \geq 1$ points in $\mathbf{R}^d$. A *convex combination* of these points is any point in $\mathbf{R}^d$ which can be expressed as an $\mathbf{R}$-linear combination

$$x = p_1 \mathbf{u}_1 + \cdots + p_M \mathbf{u}_M,$$

where each coefficient p_m is nonnegative and the sum $p_1 + \cdots + p_M$ of all these coefficients equals 1. The *convex hull* of this collection $\mathbf{u}$ is the set of all such convex combinations,[1] and we denote this subset of $\mathbf{R}^d$ by $\mathrm{Conv}(\mathbf{u})$. Now let $\mathbf{v} = \{\mathbf{v}_1, \ldots, \mathbf{v}_N\}$ be another collection of points in $\mathbf{R}^d$, and assume that we are given a map $\eta : \mathbf{u} \to \mathbf{v}$. Then, η provides a standard and fairly obvious recipe for concocting a map $\bar{\eta} : \mathrm{Conv}(\mathbf{u}) \to \mathrm{Conv}(\mathbf{v})$ as follows:

$$\bar{\eta}(p_1 \mathbf{u}_1 + \cdots + p_M \mathbf{u}_M) = p_1 \eta(\mathbf{u}_1) + \cdots + p_M \eta(\mathbf{u}_M).$$

If we restrict our attention to the subcollection $\mathbf{u}'$ of $\mathbf{u}$ and let η be the inclusion map $\mathbf{u}' \to \mathbf{u}$, we immediately see that $\mathrm{Conv}(\mathbf{u}') \subset \mathrm{Conv}(\mathbf{u})$. The d-dimensional *standard simplex* $\Delta^d \subset \mathbf{R}^{d+1}$ is defined to be $\mathrm{Conv}(\mathbf{e}_1, \ldots, \mathbf{e}_{d+1})$, the convex hull of the basis elements of $\mathbf{R}^{d+1}$. Equivalently,

$$\Delta^d = \left\{ (x_1, \ldots, x_{d+1}) \mid \text{each } x_j \geq 0 \text{ and } x_0 + \cdots + x_{d+1} = 1 \right\}.$$

For example, Δ^2 is the two-dimensional triangle determined by the standard basis vectors $\mathbf{e}_1 = (1, 0, 0)$, $\mathbf{e}_2 = (0, 1, 0)$ and $\mathbf{e}_3 = (0, 0, 1)$ in three-dimensional Euclidean space.

Let K be a simplicial complex with $d + 1$ vertices, which we order as $\{v_1, \ldots, v_{d+1}\}$. We will construct a concrete subset $|K|$ of the standard simplex Δ^d, which is the canonical geometric space associated to K. For each simplex

---

[1] It is easy to work out that the convex hull of two points is the line segment connecting them, and that the convex hull of three points (which do not all lie on the same line) is the triangle containing those three points as vertices. In higher dimensions and with many more points, things become less obvious. Determining convex hulls is a fundamental problem in computational geometry.

σ in K consisting of vertices $\{v_{i_1}, \ldots, v_{i_m}\}$, we first define $|\sigma| \subset \Delta^d$ by $|\sigma| = \text{Conv}(\mathbf{e}_{i_1}, \ldots, \mathbf{e}_{i_m})$.

Definition 1. The *geometric realization* $|K| \subset \Delta^d$ of K is the union of all $|\sigma|$ as σ ranges over simplices in K.

Thus, a counterpart to K has been delineated within Δ^d as a concrete geometric object. Before being completely satisfied with this definition, however, one might wonder: *what happens if we order the vertices* $\{v_1, \ldots, v_{d+1}\}$ *differently?* In order to arrive at a satisfactory answer, we must understand when two simplicial complexes are considered equivalent; for this purpose, we turn our attention to *simplicial maps*. These maps will require a domain and a range, so let K and L be simplicial complexes with vertex sets U and V, respectively.

Definition 2. A *simplicial map* $\phi : K \to L$ assigns to each vertex u in U a vertex $\phi(u)$ in V so that the image of each simplex $\sigma \in K$ constitutes a simplex $\phi(\sigma) \in L$.

Here, by $\phi(\sigma)$ we mean the set of vertices in V obtained by mapping each vertex of σ by ϕ into V. We have already seen examples of simplicial maps: if $K \hookrightarrow L$, then the map sending each vertex of K to itself as a vertex of L is simplicial. It turns out that any simplicial map $\phi : K \to L$ induces a continuous function of geometric realizations, which we denote by $|\phi| : |K| \to |L|$. This map acts exactly as one would expect. Namely, we note first that $|K|$ is a union of realizations of simplices $|\sigma|$ where $\sigma \in K$, so it suffices to understand how each individual $|\sigma|$ is mapped by $|\phi|$. Since ϕ maps the vertices of σ into the vertices of its image $\phi(\sigma)$, the map $\bar{\phi}$ linearly maps the convex hull $|\sigma|$ into the convex hull $|\phi(\sigma)|$. We define the function $|\phi|$ to be that transformation from $|K|$ to $|L|$ which acts on each $|\sigma| \subset |K|$ as the linear map $\bar{\phi}$. Thus, although $|\phi| : |K| \to |L|$ itself may not be a linear map, its action on each convex piece $|\sigma|$ of $|K|$ is linear. For this reason, the continuous maps between realizations induced by simplicial maps are often called *piecewise-linear* maps.

The reader is warned that arbitrary simplicial maps do not preserve dimension: one might have $\dim \phi(\sigma) < \dim \sigma$ if ϕ is not one-to-one on the vertices of σ. On the other hand, if ϕ is a *bijection* – a map that associates each vertex of U to a single vertex of V and vice versa – then not only are dimensions preserved, but also the net effect of mapping K into L via ϕ is essentially that of relabeling the vertices. In such a case, K and L are called *isomorphic* and we write $K \simeq L$. Although the geometric realizations $|K|$ and $|L|$ might disagree in terms of exactly how they sit in $\mathbf{R}^d$, they are topologically (and indeed, geometrically) equivalent because an invertible linear transformation of $\mathbf{R}^d$ (i.e., an invertible matrix) maps $|K|$ to $|L|$, with its inverse taking $|L|$ back into $|K|$. More precisely, any simplicial map $\phi : K \to L$ induces a map from the basis of $\mathbf{R}^{\#U}$ to that of $\mathbf{R}^{\#V}$ as follows:

$$\text{basis element} \xleftrightarrow{\;\simeq\;} \text{vertex of } K \xrightarrow{\;\phi\;} \text{vertex of } L \xleftrightarrow{\;\simeq\;} \text{basis element}.$$

Following this diagram from left to right produces a matrix which maps $\Delta^{\#U-1}$ to $\Delta^{\#V-1}$ so that the image of $|K|$ is contained inside $|L|$. In the special case where ϕ is a bijection of vertices, this matrix is invertible. It is in this sense that simplicial complexes (up to equivalence by isomorphism) are uniquely associated with their geometric realizations (up to equivalence by invertible linear transformations).

3 Constructing Filtrations Around Points

The process of conducting experiments and collecting data is, by its very nature, the crux of all experimental science. Experimental data can take many forms, including text, images, and even video. For our purposes, we will restrict our attention to a very specific form of data: a *point cloud*. By a point cloud, we simply mean a finite collection P of points in $\mathbf{R}^d$ for some suitable dimension d, and make no further assumptions regarding the nature of P. We would like to remark here that it is possible to construct faithful point cloud representations of just about any type of data, although it may not be advantageous to do so because the dimension d might become enormous.

A first step towards applying topological machinery to a point cloud is to construct a filtration of a simplicial complex whose vertex set can in some way be identified with P. We will discuss two standard filtrations that may be constructed around point clouds. Along the way, we will try to highlight their relative advantages and disadvantages.

The largest possible simplicial complex with vertex set P is, of course, the *complete* simplicial complex, where every possible subset of P constitutes a simplex. We will denote this complex by K_P throughout this section.[2] All the filtrations that we encounter here will be – either implicitly or explicitly – filtrations of K_P.

There are many notions of *distance* that one can reasonably impose on $\mathbf{R}^d$. For any $p \geq 1$, we can consider the p-distance

$$\mathbf{d}_p(x, y) = \sqrt[p]{\sum_{m=1}^{d} |x_m - y_m|^p},$$

so that the familiar Euclidean distance is recovered when one sets $p = 2$. Another option is the max-distance,

$$\mathbf{d}_\infty(x, y) = \max_{1 \leq m \leq d} \{|x_m - y_m|\}.$$

---

[2] In fact, K_P consists of a single $(\#P - 1)$-dimensional simplex along with all its faces!

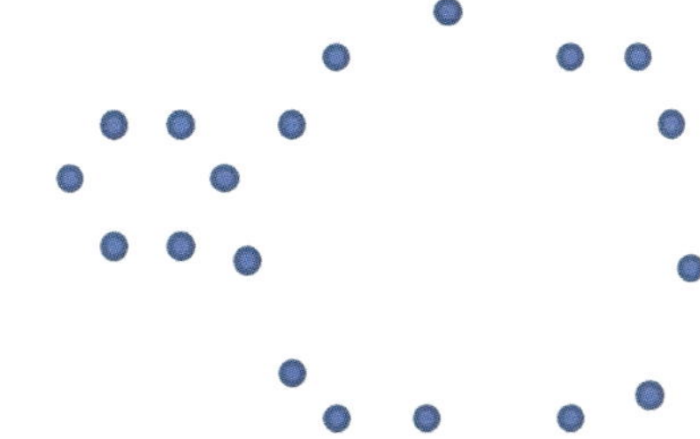

Fig. 4 An example of a small point cloud sitting in two-dimensional Euclidean space. Note that the points appear to have the shape of two *circles*, one larger than the other

The filtrations that one constructs around $P \subset \mathbf{R}^d$ depend on which notion of distance is chosen. In order to provide the most flexibility, we will simply denote the distance we use by $\mathbf{d}$ and leave the explicit choice to the reader.

For any positive real number $r \geq 0$ and a point $x \in \mathbf{R}^d$, we define the *ball of radius r around x* as

$$B_r(x) = \left\{ y \in \mathbf{R}^d \mid \mathbf{d}(x, y) < r \right\}.$$

The shape of this ball depends on the distance $\mathbf{d}$. The reason for calling this type of set a "ball" becomes clear when one uses the standard distance $\mathbf{d}_2$.

As a running example, we will consider a toy example of a point cloud in $\mathbf{R}^2$ as shown in Fig. 4. The exact coordinates of each point are relatively unimportant; we are only seeking *qualitative* information. Thus, what we will focus on here is the fact that the point cloud appears to contain two distinct loops, with the one on the right-hand side having a larger diameter than the other. In our running example, we will use the distance $\mathbf{d}_2$.

3.1 The Vietoris–Rips Filtration

Let $P \subset \mathbf{R}^d$ be our point cloud. One can compute all the pairwise distances $\mathbf{d}(p, p')$ between pairs of points p and p' in P. This data structure – consisting of P along with the pairwise distances – suffices to construct the Vietoris–Rips filtration (Fig. 5). At any given *scale* $\epsilon \geq 0$, we define the simplicial subcomplex $\mathscr{V}_\epsilon K_P$ of the complete complex K_P as follows. The vertex set is P, and each simplex σ in $\mathscr{V}_\epsilon K_P$ consists of a subcollection of vertices so that the pairwise distance between any two is less than ϵ. Let $\sigma \subset P$ be a subcollection of points $(p_1, \ldots, p_m)$. Restricting the indices i and j to $\{1, \ldots, m\}$, we have

$$\sigma \text{ is a simplex in } \mathscr{V}_\epsilon K_P \text{ if } \mathbf{d}(p_i, p_j) < \epsilon \text{ for all } i, j,$$

or equivalently,

$$\sigma \text{ is a simplex in } \mathscr{V}_\epsilon K_P \text{ if } B_{\epsilon/2}(p_i) \cap B_{\epsilon/2}(p_j) \neq \emptyset \text{ for all } i, j.$$

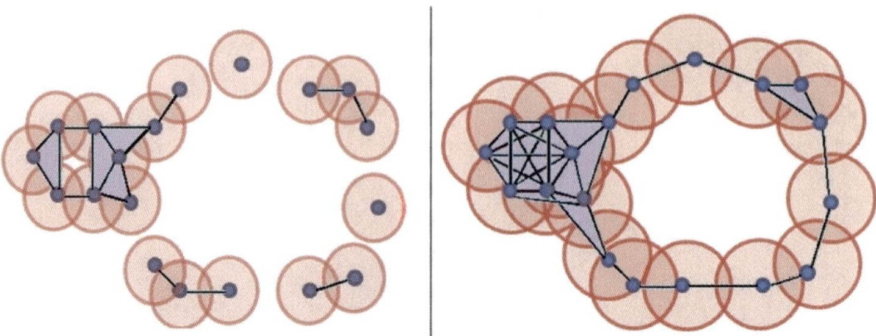

Fig. 5 Two stages of the Vietoris–Rips filtration around the point cloud from Fig. 4. The scale ϵ increases from *left* to *right*, and the balls of radius ϵ have been shown underlying the simplices. The smaller loop is captured faithfully at the smaller ϵ value, and the larger loop is captured at the larger ϵ value. But *no single scale captures both*!

Here $\cap$ stands for the intersection of sets.

It is easy to see that for any value of ϵ, our definition of $\mathscr{V}_\epsilon$ yields a genuine simplicial complex. After all, if σ is a simplex and τ is a face of σ, then the set of all pairwise distances between vertices of τ is contained in the set of the corresponding pairwise distances of σ's vertices. On the other hand, we can also immediately check that for $\delta > \epsilon$, we have $\mathscr{V}_\epsilon K_P \hookrightarrow \mathscr{V}_\delta K_P$ because if all pairwise distances are less than ϵ, they are also less than δ.

We define the function $g_{\mathscr{V}} : K_P \to \mathbf{R}$ as follows. For any simplex σ in K_P,

$$g_{\mathscr{V}}(\sigma) = \max_{p,q \text{ in } \sigma} \{\mathbf{d}(p,q)\}.$$

Whenever $\sigma \prec \tau$, we obtain $g_{\mathscr{V}}(\sigma) \leq g_{\mathscr{V}}(\tau)$ because we are taking the maximum over a larger set. Thus, g is monotone, and the following definition makes sense by Proposition 2.

Definition 3. The *Vietoris–Rips filtration* around $P \subset \mathbf{R}^d$ is the sublevelset filtration of $g_{\mathscr{V}}$.

We place the pairwise distances between points in P in ascending order, $0 \leq \epsilon_1 \leq \cdots \leq \epsilon_N$, and note that we have

$$\mathscr{V}_{\epsilon_1} K_P \hookrightarrow \mathscr{V}_{\epsilon_2} K_P \hookrightarrow \cdots \hookrightarrow \mathscr{V}_{\epsilon_N} K_P = K_P.$$

In practice, one stops well short of constructing the Vietoris–Rips filtration all the way up to ϵ_N, unless the number of points in P is very small. The reason for this is simple: the complete complex K_P contains as many simplices as there are nonempty subsets of P, so its cardinality is $2^{\#P} - 1$. Even for a tiny cloud containing only 40 points, building the full Vietoris–Rips filtration requires storing well over a *trillion* simplices in system memory!

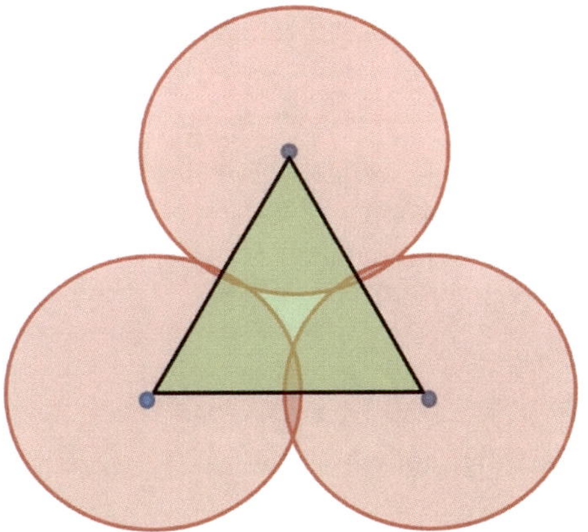

Fig. 6 A collection of balls in the plane with nonempty pairwise intersection but no triple intersection. The union of these balls clearly encloses a hole, which the overlaid Vietoris–Rips filtration fails to capture at the current radius

Advantages. Pairwise distances are easily *computable* in most settings, so, at least in principle, it is very easy to determine the scale at which a given simplex joins the Vietoris–Rips filtration. Since one only requires knowledge of pairwise distances, this filtration is extremely *flexible* in the sense that one can construct it around extremely general data types. For instance, consider a situation where the data arises from measuring correlations between various states of a complex system. In this case, it may not be natural to try to embed these states as points in some $\mathbf{R}^d$. However, a knowledge of the pairwise correlations alone is enough to construct the Vietoris–Rips complex!

Disadvantages. As we have already discussed, the Vietoris–Rips filtration is liable to become *gigantic* in terms of the number of simplices because its size scales exponentially with the number of points. Moreover, there is no control over the *dimensions* of simplices that are built, even for small values of the scale ϵ: if there are 20 points with pairwise distances all less than ϵ, then the 19-dimensional simplex containing those points will belong to $\mathcal{V}_\epsilon K_P$ even if those points are sitting in two-dimensional space! A subtler issue with these filtrations is that they are merely *approximations* to the structure of the underlying space which do not recover its structure accurately at each scale ϵ. It is easy to construct – at least with the distance $\mathbf{d}_2$ – three balls so that any pair intersects, but there is no common point in the intersection of all three, thus forming a hole (see Fig. 6). However, the geometric realization of the resulting Vietoris–Rips filtration at the given scale fails to capture that hole, because it contains the two-dimensional simplex spanning the ball centers.

A description of efficient algorithms for constructing Vietoris–Rips filtrations may be found in [41]. Most persistent-homology software packages (e.g., [29]) contain implementations of these algorithms.

3.2 *The Čech Filtration*

Letting $P \subset \mathbf{R}^d$ be our point cloud and K_P the complete simplicial complex with vertex set P, we define a simplicial subcomplex $\mathscr{C}_\epsilon K_P$ of K_P at each scale $\epsilon > 0$ in the following way. A subcollection $\sigma \subset P$ of points forms a simplex of $\mathscr{C}_\epsilon K_P$ if there exists some point x in $\mathbf{R}^d$ whose distance[3] from each vertex of σ is at most ϵ. More precisely, let $\sigma = (p_1, \ldots, p_m)$. Then,

$$\sigma \text{ is a simplex in } \mathscr{C}_\epsilon K_P \text{ if } \mathbf{d}(x, p_i) < \epsilon \text{ for all } i \text{ and some fixed } x,$$

or, equivalently,

$$\sigma \text{ is a simplex in } \mathscr{C}_\epsilon K_P \text{ if the intersection } \bigcap_{j=1}^{m} B_\epsilon(p_i) \neq \emptyset.$$

It is apparent that the construction of Čech filtrations depends crucially on the following computation: *given a collection σ of points in $\mathbf{R}^d$, what is the smallest radius r so that there exists some point x in $\mathbf{R}^d$ whose distance from each point in σ is less than r?* Discrete and computational geometers often refer to this as the *smallest enclosing ball* problem: after all, the ball of radius r around x encloses all the points in σ and, by definition, it must be the smallest ball to do so. Although there are various algorithms available to compute this minimal ball (some sacrifice exactness for speed), in general (for large point sets sitting in high dimensions) this is a complicated, nontrivial problem. Certainly, one requires a lot more computational muscle than the simple pairwise distance calculations that must be performed for constructing a Vietoris–Rips filtration.

Consider the function $g_{\mathscr{C}} : K_P \to \mathbf{R}$ defined on the simplex $\sigma = (p_1, \ldots, p_m)$ by

$$g_{\mathscr{C}}(\sigma) = \min \left\{ r \geq 0 \mid \text{there is an } x \text{ in } \mathbf{R}^d \text{ with } \mathbf{d}(x, p_j) < r \text{ for } 1 \leq j \leq m \right\}.$$

It is clear that $g_{\mathscr{C}}$ is monotone; if $\tau \prec \sigma$ and some ball $B_r(x)$ contains all the vertices of σ, then it also contains the subset of vertices which belong to τ, and hence the smallest enclosing ball for τ can have radius no larger than r.

Definition 4. The *Čech filtration* around $P \subset \mathbf{R}^d$ is the sublevelset filtration of $g_{\mathscr{C}}$.

Since there are only finitely many simplices in K_P, this function $g_{\mathscr{C}}$ assumes only finitely many values. Listing them in increasing order as $0 = \epsilon_0 \leq \epsilon_1 \leq \cdots \leq \epsilon_N$, one obtains

---

[3]This x is not necessarily a point in the cloud P, so typically the Čech filtration cannot be built from knowledge of pairwise distances alone!

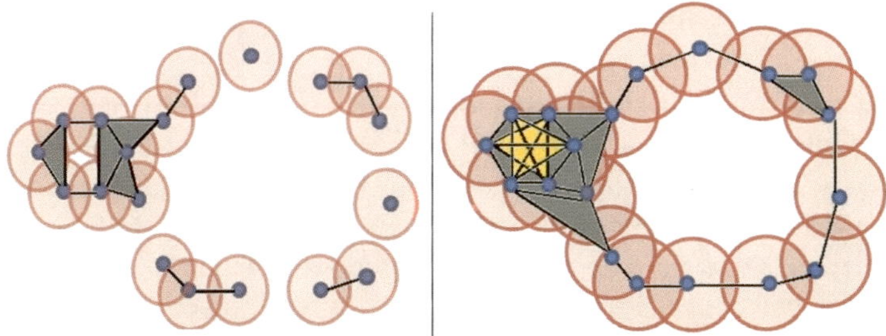

Fig. 7 Two stages of the Čech filtration of the point cloud of Fig. 4. Note that there are fewer simplices of dimension 2 and above when compared with Fig. 5 at the larger scale. For instance, the two differently colored simplices are *not* present in the Čech filtration, although they are present in the Vietoris–Rips filtration at the same scale

$$\mathscr{C}_{\epsilon_1} K_P \hookrightarrow \mathscr{C}_{\epsilon_2} K_P \hookrightarrow \cdots \hookrightarrow \mathscr{C}_{\epsilon_N} K_P = K_P.$$

The fact that higher-order intersections are taken into account when one is building the Čech filtration incurs a computational burden, but there is a substantial payoff. In particular, it follows from a result known as the *nerve theorem* that the geometric realization $|\mathscr{C}_\epsilon K_P|$ is topologically equivalent[4] to the union of all balls $B_\epsilon(p)$, where p ranges over the points in P.

Advantages. At each scale ϵ, the Čech filtration is *faithful* to the topology of the union of balls. In particular, keeping track of higher-order intersections allows one to bypass the issue highlighted in Fig. 6: the two-dimensional simplex concerned will not enter the Čech filtration until the scale where all three balls intersect, at which point there is no loop. For the same reason, the Čech filtration at a given scale is typically much *smaller*, in the sense that it contains fewer simplices than the Vietoris–Rips filtration at the same scale: see Fig. 7.

Disadvantages. As we have already noted, the *complexity* of the enclosing ball problem makes it difficult to construct Čech filtrations around point clouds in dimensions exceeding 3. As with the Vietoris–Rips filtration, a cluster of nearby points produces a simplex of high *dimension* regardless of the ambient dimension d.

Algorithms to construct the Čech filtration are described in [11], and an implementation is available as part of [24].

---

[4]This equivalence is up to a fundamental topological invariant known as *homotopy*.

3.3 Other Filtrations

While the two filtrations mentioned above are the most common ones encountered in practice, there are several other approaches to imposing simplicial structures on point clouds. In particular, the *witness complex* filtration [12] attempts to reduce the number of simplices by preprocessing the point cloud P itself as follows. We fix an acceptable "fuzz" parameter $\delta > 0$, and restrict our attention to a subset $P' \subset P$ of *landmark* points so that no two are within δ of each other. This preprocessing allows us to reduce the dimension (and hence the number) of simplices which appear at each scale $\epsilon > \delta$ in either the Vietoris–Rips or the Čech filtration. This computational advantage is not without a price, however: to the best of our knowledge, there are no explicit results about how faithfully a witness complex represents the topology of the underlying union of balls.

A drastically different approach, which is especially useful in low dimensions, involves the use of filtered *alpha complexes* [15]. Although these complexes require even more computational-geometry muscle to construct than the Čech filtration, the benefits are immense. The nerve theorem applies in the context of alpha complexes, so they are also topologically faithful to the underlying union of balls, like Čech filtrations. At the same time, the dimension of the simplices encountered in an alpha complex never exceeds d, the ambient Euclidean dimension!

4 Homology and Its Computation

Throughout the preceding sections, we have discussed various topological *features* – such as loops and cavities – which appear in geometric realizations of simplicial complexes or in the context of point clouds thickened into balls by some scale ϵ. In order to precisely understand the objects which encode and catalog such features, we must turn to algebra. Any reader who experiences moral qualms about our descent from the Olympus of geometric shapes to the Hades of algebraic formalism stands in distinguished company:

> Algebra is the offer made by the devil to the mathematician. The devil says: "I will give you this powerful machine, it will answer any question you like. All you need to do is give me your soul: give up geometry and you will have this marvellous machine."
>
> Sir Michael Atiyah

4.1 The Linear Algebra of Holes

The "marvellous machine" called *homology* detects "holes" of all dimensions by using linear algebra. It associates to each simplicial complex K a collection of algebraic objects $\mathsf{H}_d(K)$ called *homology groups*, where d ranges over the dimensions of the simplices encountered in K. Given a simplicial map $\phi : K \to L$, homology produces *group homomorphisms* $\phi_d^* : \mathsf{H}_d(K) \to \mathsf{H}_d(L)$. The type of groups and homomorphisms that one obtains depends on the choice of some underlying *coefficient system*. Here, we will use the real numbers $\mathbf{R}$. In this setting, each homology group is just some Euclidean space and each homomorphism a matrix with entries in $\mathbf{R}$.

Let K be a simplicial complex with ordered vertices. What this means for our purposes is that the vertices of any simplex σ can be uniquely written in some ascending order $(v_0, \ldots, v_d)$. The d-dimensional *chain group* $\mathsf{C}_d(K)$ of K consists of $\mathbf{R}$-linear combinations of d-dimensional simplices. Thus, a typical element of $\mathsf{C}_d(K)$ – called a d-dimensional *chain* – is $a_1\sigma_1 + \cdots + a_m\sigma_m$, where the a's are real numbers and the σ's are d-dimensional simplices. Clearly, this chain group is equivalent to $\#K_d$-dimensional Euclidean space: just use the d-dimensional simplices as a basis. Let $\sigma = (v_0, \ldots, v_d)$ be such a basis element, and for each j in $\{0, \ldots, d\}$ let σ_j be that $(d-1)$-dimensional proper face of σ which contains all the vertices except v_j. Now, the *boundary* of σ is a $(d-1)$-dimensional chain given by the alternating sum of these faces:

$$\partial_d(\sigma) = \sigma_0 - \sigma_1 + \cdots + (-1)^d \sigma_d .$$

Thus, ∂_d defines a linear transformation $\mathsf{C}_d(K) \to \mathsf{C}_{d-1}(K)$, and hence may be thought of as a matrix once we order the simplices into a basis. We define the d-dimensional *cycle group* $\mathsf{Z}_d(K)$ to be the subspace corresponding to the kernel of this matrix in $\mathsf{C}_d(K)$, and the $(d-1)$-dimensional *boundary group* $\mathsf{B}_{d-1}(K)$ is the image of this matrix as a subspace of $\mathsf{C}_{d-1}(K)$. The elements of $\mathsf{Z}_d(K)$ and $\mathsf{B}_d(K)$ are called the d-dimensional *cycles* and *boundaries*, respectively. It can be checked that each d-dimensional boundary is also a cycle.[5] Now, the d-dimensional *homology group* is defined as the quotient

$$\mathsf{H}_d(K) = \frac{\mathsf{Z}_d(K)}{\mathsf{B}_d(K)}.$$

Thus, we are interested in cycles, but do not distinguish between two cycles if they are related by a boundary. That is, we *partition* the cycles x from $\mathsf{Z}_d(K)$ into *homology classes* $[x]$, with $[x] = [y]$ whenever $x - y$ lies in $\mathsf{B}_d(K)$.

To see why we care about this quotient, let us go back to the complex of Example 1. Observe that the loop formed by ab, ac, bc corresponds to the algebraic cycle $x = ab + bc - ac$, whose boundary is 0. So far, so good. But, algebraically, even ab, ae, be forms a "loop": let $y = ab + be - ae$, and check that $\partial_1(y) = 0$. The difference between these cycles – transparent to the eye but opaque to the algebra at this point – is the presence of abe which fills up the latter cycle. In order to make

---

[5]To see why this is the case, note that the composition $\partial_d \circ \partial_{d+1}$ is the zero map from $\mathsf{C}_{d+1}(K)$ to $\mathsf{C}_{d-1}(K)$ for each dimension d.

the chain algebra recognize this fill-up, we note that $\partial_2(abc) = ab + bc - ac$. In the quotient space, this cycle y therefore ends up in the trivial homology class $[0]$. This is why algebraic cycles alone are not enough; we need to quotient by the boundaries of higher simplices.

4.2 Smith Normal Form and Betti Numbers

Staying with the simplicial complex K of Example 1, let us see what it takes to compute $H_0(K)$. First, we order the zero- and one-dimensional simplices of K in some consistent way. For convenience, we may choose the alphabetical order, and hence obtain the following sequences of cells:

$$K_0 = (a, b, c, d, e, f) \text{ and } K_1 = (ab, ac, ae, bc, bd, be, bg, cd, cg, dg, ef).$$

Next, we express the boundary operator $\partial_1 : C_1(K) \to C_0(K)$ as a matrix M_1 in our chosen basis. For instance, in the column for ac and the row for a, one finds the dot product $\langle \partial_1(ac), a \rangle = -1$, which simply extracts the coefficient of a in the boundary of ac. Proceeding in this fashion yields the following matrix:

$$M_1 = \begin{array}{c}
 \\
a \\
b \\
c \\
d \\
e \\
f \\
g
\end{array}
\begin{array}{c}
\begin{array}{ccccccccccc}
ab & ac & ae & bc & bd & be & bg & cd & cg & dg & ef
\end{array} \\
\left[\begin{array}{ccccccccccc}
-1 & -1 & -1 & 0 & 0 & 0 & 0 & 0 & 0 & 0 & 0 \\
1 & 0 & 0 & -1 & -1 & -1 & -1 & 0 & 0 & 0 & 0 \\
0 & 1 & 0 & 1 & 0 & 0 & 0 & -1 & -1 & 0 & 0 \\
0 & 0 & 0 & 0 & 1 & 0 & 0 & 1 & 0 & -1 & 0 \\
0 & 0 & 1 & 0 & 0 & 1 & 0 & 0 & 0 & 0 & -1 \\
0 & 0 & 0 & 0 & 0 & 0 & 0 & 0 & 0 & 0 & 1 \\
0 & 0 & 0 & 0 & 0 & 0 & 1 & 0 & 1 & 1 & 0
\end{array}\right]
\end{array}.$$

Using standard row and column operations (with coefficients in $\mathbf{R}$), we can put M_1 in *Smith normal form*, so that the off-diagonal entries are all zero, and the diagonal contains only zeros and ones. The number of zero entries in the diagonal of the Smith normal form is then equal to the *rank* of $H_0(K)$ as a vector space over the real numbers. One can repeat this process for all dimensions $d \geq 1$: order the cells, generate a matrix representation M_d of ∂_d in the chosen basis, and compute its Smith normal form. The number of zero entries in the diagonal of M_d's Smith normal form is called the $(d - 1)$-th *Betti number* of K, and it equals the rank of $H_{d-1}(K)$ as a Euclidean space. Keeping track of the change-of-basis matrices of the row and column operations also produces an explicit basis for $H_{d-1}(K)$ in terms of the chains in $C_{d-1}(K)$.

One may ask: *what does it all mean?* The answer is easy in low dimensions: the zero-, one-, and two-dimensional Betti numbers count the *connected components, tunnels, and cavities*, respectively, of the underlying simplicial complex.[6] In higher dimensions, the answer is subtler because we lose the ability to visualize geometry. But in any case, the Betti numbers of a simplicial complex provide computable *topological statistics* of that complex. The structure encoded by the actual groups (not just the Betti numbers) is much more intricate, but it should be clear (at least in principle) that knowledge of those groups as quotients of chains enables one to actually find components, tunnels, cavities, and their higher-dimensional analogs as linear combinations of simplices.

The situation is very similar for simplicial maps $\phi : K \to L$. Since ϕ sends simplices of K to simplices of L, for each dimension d it induces a *chain map* $\phi_d^{\#} : \mathsf{C}_d(K) \to \mathsf{C}_d(L)$ determined by the following action on the basis elements. Given $\sigma \in K_d$, we define

$$\phi_d^{\#}(\sigma) = \begin{cases} \phi(\sigma) & \text{if } \dim \phi(\sigma) = d, \\ 0 & \text{otherwise.} \end{cases}$$

One can check that $\phi_d^{\#}$ sends $\mathsf{Z}_d(K)$ to $\mathsf{Z}_d(L)$, and likewise for boundaries. Thus, $\phi_d^{\#}$ descends to a map $\mathsf{H}_d(K) \to \mathsf{H}_d(L)$ of quotient spaces, which is our homomorphism ϕ_d^{*}. More precisely, the following assignment of homology classes is well defined in the sense that it never sends two members of the same homology class in K to different homology classes in L:

$$\phi_d^{*}([x]) = [\phi_d^{\#}(x)].$$

From a computational perspective, one constructs a matrix representation of $\phi_d^{\#}$ and computes its Smith normal form in order to explicitly construct ϕ_d^{*}.

For a classical and theoretical account of simplicial homology, one can turn to the canonical algebraic-topology texts [28, 36]. But for a much more computational approach to homology (with cubical rather than simplicial complexes!), the reader is invited to consult [22]. There are highly optimized software libraries [29, 37, 38] for computing homology groups of various types of complexes.

4.3 Persistent Homology, Diagrams, and Stability

Suppose we start with a simplicial complex, and add a single extra vertex to it, disconnected from everything else. This change effectively increments the dimension of the zero-dimensional homology group by 1. One can easily construct examples where removing a single simplex also changes the dimensions drastically.

---

[6]So, there is precisely one zero in the diagonal of the Smith normal form of M_1, since K has only one connected component.

In this sense, the homology of a complex is not very stable to small changes in that complex. The antidote to this lack of stability is provided by *persistent homology.*

Persistent homology is to filtrations what homology is to simplicial complexes. Consider a filtration $\mathscr{F}$ of a simplicial complex K as shown,

$$\emptyset = \mathscr{F}_0 K \hookrightarrow \mathscr{F}_1 K \hookrightarrow \cdots \hookrightarrow \mathscr{F}_M K,$$

and note that each inclusion corresponds to a simplicial map of simplicial complexes, so one may apply the homology machine to get a sequence of Euclidean spaces connected by matrices for each dimension d:

$$\mathsf{H}_d(\mathscr{F}_1 K) \xrightarrow{\phi_d^{1\to 2}} \mathsf{H}_d(\mathscr{F}_2 K) \xrightarrow{\phi_d^{2\to 3}} \cdots \xrightarrow{\phi_d^{(M-1)\to M}} \mathsf{H}_d(\mathscr{F}_M K).$$

This structure is called a *persistence module.* The horizontal maps of homology groups are induced by chain maps arising from simplicial inclusions $\mathscr{F}_m K_d \hookrightarrow \mathscr{F}_{m+1} K_d$. Let us write $\phi^{1\to 3}$ to denote the obvious matrix product $\phi^{2\to 3} \cdot (\phi^{1\to 2})$, which gets us from the first to the third Euclidean space in our persistence module and so forth. These horizontal matrices allow one to track homological features (components, tunnels, cavities, etc.) across the entire filtration. The *p-persistent d-dimensional homology group* of the subcomplex $\mathscr{F}_m K$ is defined as the following subspace of $\mathsf{H}_d(\mathscr{F}_{m+p})$:

$$\mathsf{H}_d^p(\mathscr{F}_m K) = \phi_d^{m\to m+p}(\mathsf{H}_d(\mathscr{F}_m K)).$$

The basic idea behind this formulation is simple. Each homology class $[x]$ living in the d-dimensional homology group of $\mathscr{F}_m K$ is included into the d-dimensional homology group of $\mathscr{F}_{m+p} K$ by a string of maps on homology groups induced by simplicial inclusions. However, $\mathscr{F}_{m+p}$ contains more simplices than $\mathscr{F}_m K$ in general, so there might be a collection of $(d+1)$-dimensional simplices which fill out this cycle by making it a boundary. If this is not the case, then x has survived the journey from $\mathscr{F}_m K$ to $\mathscr{F}_{m+p} K$ safely. Otherwise, x must have met its demise at some stage q occurring before p. In the latter case, it corresponds to the trivial element $[0]$ in the homology group of $\mathscr{F}_{m+p} K$.

In order to compute homology, we had to put matrix representations of boundary operators into Smith normal form using row and column operations with coefficients in $\mathbf{R}$. Computing persistent homology groups requires a similar calculation, except that we now perform these operations over *polynomials* in one variable with coefficients in $\mathbf{R}$. Using these techniques (see the canonical reference [42, Sect. 4.2] for an explicit algorithm), one can compute for each nontrivial homology class $[x]$ in $\mathsf{H}_d(\mathscr{F}_m K)$ an unambiguous interval $[b_x, d_x)$, where the *birth* $b_x \leq m$ and the *death* $d_x > m$ are defined as follows:

- b_x is the smallest ℓ such that there is some homology class $[y]$ in $\mathsf{H}_d(\mathscr{F}_\ell K)$ with $[\phi_d^{\ell\to m}(y)] = [x]$, and

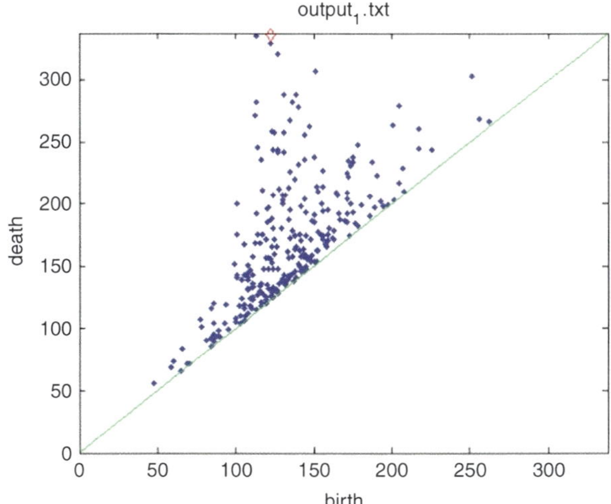

Fig. 8 A sample persistence diagram generated by the Perseus software package [29]. Births are plotted along the *horizontal axis* and deaths along the *vertical axis*. The points near the diagonal correspond to homology generators which do not persist across a large section of the filtration, and hence correspond to unstable or noisy features. On the other hand, the *dots* far from the diagonal correspond to robust features with long lifespans

- d_x is the smallest n such that $[\phi_d^{m \to n}(x)]$ is the trivial homology class $[0]$ in $\mathsf{H}_d(\mathscr{F}_n K)$.

This collection of *persistence intervals* $[b_x, d_x)$ over all such x is called the d-dimensional *persistence diagram* of the filtration $\mathscr{F}$, and it can be easily visualized as a two-dimensional cluster of points (see Fig. 8). For each x, the length $(d_x - b_x)$ measures the *lifespan* of the homology class $[x]$ across the filtration. The persistence diagram is the filtered analog of the Betti numbers in the following sense: the d-dimensional Betti number of $\mathscr{F}_m K$ is simply the number of d-dimensional persistence intervals which contain m.

Stability. There is a well-defined notion of distance between persistence diagrams, called the *bottleneck distance*. It is known [8] that the persistence diagram is stable to fluctuations in the filtration. In particular, consider a point cloud P in Euclidean space and a "noisy" version P', which is another point cloud obtained by perturbing each point of P by some distance less than a fixed $\gamma > 0$. Then, one can prove that the bottleneck distance between the dimension-d persistence diagrams of the Čech or Vietoris–Rips filtrations of P and P' is smaller than γ for every d. In this sense, the output persistence diagram is no more noisy than the input point cloud.

It is crucial to note that this stability result is a one-way street. That is, *if P and P' are near each other*, then their persistence diagrams will also be close. But it would be wrong to conclude that P and P' are close if their persistence diagrams are similar. Thus, having similar persistent homology only allows one to *conjecture* the similarity of the underlying datasets; however, having different persistent homology actually furnishes a solid *proof* that the two datasets are topologically distinct. Thus, persistent homology is better at telling things apart than at confirming their similarity.

There are various excellent resources for the persistent-homology neophyte; see [5, 6, 13, 16–18] and the references therein for many more details. The reader may also be relieved to know that using persistent homology does not require a personal desire to compute Smith normal forms of huge matrices by hand: efficient software is available for this purpose [24, 29].

5   Applications to Biological Datasets

Having established the basics of simplicial complexes and their homology, we would like to highlight some particularly appealing instances of topological inference – that is, inference based on topological techniques – from biological datasets. Selecting the right filtration to impose on a point cloud is a bit of an art form: even choosing an expedient distance function between data points requires highly specialized knowledge about the data itself, as well as a genuine understanding of the desired features which one wishes to investigate. In the absence of a general recipe that fits all possible data, the next best thing is a host of successful and interesting examples which the reader can use as signposts in his or her personal quest to build a convenient filtration.

5.1   Identification of Breast Cancer Subtypes

Breast cancer is one of the most widespread and most frequently occurring types of cancer. Since there are several variants of this cancer, considerable efforts have been made to distinguish these from each other in the search for specialized and effective treatments.

5.1.1   The Discovery of c-MYB+

A new subtype of breast cancer was detected in [31] by clustering methods acting on a filtered simplicial complex built using microarray data.

The data. A *microarray* [2] is a thin glass slide with distinguished regions – called *features* – onto which DNA molecules can attach in an orderly fashion. Using these slides, it is possible to measure efficiently as *patterns* the differences in expression between two sets of genes (from a common cell) which have been kept under different conditions. In [30], a framework called Disease-Specific Genomic

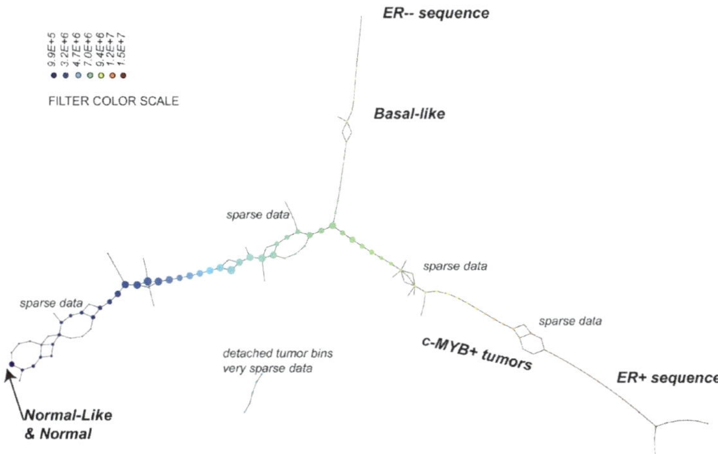

Fig. 9 Progression Analysis of Disease (PAD) results from [31] produced by Mapper [34]: the data points correspond to tumors, and their colors represent the order of magnitude of deviation from normal as measured by DSGA: *red* tumors have the largest deviation

Analysis (DSGA) was introduced, which highlights differences in expression patterns of microarrays of diseased tissue relative to a continuous range of normal phenotypes. The input data was precisely the result of DSGA performed over a sufficiently large class of normal and diseased tissues.

The complex. Nicolau et al. [30] constructed the complete simplicial complex K whose vertices T correspond to a set of tumors, and defined a function $g : T \to \mathbf{R}$ derived from the distance of each tumor from some large collection of normal phenotype tissue as yielded by regular DSGA analysis. The precise details of this distance function may be found in [31, Sect. 1.3]. Associating each simplex to the highest g-value encountered among its vertices extended g to all of K. The sublevelset filtration of g was then fed into the clustering tool Mapper [34].

The results. As shown in Fig. 9, Mapper revealed an intrinsic structure of the space of breast cancer transcriptional data that remained undetected by common clustering methods. Without any clinical or biological input except for DSGA, the construction of a suitable filtered simplicial complex followed by clustering enabled the detection of a new, unique subgroup of breast cancers called c-MYB+. These cancers are estrogen receptor-positive (ER+), and have high levels of x-MYB and low levels of innate inflammatory genes. Perhaps most importantly, there is a 100 % survival rate and no metastasis. This type of cancer does not fit into the standard classification of Luminal A/B and Normal-like subtypes of ER+ breast cancers obtained by ordinary clustering analysis.

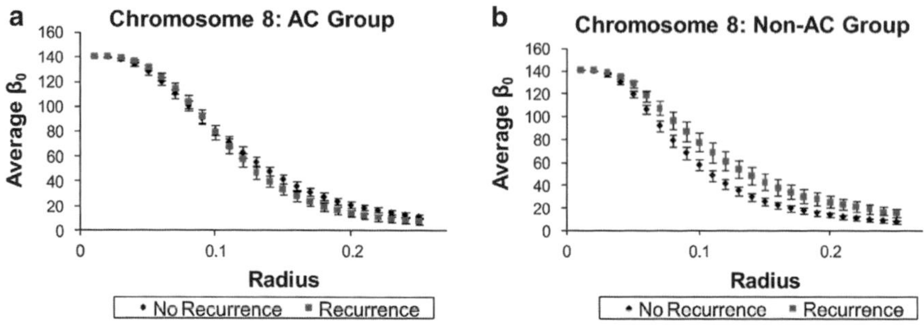

Fig. 10 Chromosome 8 plots of average Betti numbers β_0 in dimension 4 calculated for recurrent and nonrecurrent data, for radii between 0.01 and 0.25. (**a**) Patients treated with chemotherapy (AC group). (**b**) Patients not treated with chemotherapy (non-AC group). Non-AC patients have significantly higher β_0 values in the recurrent population

5.1.2 Distinguishing Between Recurrent and Nonrecurrent Subtypes

Dewoskin et al. [14] established that topological methods can partially differentiate those breast cancer subtypes which have a high recurrence rate from those which do not.

The data. *Comparative genomic hybridization* (CGH) is a method which detects chromosomal aberrations [33]. DNA from a tumor sample and from a normal reference sample are given different fluorescent labels and cohybridized onto a thin glass surface in a regular pattern. The fluorescent intensity of each region measures the differences (either *amplifications* or *deletions*) between the two samples as the logarithm of a ratio. The starting point of this analysis, therefore, is an ordered list

$$\ell = (\ell_1, \ell_2, \ldots, \ell_N)$$

of these logarithms of ratios of intensities.

The complex. One chooses an *embedding dimension d*, and creates points in $\mathbf{R}^d$ by sliding a window of width d along the list ℓ as follows. The first point is $(\ell_1, \ldots, \ell_d)$, the second one is $(\ell_2, \ldots, \ell_{d+1})$ and so forth. This creates a point cloud $P_d(\ell) \subset \mathbf{R}^d$. Although this point cloud does not retain precise knowledge of *where* the tumor DNA differs from the normal DNA, the pairwise distances are preserved and similar regions are mapped near the origin in $\mathbf{R}^d$. If the intensities are similar, then their ratio is close to 1, and hence the logarithm of the ratio is near 0. The Vietoris–Rips filtration was constructed around the point cloud $P_d(\ell)$ for various choices of dimension d.

The results. For $d = 4$, the *average zero-dimensional Betti numbers* (Fig. 10) over all of the Vietoris–Rips subcomplexes for chromosomes 8 and 11 clearly

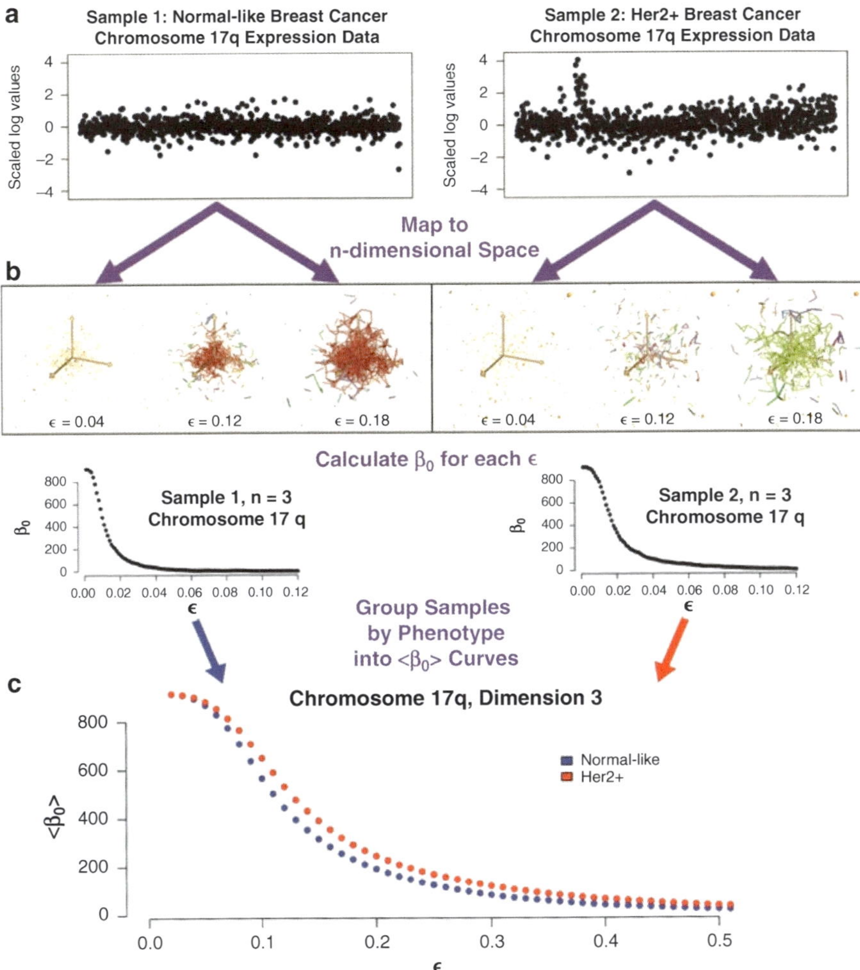

Fig. 11 Outline of the method used in [1]. (**a**) Gene expression for chromosome 17q for patients with two different types of breast cancer. (**b**) Point clouds and plots of β_0. (**c**) Plots associated with different sets of patients for a window size equal to 3. The final steps include statistical analysis for combining and correcting values

distinguished between recurrent and nonrecurrent patients who did not receive anthracycline-based chemotherapy after surgery. This method reproduced results presented in [7]. See Fig. 11 for a pictorial summary.

5.1.3 Drawing Finer Distinctions with Persistent Homology

Arsuaga et al. [1] used persistent homology instead of Betti number averages, and, starting with the same data and complex, extended the results of [14]. For an embedding dimension $d = 3$, analyzing zero-dimensional persistence diagrams had partial success in differentiating various subtypes of breast cancer. In particular, it was possible to differentiate between cancers with varying disease progression: the less aggressive types included Normal-like and Luminal A, whereas the more aggressive types were Luminal B, Basal, and Her2. The zero-dimensional persistence diagrams could differentiate intrinsic subtypes such as Basal-like and Her2 further within the class of aggressive cancers. The persistence intervals suggest that Luminal B has features in common with both the Her2 and the Basal-like subtypes.

In the future work, Arsuaga et al. hope to relate these results to cancer recurrence predictions and use the full strength of persistent homology. The fundamental question is that of which properties of breast cancers – if any – are captured by higher-dimensional homology groups. The ultimate goal is to gain insight into the periodicity of disease progression and hence select the most effective treatments.

5.2 Analysis of Neural Structures

A fundamental question arising from investigations of the brain's perception mechanisms is how a physical environment is mapped into the visual cortex, and how the resulting mental maps are used by the hippocampus for spatial navigation.

5.2.1 Activity Patterns in the Visual Cortex

The central thesis of [23] is that spontaneous cortical states resemble the patterns in oriented stimuli, i.e., that they have the same topology. The work of Singh et al. [35], described below, provides supporting evidence for this claim.

The data. The basic data consisted of multielectrode recordings from the primary visual cortex of a macaque[7] in two different settings: spontaneous activity when both eyes were closed, and natural image stimulation when one eye was open and exposed to a video sequence.

The complex. The recorded data was split into 10-s segments, and the five neurons with the highest firing rates were selected. The spike trains were binned into 50-ms intervals, so that each segment corresponded to 200 points. Finally, a witness

---

[7]*Simia inuus*, an Old World monkey.

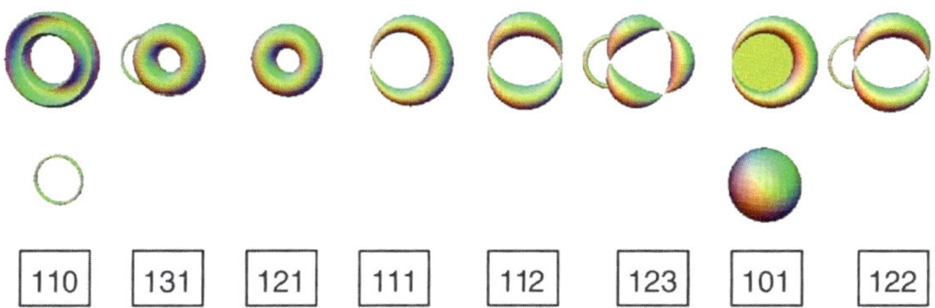

| 110 | 131 | 121 | 111 | 112 | 123 | 101 | 122 |

Fig. 12 Different topological signatures obtained in the experiment [35]. The *top row* contains examples of complexes, with the prescribed Betti number sequence signature $(\beta_0, \beta_1, \beta_2)$ shown below the corresponding complex

complex approximation to the Vietoris–Rips filtration was built around 35 landmark points.

The results. Singh et al. [35] computed the persistence intervals in dimensions 0, 1, and 2, at various scales in the Vietoris–Rips filtration and referred to such strings of Betti numbers as *signatures*. Several different Betti number sequences observed during the experiment are shown in Fig. 12, along with a sample complex whose homology exhibits those Betti numbers. Figure 13 shows histograms of Betti number distributions obtained for spontaneous and natural image stimulation, where the Betti numbers follow the same order as in Fig. 12. The features present for higher threshold scales correspond to more persistent features of the Vietoris–Rips filtration built around the data.

The experiment showed that the homology of a circle and sphere dominated the data, although the circle was much more prevalent during natural image stimulation than during spontaneous activity. The main difference between the two experimental settings appeared at lower thresholds, where the spontaneous activity exhibited much more diverse topological structures.

5.2.2 Activity Patterns in the Hippocampus

The hippocampus is a part of the brain which contains *place cells* – neurons that can detect location – clustered into regions called *place fields*. The hippocampus plays a central role in an animal's ability to navigate in its environment. However, the process by which visual data is converted into a spatial map in the brain remains mysterious. Dabaghian et al. [10] worked under the hypothesis that the topology of the map obtained from the place cells in the brain matches the topological features of the environment. That is, they conjectured that the brain does not have access to the geometric information and that it converts neural signals into a spatial map (similar to a subway map) of the surroundings based only on the spiking activity of the place cells [9, 19] and on the connectivity and adjacency information. They also

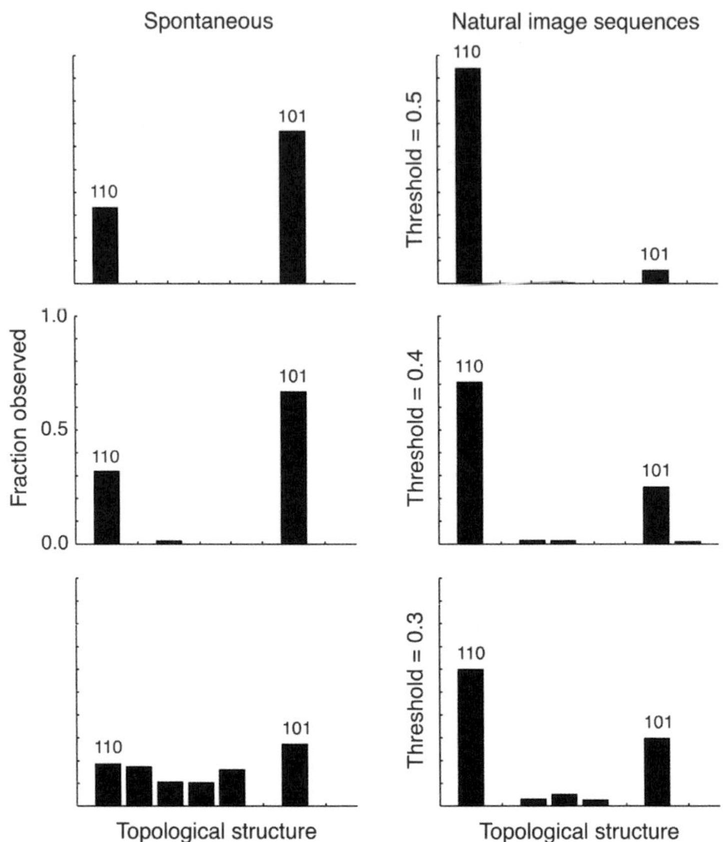

Fig. 13 Histograms of Betti number distributions obtained in the spontaneous and natural image stimulation phases of neural activity. The thresholds correspond to different Vietoris-Rips scale values. Higher thresholds correspond to more persistent features of the data. For lower threshold values, the spontaneous activity exhibits diverse topological structures, while natural image stimulation is still dominated by the homology of a circle and a sphere

assumed that the hippocampus constructs the connectivity map based on the place cell cofiring patterns.

For example, consider a rat running through a maze. As it begins to explore the environment, place fields in its hippocampus become active: in the beginning, they are be disconnected, but over time, as various navigation routes are explored, the connectivity of the active regions increases and eventually holes begin to appear.

The data. To model the activity of the place cells in a computer simulation, the authors of [10] considered the firing rate f; the size of the place field s (the part of the hippocampus that is activated when a place cell fires), of ellipsoidal shape; and the number of cells N.

The complex. A simplicial complex K was constructed as follows. Each place field was a vertex, and a d-dimensional simplex σ of K consisted of $(d-1)$ place fields which fired simultaneously during the experiment. Let $\|\sigma\|$ denote the total number of place cells involved in the simultaneous firing. The monotone function $g : K \to \mathbf{R}$ defined by

$$g(\sigma) = 1 - c\sqrt{\frac{\|\sigma\|}{N}}$$

for any $c > 0$ provides a measure of the *dissimilarity* between the place fields which form the vertices of σ. A positive c was chosen, and sublevelsets of g were used to generate a filtration of K. This is precisely the simplicial model presented in [9].

The results. The most amazing result obtained by computing persistent homology was that, if one ignored features with very small lifespans, then the homology of K was the same as the homology of the environment. The results for different experimental conditions are summarized and explained in Fig. 14, from [10]. The top row (i) shows three different experimental configurations of the environment, but we note that (B) and (C) are topologically the same. The second row (ii) contains the mean map formation times; each dot represents a place cell with a certain (f, s, N), and the size of the dot represents the percentage of trials in which this state produced the correct outcome. The color range denotes the time needed to form the map, blue denoting a short time and red almost the whole time period. Note how the third scenario (C) contains a preponderance of blue dots, which means that it was much easier for a rat to map this configuration rather than (B), even though they are topologically indistinguishable.

5.2.3 Terminal Ganglia of Crickets

The cricket *Acheta domesticus* uses hairs on its rear appendage (called a *cercus*) to detect changes in its environment. The hairs are connected via nerve endings called *afferent terminals* to the *terminal ganglion*, one of the three dense neural centers present in the cricket's body. These hairs are broadly classified as *proximal* and *distal*, depending on their distance from the ganglion. The proximal hairs are further divided into *long, medium*, and *short* categories, whereas the distal hairs are always long. Each hair has an *orientation*, a preferred direction to which it is most sensitive.

The afferent terminals of hairs with different orientations are in different places in the terminal ganglion. Hence, the cricket's response to an external stimulus depends on the region in the terminal ganglion which is excited by the stimulus, and this region depends on the direction of the stimulus. A natural question is to determine whether there is a similar dependence for spatial stimuli: i.e., whether different spatial stimuli correspond to a spatial segregation of the terminal ganglion. This would imply that the projections of the long, medium, and short hairs in the terminal

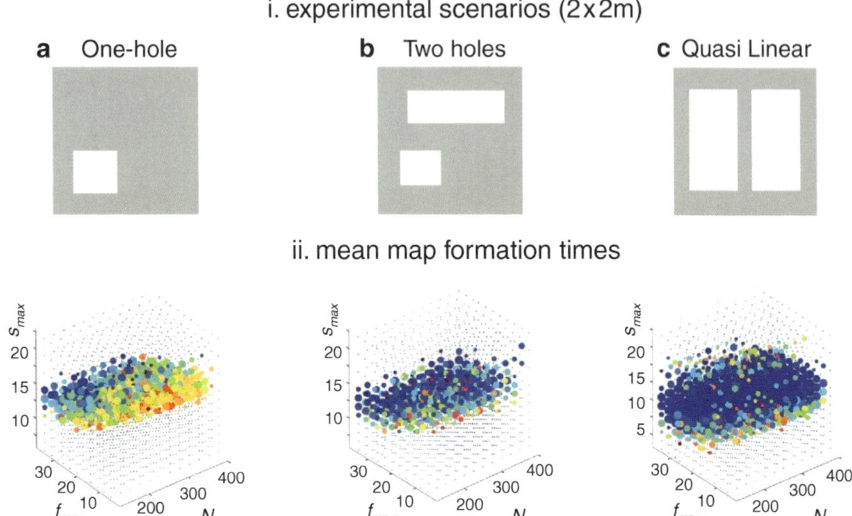

Fig. 14 (i) Three different experimental configurations of the environment: (*B*) and (*C*) are topologically identical. (ii) Point cloud approximations that reveal mean map formation times for each space configuration. Each *dot* represents a hippocampal state as defined by the three parameters (f, s, N); the size of the *dot* reflects the proportion of trials in which a given set of parameters produced the correct outcome. The color of the *dot* reflects the mean time taken over ten simulations: *blue* denotes a short time, whereas *red* stands for almost the entire period. The maximum observed time was 4.3 min for configuration (*A*), 11.7 min for (*B*), and 9.3 min for *C*

ganglion are concentrated in different regions of the terminal ganglion. Since the structure of afferent terminals and their attachment to the terminal ganglion is rather complicated, this question remained open until 2012. Recently, however, a positive answer was provided by Brown and Gedeon [3] using topological tools.

The data. The data came from experiments on afferent terminals [20, 21, 32], with the data points representing the three-dimensional locations of terminal endings in ganglia across a large number of crickets. This data was preprocessed via various standard methods, including Gaussian mixture models and nearest-neighbor techniques. The data for the afferent terminals of long, short, and medium hairs was isolated into three separate point clouds.

The complex. The authors of [3] constructed a Vietoris–Rips filtration around each point cloud, but used the distance $\mathbf{d}_1$. The scale-thickened version of such a complex consists of cubes rather than balls. The reason for doing this involved the large size of the dataset: cubes require less memory to store on a computer than do simplices, and there is a parallel theory of cubical homology.

The results. The authors of [3] compared the persistent homology of the initial point clouds for the long, medium, and short hairs separately, then for all pairwise unions, and finally for the complete dataset. The results are shown in Fig. 15 as

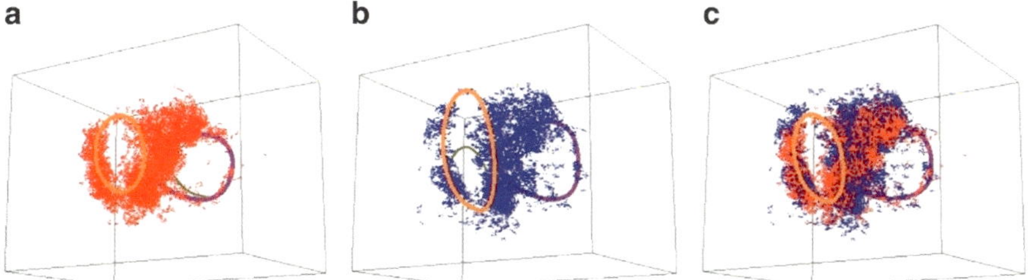

Fig. 15 Experimental data for (**a**) the short hairs, (**b**) the medium hairs, and (**c**) the medium and short hairs combined, together with the generators of the first homology. (**a**) and (**b**) have three persistent generators (*orange, purple*, and *gray*), but the last generator is filled up in the combined set

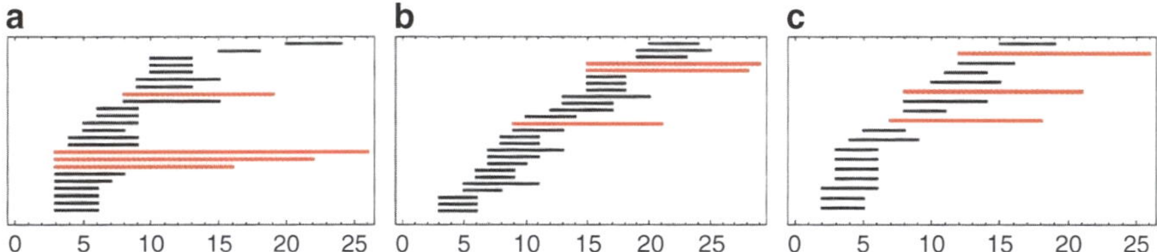

Fig. 16 Barcodes of dimension 1: β_1 persistence intervals of length more than two for the reduced datasets for the proximal hairs. (**a**) Long hairs; (**b**) medium hairs; (**c**) short hairs. Persistent generators are shown in *red*

barcodes, where each bar is drawn from the birth scale to the death scale and its length represents the lifespan of the corresponding homological feature. The persistence of the cubical filtration complex was computed using the software package cubPersistenceMD [25–27]. The persistent homologies of the individual point clouds and their unions were found to be significantly different.

As an example of the methodology, we compare the union of the short and medium proximal hairs. There are three persistent generators in each point cloud when the clouds are viewed individually. However, when one considers the combined point cloud consisting of data from both short and medium hairs, then only two of these three persistent generators remain (see Fig. 15). Computations reveal that one of the persistent generators for the medium set is filled by the terminals from the short hairs (Fig. 16).

In light of these observations, one can conclude that the nerve endings connected to the hairs are actually concentrated at different places in the terminal ganglion. Thus, there is the potential for downstream neurons to use information from the hairs. The precise nature of how these neurons synapse with the nerve endings in the terminal ganglion is unknown, and is currently under investigation.

Acknowledgements The authors thank Robert Ghrist for useful discussions, feedback and support. VN's work is supported by federal contracts FA9550-12-1-0416 and FA9550-09-1-0643. RS is grateful for the generous support by NSF 0935165 FA9550-12-1-0416 grants.

References

1. J. Arsuaga, N. Baas, D. DeWoskin, H. Mizuno, A. Pankov, C. Park, Topological analysis of gene expression arrays identifies high risk molecular subtypes in breast cancer. Applicable Algebra in Engineering, Communication and Computing. Special issue on Computer Algebra in Algebraic Topology and Its Applications. **23**, 3–15 (2012)
2. M. M. Babu, Introduction to microarray data analysis, in *Computational Genomics*, ed. by R. Grant (Taylor & Francis, 2004)
3. J. Brown, T. Gedeon, Structure of the afferent terminals in terminal ganglion of a cricket and persistent homology. PLoS ONE **7**(5), e37278 (2012)
4. G. Carlsson, Topology and data. Bull. Am. Math. Soc. (N.S.) **46**(2), 255–308 (2009)
5. G. Carlsson, V. de Silva, Zigzag persistence. Found. Comput. Math. **10**(4), 367–405 (2010)
6. G. Carlsson, V. de Silva, D. Morozov, Zigzag persistent homology and real-valued functions, in *Proceedings of the 25th Annual Symposium on Computational Geometry*, Aarhus (ACM, 2009), pp. 247–256
7. J. Climent, P. Dimitrow, J. Fridlyand, J. Palacios, R. Siebert, D.G. Albertson, J.W. Gray, D. Pincel, A. Lluch, J.A. Martinez-Climent, Deletion of chromosome 11q predicts response to anthracycline-based chemotherapy in early breast cancer. Cancer Res. **67**, 818–826 (2007). PMID: 17234794
8. D. Cohen-Steiner, H. Edelsbrunner, J. Harer, Stability of persistence diagrams. Discret. Comput. Geom. **37**(1), 103–120 (2007)
9. C. Curto, V. Itskov, Cell groups reveal structure of stimulus space. PLoS Comput. Biol. **4**, e1000205 (2008)
10. Y. Dabaghian, F. Memoli, L. Frank, G. Carlsson, A topological paradigm for hippocampal spatial map formation using persistent homology. PLoS Comput. Biol. **8**(8), e1002581 (2012)
11. S. Dantchev, I. Ivrissimtzis, Efficient construction of the Čech complex. Comput. Graph. **36**(6), 708–713 (2002)
12. V. de Silva, G. Carlsson, Topological estimation using witness complexes, in *SPBG'04 Proceedings of the First Eurographics Conference on Point-Based Graphics*, Zurich, 2004, pp. 157–166
13. V. de Silva, R. Ghrist, Coverage in sensor networks via persistent homology. Algebr. Geom. Topol. **7**, 339–358 (2007)
14. D. Dewoskin, J. Climent, I. Cruz-White, M. Vazquez, C. Park, J. Arsuaga, Applications of computational homology to the analysis of treatment response in breast cancer patients. Topol. Appl. **157**(1), 157–164 (2010)
15. H. Edelsbrunner, The union of balls and its dual shape. Discret. Comput. Geom. **13**, 415–440 (1995)
16. H. Edelsbrunner, J. Harer, *Computational Topology: An Introduction* (American Mathematical Society, Providence, 2010)
17. H. Edelsbrunner, D. Letscher, A. Zomorodian, Topological persistence and simplification. Discret. Comput. Geom. **28**, 511–533 (2002)
18. R. Ghrist, Barcodes: the persistent topology of data. Bull. Am. Math. Soc. (N.S.) **45**(1), 61–75 (2008)
19. B. Igelnik, *Computational Modeling and Simulation of Intellect: Current State and Future Perspectives*, vol. 655 (Information Science Reference, Hershey, 2011). xxix
20. G. Jacobs, F. Theunissen, Functional organization of a neural map in the cricket cercal sensory system. J. Neurosci. **16**, 769–784 (1996)
21. G. Jacobs, F. Theunissen, Extraction of sensory parameters froma neural map by primary sensory interneurons. J. Neurosci. **20**, 2934–2943 (2000)
22. T. Kaczynski, K. Mischaikow, M. Mrozek, *Computational Homology* (Springer, New York, 2004)
23. T. Kenet, D. Bibitchkov, M. Tsodyks, A. Grinvald, A. Arieli, Spontaneously emerging cortical representations of visual attributes. Nature **425**, 954–956 (2003)

24. D. Morozov, Dionysus software library, http:www.mrzv.org/software/dionysus
25. M. Mrozek, Homology software website, http://www.ii.uj.edu.pl/,mrozek/software/homology.html
26. M. Mrozek, B. Batko, Coreduction homology algorithm. Discret. Comput. Geom. **41**, 96–118 (2009)
27. M. Mrozek, T. Wanner, Coreduction homology algorithm for inclusions and persistent homology. Comput. Math. Appl. **60**(10), 2812–2833 (2010)
28. J. R. Munkres, *Elements of Algebraic Topology* (Addison-Wesley, 1984)
29. V. Nanda, Perseus: the persistent homology software, http://www.math.rutgers.edu/~vidit
30. M. Nicolau, R. Tibshirani, A. Børresen-Dale, S.S. Jeffrey, Disease-specific genomic analysis: identifying the signature of pathologic biology. Bioinformatics **23**(8), 957–965 (2007)
31. M. Nicolau, A.J. Levine, G. Carlsson, Topology based data analysis identifies a subgroup of breast cancers with a unique mutational profile and excellent survival. PNAS **108**(17), 7265–7270 (2011)
32. S. Paydar, C. Doan, G. Jacobs, Neural mapping of direction and frequency in the cricket cercal sensory system. J. Neurosci. **19**, 1771–1781 (1999)
33. D. Pinkel, D. G. Albertson, Array comparative genomic hybridization and its applications in cancer. Nat. Genet. **37**, S11–S17 (2005)
34. G. Singh, F. Mémoli, G. Carlsson, Topological methods for the analysis of high dimensional data sets and 3D object recognition, in *Eurographics, Symposium on Point-Based Graphics*, Prague, 2007
35. G. Singh, F. Mémoli, T. Ishkhanov, G. Sapiro, G. Carlsson, D. Ringach, Topological analysis of population activity in visual cortex. J. Vis. **8**(8), article 11 (2008)
36. E. H. Spanier, *Algebraic Topology* (McGraw-Hill, New York, 1966)
37. The CAPD group, *CAPD::RedHom*, http://redhom.ii.uj.edu.pl
38. The Computational HOMology Project, *CHOMP*, http://chomp.rutgers.edu
39. The Protein Data Bank, http://www.rcsb.org
40. J. C. Venter, M.D. Adams et al., The sequence of the human genome. Science **291**(5507), 1304–1351 (2001)
41. A. Zomorodian, Fast construction of the Vietoris-Rips complex. Comput. Graph. **34**, 263–271 (2010)
42. A. Zomorodian, G. Carlsson, Computing persistent homology. Discret. Comput. Geom. **33**, 249–274 (2005)

Part II
Molecular Arrangements and Structures

Combinatorial Insights into RNA Secondary Structure

Christine Heitsch and Svetlana Poznanović

Abstract The interaction of discrete mathematics with molecular biology advances our understanding of important sequence/structure/function relationships. By their nature, biological sequences are often abstracted to combinatorial objects namely strings over finite alphabets and their representation as graphs or formal languages. As described in this chapter, results based on these mathematical abstractions have been used to count, compare, classify, and otherwise analyze RNA secondary structures. In this way, they provide important insights into the base pairing of RNA sequences, thereby advancing our understanding of RNA folding.

1 Introduction

RNA has long been known to mediate the production of proteins from DNA. Moreover, RNA molecules are now understood to perform many vital regulatory and catalytic functions [21, 28, 33], other than the well-known roles of messenger, transfer, and ribosomal RNA. Like DNA, RNA molecules are nucleotide sequences. Unlike the canonical double-stranded DNA helix, most RNA molecules are single-stranded and fold into different structures via intra-sequence base pairings. Like proteins, the 3D structure of an RNA molecule has functional significance. Unlike many proteins, the size of RNA molecules, particularly for longer sequences such as viral genomes, poses challenges for the experimental determination of

C. Heitsch (✉)
School of Mathematics, Georgia Institute of Technology, Atlanta, GA 30332-0160, USA
e-mail: heitsch@math.gatech.edu

S. Poznanović
Department of Mathematical Sciences, Clemson University, Clemson, SC 29634-0975, USA
e-mail: spoznan@clemson.edu

N. Jonoska and M. Saito (eds.), *Discrete and Topological Models in Molecular Biology*, 145
Natural Computing Series, DOI 10.1007/978-3-642-40193-0_7,
© Springer-Verlag Berlin Heidelberg 2014

three-dimensional structures.[1] Hence, understanding RNA folding remains a fundamental problem in molecular biology.

Experimental methods for RNA structure determination are an active area of research. To a first approximation, though, the structure of an RNA molecule can be understood from its 2D configuration. RNA folding is hierarchical [11, 90], with intermediate states characterized by their dimensionality. The sequence is the 1D, or primary, structure. Via the noncrossing pairing of complementary bases, RNA is said to fold into a 2D, or secondary, structure. The runs of stacked base pairs, known as helices, are interspersed by single-stranded regions, called loops. For instance, the sequence $g^6 a^4 c^6 = ggggg\,aaaa\,ccccc$ folds into a structure with one helix and one loop, known as a hairpin. A variety of tertiary interactions, such as pseudoknot formation and RNA–protein binding, govern the arrangement of secondary structures in three dimensions. Since many aspects of these 3D interactions are still not well understood, quantitative modeling and analysis of RNA folding has focused mainly on secondary-structure prediction [59, 80, 102].

Given a suitable multiple sequence alignment, a common set of base pairs can be inferred from the mutual information in covarying positions [14, 67]. This is the current gold standard for RNA secondary-structure determination, especially for ribosomal sequences. For too many others, however, a sufficiently informative multiple sequence alignment is not available, and other quantitative methods must be used to predict RNA base pairing. A basic biological premise is that molecules fold to minimize the overall free energy. Hence, the majority of current RNA prediction programs are based on discrete optimization using a nearest-neighbor thermodynamic model (NNTM).

These methods are computationally similar to DNA sequence alignment, involving a recursive formulation of the optimal solution which can be solved efficiently using dynamic programming. In fact, one of the first approaches [68] was simply to maximize the number of matchings, that is, of complementary base pairs. This is often called the Nussinov model for RNA base pairing. More sophisticated objective functions soon followed [95, 105], and have continued to evolve over the years; the Turner99 parameter set for the NNTM available in the Nearest Neighbor DataBase (NNDB) [92] has over 8,000 loop energy parameters.[2]

The optimization methods themselves have also been refined and extended. The original approach predicted a single minimum-free-energy (MFE) secondary

---

[1]On March 5, 2013, the PDB [7] contained 82,105 protein and 2,510 nucleic acid structures. Concurrently, the Rfam database [12] listed 60 families of RNA molecules with at least one member having a three-dimensional structure available in the PDB. Of those, five-sixths had an average length below 250 nucleotides. The only families with an average length exceeding 400 nucleotides were three ribosomal RNA ones (archaeal, bacterial, and eukaryotic). It is worth noting that a high-resolution structure of the *E. coli* ribosome was first published only in 2005 [82].

[2]Nearly all of these are for the special cases of small internal loops, denoted by the number of single-stranded bases on each side as 1×1, $2 \times 1/1 \times 2$, and 2×2. The same special cases are included in the Turner04 parameters. It is worth noting that the number of parameters for formal-language models can also be quite large, depending on the grammar.

structure, and this is still the default output for widely used prediction programs such as UNAfold [58]/mfold [104] and RNAfold [43]. However, it has long been recognized that this optimization problem is "ill-conditioned" [102], and hence that it is important to investigate alternative low-energy secondary structures. These suboptimal configurations can be generated as representative of different substructures [103], exhaustively computed within a given energy window [97], or sampled efficiently [23] from the Gibbs/Boltzmann distribution according to the partition function [61], which also allows the calculation of base pair probabilities.

An alternative to these physics-based methods is an approach grounded in formal language theory. In this formulation, the noncrossing pairings of an RNA secondary structure are generated by a stochastic context-free grammar (SCFG), which, when coupled with phylogenetic information from a multiple sequence alignment, can be used to predict the common base pairings for a set of RNA sequences [52, 53].

However, there remain questions about RNA folding which these computational prediction methods are not well suited to answer. Hence, researchers have turned to combinatorial methods and models to count, compare, classify, and otherwise analyze RNA secondary structures. By their nature, biological sequences are often abstracted to discrete mathematical objects, namely strings over finite alphabets and their representation in graph-theoretic or formal language terms. As we describe below, this combinatorial abstraction has been particularly effective for RNA folding since it captures important aspects of the discrete base pairings.

2 Secondary Structures as Matchings: Enumeration

The primary structure of an RNA molecule is a biochemical sequence of four nucleotides, represented as a string over the alphabet $\{a, c, g, u\}$ with an orientation from left ($5'$) to right ($3'$). We denote the pairing of the ith base with the jth as (i, j), for $i < j$. The separation along the sequence between the two base-paired nucleotides, that is, $j - i - 1$, is known as the "contact distance" [27], denoted here by $c(i, j)$. A set of base pairs containing (i, j) and (i', j') is a secondary structure provided that $i = i'$ if and only if $j = j'$ and that if $i < i'$ then either $i < j < i' < j'$ or $i < i' < j' < j$. The first condition prohibits base triples, and the second constrains base pairings to be noncrossing, that is, pseudoknot-free. Both pseudoknots and base triples are known to occur in RNA molecules; however, they are typically classified as belonging to the tertiary-structure interactions. Thus, if an RNA sequence is pictured as points on a line (or a circle), the secondary structure corresponds to a noncrossing matching.

It is natural to ask for bounds on the number of possible secondary structures for a sequence of length n. Under the assumption that all pairings (i, j) with $c(i, j) \geq 1$ are allowed, there is a recursive formula [95] for the number of secondary structures $S_1(n)$. More generally, assuming that $c(i, j) \geq m \geq 1$, there is a generating function for $S_m(n)$, and first-order asymptotics can be derived for $m = 1$ using the "folklore" theorem of Bender. This yields the oft-cited exponential growth formula for the number of secondary structures [87]:

$$S_1(n) \sim \sqrt{\frac{15 + 7\sqrt{5}}{8\pi}}\, n^{-3/2} \left(\frac{3 + \sqrt{5}}{2}\right)^{n} \text{ as } n \to \infty.$$

If base pairs $(i, i+1)$ were allowed, then $S_0(n)$ would be the Motzkin numbers [26]. However, such pairings are biophysically infeasible. In reality, m should be at least 3, if not 4, but solving the functional equation analytically when $m \geq 3$ is significantly more challenging. In these cases, techniques from analytic combinatorics guarantee that there is a precisely defined asymptotic limit, which can then be approximated numerically.

It is also natural to ask for the growth rates of various types of substructures, which can be broadly categorized as helices and loops. A base pair (i', j') is *stacked* on (i, j) if $i' = i + 1$ and $j' = j - 1$, and a *helix* is a run of stacked base pairs. If (i, j) has no stacking base pair, then it closes a *loop of degree d*, which contains $k \geq 1$ unpaired nucleotides and $d \geq 0$ base pairs. (This definition of degree is consistent with that for rooted trees [86], where only the children of a vertex are counted, and it distinguishes the number of enclosed base pairs from the closing one.) A nucleotide i' (or base pair (i', j')) is in the loop closed by (i, j) if $i < i' (< j') < j$ and there exists no other base pair (i'', j'') for which $i < i'' < i' (< j') < j'' < j$. When $d = 0$, the loop is a *hairpin* and, realistically, $k = c(i, j) \geq 3$. When $d = 1$, the loop contains a single base pair (i', j'). If $k = i' - i - 1 \geq 1$, then the loop is a *left bulge*; if $k = j - j' - 1 \geq 1$, it is a *right bulge*. Otherwise, there are unpaired bases on both sides of an *internal loop*. A multiloop (or multibranch loop or junction) has degree $d \geq 2$. Finally, any remaining nucleotides or base pairs not enclosed by a base pair (i, j) belong to an *exterior loop*.

The growth rate of substructures was first addressed numerically in [35], where various statistical properties were computed for random sequences with lengths up to $n = 100$. These included the mean numbers of base pairs and of helices/loops, which grew linearly with sequence length, as well as the average loop degree, helix length, and loop size, which converged to "almost constant" values (given as 1.82, 4.57, and 5.42, respectively)[3] with increasing sequence length. These values were calculated for random sequences of length $n = 500$, and were shown to agree well with the values for natural sequences.

Subsequent work [44] gave exact asymptotics for these and other characteristics of RNA secondary structures. The methods paralleled the original enumeration results for secondary structures by giving recursion formulas and then first-order asymptotics for the distribution of different substructures using methods from analytic combinatorics, in particular Darboux's theorem. Thus, for instance, the

---

[3]The degree of a loop here is the total number of base pairs in the loop.

ratio of the expected number of helices N_n to the expected number of secondary structures S_n of length n with hairpins of length at least m is

$$\frac{N_n}{S_n} \sim \frac{(1-\alpha)^2(1+\alpha)}{2+m-2m\alpha}n$$

where $\alpha > 0$ is real, and arises from the only singularity of the generating function on its radius of convergence when the generating function is expressed in the form of Darboux's theorem. Hence, the growth rate of the average number of helices is linear in the sequence length.

While noncrossing matchings are the most immediate combinatorial abstraction of RNA secondary structures, the most broadly used one has been trees. A *plane tree* (also known as a *linear tree*) is a rooted tree whose subtrees are linearly ordered. Hence, the linear ordering respects the $5'$–$3'$ orientation of an RNA secondary structure, and the root recognizes the ends of the sequence. Under a bijection which maps unpaired nucleotides to leaves, and base pairs to internal vertices [79], it was shown that the number of secondary structures for a sequence of length n with k base pairs is

$$\frac{1}{k}\binom{n-k}{k+1}\binom{n-k-1}{k-1}.$$

The impact of these enumerative results is to highlight the exponentially large number of possible secondary structures for a given sequence, as well as the expected growth rate for different classes of substructures, such as the number of helices.

3 Secondary Structures as Trees: Comparison, Classification, Motifs, and Analysis

The bijection given above is not the only mapping from RNA secondary structures to plane trees. These graphical objects are also very useful as a more coarse-grained model of RNA loops and helices.

Initially, plane trees were used as an abstraction of RNA base pairing to compare predicted secondary structures of closely related sequences. In the first approach [84], each loop was represented as one of four types of labeled vertices ("H", "I", "B", or "M"), and the helices were edges connecting the loops. A similar representation was used in [55], except that each vertex and edge was additionally annotated with the number of single-stranded nucleotides or base pairs, respectively. In the former approach, the trees were converted to strings using parentheses to indicate subtrees/nesting relations, which were further simplified to retain only the branching information. These were then compared using a string homology program developed for DNA sequence alignments to determine similar structures (Fig. 1).

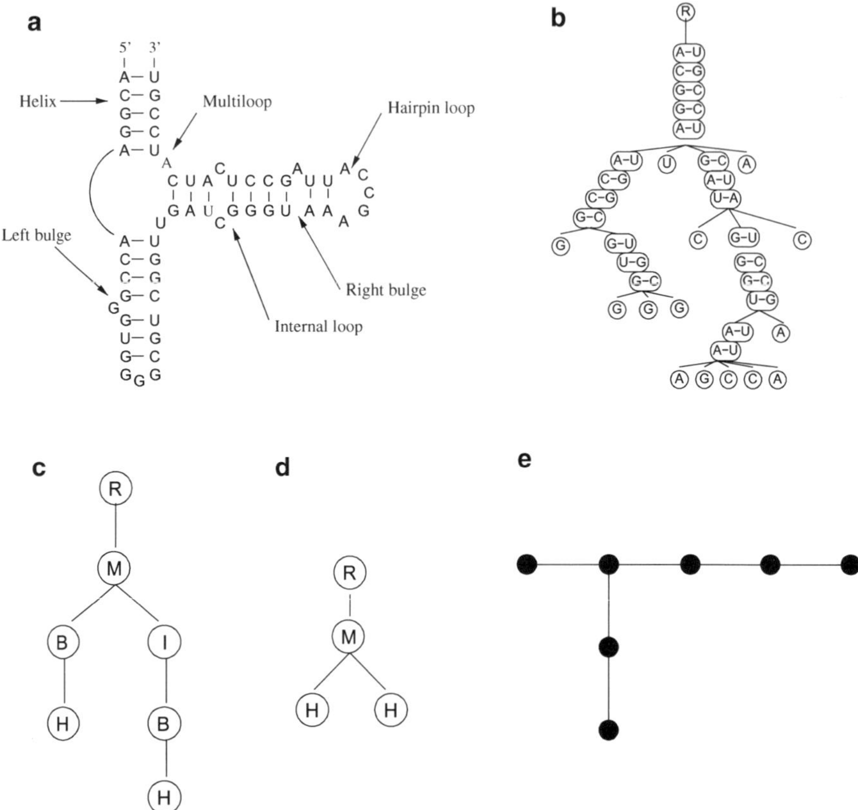

Fig. 1 An RNA secondary structure and its associated tree representations, ranging from most to least detailed

In contrast, the latter approach compared trees by introducing a value, the ratio of the number of identical nodes to the total number of nodes in the two trees, as an approximation to the tree edit distance. As with strings, the edit distance is the minimum-cost sequence of operations (insertion, deletion, and substitution of nodes) needed to transform one tree into the other [101]. Likewise, trees can also be compared by finding an optimal alignment [48]. In contrast to strings, though, the alignment distance between two trees may be greater than the editing cost, since all insertions must occur before any deletions. (See [8] for a comprehensive survey of results on tree edits.)

An algorithm for comparing RNA secondary structures as trees using edit operations was given in [85]. A feature of note is that the cost function could accommodate varying levels of resolution of the tree abstraction. The most coarse-grained would be the model introduced in [84], followed by that of [55], and down through further refinements to the nucleotide level [79]. Once all the pairwise distances had been computed, the next step was to cluster similar structures together, resulting in a taxonomy of predicted secondary structures for the 100 tRNA test sequences.

To facilitate the comparison of larger RNA secondary structures, tree edits were generalized by introducing the additional operations of node fusion and edge fusion [2]. These address some of the biological complexity not captured by the classical tree edit operations. Although string edit operations capture important aspects of DNA sequence evolution, for RNA sequences it is the structure, and not the base composition, which is most strongly conserved. These new operations take this into account by allowing helices to merge (edge fusion) or be disregarded (node fusion), with an appropriate cost function. This approach was then extended to a multiscale method for comparing RNA secondary structures [1].

The methods above were primarily focused on comparing predicted secondary structures for *different* but related sequences, such as the small subunits of two ribosomal RNAs. In contrast, there is the problem of comparing different possible secondary structures for the *same* sequence. Recall that the space of possible secondary structures is very large, growing exponentially as a function of sequence length. Moreover, numerical evidence indicates that the number of low-energy structures also grows exponentially as a function of the window size above the minimum free energy. Although exhaustive generation of suboptimal configurations is possible [97], the space of potential secondary structures is typically explored through sampling, either deterministically [103] or stochastically [23, 61]. Suboptimal structures are frequently compared using a "base pair" metric. Often, this is just the symmetric difference between the two sets of base pairs, but more sophisticated approaches are also possible, along with other options such as the mountain metric [62].

No matter what the metric, however, the challenge of making sense of a large number of different secondary structures remains. One approach [24] has been to use standard techniques to cluster secondary structures sampled according to the Gibbs/Boltzmann distribution and the unadorned base pair metric.

A more combinatorial alternative is provided by the RNAshapes program [38]. Although it can be defined more formally, the *shape* of an RNA secondary structure is simply the most abstract tree representation, where two or more helices connected by bulges or internal loops are represented by a single edge. Hence, the shape of an RNA secondary structure is its branching pattern, which corresponds to a plane tree with no vertices of degree 1, except possibly the root. Each shape has an associated *shrep*, which is the lowest-energy secondary structure with that branching configuration. As with suboptimal-structure prediction, it is possible to compute efficiently the shapes of all secondary structures within a given energy window of the MFE, and their associated shreps. For the sequences tested, the native secondary structures were always among the shreps identified.

The classification of RNA secondary structures into shapes substantially reduces the number of distinct folds under consideration, allowing a broader overview of the RNA configuration space. (If a more detailed perspective is desired, the shape of an RNA secondary structure can be represented at different levels of abstraction, as with the tree model.) Furthermore, many of the enhancements to thermodynamic optimization for RNA secondary-structure prediction – in particular, the ability to compute probabilities and to sample stochastically – extend to this more coarse-grained representation [93]. In theory, this allows a complete probabilistic analysis of the configuration space by computing the probability of a shape as the sum of all structures with that branching signature. In practice, this computation is currently feasible only for shorter sequences such as tRNA (76 nt) or the HIV-1 leader (281 nt). This is due to the fact that the growth rate of shapes is still exponential in the sequence length, albeit at a slower rate than that of the secondary structures themselves.

Under the assumption that any two bases can pair, the asymptotic enumeration of RNA shapes of length n, where n must be even, was given [57] as

$$\sqrt{\frac{6}{\pi}}\, n^{-3/2} \sqrt{3}^{\,n}.$$

Furthermore, the number of different possible shapes for a sequence of length n is also exponential in n, asymptotically $2.44251 \cdot n^{-3/2} \cdot 1.32218^n$. The proof methods used a context-free grammar, another combinatorial representation for RNA secondary structures, and techniques from analytic combinatorics. Since this type of approach has yielded a number of different insights into RNA folding, it will be addressed in Sect. 4.

Even when RNA secondary structures have different shapes, they may still share common substructures. Hence, the identification of motifs is an important problem in the study of RNA folding. This has been approached combinatorially using trees, which in this case are unrooted and unordered, or more generally using "dual graphs" to capture pseudoknotted RNA base pairings, as introduced in [36]. In this context, theoretical bounds on the number of pseudoknot-free RNA motifs are provided by known enumeration results, such as Cayley's formula for labeled trees. It was found that known RNA secondary structures correspond to only a small subset of the possible motifs, indicating that the "missing" ones would be novel RNA motifs if found. These results and subsequent work have been very fruitful in identifying candidates for novel "RNA-like" motifs, especially for their potential in a more combinatorial approach to aid in the design of such structures. To aid in the search for new RNA motifs, the RNA-As-Graphs database was developed [37], and is freely available online.

A similar graph-theoretical approach was used in [41] to distinguish "RNA-like" trees from those that are very unlikely to be RNA structural motifs. These results used the *domination number* of a graph, which is the minimum size of a dominating set. A subset of vertices $D \subset V$ is dominating if every vertex in $V \setminus D$ is adjacent to one in D. It was shown that this graph invariant captures biologically meaningful differences between different candidate RNA motifs.

A more general graph-theoretic method for identifying common motifs in a set of n unaligned RNA sequences was given in [47]. This approach is based on first identifying common helices across multiple sequences, and then constructing an n-partite weighted graph where common helices, with a similarity score exceeding some threshold, are connected by edges weighted with their similarity. Putative motifs, which can include pseudoknots, are found by identifying maximum cliques of a large enough size. Although the worst-case complexity is exponential in the number of sequences, the observed run times are much less, owing to the sparsity of the graph. A more recent algorithm [25] uses graph theory to identify tertiary motifs in three-dimensional RNA structures.

In addition to comparing RNA secondary structures, classifying them according to common branching signatures, and identifying motifs in them, plane trees can be used as a combinatorial model of RNA base pairing.

For instance, the question of "inverse folding," that is, of designing an RNA sequence with a particular MFE secondary structure, is of interest in applications ranging from biomedical therapeutics to biomolecular computation [20, 91]. To appreciate the challenges, consider an arbitrary plane tree T with n edges, which represents the desired configuration. In a thermodynamic model where an optimal secondary structure has the maximum number of stacked base pairs, the sequence $R = (g^6 a^4 c^6 a^4)^n$ has an optimal set of $5n$ base pairs (i, j) stacked on another base pair $(i - 1, j - 1)$, with the desired arrangement of helices and loops specified by T. Hence, we say that R has an optimal secondary structure corresponding to T. However, any plane tree with n edges corresponds to an optimal secondary structure for R, since there are

$$C_n = \frac{1}{n+1} \binom{2n}{n}$$

(the nth Catalan number) different ways of pairing the poly-g and poly-c segments which maximize the base pair stacking. Hence, R does not correspond uniquely to T.

To achieve the desired configuration, then, we must strengthen the association between the segments which are intended to pair with each other. In the above thermodynamic model which maximizes base pair stacking, this can be achieved. Starting at the root, we walk around the boundary of T, recording a string s of edge labels. We label each edge first with a distinct integer $k \geq 3$ (to ensure stable helices) and then with a complementary $\bar{k}$ on the return trip. We then expand the string s into nucleotides by replacing k by $g^k a^4$ and $\bar{k}$ by $c^k a^4$. The resulting sequence has a unique optimal secondary structure corresponding to T.

Although n different labels are sufficient, it is natural to ask whether they are necessary. From a design perspective, the fewer distinct labels needed, the more realistic the prospect becomes under less simplified thermodynamic models. In particular, for a given T, what is the minimum number of integers k needed to produce a string s which, when expanded into nucleotides, has a unique optimal secondary structure with the desired arrangement of helices and loops? That is, what is the minimum $k \geq 3$ needed to produce an s corresponding uniquely to T with the maximum number of stacked base pairs?

The answer depends exactly on the degree of branching in T. First, consider the "star" tree T with n edges, which has the maximum degree of branching, with a root vertex and n leaves. Then $\lceil n/2 \rceil$ integers k are necessary and sufficient to produce a string $3\bar{3}3\bar{3}4\bar{4}4\bar{4}\ldots$, which corresponds uniquely to T under the association of $(k, \bar{k})$ and $(\bar{k}, k)$ for integers $1 \leq k - 2 \leq \lceil n/2 \rceil$. This follows from the fact that there is exactly one noncrossing matching of $k\,k\,k\,k$; however, there are two distinct ways of pairing up $k\,\bar{k}\,k\,\bar{k}$.

In general, let m be the maximum number of edges incident on a vertex v in T. Then $\lceil m/2 \rceil$ integers are necessary to label the edges of v according to a preorder walk, where the label on the second transversal of an edge is the complement of the first and where each label is used at most once when entering a vertex. Moreover, if each vertex is sequentially labeled appropriately, beginning with the root, these integers are sufficient to produce a string which corresponds uniquely to T.

This analysis suggests that local constraints are necessary and sufficient for the formation of a global structure. In principle, then, any branched structure could be encoded in an RNA sequence, provided that the helices associate uniquely with their correct pairing partner. This insight was the basis for a combinatorial approach to the design problem [42], in contrast to the more common heuristic local search strategies [4, 13, 43].

Furthermore, analysis of this combinatorial model points to the importance of branching, especially the degree of branching, in RNA secondary structures. Results following from [36] have shown that there are many branching topologies that have not (yet) been observed in known RNA molecules. For instance, the RNA-As-Graphs Database [37] lists no known examples of secondary RNA structures with the star topology, having a central loop radiating n helices, when $n = 7, 8, 9$. Furthermore, the structures in RNA STRAND [5] exhibit a low degree of branching overall, with a mean degree for multiloops of 2.66 (and a standard deviation of 0.90), as do the predicted secondary structures for sequences such as 23S ribosomal RNA and picornaviral genomes [6].

What then determines the degree of branching of RNA secondary structures? The MFE secondary structure for an RNA molecule is understood to be an optimal balance between helices and loops, where base pair stacking is thermodynamically beneficial and the formation of loops is necessarily "energetically unfavorable" [97]. However, there are different types of loops, with different functions and parameters for their free energies in the thermodynamic model. What, if anything, can be said about the trade-offs within the different types of loops?

The energy of a loop in the NNTM is a function of the number (and type) of base pairs and of single-stranded nucleotides [92]. A comparison of the length-dependent initiation parameters indicates that, in general, hairpins are the energetically most expensive class, followed by internal loops and bulges, and finally by multiloops. (The exterior loop is treated separately from the others.) Of particular note is the fact that the free energy of multiloops can be increasingly negative as a function of the loop degree because of the favorable dangling ends and terminal mismatches, that is, because of the single-stranded stacking interactions with adjacent base pairs. This suggests that a high degree of branching could be locally favorable. How do we reconcile this with the low degree of branching in both known and predicted RNA secondary structures?

To address this apparent contradiction, let $\mathcal{T}_n$ be the set of plane trees with n edges. Since the trees are rooted, the degree of a vertex is defined to be the number of children [86]. Hence, a leaf has degree 0. The *type* of a plane tree $T \in \mathcal{T}_n$ is the sequence $(d_0, d_1, \ldots, d_{n-1}, d_n)$, where d_i is the number of vertices of degree i. Furthermore [86], a sequence $(d_0, d_1, d_2, \ldots, d_n)$ of nonnegative integers is the type of a plane tree with $n + 1$ vertices if and only if

$$\sum_{i=0}^{n} d_i = n + 1 \text{ and } \sum_{i=0}^{n}(i - 1)d_i = -1.$$

Consider again the combinatorial sequence $R = (g^6 a^4 c^6 a^4)^n$ whose secondary structures that maximize the number of stacked base pairs are in bijection with $\mathcal{T}_n$. To simplify the thermodynamics even further, let $R' = (g^5 c a^4 g c^5 a^4)^n$ so that structures with the maximum number of stacks have the property that all enclosed base pairs are g–c and all closing base pairs are c–g. Then, using the Turner99 parameters [92], the total free energy of the stacks is equal for all these structures, while the free energy of the hairpins is 4.10 kcal/mol, that of the internal loops is 2.3 kcal/mol, that of the branching loops with d enclosed base pairs is $3.4 - 1.5(d + 1)$, and that of the exterior loop with d enclosed base pairs is $-1.9d$. Clearly, the free energy of the loops is increasingly negative as a function of the degree d. Yet, for a given n, the MFE structure has a low degree of branching, with $d \leq 2$, as expected.

The key to understanding this lies in the fact that, in our combinatorial model, the energy of a hairpin loop can be assigned to the branching loop which created it. Using a network-flow-type analysis, an exterior loop of degree d creates d hairpins and a branching loop of degree d creates $d - 1$ new hairpins in addition to the incoming one, which it propagates. An internal loop creates no new hairpins, merely propagating the incoming one.

The MFE secondary structure for R' has the maximum number of stacked base pairs, and the difference in free energy between the MFE structure and other secondary structures for R' with $5n$ stacks is exactly the sum of the loop energies. Disregarding the special energy function for the exterior loop for the moment, these are given by the function

$$a_0 d_0 + a_1 d_1 + \sum_{i=2}^{n}[c_2 + a_2(i + 1)]d_i, \tag{1}$$

where $a_0 = 4.1$, $a_1 = 2.3$, $c_2 = 3.4$, and $a_2 = -1.5$. Using the results for plane trees given above, the optimum can be calculated by considering the cases $d_0 = 1$ with $d_1 = n$, and $2 \leq d_0 \leq n$ with $n - 2d_0 + 2 \leq d_1 \leq n - d_0$ for $0 \leq d_1 < n$ separately. The conclusion is that the maximally stacked secondary structures for R' which minimize the associated loop energies correspond to plane trees with type $((n+2)/2, 0, n/2, 0, \ldots, 0)$ when $n > 2$ is even and the type $((n+1)/2, 1, (n-1)/2, 0, \ldots, 0)$ when $n > 3$ is odd. The analysis when the exterior loop is treated with its separate function, is considerably more involved; however, the conclusion remains the same.

Thus, in this combinatorial model, the energetically most favorable configurations are those which maximize the number of multiloops, but which keep the degree of branching to a minimum. Hence, this analysis suggests that although branching in RNA secondary structures may be locally favorable, it is globally balanced by the cost of increasing numbers of hairpins. This trade-off has been analyzed stochastically in [6], where a large-deviation principle with an explicit rate function was given, and parametrically in [46], by constructing a polytope of RNA secondary structures and its normal fan.

As we have seen, representing RNA secondary structures as trees has enabled researchers to compare structures, to classify them, to identify important motifs, and, finally, to analyze their branching. As fruitful as this combinatorial abstraction has been, however, it is certainly not the only possible abstraction. In particular, interpreting RNA base pairing as a formal language and analyzing it using (stochastic) context-free grammars provides another compelling illustration of the importance of discrete mathematics in molecular biology.

4 Secondary Structures as SCFGs: Asymptotics, Prediction, Homology, and Statistics

Modeling the base pairing of an RNA sequence as a stochastic context-free grammar has enabled researchers to derive asymptotics for structural motifs. This combinatorial formulation has also proven very useful in secondary-structure prediction approaches which do not rely on a thermodynamic model, and especially for predicting a consensus secondary structure by comparative analysis using phylogenetic information. Furthermore, SCFGs have been used for structural-homology recognition in database searches, and as statistical models for evaluating the prediction accuracy of thermodynamic optimization methods.

In the 1980s, several scientists began applying Chomsky's linguistic methods to molecular biology [83]. The first results established that biological sequences can be modeled using regular grammars; since then, hidden Markov models (HMMs), which are their stochastic extensions, have been extensively used in biological sequence analysis. The usual assumption in biological sequence analysis algorithms is that the positions in the strings are uncorrelated, i.e., the identity of the nucleotide in one position has no effect on the identity of another nucleotide. However, this assumption breaks down in RNA sequence analysis because the secondary structure produces long-distance correlations between nucleotides that base-pair. These correlations cannot be modeled using regular grammars and, instead, a larger class of grammars, called context-free grammars (CFGs), needs to be used. In this section, we review how CFGs and their probabilistic versions are used to model and predict RNA secondary structure.

A formal grammar is a quadruple $\mathcal{G} = (\Sigma, N, S, R)$, where Σ is a finite alphabet of terminal symbols, N is a finite set of nonterminals, $S \in N$ is the start nonterminal, and R is a finite set of production rules that specify how sequences that contain nonterminals can be rewritten by expanding the nonterminals to new subsequences. For RNA sequences, the terminals are $\Sigma = \{a, c, g, u\}$. A grammar is *context-free* if all production rules are of the form $V \to \alpha$, where V is a single nonterminal symbol and α is a string of terminal and nonterminal symbols (α may be the empty string ϵ). An example of a CFG [52] for modeling RNA secondary structures with at least two unpaired bases in each hairpin loop is the grammar with nonterminals $N = \{S, L, F\}$, terminals $\Sigma = \{a, c, g, u\}$, and the following rules:

$$S \to LS \mid L \qquad\qquad\qquad\qquad\qquad\qquad \text{loops,}$$

$$L \to dFd' \qquad (d, d') \in \Sigma^2 \qquad\qquad \text{base pair starting a helix,}$$

$$L \to t \qquad\qquad t \in \Sigma \qquad\qquad \text{unpaired nucleotide in a loop,}$$

$$F \to LS \qquad\qquad\qquad\qquad\qquad\qquad \text{termination a helix,}$$

$$F \to dFd' \qquad (d, d') \in \Sigma^2 \qquad\qquad \text{base pair extending a helix.}$$

The vertical bar here serves as a separator between two productions. The parse

$$S \to LS \to aFuS \to acFguS \to acLSguS \to acaSguS$$

$$\to acaLguS \to acaaguS \to acaaguL \to acaaguc$$

corresponds to the structure with primary sequence $acaaguc$ and base pairs $(1, 6)$ and $(2, 5)$.

CFGs coupled with analytic combinatorics can be used as an efficient tool for deriving the asymptotics of the expected number of motifs in secondary structures as a function of sequence length. For example, Clote et al. [19] computed the expected $5'$–$3'$ distance over all secondary structures with n nucleotides in which all hairpins have at least θ unpaired bases, and showed that it does not depend on n. The $5'$–$3'$ distance in a secondary structure on $a_1, \ldots, a_n$ is the minimum path length from a_1 to a_n in the graphical representation of the structure as a matching of n points on a line to which the edges (a_i, a_{i+1}), $1 \le i \le n - 1$, are added. Clote et al. constructed a CFG which generates strings over $\{\circ, (,)\}$, in which the parentheses are correctly matched and each pair of corresponding parentheses is at least θ positions apart. The grammar productions were defined so that exterior nucleotides could be distinguished from nucleotides that are not in the exterior loop. Using the DSV methodology [81] Clote et al. obtained a functional equation for the generating function $S_\theta(z, u)$ of secondary structures with respect to the sequence length and the number of exterior bases in the structure. The average $5'$–$3'$ distance (and its higher moments if desired) can be computed from the asymptotics of the coefficients of $S_\theta(z, u)$ and its derivatives. The same techniques can be used to find the expected $5'$–$3'$ distance in a structure over a sequence compatible with a stickiness p. The *stickiness* parameter p represents the probability that any two positions i, j can base-pair. For example, for sequences with mononucleotide frequencies p_A, p_C, p_G, p_U, in which only canonical base pairs are allowed, the stickiness is $p = 2(p_A p_U + p_G p_U + p_G p_U)$ [44]. The asymptotic number of canonical (without isolated base pairs) and saturated (no additional base pair can be added without violating the noncrossing condition) structures with n nucleotides was derived in [18] using the same method.

Adding a stochastic structure to a grammar allows the development of probabilistic models, which can be used for a variety of computational problems related to RNA. These include secondary-structure prediction for a single sequence, consensus secondary-structure prediction by comparative analysis, and structural homology recognition in database searches. Formally, a stochastic CFG is a quintuple $\mathcal{G} = (\Sigma, N, S, R, P)$, where P is a probability function that assigns a probability $P(V \to \alpha)$ to every production rule $V \to \alpha$ such that $\sum_\alpha P(V \to \alpha) = 1$ for every nonterminal V [31]. The SCFG induces a probability distribution over the set of parses if one allows "infinite parses" as well: the probability of each parse is the product of the probabilities of the rules applied to produce it.

An SCFG that can be used for RNA structure prediction generates a language of RNA sequences, and productions are chosen so that parses of a string correspond to secondary structures. Ideally, this correspondence is one-to-one and the optimal parse of a sequence can be interpreted as the optimal structure. Such grammars are called *structurally unambiguous*. A secondary structure for a single nucleotide sequence can then be predicted using two methods. The first one predicts the most probable parse of the sequence. This can be done efficiently using the Cocke–Younger–Kasami (CYK) algorithm [31, 99], which computes the probability of the most likely derivation, and traceback can be used to find the most likely structure. The CYK algorithm is analogous to the Viterbi algorithm for HMMs and was originally designed for SCFGs in Chomsky normal form, but other parsers [15, 32, 88] can deal with any SCFG. The second method applies posterior decoding using base-pair probability matrices. These can be computed using the inside–outside algorithm [54]. The method then predicts the structure with the maximum expected number of correct positions via dynamic programming.

The idea of using formal language methods for RNA secondary-structure prediction goes back to Sakakibara et al. [74] and Eddy and Durbin [34]. Since then, a number of prediction methods that combine thermodynamics and evolutionary information have been designed [9, 10, 16, 45, 49, 60, 76]. One of the most successful of the fully developed probabilistic methods that predict a consensus structure for a given alignment of sequences was developed by Knudsen and Hein [52] and implemented in Pfold [53]. For a single sequence, the method predicts the most likely structure for the SCFG, while for multiple sequences it maximizes the joint probability of the structure and the alignment given the maximum likelihood estimate of the phylogenetic tree relating the sequences. This model can be readily extended to include experimental information about the structure obtained from chemical probes [89]. It is worth mentioning here that chemical and enzymatic-probing data have been integrated into the thermodynamic model as well [22, 94, 96, 100].

Designing SCFG-based methods for structure prediction requires constructing a grammar with appropriate production rules and obtaining good probability parameters. A good design would be one that captured the important statistics of structural features with as much biological realism as possible. However, it must also be simple enough, with a reasonable number of parameters that could be trained on existing data. The space of SCFGs that can be used for RNA secondary-structure modeling is quite large. However, the currently dominant ones seem to have been constructed using the intuition of the researchers [29]. The seemingly arbitrary choice of grammars and the issue of grammar comparison was addressed in [29], where several grammars were compared. However, the set of grammars tested was by no means exhaustive. It is only recently that automated search techniques have been employed to find potentially more effective grammars [3]. The search was done on grammars in which the production rules are of the type $T \rightarrow UV, T \rightarrow ., T \rightarrow (U)$, for nonterminals T, U, V. The number of such grammars with n nonterminals is $2^{n^3+n^2+n}$, which allows exhaustive search for $n = 2$. Larger grammars were explored using an evolutionary algorithm. The search proved effective in finding grammars that have strong predictive accuracy, as good as or slightly better than those designed manually. The probabilities in a grammar are usually taken to be the maximum-likelihood probabilities obtained from a selected set of trusted secondary structures. If the grammar is structurally unambiguous, these can easily be found by counting the frequencies of the productions used in parsing the structures. Otherwise, they can be estimated using the inside–outside algorithm, analogously to the forward–backward algorithm for HMMs.

Note that even though the sum of the probabilities of all productions starting with a fixed nonterminal is 1, this may not define a probability distribution on the whole language. This happens if a parse starting from S does not end in finitely many steps with probability 1. Such grammars are called inconsistent. A simple example is the grammar with one nonterminal and one terminal symbol, and production rules $S \rightarrow SS$ and $S \rightarrow a$ [17]. Let p be the probability of the first production, let $1 - p$ be the probability of the second one, and let P_h be the total probability of all the parse trees with depth less than or equal to h. Then $P_1 = 1 - p$ (corresponding to the parse $S \rightarrow a$) and $P_{h+1} = pP_h^2 + 1 - p$ for $h > 1$. It is not difficult to show that P_h is nondecreasing and converges to $\min\{1, (1 - p)/p\}$. This means that proper probability over all parses is obtained if and only if $p \leq \frac{1}{2}$. Interestingly, SCFGs with inside–outside parameters obtained from a finite training set are always consistent [75]. However, even for inconsistent grammars, it always makes sense to consider the probability distributions induced by the SCFG on strings of fixed length n by taking appropriate conditional probabilities. Namely, the probability of a structure with n nucleotides is the sum of the probabilities of all its parses divided by the total probability of a parse to finish with a string of length n.

It then becomes natural to ask how, given a CFG, the induced probability distribution changes with a change in the probability parameters. Motivated by Dowell and Eddy [29], where the Pfold grammar was benchmarked against eight other SCFGs for secondary-structure prediction, with the conclusion that it achieves the best prediction accuracy, the present authors analyzed this grammar in detail in [69]. Specifically, we analyzed the distribution of the numbers of base pairs, helices, and various types of loops in structures over a sequence of length n. Using the DSV methodology and analytic combinatorics, we proved that the distributions of these motifs converge to a Gaussian, and we computed the expected number of motifs as a function of n. As a surprising consequence, we obtained some relations between the expected numbers of motifs which do not depend on the probability parameters of the SCFG which define the distributions. For example, the expected number of helices is always four times larger than the expected number of multiloops. This provided a mathematical explanation of a fact that was observed previously by Knudsen [51], that the Pfold grammar predicted the structure of tRNA much more accurately than that of 5S rRNA. Namely, the tRNA structure is known to resemble a cloverleaf – it has one multiloop with four helices – whereas the helix-to-multiloop ratio in the 5S structures is about 8. We also showed that the expected $5'$–$3'$ distance is bounded by a constant that does not depend on the sequence length. As already mentioned, this phenomenon was proved by Clote for the homopolymer model and for structures with stickiness p [19], and it has been experimentally observed for the thermodynamic model [98].

Besides providing a simple method for secondary-structure prediction using the CYK algorithm, SCFGs can also be a basis for more complex prediction approaches. For example, a stochastic sampling algorithm can be designed as a probabilistic counterpart to Sfold using SCFGs [65, 78]. Moreover, modeling secondary structure with an SCFG has the advantage that it provides a way of designing a probabilistic model that combines information from different scoring systems that otherwise do not combine naturally. Pfold was one example; it computed the optimal structure for a given alignment. Another example of such an application is the design of a consensus structure prediction algorithm for two homologous sequences which are not a priori aligned. Such an algorithm should also combine evolutionary and thermodynamic information. However, it is not clear what score should be optimized by the algorithm so that a mathematically optimal structure is most likely to represent the biologically correct structure. The initial algorithms used an ad hoc additive combination of alignment-scoring matrices and base pair maximization [40], or optimized the sum of two structures' predicted free energies, while using an ad hoc pseudoenergy penalty for insertions/deletions [60].

Dowell and Eddy [30] developed an algorithm for simultaneous folding and alignment of two sequences using "pairSCFGs". These extend traditional SCFGs for RNA modeling by emitting two correlated sequences instead of just one, and an RNA sequence is also allowed to contain a gap to accommodate insertions and deletions. Good pairSCFGs can be constructed from structurally unambiguous traditional SCFGs, albeit with some modifications to ensure that the design is also alignment-unambiguous. This is an additional desired property of pairSCFGs, which guarantees that the optimal parse corresponds to the optimal alignment and structure.

In addition to prediction, SCFGs can be also be used for measuring the similarity of a sequence to a given family of aligned homologous sequences with a known structure. This is applicable in situations where one is presented with a multiple alignment of an RNA sequence family with known secondary structure and wants to search a sequence database for homologs that significantly match both the sequence and the structure of the query. An SCFG built to capture the consensus structure of an alignment with position-specific scores is called a profile SCFG, or a covariance model, and was first described by Eddy and Durbin [34]. It is built around the consensus structure tree so that each node corresponds to one or more nonterminals. One of those nonterminals corresponds to the type of the node (a base pair, unpaired bases on the $5'$ or $3'$ side of a helix, or a bifurcation), and other nonterminals are used to model deletions and insertions in the target sequence with respect to the consensus. Profile-SCFG-based search for homologs of a single RNA molecule with a known secondary structure is also possible [50]. A collection of RNA sequence families, represented by multiple sequence alignments and covariance models, is freely available from the online database Rfam [39].

Finally, we mention an SCFG-based statistical method for estimating the accuracy of structures predicted by other models, such as the MFE structure, proposed by Nebel [63,64]. This method uses a kind of distance function defined in terms of secondary-structure motifs and their lengths to compare the MFE structure with the expected values for the model, which are supposed to describe the set of structures used for the training of the grammar probabilities and can be obtained using analytic combinatorics.

5 Conclusion

As we have illustrated, there are several combinatorial structures that have been used as models for RNA base pairing. Historically, matchings were the first one used, and were very appropriate for obtaining bounds on and asymptotic estimates of the size of the space of secondary structures. Trees, on the other hand, are well-studied mathematical objects that are useful for capturing more coarse-grained information and for multiscale comparison of secondary structures. SCFGs are attractive because they are readily extendable to allow incorporation of different types of information in applications such as secondary-structure prediction. In other words, all of these different models emerged as the right tool for answering some relevant biological and computational questions, and their future use will depend on how easily they can be extended to model biological reality.

For example, although this review does not address pseudoknots, much has been done in understanding their combinatorics. A pseudoknot is formed by base pairs (i, j) and (i', j') such that $i < i' < j < j'$. Following directly from the definition, pseudoknotted structures can be modeled as matchings which, when drawn on a circle, have some base pairings that cross. Appropriate restrictions on the type of crossings allowed can be introduced to reflect the biological complexity of pseudoknots. Hence, matchings provide a model which permits an in-depth combinatorial and probabilistic analysis of pseudoknots, including their uniform generation [70]. On the other hand, using CFGs for general pseudoknot modeling is challenging because arbitrary nonnested correlations cannot be generated using a single CFG. The next category of languages in the Chomsky hierarchy, the context-sensitive ones, are not attractive because their parsing is an NP-complete problem. However, modeling pseudoknotted structures using SCFGs becomes possible if some restrictions on the complexity of the pseudoknots are assumed. For example, Rivas and Eddy [72, 73] have described a class of extended SCFGs that can model a restricted class of pseudoknots; see also [56, 66, 71, 77].

In addition to pseudoknots, there are many open problems in RNA folding which may be amenable to combinatorial modeling and analysis. The foremost remains that accurate prediction of RNA secondary structures, particularly of longer sequences with many thousands of nucleotides. Related problems include the incorporation of kinetic effects into secondary-structure prediction algorithms, the identification of structural motifs, the design of RNA secondary structures, and the challenge of extending our understanding beyond base pairing to the modeling, analysis, and prediction of three-dimensional RNA molecular structures.

References

1. J. Allali, M.F. Sagot, A multiple graph layers model with application to RNA secondary structures comparison, in *String Processing and Information Retrieval*, ed. by M. Consens, G. Navarro. Lecture Notes in Computer Science, vol. 3772 (Springer, Berlin, 2005), pp. 348–359
2. J. Allali, M.F. Sagot, A new distance for high level RNA secondary structure comparison. IEEE/ACM Trans. Comput. Biol. Bioinform. **2**(1), 3–14 (2005)
3. J.W. Anderson, P. Tataru, J. Staines, J. Hein, R. Lyngsø, Evolving stochastic context–free grammars for RNA secondary structure prediction. BMC Bioinformatics **13**(1), 78 (2012)
4. M. Andronescu, A.P. Fejes, F. Hutter, H.H. Hoos, A. Condon, A new algorithm for RNA secondary structure design. J. Mol. Biol. **336**(3), 607–624 (2004)
5. M. Andronescu, V. Bereg, H.H. Hoos, A. Condon, RNA STRAND: The RNA secondary structure and statistical analysis database. BMC Bioinform. **9**(340) (2008)
6. Y. Bakhtin, C.E. Heitsch, Large deviations for random trees and the branching of RNA secondary structures. Bull. Math. Biol. **71**(1), 84–106 (2009)
7. F.C. Bernstein, T.F. Koetzle, G.J. Williams, E.E. Meyer Jr., M.D. Brice, J.R. Rodgers, O. Kennard, T. Shimanouchi, M. Tasumi, The protein data bank: a computer-based archival file for macromolecular structures. J. Mol. Biol. **112**(3), 535–542 (1977)
8. P. Bille, A survey on tree edit distance and related problems. Theor. Comput. Sci. **337**(1–3), 217–239 (2005)
9. E. Bindewald, B. Shapiro, RNA secondary structure prediction from sequence alignments using a network of k-nearest neighbor classifiers. RNA **12**(3), 342–352 (2006)

10. D. Bouthinon, H. Soldano, A new method to predict the consensus secondary structure of a set of unaligned RNA sequences. Bioinformatics **15**(10), 785–798 (1999)
11. P. Brion, E. Westhof, Hierarchy and dynamics of RNA folding. Annu. Rev. Biophys. Biomol. Struct. **26**, 113–137 (1997)
12. S.W. Burge, J. Daub, R. Eberhardt, J. Tate, L. Barquist, E.P. Nawrocki, S.R. Eddy, P.P. Gardner, A. Bateman, Rfam 11.0: 10 years of RNA families. Nucleic Acids Res. **41**(Database issue), D226–D232 (2013)
13. A. Busch, R. Backofen, INFO-RNA – a fast approach to inverse RNA folding. Bioinformatics **22**(15), 1823–1831 (2006)
14. J.J. Cannone, S. Subramanian, M.N. Schnare, J.R. Collett, L.M. D'Souza, Y. Du, B. Feng, N. Lin, L.V. Madabusi, K.M. Müller, N. Pande, Z. Shang, N. Yu, R.R. Gutell, The Comparative RNA Web (CRW) Site: an online database of comparative sequence and structure information for ribosomal, intron, and other RNAs. BMC Bioinform. **3**(1), 2 (2002)
15. J. Chappelier, M. Rajman, A generalized CYK algorithm for parsing stochastic CFG, in *First Workshop on Tabulation in Parsing and Deduction (TAPD98)*, Paris, 1998, pp. 133–137. Citeseer
16. J. Chen, S. Le, J. Maizel, Prediction of common secondary structures of RNAs: a genetic algorithm approach. Nucleic Acids Res. **28**(4), 991–999 (2000)
17. Z. Chi, S. Geman, Estimation of probabilistic context-free grammars. Comput. Linguist. **24**(2), 299–305 (1998)
18. P. Clote, E. Kranakis, D. Krizanc, B. Salvy, Asymptotics of canonical and saturated RNA secondary structures. J. Bioinform. Comput. Biol. **7**(05), 869–893 (2009)
19. P. Clote, Y. Ponty, J. Steyaert, Expected distance between terminal nucleotides of RNA secondary structures. J. Math. Biol. **65**, 1–19 (2012)
20. A. Condon, Problems on RNA secondary structure prediction and design, in *Automata, Languages and Programming*, ed. by J.C.M. Baeten et al. Lecture Notes in Computer Science, vol. 2719 (Springer, Berlin, 2003), pp. 22–32
21. J. Couzin, Breakthough of the year: small RNAs make big splash. Science **298**(5602), 2296–2297 (2002)
22. K.E. Deigan, T.W. Li, D.H. Mathews, K.M. Weeks, Accurate SHAPE-directed RNA structure determination. Proc. Natl. Acad. Sci. **106**(1), 97–102 (2009)
23. Y. Ding, C.E. Lawrence, A statistical sampling algorithm for RNA secondary structure prediction. Nucleic Acids Res. **31**(24), 7280–7301 (2003)
24. Y. Ding, C.Y. Chan, C.E. Lawrence, RNA secondary structure prediction by centroids in a Boltzmann weighted ensemble. RNA **11**, 1157–1166 (2005)
25. M. Djelloul, A. Denise, Automated motif extraction and classification in RNA tertiary structures. RNA **14**(12), 2489–2497 (2008)
26. R. Donaghey, L.W. Shapiro, Motzkin numbers. J. Comb. Theory Ser. A **23**(3), 291–301 (1977)
27. K.J. Doshi, J.J. Cannone, C.W. Cobaugh, R.R. Gutell, Evaluation of the suitability of free-energy minimization using nearest-neighbor energy parameters for RNA secondary structure prediction. BMC Bioinform. **5**(1), 105 (2004)
28. J.A. Doudna, Structural genomics of RNA. Nat. Struct. Biol. **7**, 954–956 (2000)
29. R. Dowell, S. Eddy, Evaluation of several lightweight stochastic context-free grammars for RNA secondary structure prediction. BMC Bioinformatics **5**(1), 71 (2004)
30. R. Dowell, S. Eddy, Efficient pairwise RNA structure prediction and alignment using sequence alignment constraints. BMC Bioinformatics **7**(1), 400 (2006)
31. R. Durbin, S. Eddy, A. Krogh, G. Mitchison, *Biological Sequence Analysis: Probabilistic Models of Proteins and Nucleic Acids* (University Press, Cambridge/New York, 1998)
32. J. Earley, An efficient context-free parsing algorithm. Commun. ACM **13**(2), 94–102 (1970)
33. S. Eddy, Noncoding RNA genes. Curr. Opin. Genet. Dev. **9**(6), 695–699 (1999)
34. S. Eddy, R. Durbin, RNA sequence analysis using covariance models. Nucleic Acids Res. **22**(11), 2079–2088 (1994)

35. W. Fontana, D. Konings, P.F. Stadler, P. Schuster, Statistics of RNA secondary structures. Biopolymers **33**(9), 1389–1404 (1993)
36. H.H. Gan, S. Pasquali, T. Schlick, Exploring the repertoire of RNA secondary motifs using graph theory; implications for RNA design. Nucleic Acids Res. **31**(11), 2926–2943 (2003)
37. H.H. Gan, D. Fera, J. Zorn, N. Shiffeldrim, M. Tang, U. Laserson, N. Kim, T. Schlick, RAG: RNA-as-graphs database–concepts, analysis, and features. Bioinformatics **20**(8), 1285–1291 (2004)
38. R. Giegerich, B. Voß, M. Rehmsmeier, Abstract shapes of RNA. Nucleic Acids Res. **32**(16), 4843–4851 (2004)
39. S. Griffiths-Jones, A. Bateman, M. Marshall, A. Khanna, S. Eddy, Rfam: an RNA family database. Nucleic Acids Res. **31**(1), 439–441 (2003)
40. J. Havgaard, R. Lyngsø, J. Gorodkin, The FOLDALIGN web server for pairwise structural RNA alignment and mutual motif search. Nucleic Acids Res. **33**(suppl 2), W650–W653 (2005)
41. T. Haynes, D. Knisley, E. Seier, Y. Zou, A quantitative analysis of secondary RNA structure using domination based parameters on trees. BMC Bioinform. **7**, 108 (2006)
42. C.E. Heitsch, A. Condon, H.H. Hoos, From RNA secondary structure to coding theory: a combinatorial approach, in *DNA8: Revised Papers from the 8th International Workshop on DNA Based Computers*, Sapporo, ed. by A.O.M. Hagiya. Lecture Notes in Computer Science, vol. 2568 (Springer, London, 2003), pp. 215–228
43. I.L. Hofacker, W. Fontana, P.F. Stadler, L.S. Bonhoeffer, M. Tacker, P. Schuster, Fast folding and comparison of RNA secondary structures. Monatsh. Chem. **125**(2), 167–188 (1994)
44. I.L. Hofacker, P. Schuster, P.F. Stadler, Combinatorics of RNA secondary structures. Discret. Appl. Math. **88**(1–3), 207–237 (1998)
45. I. Hofacker, M. Fekete, P. Stadler, Secondary structure prediction for aligned RNA sequences. J. Mol. Biol. **319**(5), 1059–1066 (2002)
46. V. Hower, C.E. Heitsch, Parametric analysis of RNA branching configurations. Bull. Math. Biol. **73**(4), 754–776 (2011)
47. Y. Ji, X. Xu, G.D. Stormo, A graph theoretical approach to predict common RNA secondary structure motifs including pseudoknots in unaligned sequences. Bioinformatics **20**(10), 1591–1602 (2004)
48. T. Jiang, L. Wang, K. Zhang, Alignment of trees – an alternative to tree edit. Theor. Comput. Sci. **143**(1), 137–148 (1995)
49. V. Juan, C. Wilson, RNA secondary structure prediction based on free energy and phylogenetic analysis. J. Mol. Biol. **289**(4), 935 (1999)
50. R. Klein, S. Eddy, Rsearch: finding homologs of single structured RNA sequences. BMC Bioinformatics **4**(1), 44 (2003)
51. M. Knudsen, Stochastic context-free grammars and RNA secondary structure prediction. Ph.D. thesis, Aarhus Universitet, Datalogisk Institut, 2005
52. B. Knudsen, J. Hein, RNA secondary structure prediction using stochastic context-free grammars and evolutionary history. Bioinformatics **15**(6), 446–454 (1999)
53. B. Knudsen, J. Hein, Pfold: RNA secondary structure prediction using stochastic context-free grammars. Nucleic Acids Res. **31**(13), 3423–3428 (2003)
54. K. Lari, S. Young, The estimation of stochastic context-free grammars using the inside-outside algorithm. Comput. Speech Lang. **4**(1), 35–56 (1990)
55. S.Y. Le, R. Nussinov, J.V. Maizel, Tree graphs of RNA secondary structures and their comparisons. Comput. Biomed. Res. **22**(5), 461–473 (1989)
56. F. Lefebvre, An optimized parsing algorithm well suited to RNA folding. Proc. Int. Conf. Intell. Syst. Mol. Biol. **3**, 220–230 (1995)
57. W.A. Lorenz, Y. Ponty, P. Clote, Asymptotics of RNA shapes. J. Comput. Biol. **15**(1), 31–63 (2008)
58. N.R. Markham, M. Zuker, UNAFold: software for nucleic acid folding and hybridization, in *Bioinformatics: Structure, Function, and Applications*, ed. by J.M. Keith. Methods in Molecular Biology, vol. 453 (Humana Press, Totowa, 2008), pp. 3–31

59. D.H. Mathews, Revolutions in RNA secondary structure prediction. J. Mol. Biol. **359**(3), 526–532 (2006)
60. D. Mathews, D. Turner et al., Dynalign: an algorithm for finding the secondary structure common to two RNA sequences. J. Mol. Biol. **317**(2), 191 (2002)
61. J.S. McCaskill, The equilibrium partition function and base pair binding probabilities for RNA secondary structure. Biopolymers **29**(6–7), 1105–1119 (1990)
62. V. Moulton, M. Zuker, M. Steel, R. Pointon, D. Penny, Metrics on RNA secondary structures. J. Comput. Biol. **7**(1), 277–292 (2000)
63. M.E. Nebel, *On a Statistical Filter for RNA Secondary Structures* (Johann-Wolfgang-Goethe-University, Institut für Informatik, Frankfurt, 2002)
64. M.E. Nebel, Identifying good predictions of RNA secondary structure, in *Pacific Symposium on Biocomputing*, Lihue, 2003, vol. 9, ed. by R.B. Altman, A.K. Dunker, L. Hunter, T.E. Klein, pp. 423–434
65. M.E. Nebel, A. Scheid, Evaluation of a sophisticated SCFG design for RNA secondary structure prediction. Theory Biosci. **130**(4), 313–336 (2011)
66. M.E. Nebel, F. Weinberg, Algebraic and combinatorial properties of common RNA pseudo-knot classes with applications. J. Comput. Biol. **19**(10), 1134–1150 (2012)
67. H.F. Noller, C.R. Woese, Secondary structure of 16S ribosomal RNA. Science **212**(4493), 403–411 (1981)
68. R. Nussinov, G. Pieczenik, J.R. Griggs, D.J. Kleitman, Algorithms for loop matchings. SIAM J. Appl. Math. **35**(1), 68–82 (1978)
69. S. Poznanović, C. Heitsch, Asymptotic distribution of motifs in a stochastic context-free grammar model of RNA folding (2012, preprint). arXiv:1204.3670
70. C. Reidys, *Combinatorial Computational Biology of RNA: Pseudoknots and Neutral Networks* (Springer, New York/London, 2010)
71. C.M. Reidys, F.W.D. Huang, J.E. Andersen, R.C. Penner, P.F. Stadler, M.E. Nebel, Topology and prediction of RNA pseudoknots. Bioinformatics **27**(8), 1076–1085 (2011)
72. E. Rivas, S.R. Eddy, A dynamic programming algorithm for RNA structure prediction including pseudoknots. J. Mol. Biol. **285**(5), 2053–2068 (1999)
73. E. Rivas, S. Eddy, The language of RNA: a formal grammar that includes pseudoknots. Bioinformatics **16**(4), 334–340 (2000)
74. Y. Sakakibara, M. Brown, R. Hughey, I. Mian, K. Sjölander, R. Underwood, D. Haussler, Stochastic context-free grammers for tRNA modeling. Nucleic Acids Res. **22**(23), 5112–5120 (1994)
75. J. Sánchez, J. Benedí, Consistency of stochastic context-free grammars from probabilistic estimation based on growth transformations. IEEE Trans. Pattern Anal. Mach. Intell. **19**(9), 1052–1055 (1997)
76. D. Sankoff, Simultaneous solution of the RNA folding, alignment and protosequence problems. SIAM J. Appl. Math. **45**(5), 810–825 (1985)
77. C. Saule, M. Régnier, J.-M. Steyaert, A. Denise, Counting RNA pseudoknotted structures. J. Comput. Biol. **18**(10), 1339–1351 (2011)
78. A. Scheid, M.E. Nebel, Statistical RNA secondary structure sampling based on a length-dependent SCFG model. Technical report, University of Kaiserslautern, 5, 2012
79. W.R. Schmitt, M.S. Waterman, Linear trees and RNA secondary structure. Discret. Appl. Math. **51**(3), 317–323 (1994)
80. P. Schuster, P.F. Stadler, A. Renner, RNA structures and folding: from conventional to new issues in structure predictions. Curr. Opin. Struct. Biol. **7**(2), 229–235 (1997)
81. M.P. Schützenberger, On context-free languages and push-down automata. Inf. Control **6**, 246–264 (1963)
82. B.S. Schuwirth, M.A. Borovinskaya, C.W. Hau, W. Zhang, A. Vila-Sanjurjo, J.M. Holton, J.H.D. Cate, Structures of the bacterial ribosome at 3.5 Å resolution. Science **310**(5749), 827–834 (2005)
83. D. Searls, The language of genes. Nature **420**(6912), 211–217 (2002)

84. B.A. Shapiro, An algorithm for comparing multiple RNA secondary structures. Comput. Appl. Biosci. **4**(3), 387–393 (1988)
85. B.A. Shapiro, K. Zhang, Comparing multiple RNA secondary structures using tree comparisons. Comput. Appl. Biosci. **6**(4), 309–318 (1990)
86. R.P. Stanley, *Enumerative combinatorics. Vol. 2.* Cambridge Studies in Advanced Mathematics, vol. 62 (Cambridge University Press, Cambridge, 1999)
87. P.R. Stein, M.S. Waterman, On some new sequences generalizing the Catalan and Motzkin numbers. Discret. Math. **26**(3), 261–272 (1979)
88. A. Stolcke, An efficient probabilistic context-free parsing algorithm that computes prefix probabilities. Comput. Linguist. **21**(2), 165–201 (1995)
89. Z. Sükösd, B. Knudsen, J. Kjems, C. Pedersen, Ppfold 3.0: fast RNA secondary structure prediction using phylogeny and auxiliary data. Bioinformatics **28**, 2691–2692 (2012)
90. I. Tinoco Jr., C. Bustamante, How RNA folds. J. Mol. Biol. **293**(2), 271–281 (1999)
91. B.J. Tucker, R.R. Breaker, Inventing and improving ribozyme function: rational design versus iterative selection methods. Curr. Opin. Struct. Biol. **15**(3), 342–348 (2005)
92. D.H. Turner, D.H. Mathews, NNDB: the nearest neighbor parameter database for predicting stability of nucleic acid secondary structure. Nucleic Acids Res. **38**, D280–D282 (2010)
93. B. Voß, R. Giegerich, M. Rehmsmeier, Complete probabilistic analysis of RNA shapes. BMC Biol. **4**(1), 5 (2006)
94. S. Washietl, I.L. Hofacker, P.F. Stadler, M. Kellis, RNA folding with soft constraints: reconciliation of probing data and thermodynamic secondary structure prediction. Nucleic Acids Res. **40**(10), 4261–4272 (2012)
95. M.S. Waterman, Secondary structure of single-stranded nucleic acids, in *Studies in Foundations and Combinatorics*, ed. by G.-C. Rota. Advances in Mathematics. Supplementary Studies, vol. 1 (Academic, New York, 1978), pp. 167–212
96. K. Weeks, Advances in RNA structure analysis by chemical probing. Curr. Opin. Struct. Biol. **20**(3), 295–304 (2010)
97. S. Wuchty, W. Fontana, I.L. Hofacker, P. Schuster, Complete suboptimal folding of RNA and the stability of secondary structures. Biopolymers **49**(2), 145–165 (1999)
98. A. Yoffe, P. Prinsen, W. Gelbart, A. Ben-Shaul, The ends of a large RNA molecule are necessarily close. Nucleic Acids Res. **39**(1), 292–299 (2011)
99. D. Younger, Recognition and parsing of context-free languages in time n^3. Inf. Control **10**(2), 189–208 (1967)
100. K. Zarringhalam, M.M. Meyer, I. Dotu, J.H. Chuang, P. Clote, Integrating chemical footprinting data into RNA secondary structure prediction. PloS ONE **7**(10), e45160 (2012)
101. K. Zhang, D. Shasha, Simple fast algorithms for the editing distance between trees and related problems. SIAM J. Comput. **18**(6), 1245–1262 (1989)
102. M. Zuker, RNA folding prediction: the continued need for interaction between biologists and mathematicians, in *Some Mathematical Questions in Biology – DNA Sequence Analysis* (New York, 1984). Lectures on Mathematics in the Life Sciences, vol. 17. (American Mathematical Society, Providence, 1986), pp. 87–124
103. M. Zuker, On finding all suboptimal foldings of an RNA molecule. Science **244**(4900), 48–52 (1989)
104. M. Zuker, Mfold web server for nucleic acid folding and hybridization prediction. Nucleic Acids Res. **31**(13), 3406–3415 (2003)
105. M. Zuker, P. Stiegler, Optimal computer folding of large RNA sequences using thermodynamics and auxiliary information. Nucleic Acids Res. **9**(1), 133–148 (1981)

Redundant and Critical Noncovalent Interactions in Protein Rigid Cluster Analysis

Naomi Fox and Ileana Streinu

Abstract A protein's fold is held together by weak noncovalent interactions, known to break and form during naturally occurring fluctuations. Rigidity analysis leverages connectivity information about these interactions, calculated from PDB structural data, and computes a decomposition of the molecule into groups of atoms, called rigid clusters, that tend to remain together during such local motions. A crucial question in the application of this technique is how robust the results of rigidity analysis are to small variations in the noncovalent network. *If any particular interaction within a cluster were to break, would the cluster remain rigid, would it "shatter" into many smaller clusters, or would the flexibility increase but only negligibly?* In this chapter, we overview the mathematical principles underlying rigidity analysis, and propose a method for classifying the interactions which are *redundant* or *critical* for a computed cluster decomposition. We also measure the change in cluster size upon the interaction's removal, which we refer to as its *criticality value*. In addition, we propose a new method for assigning scores to the rigid clusters based on the fraction of interactions that are redundant. We demonstrate this classification scheme on a data set of multiple conformations of 16 proteins. We have found that typically the dominant rigid clusters do not contain highly critical interactions, yet, when such interactions exist, they tend to be concentrated around the active site. In our case studies, we have found that removal of these interactions results in functionally relevant changes in rigidity. We present

N. Fox
Department of Computer Science, University of Massachusetts Amherst, 140 Governors Drive, Amherst, MA 01002, USA
e-mail: fox@cs.umass.edu

I. Streinu (✉)
Department of Computer Science, Smith College, Northampton, MA 01063, USA

Department of Computer Science, University of Massachusetts Amherst, 140 Governors Drive, Amherst, MA 01002, USA
e-mail: istreinu@smith.edu; streinu@cs.umass.edu

N. Jonoska and M. Saito (eds.), *Discrete and Topological Models in Molecular Biology*, 167
Natural Computing Series, DOI 10.1007/978-3-642-40193-0_8,
© Springer-Verlag Berlin Heidelberg 2014

our results on the redundancy and presence of critical interactions on benchmarking data sets, with case studies on adenylate kinase, dihydrofolate reductase, DNA polymerase β, HIV-1 protease, and cytochrome-c. We also provide survey results on a larger data set of 150 proteins. These methods have been implemented in the KINARI-Redundancy server, publicly available from the KINARI-Web site (http://kinari.cs.umass.edu).

1 Introduction

Atomic fluctuations are essential for protein function, because they permit the structure to adjust to the binding of another molecule [32]. The native state is stabilized by weak noncovalent interactions, namely hydrogen bonds (H-bonds) and hydrophobic interactions, which break and form frequently during these fluctuations. When existing weak interactions are broken, the released atomic groups can make new interactions of comparable energy, potentially resulting in conformational rearrangement. Protein structures continuously fluctuate about the equilibrium conformation observed in X-ray crystallography and NMR experiments. Therefore, when developing methods that rely on Protein Data Bank (PDB) structural data to predict protein rigidity and flexibility, it is crucial to assess how these fluctuations may affect the results.

Rigidity analysis is a computationally efficient method for predicting rigid and flexible regions in proteins. Examples of systems for this include MSU-FIRST (now Proflex) [20], ASU-FIRST [4], and our own KINARI [9]. A typical application to a 100 residue protein takes seconds, permitting large data sets to be analyzed. Rigidity analysis has been used to show the change in flexibility between different conformations and complexes, such as HIV-1 protease [20] and the Ras–Raf complex [13].

Figure 1 depicts the step-by-step processing performed by KINARI. The protein is first modeled as a body–bar–hinge framework, and then a special multigraph is built, where each body is assigned a vertex, each hinge is assigned five edges, and each bar is assigned one edge. This graph serves as input to a pebble game algorithm, which determines the components in terms of $(6, 6)$-*graph sparsity*, a concept introduced in [27]. The output of the pebble game is then converted back in terms of clusters of atoms within the protein. An educational site (http://linkage.cs.umass.edu/pg/) and a video [28] introducing these mathematical concepts to a general audience are linked to from the KINARI-Web site.

A simplifying assumption of this approach is that the set of interactions is static. Yet, as demonstrated in molecular dynamics simulations, noncovalent interactions break and form rapidly, typically over nanoseconds [26]. An open question concerns the sensitivity and robustness of the rigidity results obtained from a single conformation, typically taken from the PDB. When using a rigidity analysis system, how confident should we be in the resulting rigid cluster decomposition? *If any particular interaction within a cluster were to break, would the cluster remain rigid,*

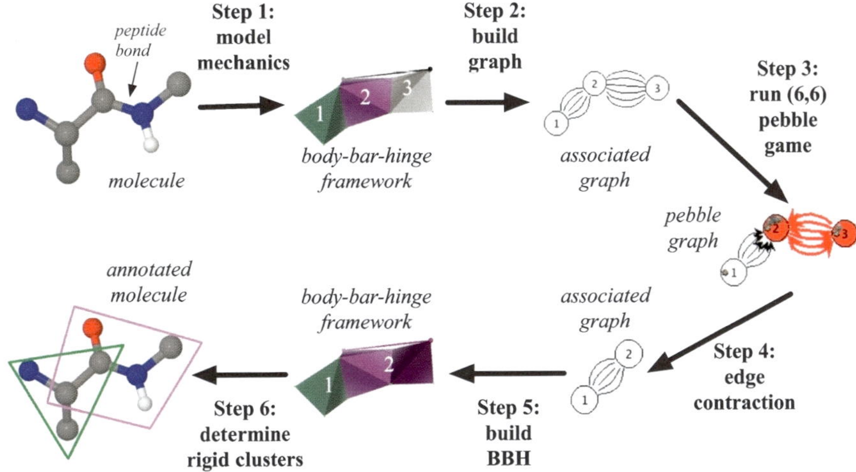

Fig. 1 Steps of protein rigidity analysis. "BBH" stands for the calculated body–bar–hinge mechanical model that captures the interconnectivity of the resulting rigid clusters

would it "shatter" into many smaller clusters, or would the flexibility increase but only negligibly?

In this chapter, we present our investigation of the prevalence of redundant and rigidity-critical interactions:

1. *Redundant interactions.* How much redundancy is built into the network of interactions which hold the rigid clusters together? What is the tolerance of a cluster to the loss of any particular interaction?
2. *Rigidity-critical interactions.* How prevalent are nonredundant (*critical*) inter-actions? These are the interactions which, when broken, cause a nonnegligible change in flexiblity that may affect function. How much will a cluster's size decrease when a critical interaction breaks?

We address these questions by proposing a method to classify noncovalent interactions. Based on their individual contribution to the rigidity of the cluster, they are labeled as either *redundant* or *critical*. In addition, we describe a method for scoring clusters using the classification. The *criticality value* of an interaction is the change in cluster size upon the interaction's removal. We characterize the typical occurrence of redundant and critical interactions with an evaluation on a benchmark data set of over 120 proteins. We show with case studies that interactions with criticality values greater than or equal to 0.10 tend to be concentrated in the same local region and their removal causes functionally relevant changes in rigidity. We have made these methods available from the KINARI-Web server (http://kinari. cs.umass.edu) [9].

2 Background

Protein rigidity analysis processes molecular data and provides a coarse-grained representation capturing a few essential mechanical properties of the molecule. The coarse-grained representation directly provides flexibility information, and this can, in principle, be subsequently leveraged in motion generation methods [12]. In this section, we describe the details of the steps undertaken by KINARI, as depicted in Fig. 1. First, we provide a brief introduction to the mathematical rigidity theory on which the software relies. Then we discuss how the protein modeling is performed.

2.1 Rigidity Theory

A *body–bar–hinge framework* is a 3D mechanical structure made from rigid bodies, pairs of which are connected through hinges and bars. The hinges admit only a rotation of the two incident bodies around the hinge axis. The fixed-length bars connect the bodies at universal joints which allow full rotational freedom. If the only motions of the framework are the trivial rigid motions (those which move the whole system rigidly, maintaining all the pairwise distances between all points), then the framework is said to be rigid. Otherwise, it is flexible. Figure 1 shows an example of a body-bar-hinge framework created in Step 1, composed of three bodies connected through two hinges and one bar.

In isolation, a rigid body has six trivial degrees of freedom (DOFs): all rigid motions can be generated by translations along, and rotations about, the x, y, and z axes. Two disconnected rigid bodies have a total of 12 DOFs; k disconnected rigid bodies have a total of $6k$ DOFs. To a body–bar–hinge framework we associate a multigraph (called the Tay graph), in which a body is represented by a vertex. Then, each bar between two bodies is represented by an edge between the corresponding vertices in the Tay graph, and a hinge is represented by five edges. The intuition behind this association is that when two bodies are connected by a bar, one DOF is removed. If two rigid bodies are connected at a hinge joint, five DOFs are removed. Adding additional bars between the two bodies can remove up to six DOFs, at which point the two rigid bodies are rigidly attached and form a single rigid body. The remaining trivial six DOFs cannot be removed by connecting the bodies by additional bars or hinges.

A simple counting rule, due to Tay [34] (see also [35]) and rigorously proven to be valid by *Tay's theorem*, can be used on a multigraph associated to a body–bar–hinge framework to determine the rigidity and the DOFs of the framework.

Theorem 1. *Tay's theorem: Theorem for 3D body–bar–hinge framework (Tay, Whitely). A multigraph G, with n vertices and m edges, is the graph of a generic minimally rigid body–bar–hinge framework iff any subset of n' vertices in G spans at most $6n' - 6$ edges and $m = 6n - 6$.*

This theorem gives a combinatorial condition for a *generic* body–bar–hinge framework to be minimally rigid. *Generic* means that it holds for most of the geometric realizations of the graph. However, it is possible that some very special situations may still be flexible, owing to specific geometric dependencies which cannot be detected by combinatorics only. *Minimal rigidity* means that the framework is rigid, but the removal of any constraint equivalent to one bar will turn it into a flexible structure.

For body–bar–hinge frameworks, the (6,6)-pebble game algorithm of [27] run on the associated graph determines if the framework is generically minimally rigid and, if not, computes its rigid components, DOFs, and overconstraints. The pebble game algorithm runs in time $O(n^2)$, where n is the number of vertices in the multigraph [27]. Internally, the algorithm operates on an auxiliary *directed* multigraph, called the pebble-game graph, as follows. The graph is initialized by placing six pebbles on each vertex, representing the six DOFs contributed by each body. Then edges are considered one at a time, in an arbitrary order. An attempt is made to insert the edge into the pebble-game graph. The acceptance rule is that at least seven pebbles must be present on the vertices that mark its endpoints. Special rules are applied in order to collect seven pebbles from neighbors, following the existing directed edges and a depth-first search strategy, but if this is not possible, then the edge is rejected and declared "overconstrained". Otherwise, the edge is declared "independent" and is inserted into the pebble-game graph. As this is done, a pebble is removed from one of the two endpoints, and the newly inserted edge is oriented away from the endpoint which has just lost a pebble. Figure 2 shows the execution of the pebble game on the example graph of Fig. 1. The algorithm also maintains collections of edges that cannot be extended further, and these correspond to rigid components (shown in red in the last snapshot in Fig. 2).

2.2 Modeling Molecules for Rigidity Analysis

We focus now on the modeling core of our software package KINARI. We use *modeling* to refer to the process and the rules for associating a body–bar–hinge framework to a molecule. The bodies, made of rigid groups of atoms, are determined first, then bar and hinge constraints are placed between them, depending on the type of bond. Figure 3 illustrates how bodies are determined from atoms connected by covalent bonds. Each multivalent atom, together with its bonded neighbors, forms a rigid body (outlined in the figure by blue and green pipes). When two bodies overlap, the overlap consists of two bonded atoms; they determine an axis which acts as a hinge (Fig. 3b). When two atoms are connected by a nonrotatable bond, such as a peptide or double covalent bond, an additional bar is placed between the two bodies in the mechanical model to lock the hinge, prohibiting rotation. Stabilizing noncovalent interactions, namely hydrogen bonds ("H-bonds") and hydrophobic interactions, can in principle be modeled in many ways, depending on how "strong" we believe the interaction to be. This is done by placing between one and six

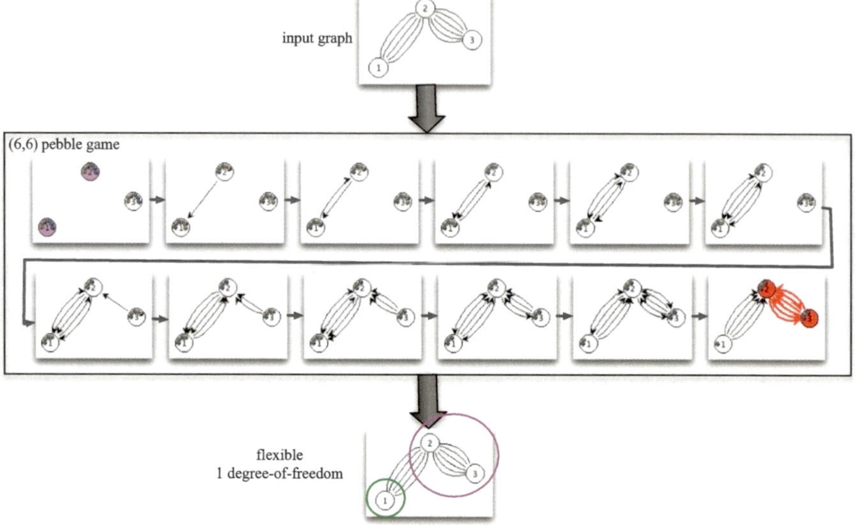

Fig. 2 Execution of the pebble game algorithm for determining rigidity of a 3D body–bar–hinge framework. The pebble game is run on an auxiliary directed graph. At the start of the game, six pebbles are placed on each vertex, representing the six degrees of freedom contributed by each body. When an edge is placed, a pebble is removed. In this example, vertices 2 and 3 have been found to belong to a single component (The images were generated with the Java applet from [28])

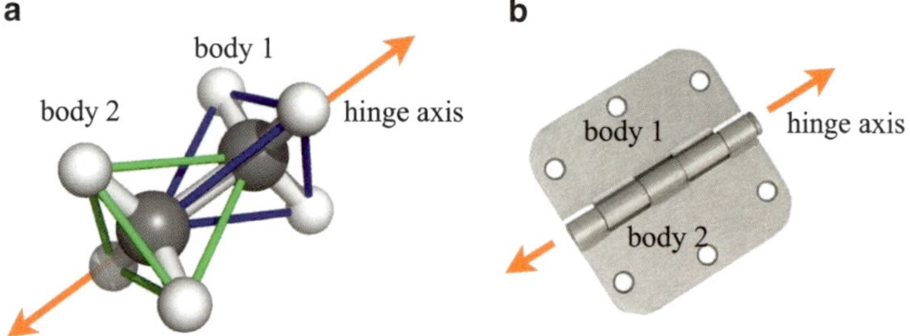

Fig. 3 Modeling the mechanics of a molecule. (**a**) Each body is composed of a carbon atom (*gray*) and its covalently connected neighbors. The two bodies are outlined in *green* and *blue*. Note that each C atom belongs to both bodies. The C–C bond is mechanically equivalent to the hinge shown in (**b**)

bars in the model. In KINARI, the default model used for H-bonds is the same as for covalent bonds. Hydrophobic interactions are modeled with the heuristic of ASU-FIRST [4], placing an interaction between C–C, C–S, or S–S pairs when their van der Waals surfaces are within a cutoff distance of 0.25 Å. By default, for each hydrophobic interaction, two bars are placed into the mechanical model. Since this kind of modeling may sometimes give biologically unrealistic results (e.g., by producing a protein that appears to be more rigid than is known from experimental data), KINARI allows the modeling of noncovalent interactions to be changed by the user, either at the level of individual interactions or using a cutoff value which eliminates those interactions considered to be too weak.

Once the modeling has been performed, the algorithm for rigidity analysis described in the previous section is applied. The resulting output provides a list of groups of atoms that remain rigidly attached to each other by the given interactions; these sets of atoms form larger rigid bodies, referred to as *rigid clusters*. These clusters are maximal; no other atom can be added to expand any of them (under the given set of interactions).

3   Literature Review

The most accurate computational methods for studying protein motion are based on all-atom physical simulations, called molecular dynamics (MD) simulations. At short simulation times, these methods reveal atomic fluctuations. Longer simulation times, long enough for domain-level conformation changes, are prohibitively expensive, computationally. Therefore, other, more efficient methods, which do not rely on simulation, have been proposed to probe the rigidity and flexibility of macromolecular structures.

When two distinct conformations are available, it has been proposed that a rigid cluster decomposition can be computed by comparing them. Heuristics for accomplishing this decomposition have been developed in HingeFind [41], DynDom[16], and RigidFinder [1]. A more challenging task is to compute a rigid cluster decomposition based on a single conformation, since most of the available data in the PDB is of this nature. The pebble game rigidity analysis described in the previous section is one such method. An alternative approach is based on normal mode analysis; this examines a single structure to detect the mobility of atoms by computing correlated motions [6]. It does not directly determine information about the rigidity and flexibility of structural regions, but several heuristics have been proposed to address this issue. Normal mode analysis may show that atoms in a certain domain, such as a mobile α-helix, move collectively, but it will not show that this region is rigid and unlikely to deform while other regions, which are flexible, will permit deformation.

Various other efforts have been made to quantify local flexibility in a protein. Flexibility indices have been proposed based on B-values from PDB files and normal mode analysis [23, 25, 37].

Pebble-game-based rigidity analysis was pioneered by Jacobs, Thorpe, and collaborators at Michigan State University, and implemented in the FIRST software package (now called Proflex) [19, 20]. A second implementation of FIRST, developed in Thorpe's laboratory at Arizona State University, was based on a different underlying model and a variation of the pebble game, and made available on the Flexweb server (http://flexweb.asu.edu) [4]. To distinguish between the two systems, we refer to the two versions as MSU-FIRST and ASU-FIRST. The FRODA method was included in ASU-FIRST to generate motions by moving rigid clusters and maintaining chemical constraints [38]. Recently, we have developed a comprehensive library for rigidity analysis of molecular structures, together with a Web application, KINARI-Web (http://kinari.cs.umass.edu) [9]. The modular design of our software allows easy extensions and tool development. A specific feature is the inclusion of several modeling options, allowing more freedom in exploring biological hypotheses and future benchmarking experiments. We have also recently released a library, KINARI-Lib, which contains an implementation of core data structures and algorithms for rigidity analysis [10]. A rigorous mathematical and computational presentation of pebble game algorithms is available [27].

One of the main differences between KINARI and ASU-FIRST is accuracy in modeling. KINARI builds a mechanical model where rigid bodies of atoms overlap on rotatable bonds behaving like hinges, as shown in Fig. 3. By contrast, ASU-FIRST models the protein using a special kind of multigraph where vertices represent the atoms and each edge represents the removal of a single degree of freedom between the atoms, skipping the mechanical modeling steps (steps 1 and 5 shown in Fig. 1). As a result, the rigid clusters computed by ASU-FIRST are disjoint and share no hinge joints. The rigid clusters identified by KINARI and ASU-FIRST are thus not identical, but when the same input PDB files, bonds and interactions, and modeling options are used, they will be in one-to-one correspondence. There is ongoing work on KINARI to investigate extensions that will further increase and stabilize the modeling accuracy, with the goal of obtaining a set of biologically validated modeling rules. In particular, we have used KINARI to investigate improved modeling accuracy of H-bonds and hydrophobic interactions [8]. Also, KINARI has been applied to predict deleterious effects of mutations [22] and to probe flexibility changes in crystals and biological assemblies [21].

Finally, we discuss those extensions of pebble game rigidity analysis which provide insight into cluster redundancy and deformability. The output of MSU-FIRST included a *flexibility index*, associated with each bond, to "characterize the degree of flexibility" [20]. We will describe the MSU-FIRST flexibility index in further detail in Sect. 5.4, where we compare it with our method for scoring rigid clusters. Gohlke et al. extended the MSU-FIRST flexibility index from a single protein to an MD trajectory, and used it to show changes in flexibility during protein docking [13].

A related method, made available in ASU-FIRST and on the Flexweb server, is *dilution analysis* [17, 33]. This has been interpreted as *simulated unfolding* because H-bonds are broken one by one, in order of energy. The rigid clusters of the protein are computed at each step, with the most stable part, called the *folding core*, remaining at the end. Dilution analysis was used to show that proteins undergo a rapid phase transition from rigid to floppy [17], to computationally identify the folding core of a protein [33], and to compare patterns of rigidity within homologues [11, 39].

Dilution analysis studies an ensemble of models which are hypothesized to reveal the unfolding path. Other efforts have been made to study an ensemble around the native state. The *duty cycle*, calculated from MD snaphots, the *duty cycle* is the percentage of time a particular interaction is present. It has been used as a criterion for determining which interactions to include in a rigidity analysis [26]. More recently, Gonzalez et al. proposed a heuristic method, called the virtual pebble game, for predicting the ensemble-averaged rigidity for a protein with fluctuating noncovalent interactions [14].

The method we propose here, *redundancy analysis*, has two goals. First, to assess the sensitivity of predicted clusters to small variations in the network of interactions. Second, to validate if *deformability* is an intrinsic property of protein rigid clusters. This approach is distinct from all previous extensions of rigidity analysis. Rather than studying unfolding, as was done with dilution analysis, our goal is to better understand the rigidity, stability, and flexibility properties of the protein in its native state. Since we are interested in finding explicit critical interactions, we cannot do this by sampling, which may miss them; instead, our method performs an exhaustive study of all the existing weak interactions, and classifies them.

4 Materials and Methods

In this section, we present our methodology for *redundancy analysis* of proteins. We introduce two algorithms, which are applied to a rigid cluster first for classifying redundant and critical noncovalent interactions and second for calculating a redundancy score. Later in this chapter, we will evaluate the method through case studies and surveys on several data sets. The resulting software package, KINARI-Redundancy, is available as an extension to the KINARI-Web server [9].

To apply redundancy analysis, we must preprocess the PDB file and select the modeling options. The KINARI curation tool allows the user to select atoms, ligands, chains, water molecules, etc. for the rigidity analysis. If the PDB file does not include hydrogen atoms, they can be added by the REDUCE software package [40], which is incorporated in KINARI. The curation tool then calculates the various atom–atom interactions, such as covalent bonds, hydrophobic interactions, and H-bonds, and assigns energies to the H-bonds. KINARI identifies H-bonds using the HBPLUS software package [31], and calculates the energy with the Mayo Lab function [30]. The curation tool discards the weaker H-bonds with energies smaller than a user-specified cutoff. Hydrophobic interactions are calculated using the same methodology as that developed in ASU-FIRST, and described as function H3 in the FIRST user guide [7]. For every carbon–carbon, carbon–sulfur, or sulfur–sulfur pair, if the van der Waals surfaces are within a cutoff distance of 0.25 Å, the curation tool places a hydrophobic interaction on the pair.

For each type of interaction, the user selects how to model the interaction from the available options of *hinge*, or 0 to 6 *bars*. As previously described, KINARI uses a body–bar–hinge model and the (6,6)-pebble game [27, 35]. A hinge models the particular type of constraint imposed by a covalent bond, fixing the bond length and bond-bending angle, but permitting rotation around the bond.

Algorithm: Classify redundant and critical interactions in a cluster

Input: *A single rigid cluster computed by rigidity analysis. The cluster is a protein fragment represented as a list of n atoms, covalent bonds and noncovalent interactions (hydrogen bonds, and hydrophobic interactions). Each interaction is labeled with how it is modeled (hinge or 1 to 6 bars).*

Output: *A set of interactions labeled as critical and redundant, with associated criticality values.*

Method: *For each noncovalent interaction in the cluster:*

1. *Remove interaction from the set and rerun rigidity analysis (as shown in Fig. 1).*
2. *If result is a single rigid cluster containing all n atoms, label the interaction as redundant. Otherwise, label the interaction as critical. Find the largest cluster in the result and let k be its number of atoms. Assign the interaction a criticality value of $(1 - k/n)$.*

Fig. 4 Algorithm for redundancy analysis of a rigid cluster

Once the protein has been curated and the modeling options specified, the rigidity analysis identifies the rigid clusters. Our two methods which work on each rigid cluster are described below. Also included is a description of the data set that will be used in our evaluation.

4.1 Identifying the Critical and Redundant Interactions Within a Cluster

A rigid cluster is a maximal set of atoms and of all bonds and interactions that hold them rigidly together. An interaction is *redundant* if its removal does not lead to the cluster becoming flexible. To identify the redundant interactions among the noncovalent interactions, we proceed as follows. One after another, we remove a noncovalent interaction, perform rigidity analysis, and verify if the cluster remains rigid, in which case we classify it as *redundant*. Otherwise, the interaction is classified as *critical*. Note that once an interaction has been classified, it is placed back in the cluster. This is different from the analysis performed during dilution analysis, where a removed interaction is not placed back. See Fig. 4 for a description of the classification algorithm. Another difference is that hydrophobic interactions are not involved in dilution analysis, only H-bonds, which have an associated energy.

To classify each interaction, we run the pebble game. In the worst case, this takes $O(n^2)$ time, thus the entire classification runs in cubic time.

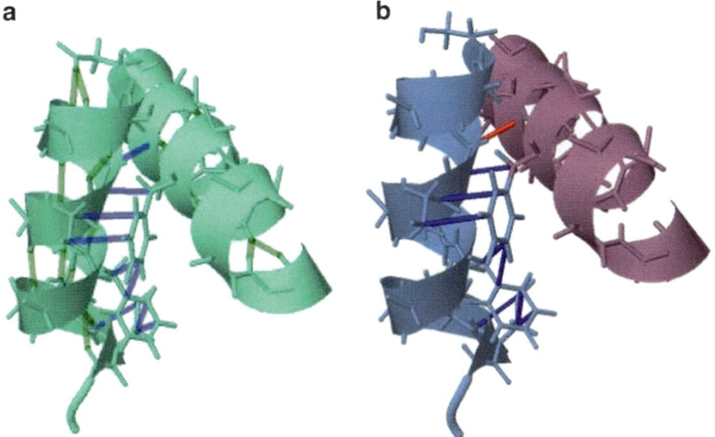

Fig. 5 (**a**) The largest rigid cluster of 1HRC, shown with H-bonds (*green*) and hydrophobic interactions (*blue*). (**b**) Removing the *red* hydrophobic interaction (with a criticality value of 0.44) from the largest rigid cluster of 1HRC causes the two α-helices in the cluster to break apart into separate rigid clusters (colors chosen at random)

To further distinguish the impact of each interaction on the overall stability of the protein, we define its *criticality value*, based on how much the cluster size is impacted. We measure the size of the rigid cluster once an interaction has been removed and the rigidity analysis rerun. The change in size of the cluster becomes the criticality value of the interaction. For example, the removal of an interaction with a criticality value of 0.10 will cause 10 % of the cluster's atoms to break off into one or more separate clusters. We are interested in the cases in which high-impact critical interactions occur.

To illustrate this, see Fig. 5, which shows the largest rigid cluster in cytochrome-c (1HRC). The redundancy of this protein will be discussed in a case study in Sect. 5.2. The cluster is composed of two α-helices bound together by hydrophobic interactions (shown in blue). The hydrophobic interaction colored in red is critical. When it is removed, each α-helix breaks off into its own rigid cluster. The criticality value of this interaction is 0.44.

4.2 Scoring of Clusters by Redundancy

We now define the *cluster redundancy score*, $\Phi(i)$ in Eq. 1 below, which requires the classification of noncovalent interactions. The set of all noncovalent interactions in cluster i is denoted by $N(i)$, and we denote the subset of $N(i)$ which are redundant by $R(i)$. Each interaction j is assigned a weight w_j, determined by how it is modeled. In the study described below, H-bonds and hydrophobic interactions had weights of 5 and 2, respectively. Theoretically, these values correspond to the maximum number of degrees of freedom that may be removed when the interaction is included in the mechanical model; in terms of our software, these values refer to the modeling of H-bonds as hinges (which place five edges in the associated Tay graph) and of hydrophobic interactions as two bars (which place two edges in the graph).

The formula defining the cluster redundancy score is

$$\Phi(i) = \frac{\sum_{j \in R(i)} w_j}{\sum_{k \in N(i)} w_k}. \tag{1}$$

If all of the noncovalent interactions within the cluster a redundant, the redundancy score is 1; when they are all critical, the redundancy score is 0.

4.3 Data Sets

We employed several different data sets in this current study.

Multiple conformations. In this first data set, we included the PDB files of proteins used in various publications for the validation of the MSU-FIRST software package [20]. The proteins were HIV-1 protease (1HHP, 1HTG), dihydrofolate reductase (1RA1, 1RX1, 1RX6), adenylate kinase (1AKY, 1DVR), and lysine–arginine–ornithine-binding (LAO-binding) protein (1LST, 2LAO). We curated the protein data using the KINARI-Web curation tool. Ligands were removed for all structures except that of adenylate kinase. Hydrogen atoms, bonds, and interactions were calculated with the default options, as described in [9]. Since 1HHP is a homodimer but only chain A is included in the PDB file, we applied a symmetry operation to compute the dimer [5]. Building the biological unit is available as an option from the KINARI-Web curation tool. We also employed a data set of 12 proteins used by the Gerstein Laboratory to validate the RigidFinder server [1]. We excluded five of the proteins in this data set that contained more than 500 residues.

Proteins with known foldons. We included a case study of cytochrome-*c*, a protein for which the foldons, intermediate structures which form during the folding process, are known [29].

Pdomain benchmark data set. In order for our survey to characterize the presence of redundant and critical interactions over a range of different proteins, we used the Pdomain Balanced Domain Benchmark 3 data set [18]. The original purpose of this data set was for benchmarking domain identification systems. We chose to use this set because of the good coverage of protein fold space. We excluded six PDB files with clusters larger than 6,000 atoms, so in total we included 121 PDB files in the analysis.

4.4 Redundancy Server

The redundancy analysis methods presented here can serve as a tool for investigating the robustness of rigidity results, in particular for users who wish to hand-edit their interaction set. We have deployed a server for redundancy analysis on the KINARI website, to accompany the KINARI-Web standard rigidity analyzer [9] and the KINARI-Mutagen tools [22]. Preprocessed examples and a video tutorial are available to facilitate use. KINARI-Redundancy provides the following functionality:

- To curate PDB data and assign modeling options, as supported by the KINARI-Web server [9].
- To color clusters according to their redundancy score.
- To examine one cluster at a time in further detail and to color critical and redundant interactions.
- To filter the set of interactions displayed. A threshold can be selected to show only interactions with higher criticality values.

5 Results and Discussion

We have applied our new methods to the data sets described in the previous section, and we present the results in the next three subsections. We then compare the redundancy score with the flexibility index of MSU-FIRST [20] and present a discussion of future applications of our method.

5.1 Analysis of Multiple Conformations

We performed redundancy analysis on the multiple-conformation data set described in Sect. 4.3. The rigidity analysis results for some of these proteins have been presented previously [20]. What we seek out here is the additional information that the *redundancy* analysis can give us about these proteins.

The table in Fig. 6 lists the results of our analysis. Most of the proteins had very few interactions with criticality values greater than or equal to 0.10. We investigate these outliers in case studies on adenylate kinase, dihydrofolate reductase, DNA polymerase β, and HIV-1 protease.

5.1.1 Adenylate Kinase

We first discuss the results on yeast adenylate kinase, a monomer known to undergo domain-level hinge motion upon ligand binding. Figure 7 shows decompositions of this monomer, depicted in the ATP-bound, open conformation (1DVR).

Protein	PDB	All		LRC		H-bonds in LRC				Hydrophobics in LRC			
		Num. Residues	Num. Atoms	Num. Atoms	max crit. val.	total	with criticality value:			max crit. val.	total	with criticality value:	
							>0	>0.10	>0.25			>0	>0.10	>0.25
MSU-FIRST data set													
HIV-1 protease	1HHP	198	3126	1802	0.02	134	36	0	0	0.01	96	54	0	0
	1HTG	198	3126	1791	0.36	134	31	14	0	0.02	87	48	0	0
Dihydrofolate Reductase	1RA1	159	2484	1683	0.11	122	37	13	0	0.1	98	37	2	0
	1RX1	159	2484	1606	0.06	122	27	0	0	0.01	110	40	0	0
	1RX6	159	2484	1461	0.1	118	21	1	0	0.1	99	43	0	0
Adenylate Kinase	1AKY	220	3469	2032	0.12	176	51	5	0	0.01	89	29	0	0
	1DVR	220	3435	1787	0.17	144	25	2	0	0.01	124	36	0	0
LAO-binding protein	1LST	238	3554	1224	0.07	112	22	0	0	0.02	35	20	0	0
	2LAO	238	3608	1289	0.09	120	25	0	0	0.03	50	27	0	0
Gerstein Lab data set													
Bungarotoxin	1IDG	74	1084	59	0.44	3	3	3	3	0	7	0	0	0
	1IDI	74	1085	51	0	0	0	0	0	0.41	8	3	3	1
Calmodulin	1CLL	147	2185	1068	0.43	109	31	20	17	0.41	47	27	9	9
	1CTR	147	2139	456	0.26	41	12	2	2	0.06	19	5	0	0
Cro repressor	5CRO	61	958	324	0.02	24	5	0	0	0.03	22	6	0	0
	6CRO	61	948	306	0.43	23	10	6	3	0.09	29	7	0	0
HIV-1 protease	3HVP	198	1534	771	0.28	48	24	7	3	0.12	77	30	4	0
	4HVP	198	1534	638	0.36	48	26	19	9	0.04	38	18	0	0
S100A6	1K9K	90	1426	841	0.09	85	21	0	0	0.02	23	7	0	0
	1K9P	90	1435	338	0.49	39	9	2	2	0.49	10	7	6	6
Alcohol Dehydrogenase	6ADH	374	1757	1757	0.08	85	24	0	0	0.1	181	55	3	0
	8ADH	374	3819	3819	0.07	284	74	0	0	0.07	243	99	0	0
Antigen 85C	1DQY	282	3360	3360	0.02	269	43	0	0	0.02	306	51	0	0
	1DQZ	282	3118	3118	0.02	254	48	0	0	0.01	219	41	0	0
Aspartate Aminotransferase	1AMA	410	3576	3576	0.05	303	62	0	0	0.02	181	48	0	0
	9AAT	410	3383	3383	0.03	317	63	0	0	0.01	147	46	0	0
Bacterio-rhodopsin	1BRD	226	1818	1818	0.04	136	16	0	0	0.02	152	57	0	0
	2BRD	226	2148	2148	0.06	208	34	0	0	0.03	146	56	0	0
DNA Polymerase Beta	2FMQ	335	3106	3106	0.29	273	74	9	3	0.29	144	64	7	4
	9ICI	335	2336	2336	0.03	147	25	0	0	0.02	222	56	0	0
Malate Dehydrogenase	1BMD	332	3136	3136	0.08	292	52	0	0	0.01	146	53	0	0
	4MDH	333	2939	2939	0.04	220	51	0	0	0.04	192	69	0	0
Adenylate Kinase	2ECK	214	444	444	0.09	34	7	0	0	0.02	29	9	0	0
	4AKE	214	549	549	0.71	43	26	18	12	0.71	38	17	4	4

Fig. 6 Prevalence of critical and redundant interactions in the largest rigid clusters (LRCs) of the MSU-FIRST and Gerstein Laboratory data sets. For each PDB files, the total number of residues and atoms are shown, as well as the size of the LRC. We classified each of the H-bonds and hydrophobics as either critical or redundant to its cluster's rigidity. The numbers of interactions with criticality values ≥ 0.0, ≥ 0.0, and ≥ 0.0 are displayed. There were no interactions with criticality values ≥ 0.50 in the data set

The domain containing the binding site is labeled as the LID domain. In early work of Jacobs et al. to validate the MSU-FIRST system, six flexible loops were detected in the open conformation (1DVR) [20]. Four of the six flexible loops were also detected in the closed conformation, 1AKY. The two loops not detected in 1AKY were those at the N- and C-terminals of the six α-helices (95 ILE to 108 GLN). The loops are labeled a–f in Fig. 7.

When the redundancy analysis was ran, two of the H-bonds (1 and 2) were found to decrease the size of the largest rigid cluster (LRC) by 17 and 12 %, respectively, and are shown in Table 1. H-bond 1 lies near the N-terminal, at the end of the parallel β-sheet, and connects the β-sheet to the f-loop, between 6 ARG O and 113 GLU H. H-bond 2 is between 105 LEU O and 110 THR H, and anchors the end of the six α-helices to the f-loop. Figure 7a shows the residues which engage in the two very

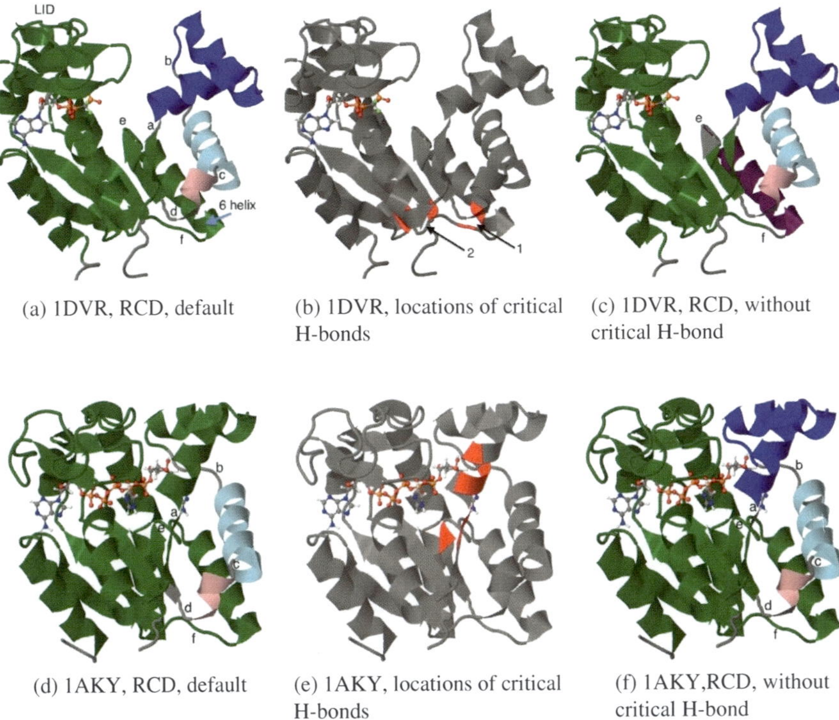

(a) 1DVR, RCD, default

(b) 1DVR, locations of critical H-bonds

(c) 1DVR, RCD, without critical H-bond

(d) 1AKY, RCD, default

(e) 1AKY, locations of critical H-bonds

(f) 1AKY, RCD, without critical H-bond

Fig. 7 Case study of adenylate kinase. (**a–c**) Open conformation 1DVR. (**a**) The rigid cluster decomposition (RCD) determined by KINARI v1.0. The *gray* regions are flexible, and each colored region is a rigid cluster. (**b**) H-bonds with criticality values of 0.17 (1) and 0.12 (2) were found in the largest rigid cluster (*green*). The residues which engage in these H-bonds are highlighted in *red*. (**c**) Upon removing H-bond 1, the e-loop and part of the f-loop become flexible. (**d**) Similarly, for H-bond 2, both the e- and the f-loops gain flexibility, although the region of flexbility in the f-loop is closer to the β-sheet than the α-helix (**d** and **e**) Closed conformation 1AKY. (**d**) Rigid cluster decomposition with default options. (**e**) Location of the five interactions which have the greatest impact on cluster size. (**f**) After removing a critical interaction

Table 1 Critical interactions in adenylate kinase (open, 1DVR). Two H-bonds (HB) and no hydrophobic interactions with criticality values ≥ 0.10 were detected

ID	Atom 1	Atom 2	Type	Energy	Criticality value
1	6 ARG O	113 GLU H	HB	−5.3	0.17
2	105 LEU O	110 THR H	HB	−0.58	0.12

critical interactions, highlighted in red. When either of these interactions is removed, the 6-alpha-helix breaks apart from the cluster and the e- and f-loops become flexible. The rest of the LRC remains intact. These two H-bonds were assigned energies of −0.58 and −5.3 kcal/mol by the Mayo Laboratory energy function [30].

Table 2 Critical interactions in largest rigid cluster of adenylate kinase closed conformation (1AKY). Five H-bonds (HB) and no hydrophobic interactions with criticality values ≥ 0.10 were detected

ID	Atom 1	Atom 2	Type	Energy	Criticality value
1	34 ALA O	38 ALA H	HB	−3.96	0.12
2	37 ASP OD1	40 ASP HH21	HB	−4.78	0.12
3	40 ARG HH12	301 AP5 O1E	HB	−2.22	0.12
4	40 ARG HH22	301 AP5 O1E	HB	−4.68	0.12
5	33 LEU O	89 LEU H	HB	−4.89	0.12

We investigated whether using a dilution analysis would identify these critical interactions [33]. In dilution, or simulated unfolding, H-bonds are removed one by one, cumulatively, in order of weakest to strongest, simulating H-bonds breaking during denaturation. Of the 146 H-bonds in the LRC, H-bond 2 was the 18th weakest, with an energy of −0.58 kcal/mol. H-bond 1, with an energy of −5.3 kcal/mol, ranked 109th, and would not have been revealed by a dilution analysis.

We next analyzed the closed conformation, 1AKY. With the default options, KINARI predicts a larger LRC than for 1DVR, the open conformation. The MSU-FIRST software package detected four flexible loop regions (b, c, d, and f) [20]. With the default options, KINARI detected three of these loops (b, c, and d). Our redundancy analysis detected five H-bonds with criticality values greater than or equal to 0.12, listed in Table 2. When any of these interactions are removed, the resulting RCD contains a flexible region in the a-loop.

To summarize, the LRCs of 1AKY and 1DVR both contained multiple interactions with criticality values greater than or equal to 0.10, and these were concentrated together in the structure. In 1DVR, the removal of either of the two very critical interactions caused the same two loops to gain flexibility concurrently. In 1AKY, the removal of any of the five very critical interactions all caused the a-loop to become flexible. These loops were determined to be important to the flexibility and mobility that are required for ADK to perform its function.

5.1.2 Dihydrofolate Reductase

1RA1, 1RX1, and 1RX6 are the structures of the open, closed, and occluded conformations, respectively, of *E. coli* dihydrofolate reductase, a small enzyme which plays an essential role in building DNA. The flexibility of the Met20 loop (residues 9–24) and the βF–βG loop (residues 116–132) near the active site plays a role in promoting the release of the product. Figure 8a shows an alignment of the three DHFR structures, demonstrating the high mobility of the Met20 loop.

The KINARI rigid cluster decompositions of the three structures show the protein to be mostly rigid, with most of the protein contained in the LRC. Each of the decompositions shows flexible regions in the βF–βG loop. In 1RX1, the closed

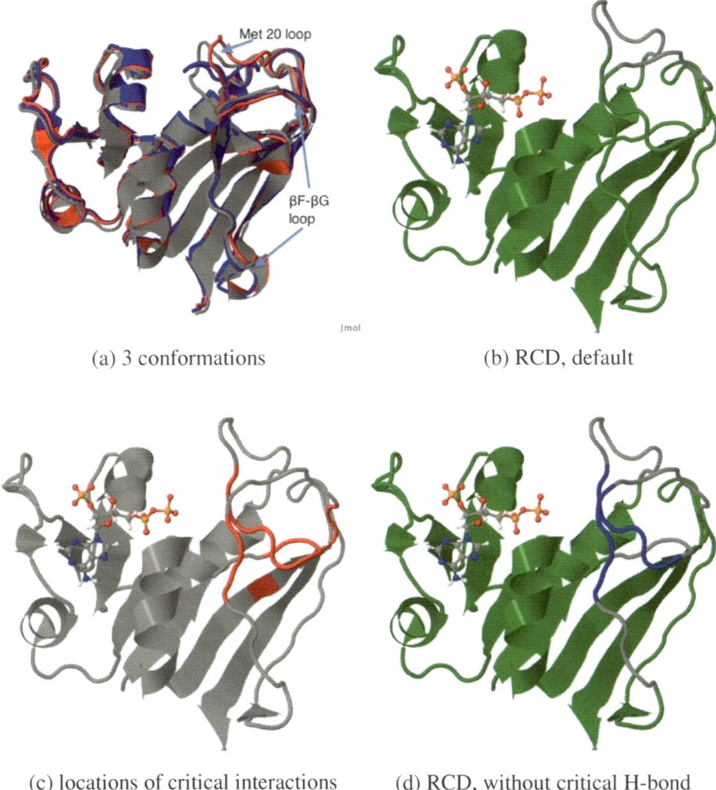

| (a) 3 conformations | (b) RCD, default |
| (c) locations of critical interactions | (d) RCD, without critical H-bond |

Fig. 8 Case study of dihydrofolate reductase. (**a**) A 3D alignment of three conformations (1RA1, 1RX1, and 1RX6) shows the high mobility of the Met20 loop. (**b**) With the default options, the rigidity results for 1RA1 show flexibility in the Met20 loop, but the βF–βG loop is almost entirely rigid. (**c**) The residues which engage in very critical interactions in the largest rigid cluster of 1RA1 are almost all within the Met20 and βF–βG loops. (**d**) After removal of a very critical interaction, a smaller cluster (*blue*) composed of part of the Met20 and βF–βG loops breaks off from the LRC, and a region in the βF–βG loop becomes flexible

conformation, the Met20 loop was determined to be locked – it is contained in the largest rigid cluster. In 1RA1 and 1RX6, the open and occluded conformations, some regions of the Met20 loop were determined to be flexible, but there was some variation between the two RCDs. These results agree with earlier results of Jacobs et al. obtained using the MSU-FIRST rigidity analysis software package [20].

Table 6 (see Sect. 5.2) includes the results of our redundancy analysis on these three conformations. The closed conformation contained no interactions with criticality values greater than or equal to 0.10, and the occluded conformation contained only one. In the open conformation (1RA1), over 10 % of the H-bonds (13 of 122) and two of the hydrophobic interactions had criticality values greater

Table 3 Critical interactions in dihydrofolate reductase (1RA1). Thirteen H-bonds (HB) and twelve hydrophobic interactions (HP) with criticality values ≥ 0.10 were detected. The type, energy (kcal/mol), and criticality value for each interaction are shown. Also, the locations of both atoms involved in the interaction, with respect to the Met20 and βF–βG loops are shown. Any of these interactions, when removed, leads to increased flexibility in the βF–βG loop, whether or not the two atoms a located within the loops. See also Fig. 8

ID	Atom 1	Atom 2	Type	Energy	Criticality value	In Met20 or βF–βG loop?
1	8 LEU H	113 LEU O	HB	−4.84	0.10	Neither
2	116 ASP H	150 ASP O	HB	−7.07	0.10	Both in βF–βG loop
3	8 LEU O	115 LEU H	HB	−5.58	0.10	Neither
4	115 ILE O	117 ILE H	HB	−1.74	0.10	One in βF–βG loop
5	9 ALA H	13 ALA O	HB	−5.77	0.10	Both in Met20 loop
6	10 VAL O	13 VAL H	HB	−2.69	0.10	Both in Met20 loop
7	10 VAL H	117 VAL O	HB	−6.33	0.10	Met20 and βF–βG loop
8	12 ARG O	125 ARG H	HB	−5.4	0.10	Met20 and βF–βG loop
9	12 ARG HE	125 ARG O	HB	−6.79	0.10	Met20 and βF–βG loop
10	14 ILE H	123 ILE O	HB	−6.91	0.11	Met20 and βF–βG loop
11	15 GLY O	122 ASP H	HB	−1.81	0.11	Met20 and βF–βG loop
12	15 GLY O	123 GLY H	HB	−2.02	0.11	Met20 and βF–βG loop
13	15 GLY H	123 GLY O	HB	−2.38	0.11	Met20 and βF–βG loop
14	11 ASP C	12 ARG CG	HP	N/A	0.10	Both in Met20 loop
15	123 THR C	124 HIS CG	HP	N/A	0.10	Met20 and βF–βG loop

than 0.10. These very critical interactions are all concentrated around the active site, adjacent to the Met20 and βF–βG loops. Table 3 lists the set of very critical H-bonds and hydrophobic interactions and their locations. The majority (7 of the 13) of very critical H-bonds connect the Met20 and βF–βG loops. The remainder of the H-bonds are in the local area of the two mobile loops. Removing any of these H-bonds increases the extent the flexible region in the Met20 or βF–βG loop. None of these very critical interactions involve the ligand.

For example, when we remove the H-bond between atoms 8 LEU H and 113 LEU O, the flexible regions of the Met20 loop and βF–βG loop increase substantially, as depicted in Fig. 8d, even though this particular H-bond does not involve residues within those loops.

The LRCs of 1RX1 (closed) and 1RX6 (occluded) contain few or no H-bonds with high criticality values. These structures already have more flexibility in the βF–βG loop.

The prevalence of very critical interactions in the active site region for the open conformation shows that the rigidity of the βF–βG loop is "hanging by a thread". These H-bonds tend to be strong, and mostly backbone–backbone H-bonds, so they are unlikely to break independently. But snipping any of these very critical interactions will cause the βF–βG loop to gain flexibility.

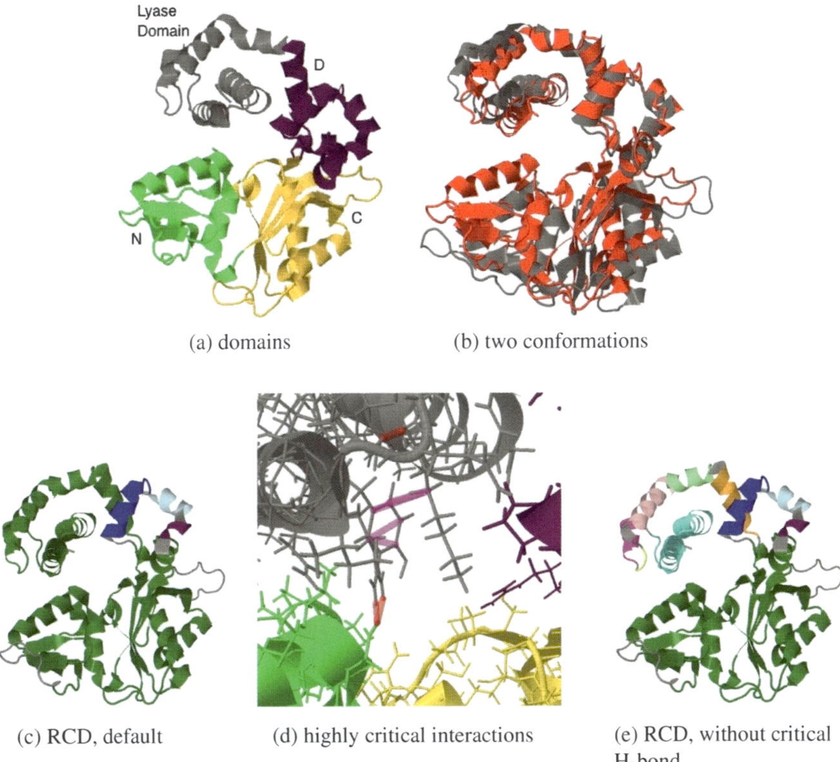

(a) domains

(b) two conformations

(c) RCD, default

(d) highly critical interactions

(e) RCD, without critical H-bond

Fig. 9 Case study of DNA polymerase β (2FMQ). (**a**) Four important functional domains. The lyase domain (*gray*) is the important catalytic site. The N domain interacts more strongly with the lyase domain in the closed conformation. (**b**) 3D structural alignment of 2FMQ (*red*) and 9ICI (*gray*). (**c**) The rigid cluster decomposition of 2FMQ shows one dominant rigid cluster (*green*) containing both the lyase and the N domains. *Gray* regions are flexible. (**d**) If any of the H-bonds (*red*) and hydrophobic interactions (*pink*) shown are removed, the lyase domain breaks off from the dominant rigid cluster, as shown in (**e**). Note that only two of these interactions actually crossbrace between the lyase domain and the other domains in the protein; the rest lie completely in the lyase domain. Therefore, the loss of an interaction within the lyase domain will cause the separation of the lyase domain from the rest of the cluster

5.1.3 DNA Polymerase β

DNA polymerase β (POLB) is a 335-residue DNA- and metal-binding enzyme, responsible for base excision repair of DNA. It is active as a monomer and is composed of an N-terminal 90-residue lyase domain connected to a C-terminal polymerase domain, composed of three subdomains [2]. The lyase domain and three subdomains are depicted in Fig. 9a. We examined the redundancy of one conformation of POLB, 2FMQ. The rigidity analysis determined that the LRC

Table 4 Critical interactions in the largest rigid cluster of DNA polymerase β (2FMQ). Three H-bonds (HB) and four hydrophobic interactions (HP) with criticality values ≥ 0.25 were detected. The type, energy (kcal/mol), and criticality value for each interaction are shown

ID	Atom 1	Atom 2	Type	Energy	Criticality value
1	26 GLU O	32 GLU H	HB	−6.98	0.29
2	40 ARG HH12	276 ARG OD2	HB	1.08	0.29
3	40 ARG HH22	276 ARG OD2	HB	−1.61	0.29
4	27 LYS CB	36 LYS CG	HP	N/A	0.29
5	27 LYS CB	36 LYS CD1	HP	N/A	0.29
6	36 TYR CE1	40 TYR CD	HP	N/A	0.29
7	36 TYR CZ	40 TYR CD	HP	N/A	0.29

contains 3,106 atoms. When redundancy analysis was performed on the cluster, 200 of 274 H-bonds were determined to be redundant and 80 of 144 hydrophobic interactions were determined to be redundant, leading to a redundancy score of 0.70. Three of the H-bonds and four of the hydrophobic interactions had criticality values of 0.29 or greater (see Table 4).

When any of these seven interactions was removed, the lyase domain broke off from the largest rigid cluster. The lyase domain did not form a single cluster, but instead shattered into many smaller clusters. For example, when we removed the H-bond between 26 GLU O and 32 GLU H and reran our redundancy analysis, we obtained the results shown in Fig. 9e. The large rigid cluster now remaining contained no interactios with criticality value greater than 0.03. The results of removing any of the other six critical interactions were similar. The 2FMQ conformation is more closed than the 9ICI conformation. We repeated the redundancy analysis on 9ICI and found that the rigid clusters were in good agreement with the clusters of 2FMQ after the removal of any of the critical interactions (as in Fig. 9e). For the largest rigid cluster of 9ICI, composed of 2,336 atoms, the maximum criticality value of any of its consituent interactions was 0.03.

To summarize, the decomposition determined by KINARI, run with the default options on the closed conformation (2FMQ), is very rigid, and the lyase domain is included in a large rigid cluster spanning other functional domains of the protein. With our redundancy analysis, we found seven interactions with very high criticality values (0.29 or greater). When any of these interactions were removed, the lyase domain decoupled from the other domains and became very flexible, better matching the rigidity results for the open conformation (9ICI).

5.1.4 HIV-1 Protease

Our analysis of two conformations of HIV-1 protease, 1HHP (open) and 1HTG (closed), uncovered interesting differences due to asymmetries in the 1HTG dimer. The PDB file for 1HHP contains only a single chain, and therefore we computed

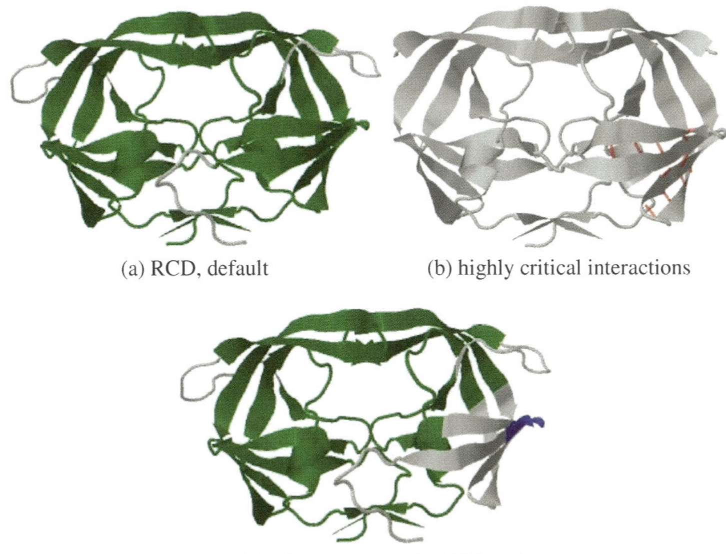

(a) RCD, default   (b) highly critical interactions

(c) RCD, without critical H-bond

Fig. 10 Ten interactions with criticality values ≥ 0.10 were found in PDB 1HTG. (**a**) Before removing any interactions. (**b**) The interactions with criticality values ≥ 0.10 in the LRC are all H-bonds in the β-sheet of chain A. (**c**) After the removal of any of these interactions, the β-sheet of chain A breaks off from the LRC and becomes flexible

the positions of the atoms in chain B, resulting in a dimer that was completely symmetric. The rigidity analysis results showed a large rigid cluster consisting of 78 of the 99 residues in each chain (residues 1–14, 19–36, 43–45, and 56–98). No interactions with a criticality value greater than 0.10 were found.

For 1HTG, which was crystallized as a dimer, both chains are already included in the PDB file. The results of the rigidity analysis of 1HTG reflect some of the asymmetries in the two chains (see Fig. 10). With the default options, the largest rigid cluster contained 81 residues from chain A (residues 9–33 and 43–98) and 90 residues from chain B (residues 1–33 and 43–98). More differences in the rigidity properties of the two chains were detected when the redundancy of the structure was analyzed. Fourteen interactions were found with criticality values greater than or equal to 0.25 (see Table 5). All of these critical interactions were found in a β-sheet of chain A. Owing to asymmetries between chains A and B, the sets of H-bonds and hydrophobic interactions were not the same in the two chains. For example, in chain B, 65 GLU CG and 68 GLY C were a distance of 3.60 Å and fell within the cutoff distance for a hydrophobic interaction. The same pair of atoms in chain B were a distance of 5.18 Å, much greater than the 3.65 Åcutoff distance, and no hydrophobic interaction was placed on them.

Table 5 Critical interactions in the largest rigid cluster of HIV-1 protease (1HTG). Fourteen H-bonds (HB) with criticality values ≥ 0.25 were detected, all in chain A. The type, energy (kcal/mol), and criticality value for each interaction are shown

ID	Atom 1	Atom 2	Type	Energy	Criticality value
1	9 PRO O	24 LEU H	HB	−7.22	0.36
2	11 VAL O	22 ALA H	HB	−5.82	0.36
3	11 VAL H	22 ALA O	HB	−5.53	0.36
4	13 ILE O	20 LYS H	HB	−5.98	0.27
5	13 ILE H	20 LYS O	HB	−5.13	0.27
6	14 LYS O	65 GLU H	HB	−6.17	0.27
7	14 LYS H	65 GLU O	HB	−3.34	0.27
8	15 ILE H	18 GLN O	HB	−6.89	0.27
9	64 ILE H	71 ALA O	HB	−6.28	0.27
10	62 ILE O	73 GLY H	HB	−4.10	0.27
11	62 ILE H	73 GLY O	HB	−5.52	0.27
12	64 ILE O	71 ALA H	HB	−5.57	0.27
13	66 ILE O	69 HIS H	HB	−0.82	0.27
14	66 ILE H	69 HIS O	HB	−4.94	0.27

5.2 Correlating Redundancy and Foldons; Case Study of Cytochrome-c

Figure 11a shows the five largest rigid clusters of cytochrome-c (1HRC). The rigidity of this protein has been previously investigated [33, 36, 39]. Here, we describe a refinement of the analysis from the point of view of redundancy (Table 6). When we focus on the largest rigid cluster (blue in Fig. 5a and shown in Fig. 5b), composed of two α-helices, we can identify 24 H-bonds (shown in green) and 10 hydrophobic interactions (shown in blue). Each α-helix is held together by H-bonds, while the hydrophobic interactions effectively "zip up" the two α-helices and hold them rigidly together. Redundancy analysis determined that 25 % of the H-bonds and 40 % of the 10 hydrophobic interactions were critical. Removing any of these critical interactions will cause the cluster to break up and become flexible. Most of the interactions that we have labeled as "critical" do not have a large impact on the cluster size. For each of the critical noncovalent interactions, we monitored how the original cluster size of 251 atoms decreased when the critical interactions were removed. For all of the H-bonds, the cluster size after removal remained at at least 86 % of its original value. For three of the four hydrophobic interactions, the cluster size remained at at least 98 %. For one hydrophobic interaction, the cluster size dropped to 56 %. Figure 5b shows how the cluster rigidity is affected when this particular interaction is removed. This demonstrates how each critical interaction may have a different degree of impact on a cluster's rigidity when it is removed.

Extensive studies have been undertaken to understand the folding kinetics of this protein. The foldons, intermediate structures which form during the folding process, and the order in which they form have been experimentally identified using HX

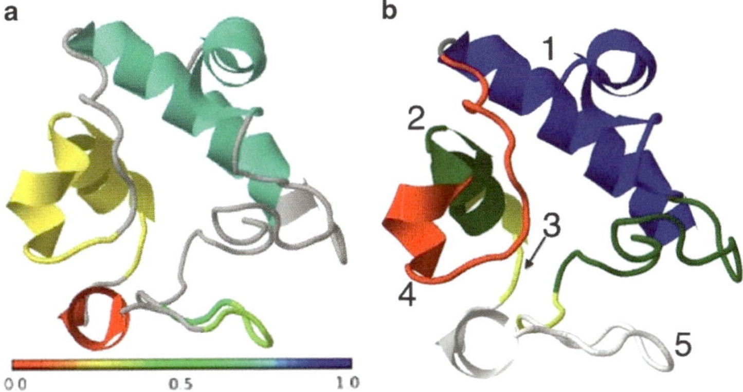

Fig. 11 Case study of cytochrome-c (1HRC), where the rigid clusters and foldons are compared. (**a**) The five largest rigid clusters of 1HRC are colored by their redundancy score, from least redundant (*red*) to most redundant (*blue*). (**b**) Experimentally determined foldons are numbered by their stepwise folding order

Table 6 Redundancy scores for the five largest rigid clusters of 1HRC, shown in Fig. 11a. Listed for each cluster are the number of atoms, the number of redundant H-bonds and the total number of H-bonds in the cluster, the corresponding numbers for the hydrophobic interactions, our calculated redundancy score, and the greatest decrease in size observed after removing an interaction

Cluster	Atoms	HB redundant/all	HP redundant/all	Score	Max. criticality value
Blue	251	18/24	6/10	0.73	0.44
Left yellow	194	2/13	6/12	0.25	0.55
Red	80	0/4	0/0	0.00	0.88
Green	35	0/1	5/6	0.59	0.40
Right yellow	30	0/1	2/3	0.36	0.34

experiments [29]. The foldons are shown in Fig. 11b in blue (residues 1–19 and 87–105, i.e., the N- and C-terminal α-helices), green (residues 60–70 and 19–36, the α-helix and V-loop), yellow (residues 37–39 and 58–61, the short two-stranded antiparallel β-sheet), red (residues 71–85, the V-loop), and white (residues 40–57, the V-loop). Although the helix and V-loop marked in green appear disconnected, the two parts engage in hydrophobic interactions between the side chains PHE 36 and LEU 64. The HEME ligand is not shown in the picture.

Visually, there is some nice agreement between these experimentally identified foldons and those determined by KINARI. In particular, the first foldon, i.e., the N- and C-terminal α-helices, correlates well with the most redundant rigid cluster found in our analysis. The α-helix of the second foldon and the entire third and fourth foldons (yellow and red) lie in the yellow cluster, which has a lower redundancy. The last foldon, white, has not been placed in a single cluster but was instead determined to be flexible and lies in a number of clusters. A nice result is that the redundancy scores calculated (Table 6) also correlate with the foldon order.

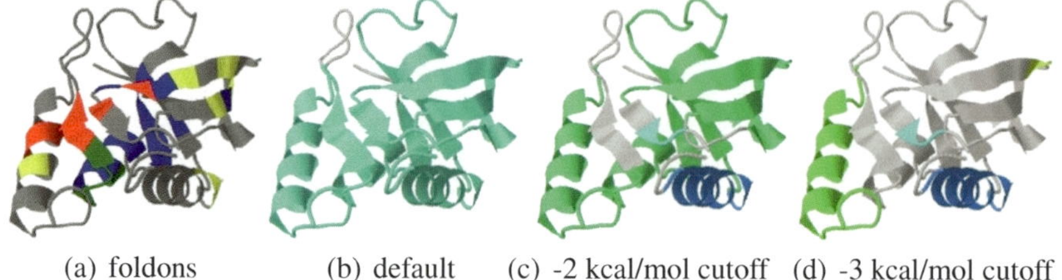

(a) foldons   (b) default   (c) -2 kcal/mol cutoff   (d) -3 kcal/mol cutoff

Fig. 12 Case study of SNase protein (1SNP). Here, we compare the experimentally determined foldons of 1SNP with the KINARI rigid cluster decompositions. (**a**) Experimentally determined foldons. (**b**) With default options, the entire protein is determined to be almost completely rigid with a redundancy score of 0.72. (**c**) After removal of the H-bonds using a cutoff of -2 kcal/mol, the minor β-strands are determined to be flexible, while the β-barrel remains in the largest rigid cluster. (**d**) With a cutoff of -3 kcal/mol, the β-barrel no longer lies in a larger rigid cluster, and two of the α-helices have remained rigid. Rigidity and redundancy analysis do not appear to give strong insight into these SNase foldons, unlike the case of cytochrome-c, which is all alpha, and for which there has been success in determining foldons

Although this case study of cytochrome-c demonstrates that KINARI with the default parameters can assist in identifying these foldons, further analysis is needed for us to know if this extends to other proteins for which foldons are known. SNase is a 149-residue mixed α/β protein with three α-helices, a major five-stranded β-barrel, and three minor β-strands. It is composed of five foldons, the first foldon of which is composed primarily of the β-barrel. Analysis of SNase (PDB file 1SNP) [3] using the default parameters revealed an almost entirely rigid structure, with no differentiation between the structural elements identified to form foldons (Fig. 12). Excluding the weaker H-bonds, which is the conventional way to tune rigidity results, leads to the β-barrel losing rigidity before other foldon regions, which does not agree with the experimental data.

A more thorough investigation to correlate foldon stability and redundancy remains to be done.

5.3   *Survey of a Pdomain Benchmark Data Set*

We calculated critical and redundant interactions for the largest rigid cluster of each protein in the Pdomain benchmark 3 data set described earlier in Sect. 4.3. To get a better sense of how redundancy and the presence of critical interactions correlate with size, we divided the 121-protein data set into three parts according to the size of the LRC: small (fewer than 500 atoms, 12 % of data set), medium (500–1,000 atoms, 20 % of data set), and large (greater than 1,000 atoms, 69 % of data set).

- *Redundancy scores.* Overall, the mean redundancy score (and standard deviation, s) was 0.74 (s $=$ 0.10). The small clusters had a lower mean redundancy, 0.66 (s $=$ 0.17), than the medium clusters, with a redundancy score of 0.73 (s $=$ 0.11), and the large clusters, with a redundancy score of 0.75 (s $=$ 0.07), showing a trend in which the redundancy score increases and the variance decreases with cluster size. Therefore, the larger rigid clusters were shown to be more robust, and less sensitive to changes in rigidity.

- *Prevalence of highly critical interactions.* Figure 13 shows the cumulative distributions of clusters containing interactions with increasing criticality values. Although virtually all of the clusters contained some critical interactions, most of the clusters did not contain interactions with criticality values greater than or equal to 0.10. Interactions with criticality values greater than or equal to 0.50 occurred in 13 % of the small and medium-sized clusters, but were quite rare in the large clusters, occurring in fewer than 4 % of them.

5.4 *Comparison with Other Techniques*

MSU-FIRST included a feature to assign each bond a *flexibility index*, using a count of the redundant constraints [20]. In the underlying bar-and-joint model of the protein used by MSU-FIRST, each bond was represented by a number of constraints: *central-force constraints* for holding bond lengths, and *external constraints* for holding bond-bending and dihedral angles. Using the MSU-FIRST pebble game, the input macromolecule was decomposed into three types of regions: *isostatically rigid*, *overconstrained*, and *underconstrained*. An isostatically rigid region was a rigid cluster in which the removal of any of the constraints would cause the cluster to become flexible. An overconstrained region was a rigid cluster in which at least one of the constraints was redundant. An underconstrained region was a region in which placing any additional bond-bending or dihedral-angle constraints would cause a rigid cluster to form.

Once the different regions had been identified, Eq. 2 below was used to calculate the flexibility index for each bond [20]. Here, H_k and F_k are the number of rotatable bonds and the number of degrees of freedom, respectively, in the kth underconstrained region. C_j and R_j are the numbers of bonds and redundant constraints, respectively, in the jth overconstrained region. The flexibility index is negative for overconstrained regions and positive for underconstrained regions. The equation for this index is

$$f_i \equiv \begin{cases} \frac{F_k}{H_k} & \text{in an underconstrained region,} \\ 0 & \text{in an isostatically rigid region,} \\ \frac{-R_j}{C_j} & \text{in an overconstrained region.} \end{cases} \tag{2}$$

The flexibility index was defined for bonds in any region of the protein, not just the rigid clusters. In order to directly compare our redundancy score, we transformed the flexibility index formula to an equation for scoring rigid clusters,

$$\Psi(i) = \frac{R_i}{C_i}. \tag{3}$$

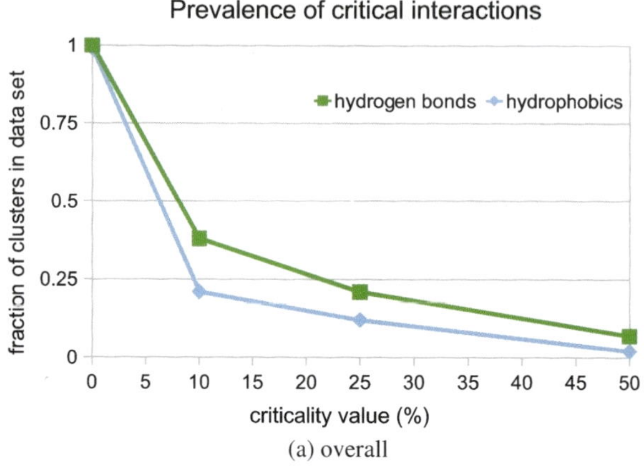

(a) overall

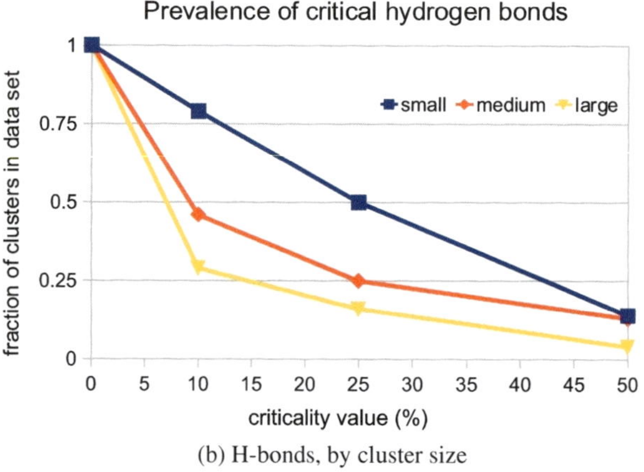

(b) H-bonds, by cluster size

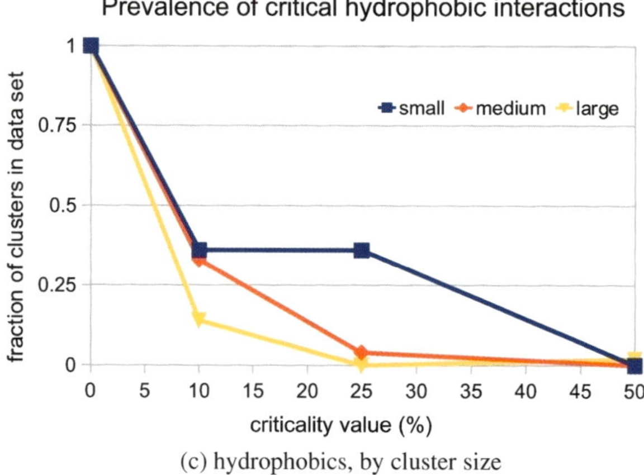

(c) hydrophobics, by cluster size

Fig. 13 Prevalence of critical interactions in the Pdomain benchmark data set. (**a**) Overall comparison of occurrence of critical H-bonds and hydrophobic interactions. (**b–c**) Occurrence of critical H-bonds and hydrophobic interactions for small, medium, and large clusters

Here, $\Psi(i)$ is the MSU-FIRST redundancy score for cluster i, and R_i and C_i are the numbers of redundant and central-force constraints in the cluster.

This scoring formula and ours (Eq. 1) are not equivalent. We can demonstrate this with a small example. A cluster consisting of a ring of five atoms connected by four single covalent bonds and one H-bond is assigned a score of 0 by our method. This score signifies that there is no redundancy in the set of noncovalent interactions and that if the H-bond were removed, the cluster would break. MSU-FIRST assigns the same cluster a score of $1/5$ (one redundant constraint and five bonds). Unlike our approach, the MSU-FIRST approach does not detect the critical interactions.

ASU-FIRST also included a flexibility index that did not use any information about redundancy. The index for bonds in rigid clusters was based on the size of the cluster [7].

6 Conclusion and Further Directions

Motivated by the need to understand the sensitivity of rigid clusters to changes in the set of noncovalent interactions, we have proposed a method for classifying the noncovalent interactions as *critical* or *redundant*. An interaction is *critical* if, when it is removed, the cluster it is contained in breaks up and becomes flexible. We have also proposed a method to score clusters using the redundancy of the noncovalent interactions. We have implemented these methods in the KINARI-Redundancy extension to KINARI, our protein rigidity analysis software package. We have provided results of our classification and scoring for the clusters obtained from a data set of PDB files. This method may be applicable to other related questions previously posed in the literature.

- *Energy.* Energy functions provide a way to compare the relative strengths of H-bonds. Different functions have been proposed that calculate the energy required to break an H-bond, using the local bond geometry [24, 30]. We may infer that the stronger interactions, the ones that require more energy to break, are the critical ones.
- *Flickering.* The *flickering* phenomenon is the forming and breaking of interactions at varying rates during the natural fluctuations of a protein about the native state [26]. The *duty cycle*, which is the percentage of time a particular interaction is present, may be used similarly to the energy, to rank interactions by how likely they are to break. An advantage of the duty cycle concept is that it can be generalized to any interaction which may break and form. This is especially valuable for hydrophobic interactions, since the associated energy is unknown.
- *Evolutionary conservation.* Using proteins within the same family with high structural conservation, it has been shown that even when there is low sequence identity, a network of hydrophobic interactions between residues is conserved [15]. The conserved interactions may be considered the "critical"

ones, and, similarly to the high-duty-cycle interactions, identifying conserved interactions does not rely on computing an energy function.

- *Other flexibility index methods.* Another type of flexibility index has been computed from the amino acid sequence using parameters derived from B-values from a training set of PDB files [23, 37]. Yet another method has used normal mode analysis to calculate the local chain deformability for each residue along the backbone [25]. This method was shown to produce results comparable to the MSU-FIRST flexibility index in a case study on 16pk kinase [25].
- *Dilution.* Dilution analysis reveals an unfolding pathway of a protein by removing H-bonds one by one and performing rigidity analysis. It would be interesting to correlate the set of H-bonds identified as critical using our method with those which cause the greatest changes in the rigidity during dilution. Dilution analysis was used as a tool to show the coordinated states of thermophilic and mesophilic protein homologues [11]. This study used measures of the rigidity properties to identify the transition point from flexible to rigid, and showed that in two-thirds of the proteins in the data set, the transition point occurred at a higher temperature in the thermophilic than in the mesophilic homologue. The classification of the critical and redundant interactions may be used as additional information to improve the order in which the bonds are removed in dilution analysis, which may lead to more consistently corresponding transition points.
- *Computational efficiency.* Our algorithm for classifying all interactions as critical or redundant, described in Sect. 4.1, takes in the worst case cubic time in the number of atoms, so the method does not scale well to proteins with more than 500 residues. We have shown that owing to the rarity of very critical interactions, a uniform sampling approach is inadequate. Because these critical interactions tend to be concentrated together, however, a targeted sampling approach may be sufficient if some knowledge of the structure is available a priori. Another reasonable approach to speeding up the algorithm would be to devise a method that leverages common intermediate states of the pebble game, so that a new run of the pebble game would not need to be performed for each interaction.

Acknowledgements The studies described in this chapter were funded by the National Institute of General Medical Sciences grant DMS-0714934 as part of the Joint Program in Mathematical Biology supported by the Directorate for Mathematical and Physical Sciences of the National Science Foundation and the National Institute of General Medical Sciences of the National Institutes of Health, and by the Mathematical Challenges grants NSF CCF-1016988 and DARPA to IS.

References

1. A. Abyzov, R. Bjornson, M. Felipe, M. Gerstein, RigidFinder: a fast and sensitive method to detect rigid blocks in large macromolecular complexes. Proteins **78**(2), 309–324 (2010)
2. W.A. Beard, S.H. Wilson, Structure and mechanism of DNA polymerase beta. Chem. Rev. **106**(2), 361–382 (2006)

3. S. Bedard, L.C. Mayne, R.W. Peterson, A.J. Wand, S.W. Englander, The foldon substructure of staphylococcal nuclease. J. Mol. Biol. **376**, 1142–1154 (2008)
4. M.V. Chubynsky, B.M. Hespenheide, D.J. Jacobs, L.A. Kuhn, M. Lei, S. Menor, A.J. Rader, M.F. Thorpe, W. Whiteley, M.I. Zavodszky, Constraint theory applied to proteins. Nanotechnol. Res. J. **2**, 61–72 (2008)
5. P. Clark, J. Grant, S. Monastra, F. Jagodzinski, I. Streinu, Periodic rigidity of protein crystal structures, in *2nd IEEE International Conference on Computational Advances in Bio and Medical Sciences (ICCABS'12)*, Las Vegas, Feb 2012
6. Q. Cui, I. Bahar (eds.), *Normal Mode Analysis: Theory and Applications to Biological and Chemical Systems* (Chapman & Hall/CRC, Boca Raton, 2006)
7. Floppy Inclusions and Rigid Substructure Topography (FIRST) 6.2.1 User Guide, Oct 2009, available online http://flexweb.asu.edu
8. N. Fox, I. Streinu, Towards accurate modeling of noncovalent interactions for protein rigidity analysis. BMC Bioinform. (2013, to appear)
9. N. Fox, F. Jagodzinski, Y. Li, I. Streinu, KINARI-Web: a server for protein rigidity analysis. Nucleic Acids Res. **39**(Web Server Issue), W177–W183 (2011)
10. N. Fox, F. Jagodzinski, I. Streinu, Kinari-lib: a C++ library for pebble game rigidity analysis of mechanical models, in *Minisymposium on Publicly Available Geometric/Topological Software*, Chapel Hill, 17–19 Jun 2012
11. H. Gohlke, S. Radestock, Exploiting the link between protein rigidity and thermostability for data-driven protein engineering. Eng. Life Sci. **8**, 507–522 (2008)
12. H. Gohlke, M.F. Thorpe, A natural coarse graining for simulating large biomolecular motion. Biophys. J. **91**(6), 2115–2120 (2006)
13. H. Gohlke, L.A. Kuhn, D.A. Case, Change in protein flexibility upon complex formation: analysis of Ras-Raf using molecular dynamics and a molecular framework approach. Proteins **56**(2), 322–337 (2004)
14. L.C. Gonzalez, H. Wang, D.R. Livesay, D.J. Jacobs, Calculating ensemble averaged descriptions of protein rigidity without sampling. PLoS ONE **7**(2), e29176 (2012)
15. K. Gunasekaran, A.T. Hagler, L.M. Gierasch, Sequence and structural analysis of cellular retinoic acid-binding proteins reveals a network of conserved hydrophobic interactions. Proteins **54**(2), 179–194 (2004)
16. S. Hayward, H.J.C. Berendsen, Systematic analysis of domain motions in proteins from conformational change: new results on citrate synthase and T4 lysozyme. Proteins Struct. Funct. Bioinform. **30**, 144–154 (1998)
17. B.M. Hespenheide, A.J. Rader, M.F. Thorpe, L.A. Kuhn, Identifying protein folding cores: observing the evolution of rigid and flexible regions during unfolding. J. Mol. Graph. Model. **21**(3), 195–20 (2002)
18. T.A. Holland, S. Veretnik, I.N. Shindyalov, P.E. Bourne, Partitioning protein structures into domains: why is it so difficult? J. Mol. Biol. **361**(3), 562–590 (2006)
19. D.J. Jacobs, Generic rigidity in three-dimensional bond-bending networks. J. Phys. A Math. Gen. **31**(31), 6653–6668 (1998)
20. D.J. Jacobs, A.J. Rader, M.F. Thorpe, L.A. Kuhn, Protein flexibility predictions using graph theory. Proteins **44**, 150–165 (2001)
21. F. Jagodzinski, P. Clark, T. Liu, J. Grant, S. Monastra, I. Streinu, Rigidity analysis of periodic crystal structures and protein biological assemblies. BMC Bioinform. (2013, to appear)
22. F. Jagodzinski, J. Hardy, I. Streinu, Using rigidity analysis to probe mutation-induced structural changes in proteins. J. Bioinform. Comput. Biol. **10**(3), 1242010-1–1242010-17 (2012)
23. P.A. Karplus, G.E. Schulz, Prediction of chain flexibility in proteins. Naturwissenschaften **72**, 212–213 (1985)
24. T. Kortemme, A.V. Morozov, D. Baker, An orientation-dependent hydrogen bonding potential improves prediction of specificity and structure for proteins and protein-protein complexes. J. Mol. Biol. **326**(4), 1239–1259 (2003)
25. J.A. Kovacs, P. Chacon, R. Abagyan, Predictions of protein flexibility: first-order measures. Proteins **56**(1), 661–668 (2004)

26. M. Kurnikova, T. Mamanova, B.M. Hespenheide, R. Straub, Protein flexibility using constraints from molecular dynamics simulations. Phys. Biol. **2**(4), S137–S147 (2005)
27. A. Lee, I. Streinu, Pebble game algorithms and sparse graphs. Discret. Math. **308**(8), 1425–1437 (2008)
28. A. Lee, I. Streinu, L. Theran, Analyzing rigidity with pebble games, in *Symposium on Computational Geometry*, College Park (ACM, New York, 2008), pp. 226–227
29. H. Maity, M. Maity, M.M.G. Krishna, L. Mayne, S.W. Englander, Protein folding: the stepwise assembly of foldon units. PNAS **102**(13), 4741–4746 (2005)
30. S.L. Mayo, B.I. Dahiyat, D.B. Gordon, Automated design of the surface positions of protein helices. Protein Sci. **6**(6), 1333–1337 (1997)
31. I. McDonald, J.M. Thornton, Satisfying hydrogen bonding potential in proteins. J. Mol. Biol. **238**(5), 777–793 (1994). hplus reference
32. G.A. Petsko, D. Ringe, *Protein Structure and Function* (New Science Press, London, 2004)
33. A.J. Rader, B.M. Hespenheide, L.A. Kuhn, M.F. Thorpe, Protein unfolding: rigidity lost. PNAS **99**, 3540–3545 (2002)
34. T.S. Tay, Rigidity of multigraphs I: linking rigid bodies in n-space. J. Comb. Theory Ser. B **26**, 95–112 (1984)
35. T.S. Tay, W. Whiteley, Recent advances in the generic rigidity of structures. Struct. Topol. **9**, 31–38 (1984)
36. M.F. Thorpe, B.M. Hespenheide, Y. Yang, L.A. Kuhn, Flexibility and critical hydrogen bonds in cytochrome c. Pac. Symp. Biocomput. 191–202 (2000)
37. M. Vihinen, E. Torkkila, P. Riikonen, Accuracy of protein flexibility predictions. Proteins **19**(2), 141–149 (1994)
38. S. Wells, S. Menor, B.M. Hespenheide, M.F. Thorpe, Constrained geometric simulation of diffusive motion in proteins. Phys. Biol. **2**, S127–S136 (2005)
39. S. Wells, J.E. Jimenez-Roldan, R.A. Romer, Comparative analysis of rigidity across protein families. Phys. Biol. **6**(4), 046005 (2009)
40. J.M. Word, J.S. Richardson, D.C. Richardson, S.C. Lovell, Asparagine and glutamine: using hydrogen atom contacts in the choice of sidechain amide orientation. J. Mol. Biol. **285**, 1735–1747 (1999)
41. W. Wriggers, K. Schulten, Protein domain movements: detection of rigid domains and visualization of hinges in comparisons of atomic coordinates. Proteins Struct. Funct. Bioinform. **29**(1), 1–14 (1998)

Modeling Autonomous Supramolecular Assembly

Meera Sitharam

Abstract Supramolecular assembly is often a remarkably robust, rapid and spontaneous process, starting from a small number of monomeric types. Although the process occurs widely in nature and is increasingly important in healthcare and engineering, it is poorly understood. Icosahedral viral shell assembly is one such outstanding example. We sketch the experimental roadblocks that necessitate mathematical and computational modeling of assembly, and list the types of experimental data available for model validation, thereby defining the models' input and output, and framing the scope of model predictions. We isolate the various factors, specifically *configurational and combinatorial entropy* that influence spontaneous supramolecular assembly, pinpointing the modeling challenges and motivating the use of *multiscale* models. We then survey existing modeling paradigms for the modeling different scales, emphasizing the newest models and paradigms developed by the author's group, geared towards not only predicting, but also intuitively explaining, analyzing and engineering assembly processes. The models leverage geometric and algebraic characteristics unique to molecular assembly (as opposed to folding), and permit provable performance guarantees together with some level of forward and backward analysis as well as a desired level of precision and refinability of prediction.

1 Motivation

Understanding supramolecular assembly is useful for many practical applications. Rational drug design is a vast area of study and requires understanding the site-specific assembly or docking of ligands with proteins and other biomolecules.

M. Sitharam (✉)
Computer and Information Sciences and Engineering, University of Florida, P.O. Box
32611-6120, Gainesville, FL, USA
e-mail: sitharam@cise.ufl.edu

N. Jonoska and M. Saito (eds.), *Discrete and Topological Models in Molecular Biology*,   197
Natural Computing Series, DOI 10.1007/978-3-642-40193-0_9,
© Springer-Verlag Berlin Heidelberg 2014

Similarly, nanoscale self-assembly of materials is a vast area of study in nanotechnology. Many viral capsids form by self-assembly of an icosahedral shell from nearly identical coat protein monomers enclosing genomic material. Understanding how to disrupt assembly permits us to target this part of the viral lifecycle using drugs and vaccines The pathophysiology of viral infections includes other parts of the viral lifecycle that involve site-specific docking and assembly. Understanding how to encourage assembly can help engineer effective viral vectors that are used as transport for gene therapy or potentially for bacteriophage virus therapy to attack specific bacteria.

Scope. In this paper, we are interested only in structures formed by direct, autonomous assembly rather than structures assembled with the aid of extraneous chaperones or scaffolding molecules that do not end up as part of the assembled structure. Furthermore, we are not interested in structures formed by multistage assembly; i.e., by various deformation and/or folding processes subsequent to the assembly of an initial structure.

1.1   Limitations of Experimental Data and Modeling Motivation

Supramolecular assembly is a rapid, economical process driven by weak interactions and non-covalent binding between the constituent molecular components. The assembly takes place spontaneously at room temperature, in solution, or in a lipid bilayer membrane. Available types of experimental data on supramolecular assembly include:

- X-ray crystallography for details of relatively large assembled structures (often possessing nontrivial symmetries, as in the case of icosahedral viruses);
- Cryo-electron microscopy and stoichiometry studies of various approximate subassembly intermediate structures and their sizes;
- Primary sequence or even NMR spectroscopy structure of the starting monomers and smaller (sub)assemblies;
- Calorimetric studies to determine dissociation energies for (sub)assemblies;
- In vitro systems to measure concentrations of various subassembly intermediates;
- Selective mutagenesis of starting monomers, and its effect in encouraging or disrupting assembly; and
- Mining a comparable database of all of the above types of data for assembly systems classified by various similarity criteria, for example, by structural or biological similarity of viruses.

Despite the above types of experimental data and exploration capabilities, supramolecular assembly processes are poorly understood partly because of their remarkable rapidity, spontaneity and robustness. Spontaneity makes it difficult to control in vitro, rapidity makes it difficult to get snapshots, and robustness (multiple pathways and insensitivity to individual interactions of the constituent molecules) makes it difficult to isolate *crucial combinations of assembly-driving interactions*—from among a combinatorial explosion of possible combinations. In addition, many of these experimental methods are labor- and resource-intensive, making blind alleys extremely expensive.

This generates a strong motivation to go beyond guesswork guided by theoretical first principles alone, and develop effective mathematical and computational models for supramolecular assembly that can inform further experimentation. On the other hand, the necessity to validate model predictions using the available experimental data and within the prevailing experimental capabilities frames the scope of our models, and defines their inputs, outputs and tuning parameters.

1.2 Prediction Tasks

Based on the previous discussion, we focus on models for the following types of prediction tasks.

- *Input:* the 3D configurations of the rigid components of the starting monomers, and the inter-component interactions (Sect. 2 describes how they are formally specified). *Output:* prediction of the terminal assembly structures and their concentrations (or probabilities).
- *Input:* as in the previous item, plus a 3D configuration of final assembly. *Output:* prediction of those atoms or monomers that are crucial for the assembly process to terminate in the given input assembly configuration.
- *Input:* as in the previous item. *Output:* prediction of minimal atomic alterations that would significantly increase probability of the assembly process terminating in the given input assembly configuration.
- *Input:* as in the previous item, additionally more than one choice of final assembly configuration. *Output:* prediction of key events such as specific intermediate subassembly configuration choices during assembly that determine which one of the final assembly configuration results.

These types of predictions cannot be made by theoretical first principles, combinatorial experimentation (trying various possibilities), and guesswork alone, even with the help of known data on similar assemblies and biological knowledge about evolutionarily conserved structures. In addition, for larger assemblies, these predictions cannot be made by direct application of standard methods such as Monte Carlo or molecular dynamics mixed with informatics style approaches for mining existing knowledge for similar assemblies.

1.3 The Methods of This Paper

The methods emphasized in this paper begin with isolating and abstracting crucial factors influencing assembly, thus motivating a *multiscale* model of assembly.

One such factor influencing supramolecular assembly at the *nanoscale* is *configurational entropy* of small assemblies at inter-monomeric interfaces, driven by weak forces and non-covalent binding. The exact computation of configurational entropy is considered a notoriously difficult problem in chemical theory and computational chemistry.

This paper describes our new modeling paradigm towards the judicious approximation of configurational entropy suited to a specific type of prediction that can be validated by mutagenesis experiments. The paradigm consists of two aspects. The first aspect is the generation of an *atlas* of the configuration space using classical Thom–Whitney stratification from algebraic geometry. The second aspect is our new theory of *convexification*, i.e., choosing parameters by which the regions of the stratification can be represented as convex regions. Both aspects are implemented as a prototype software EASAL (*efficient atlasing and search of assembly landscape*). Recent mutagenesis validation of predictions of crucial interactions for the assembly of AAV2 (Adeno Associated Virus) were based on an approximation of interface configurational entropy obtained by EASAL.

Another crucial factor influencing assembly at the *microscale* is *combinatorial entropy* in the formation of larger assemblies from smaller subassembly intermediates, especially when symmetries are present. This, too, is difficult to model or compute. Traditional approaches have been primarily based on simplified geometric approximations of the assembly constituents, and local assembly rules, together with statistical mechanics simulation heuristics that incorporate kinetics as well. Most of these methods do not provide performance guarantees, nor facilitate backward analysis of the computational model's input–output function; nor are they suited to providing intuitive, mechanistic explanations and predictions.

This paper details our approach for combinatorial entropy at the microscale, using algorithms with performance guarantees (or even generating functions), for counting assembly pathways with desired features, especially in the presence of symmetry.

1.4 Organization

In Sect. 2 we discuss factors influencing assembly, motivating a multiscale model of assembly. In Sect. 3 we discuss the crucial *nanoscale* factor that influences supramolecular assembly namely *interface configurational entropy* at inter-monomeric interfaces. We give a brief sketch of the literature tracing the long and distinguished history of the notoriously difficult problem of configurational entropy computation. We then describe our modeling paradigm for approximating interface configurational entropy: generation of an atlas of the configuration space using stratified convexification, implemented as a prototype software EASAL. In Sect. 4 we discuss the crucial *microscale* factor of that *combinatorial entropy*, which influences the number of pathways to formation of larger assemblies from smaller subassembly intermediates, especially when symmetries are present.

We briefly survey traditional approaches based on local assembly rules and statistical mechanics simulation heuristics that incorporate kinetics. We describe our approach for computing combinatorial entropy at the microscale, using algorithms (or even generating functions), for counting assembly pathways. In Sect. 5 we briefly present recent mutagenesis validation of predictions of crucial interactions for the assembly of AAV2 (Adeno Associated Virus), based on an approximation of interface configurational entropy obtained by EASAL. We conclude by highlighting the remaining challenges in Sect. 6.

2 Multiscale Model Based on Factors Influencing Assembly

First we describe an assembly system, i.e., the typical input to an assembly process. This is followed by a discussion of the key factors that influence assembly which highlights the challenges of the above prediction tasks and motivates a multiscale assembly model.

2.1 Assembly System

An input to a computational model of an assembly process is an *assembly system* consisting of the following.

- A collection of of *monomers* drawn from a small set of *monomeric types* (often just a single type). Each monomeric type is specified as a collection of *rigid molecular components*; a rigid component is in turn specified as the set of positions of the centers of their constituent *atoms*, in a local coordinate system. In many cases, an *atom* could be the representation for the average position of a *collection of atoms in an amino acid residue*. Note that an assembly *configuration* is given by the positions and orientations of the entire set of n rigid molecular components in an assembly system, relative to one fixed component. Since each rigid molecular component has 6 degrees of freedom, a configuration is a point in $6(n - 1)$-dimensional Euclidean space.
- The pairwise component of the potential energy function of the assembly system, specified as a sum of potential energy (also called enthalpy) terms between pairs of constituent atoms i and j in two different rigid components of the assembly system. The weak interactions between the rigid molecular components is captured by this potential energy function. The pairwise potential energy terms are, in turn, specified using pairwise *Lennard-Jones* and *Hard-Sphere pairwise potential energy functions*. The pairwise Lennard-Jones term is typically present only for selected pairs of atoms, i and j, one from each component, while the Hard-Sphere potentials apply to all other pairs. Both are functions of the distance $d_{i,j}$ between i and j; the former function is typically discretized

to take different constant values on 3 intervals for the distance value $d_{i,j}$: $(0, l_{i,j}), (l_{i,j}, u_{i,j}), and (u_{i,j}, \infty)$. Typically, $l_{i,j}$ is the so-called Van der Waal or steric distance given by "forbidden" regions around atoms i and j. And $u_{i,j}$ is a distance where the attractive (electrostatic or other weak) forces between the two atoms is no longer strong (typically these forces decay as the reciprocal of some power of $d_{i,j}$). Intuitively, the interval $(0, l_{i,j})$ is where the repulsive force dominates, and $(l_{i,j}, u_{i,j})$ is where the attractive force and repulsive forces are balanced, and $(u_{i,j}, \infty)$ is where neither force is strong. Over these 3 intervals, respectively, the Lennard-Jones potential assumes a very high value $h_{i,j}$, a small value $s_{i,j}$, and a medium value $m_{i,j}$. All of these *bounds* for the intervals for $d_{i,j}$, as well as the values for the Lennard-Jones potential on these intervals are *specified constants* as part of the input to the assembly model. These constants are specified for each pair of atoms i and j, i.e., the subscripts are necessary. The middle interval is called the *well*. The Hard-Sphere potentials are defined solely by the Van der Waal's forbidden distance, $l_{i,j} = u_{i,j}$.

- A non-pairwise component of the potential energy function in the form of *global potential energy* terms that capture the tethers between the rigid components within a monomer, as well as other global potential energy terms that implicitly represent the solvent (water or lipid bilayer membrane) effect [16, 23, 24]. These are specified using discrete values over intervals of the distances or angles between pairs of entire rigid components (as opposed to pairs of atoms).

It is important to note that all the above potential energy terms are *functions of the assembly configuration*.

Observe that an assembly system can alternatively be represented as a set of rigid molecular components drawn from a small set of types, together with *assembly constraints*, in the form of distance and angle intervals. These constraints define *feasible* configurations (where the pairwise inter-atoms distances are larger than $l_{i,j}$, and any relevant tether and implicit solvent constraints are satisfied). The set of feasible configurations is called the *assembly configuration space*. The *active constraint* regions of the configuration space are regions where at least one of the Lennard-Jones inter-atom distances lies in the well, i.e., the interval $(l_{i,j}, u_{i,j})$.

Note that for the prediction tasks given above, the input to the assembly model consists of an assembly system, optionally accompanied by one or more final assembly configurations.

2.2 Factors, Challenges, Multiscale

The assembly configuration space can be partitioned into regions with constant potential energy. The *free energy* of such a region is related to the probability of finding the assembly system in a configuration in the region and is dependent on both its potential energy (inversely) and on the log volume of the region (directly). The former is constant over the region, as defined, and is easy to compute for our model. Roughly speaking, the latter represents the *configurational entropy* of the region. We refer the reader to [18, 42] for a succinct exposition of the relationship between these properties.

2.2.1 Nanoscale: Interface Configurational Entropy

At equilibrium, the configuration space (complex of constant potential energy regions) partitions into potential energy *basins* representing *equilibrium* configurations.

The potential energy computation for these configurations is immediate, and the challenge is to compute the configurational entropy, i.e., volumes of these basins to determine the *stability or binding affinity* for these equilibrium configurations.

The dimension as well as geometric and topological complexity of a potential energy basin corresponding to an equilibrium assembly configuration make the computation of the basin volume challenging. If the volume is determined by sampling, it takes time *exponential* in the dimension, and each rigid component in the assembly system punishingly adds 6 to this dimension. Already for small, *interface assembly systems* that are associated with specific types of interfaces between rigid molecular components, this *interface configurational entropy* computation at the *nanoscale* is thus highly challenging.

2.2.2 Microscale: Combinatorial Entropy

For larger, microscale assemblies, this type of direct configurational entropy computation is impossible. Instead, they are treated as being recursively assembled as an interface assembly system, from a small number of stable intermediate subassemblies [35]. This recursive assembly is usually represented as an *assembly tree* whose leaves are the rigid molecular components of the assembly system, the root is the final large assembly configuration, and the internal nodes are the intermediate subassembly configurations. The overall entropy of a configuration space region of the large assembly C is a combination of:

- The entropies of its small number of constituent equilibrium subassemblies C_i;
- The *interface configurational entropy* of the assembly of the C_i's to form C;
- The *combinatorial entropy* at the *microscale* that arises from the number of different collections of subassemblies C_i that assemble to form C, heavily influenced by the symmetries of C [6, 7, 36]; and, finally,
- The *microscale kinetics* that interrelate the stability and binding affinity of different interface assembly configurations, with the concentrations of the constituent subassembly configurations.

The potential energy basins corresponding to equilibrium configurations of the large assembly system C as well as their stability and binding affinity are again

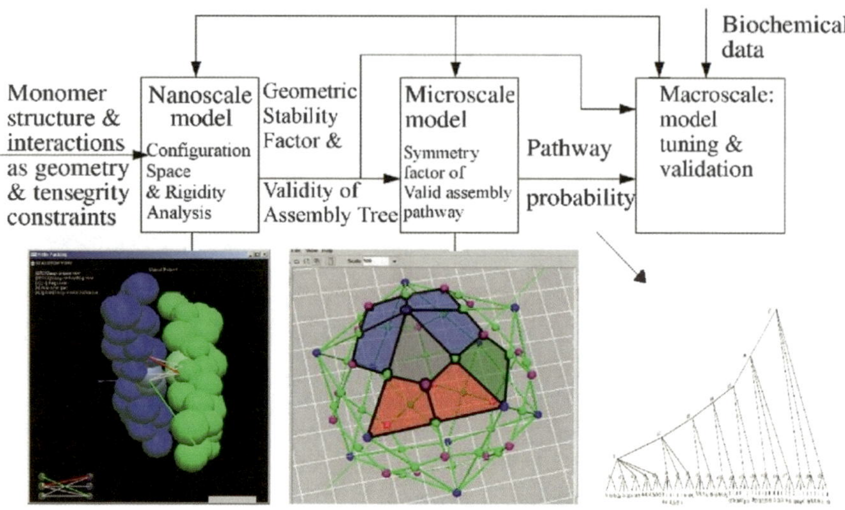

Fig. 1 Multiscale assembly model scales shown; *Left*: combinatorial entropy using nanoscale interface assembly system of 2 rigid molecular components with pair potentials; *Mid*: Large T = 1 viral, microscale assembly shown as polyhedron; *Right*: whose combinatorial entropy is given using (recursive) assembly trees

determined by the geometry and topology of the initial partition into constant potential energy regions as well as *microscale kinetics*.

The above discussion isolates the factors influencing assembly as: *potential energy*; *interface configurational entropy and nanoscale kinetics*; *combinatorial entropy and microscale kinetics*. This motivates a 3-scale model for assembly (see Fig. 1).

3 Nanoscale Models: Interface Configurational Entropy

We discuss three types of models that attempt to capture the following related properties of interface configuration space regions for small assemblies: free energy, partition function (relative probability), stability, binding affinity, configurational entropy. We refer the reader again to [18,42] for understanding the exact relationship between these properties.

3.1 Stability Based on Extent of Rigidity

In [35], rigidity was roughly equated with being an equilibrium assembly configuration (i.e., low-energy representative configuration of a potential energy basin) and a further shortcut was used to quantify the stability of an equilibrium assembly configuration, namely, the number of Lennard-Jones pairs that had to be removed in order to degenerate into a flexible configuration with many small rigid sub-assemblies. This shortcut has also been suggested by Ileana Streinu in a personal communication. However, as mentioned in the previous section, even for small assemblies, the bottleneck in computing the stability and binding affinity of equilibrium configurations is the computation of the volume of the high-dimensional potential energy basin corresponding to the equilibrium configuration, possessing a complicated geometry and topology. This rigidity-based approximation of the volume is too coarse to be effective, as demonstrated for example in trying to determine crucial interactions for AAV2 assembly as in Sect. 5, for which a different method had to be used. In particular, for that example, even a straightforward PCA or eigenvalue-based method outperformed the rigidity-based method.

3.2 Traditional Methods for Configurational Entropy and Free Energy

There has been a long and distinguished history of configurational entropy and free energy computation methods [2, 8, 12–15, 18–20], many of which use as input the configuration trajectories of molecular dynamics or Monte Carlo simulations.

All configurational entropy computations reduce to computing cartesian volumes of constant potential energy regions of a configuration space, as mentioned earlier [18,42]. Even methods that directly compute *partition integrals* (i.e., probabilities of a configuration being in a region of the configuration space) or directly compute free energy (e.g., the Mining Minima method [12]) must effectively compute volumes of configuration space regions since free energy gradients (binding affinities) are effectively based on *entropy differences* between the configuration space regions that correspond to "before" and "after" assembly. Again, the picture of configurational entropy as volume computation for configuration space regions clarifies the intrinsic nature of the two mutually compounding challenges that *any* method will have to overcome: dimensionality and topological/geometric complexity.

As mentioned earlier, accurate computation of volumes of configuration space regions cannot escape exponential dependence on dimension as long as the computation is achieved by counting samples explicitly. Sampling is often the only way to compute the volume of constant potential energy regions, since they are typically semi-algebraic sets (i.e., sets of configurations satisfying systems of quadratic inequalities, since distance is a quadratic function of the cartesian configuration). Such semi-algebraic sets have high geometric and topological complexity, going beyond just the inherent nonlinearity, even in relatively low-dimensional scenarios. In addition, during sampling, Jacobian computations are necessary to map from the "free" internal coordinates to constant potential energy regions of the cartesian configuration space. Such Jacobian computations are necessary since Lennard-Jones and Hard-Sphere pair potentials are both dependent on interatom distances, which depend quadratically (not linearly) on the cartesian coordinates of a configuration.

For instance, with two rigid molecular components, the dimension of the cartesian configuration space is just 6. However, when each component has tens of atoms, the active constraint regions induced by any standard potential landscape are complexes of nested boundaries of different (effective) dimensions. It is due to this reason that one cannot guarantee the ergodicity of Monte Carlo sampling nor give any reasonable bounds on the number of rejected samples. For both Monte Carlo and molecular dynamics, uniform sampling can only be claimed in the limit, or "if run for sufficiently long, or starting from sufficiently many initial configurations." This also causes problems for many entropy computation methods that rely on principal component analyses of the covariance matrices from a trajectory of samples in internal coordinates, followed by quasiharmonic [2] or nonparametric (such as nearest-neighbor-based) [14] estimates. Such methods generally *overestimate* the volumes of configuration space regions with high geometric or topological complexity, even when hybridized with higher-order mutual information [15], and nonlinear kernel methods, such as the Minimally Coupled Subspace approach of [13]. Ab initio methods such as [8] based on geometric algebras (Lie algebra, Grassman-Cayley algebra, etc., common in robotics) are used to give bounds or to approximate configurational entropy without relying on Monte Carlo or molecular dynamics sampling. However, it is not clear how to extend them beyond restricted assembly systems such as a chain or loop of rigid molecular components each consisting of at most three atoms, where each component is noncovalently bound to each neighboring component at exactly two sites.

There has been some research on inferring the topology of the configuration space [11, 22, 30, 38] starting from Monte Carlo and molecular dynamics samples, and using the topology to guide dimensionality reduction [41].

3.3 Approximations of Configurational Entropy via Atlas of Configuration Space

As mentioned earlier, geometric constraints on interatom distances and angles can be extracted from our potential energy function. A recent paper by the author introduces the notion of an *atlas* of a configuration space, which consists of two ingredients. The first ingredient is a *stratification* of the configuration space into *active constraint regions* 3.3.1. The second ingredient is a representation of each active constraint region by carefully chosen parameters that make the region convex (following subsection).

3.3.1 Stratification, Active Constraint Regions

Consider an assembly configuration space $\mathscr{A}$ of k rigid components, defined by a system A of assembly constraints. The configuration space has dimension

$m \leq 6(k-1)$, the number of internal degrees of freedom of the configurations since a rigid object in Euclidean 3-space has 6 rotational and translational degrees of freedom. For $k = 2$, m is at most 6 and in the presence of two tether constraints, it is at most 4.

A *Thom-Whitney stratification* [21] of the configuration space $\mathscr{A}$ (see Fig. 2) is a partition of the space into regions grouped into strata X_i of $\mathscr{A}$ that form a filtration $\emptyset \subset X_0 \subset X_1 \subset \ldots \subset X_m = \mathscr{A}$, $m = 6(k-1)$. Each X_i is a union of nonempty closed *active constraint regions* R_Q where $m - i$ inequality constraints $Q \subseteq A$ are active, meaning equality is attained and they are independent. Each active constraint set Q is itself part of at least one, and possibly many, hence l-indexed, nested chains of the form $\emptyset \subset Q_0^l \subset Q_1^l \subset \ldots \subset Q_{m-i}^l = Q \subset \ldots \subset Q_m^l$. These induce corresponding reverse nested chains of active constraint regions $R_{Q_j^l}: \emptyset \subset R_{Q_m^l} \subset R_{Q_{m-1}^l} \subset \ldots \subset R_{Q_{m-i}^l} = R_Q \subset \ldots \subset R_{Q_0^l}$. Note that here for all l, j, $R_{Q_{m-j}^l} \subseteq X_j$ is closed and j dimensional.

We represent the active constraint system for a region by an *active constraint graph* whose vertices represent the participating atoms (at least 3 in each rigid component) and edges representing the active constraints between them. Between a pair of rigid components, there are only a small number of possible active constraint graph isomorphism types since there are at most 12 contact vertices.

There could be regions of the stratification of dimension j whose number of active constraints exceeds $6(k-1) - j$, i.e., the active constraint system is overconstrained, or whose active constraints are not all independent. Dependent constraints diminish the set of realizations. For entropy calculations, these regions should be tracked explicitly, but in the present paper we do not consider these special regions in the stratification. Our regions are obtained by choosing any $6(k-1) - j$ independent active constraints.

3.3.2 Convex Representation of Active Constraint Region and Atlas

A new theory of Convex Cayley Configuration Spaces (**CCCS**) recently developed by the author [37] gives a clean characterization of active constraint graphs whose configuration spaces are convex when represented by a specific choice of so-called *Cayley parameters*, i.e., distance parameters between pairs of atoms that are inactive in the given active constraint region (see Fig. 3). Such active constraint regions are said to be *convexifiable*, and the corresponding Cayley parameters are said to be its *convexifying* parameters.

The *Atlas* of an assembly configuration space is a stratification of the configuration space into convexifiable regions. In [26] we have shown that *molecular assembly configuration spaces with 2 rigid molecular components have an atlas.* The software EASAL (Efficient Atlasing and Search of Assembly Landscapes) efficiently finds the stratification, incorporates provably efficient algorithms to choose the Cayley parameters [37] that convexify an active constraint region,

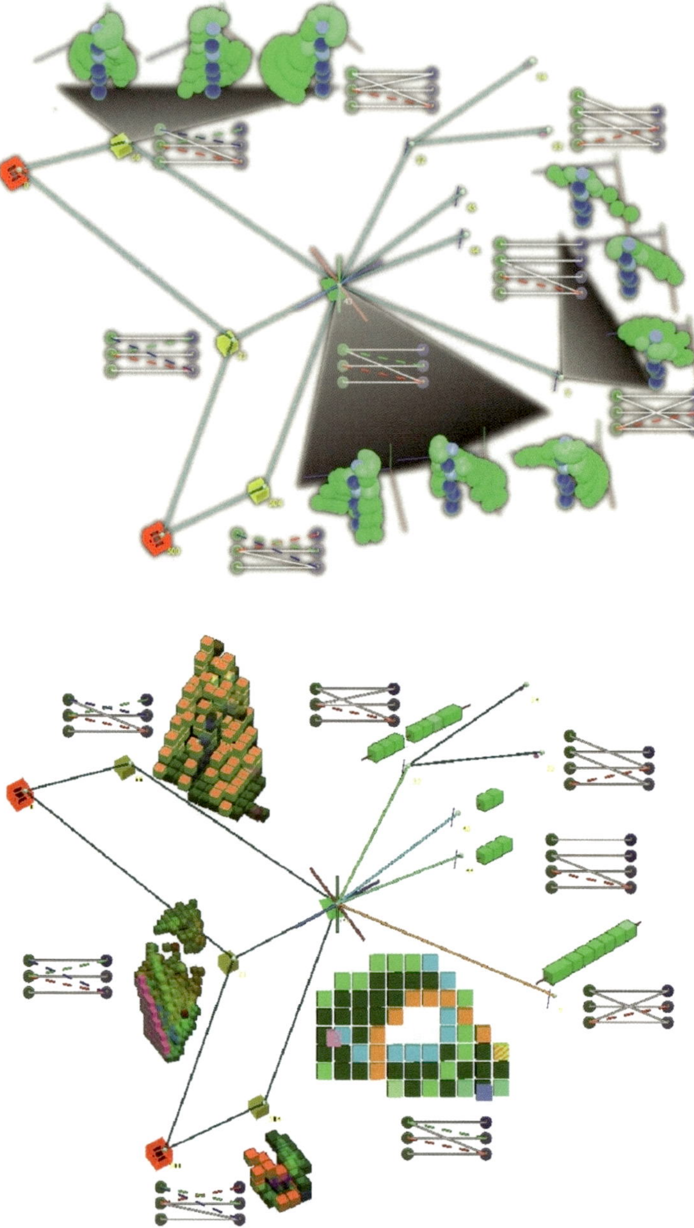

Fig. 2 *Top*: atlas portion, with active constraint regions labeled by their active constraint graphs (*dark edges*); the regions are shown as sweeps around a stationary reference molecule. *Bottom*: active constraint regions with convexifying Cayley parameters (*light edges*), which decrease with dimension, as edges are added to the active constraint graph; note intersection with the complement of a convex subregion in the center. Edges are successively added to the active constraint graphs for the child and descendant atlas regions as more constraints become active

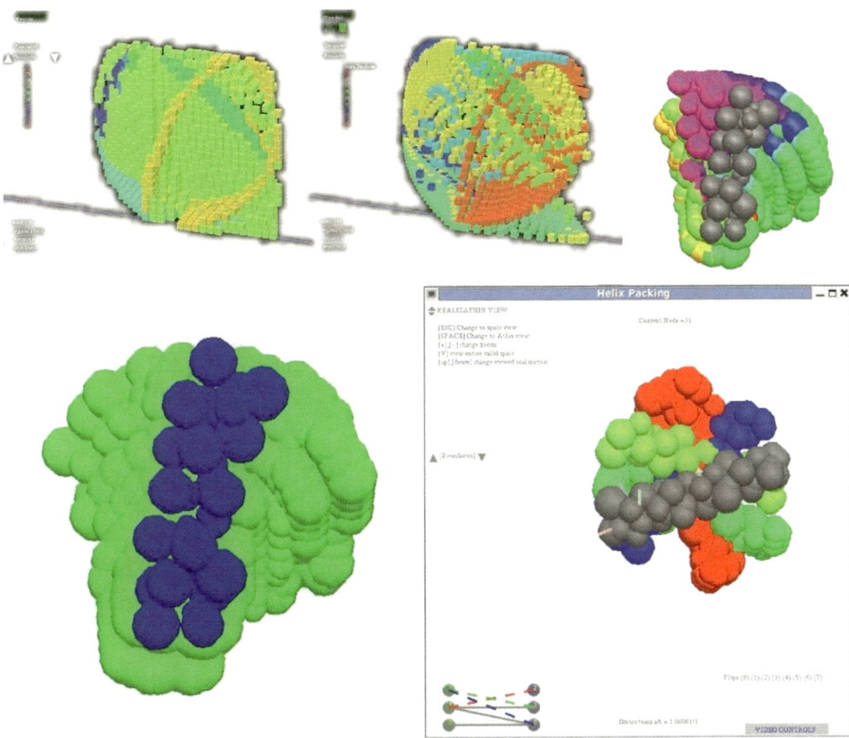

Fig. 3 *Top Left*: atlas region showing interiors and boundaries sampled in its convexifying Cayley parameters; boundary/child regions sampled in their own Cayley parameters and mapped back to the parent region's Cayley parameters (*note increase in samples*). *Top Right*: boundary/child regions sampled in their own Cayley parameters shown as sweeps around grey reference (toy) helix. *Bottom Left*: union of boundary regions sampled in parent's Cayley parameters, shown as sweep around blue reference helix (*notice (b) is bigger*) *Bottom Right*: sweep of one of the boundary regions sampled in parent's Cayley parameters is shown in *red* around gray reference helix; the sampling *misses the other colored configurations* in the same boundary region, obtained by sampling in its own Cayley parameters

efficiently computes bounds for the parametrized convex regions [9], and converts the parametrized configurations into standard cartesian configurations [29].

The key point is that EASAL is *tailored for assembly and leverages its unique properties*; in particular, even simple folding configuration spaces (e.g., the classic cycloheptane or cyclooctane) do not have atlases.

3.3.3   EASAL-Based Approximations of Configurational Entropy

There are many natural ways to approximate configurational entropy. Their efficacy depends on the particular application where they are used. We give one example here that we used for determining crucial constraints as in Sect. 5. The potential energy basins of an interface assembly system are centered around the configurations in the zero-dimensional active constraint regions of the configuration space atlas. These regions cannot be found by EASAL without finding the higher-dimensional regions of the atlas. Furthermore, each distinct configuration in such a region is rigid and could be considered an equilibrium assembly configuration with its own potential energy basin. Any configuration in a basin satisfies at least $6(n-1)$ of the input constraints (for n rigid molecular components), i.e., the corresponding interatomic distances fall within their respective Lennard-Jones wells. The number of copies of one of the configurations in a basin is the number of higher-dimensional regions of the atlas whose active constraint graphs are subgraphs of the active constraint graph of the given configuration. This is an approximate measure of the size or volume of a potential energy basin (configurational entropy associated with that basin).

4 Microscale Model: Combinatorial Entropy

As mentioned in Sect. 2, the computation of combinatorial entropy requires both (a) a count of assembly trees (defined in Sect. 2; see Fig. 4) weighted by the combined probability of their constituent stable subassemblies (in turn obtained from their free energies); and (b) microscale kinetics as described in Sect. 2.

4.1 Combinatorial Entropy via Simplified Assembly Components and Local Rules

The assembly model [34] combines both (a) and (b) above, based on the "local-rules" theory of [3–5, 33]. In addition, differentiation of these models from other similar [17, 25, 31, 32, 43] models is given in [34]. The model in [34] relies crucially on the following: (i) full-blown dynamic simulation (their approach has no static analogy for analyzing successful assembly trees alone); (ii) simple polygonal representations of monomers and an explicitly specified set of stable configurations for the subassemblies; (iii) simplified geometric interactions between monomeric types explicitly and procedurally specified as local rules. The above type of assembly model provided just the necessary level of detail to answer these kinds of questions about concentrations of subassembly configurations. However, forward and reverse analysis are difficult as are intuitive explanations as to what sets of local rules and stable subassemblies are likely to result in a given final assembly.

4.2 Combinatorial Entropy via Assembly Trees and Orbits (Pathways)

To compute the weighted sum of assembly trees for an assembly configuration A as given in (a) above, it is useful to analyze orbits of assembly trees under the action of

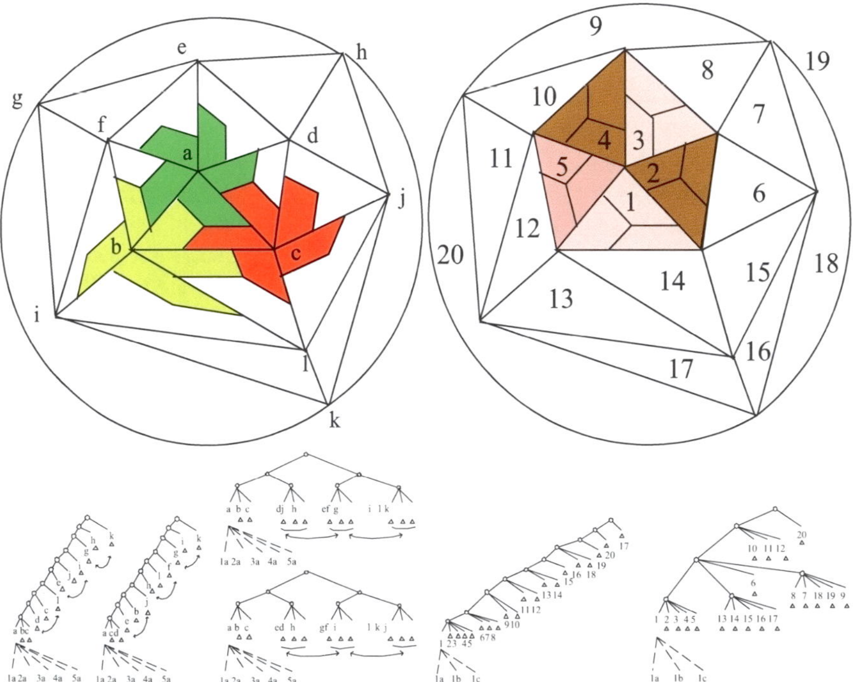

Fig. 4 *Top*: icosahedral assemblies shown as 12 pentamers, or 20 trimers; *Bottom*: Assembly Trees based on (*left*) sequential addition of pentamers, and starting with a trimer of pentamers, with bottom level triangles representing pentamers, or (*right*) sequential addition of trimers starting with a pentamer, with bottom level triangles representing trimers

automorphism group G of A, or the polyhedral graph corresponding to A. Similarly it is useful to analyze orbits of subassemblies A' under the action of G. The induced action of G on an assembly tree is obtained by the G acting on each subassembly occurring as a node in the tree. Two trees in the same orbit under this induced action represent the same assembly process (properties such as the combined probabilities of its constituent subassemblies are the same), hence we call each orbit under this action an *assembly pathway*. Similarly the orbit of a subassembly A' under G is called an *assembly type*. More generally, if any finite group G acts on a finite set S, there is an induced action of G on the set of assembly trees for S. Even if all assembly trees have equal probability of occurring, not all assembly pathways have equal probability of occurring, since the corresponding orbits have different sizes depending on their stabilizer subgroup in G. In [36], we formulated various questions about probabilities of pathways with various properties. In [6, 7] we answered some of these questions; specifically, in [7], we gave explicit generating functions for counting all pathways with a given orbit size.

5   Validation of EASAL Prediction of AAV2 Crucial Interactions

The results in this section have appeared in [40]. We started from simplified potential energy landscapes designed from the known X-ray structure of AAV2 coat protein monomers and interfaces [1, 27, 28] (data provided by Mavis Agbandje-McKenna's lab, see Fig. 2). For each of the three interfaces (2-fold, 3-fold and 5-fold), we determined the pairs of interacting atoms that are conserved in related viruses (10–20 pairs for each interface). These were used as the candidate interactions for the crucial interactions. For the mutagenesis experiment in McKenna's lab, these candidate interactions were disabled one by one, by mutating one of the atoms in the pair. The effect of the mutation on assembly efficacy was determined by measuring concentration of successfully assembled viral shells via cryo-electron microscopy. *This experiment (Bennett, 2012, unpublished manuscript) took at least 2 years.*

For EASAL's predictions, we treated monomers as single rigid components in the interface assembly systems. We used Lennard-Jones potentials for the above pairs of interacting atoms and hard spheres for the sterics of the remaining atoms. No solvent effects were considered. For each interface, for each of its interactions, the approximate interface configurational entropy was computed as described in Sect. 3, when the specific interaction was dropped. We called this the *sensitivity* of that interaction. In fact, for each of the interfaces we generated a new atlas and computed the above quantity for more than one assembly system obtained from different pairs of participating multimers—see below for a detailed description. The rationale was that the same interface drives assembly of different types of multimer-pairs during the formation of larger intermediate subassemblies. We obtained a cumulative sensitivity ranking for each interaction, over all of the relevant interface assembly systems for that interaction. *This computation took 1 week.*

The tabulated results for dimer and pentamer interfaces are given in the two parts of Table 1. The atom numbers in the first two columns are standard numbering used in the cited papers. In some cases, Atom 1 interacts with more than one partner Atom 2. Mutagenesis disables all interactions in which a mutated atom participates. Atom 1 and Atom 2 give the residue. In both cases, the highest ranked interactions (the corresponding atom pair names are given) output by EASAL indicate that assembly is most sensitive to these interactions. They were validated by mutagenesis, resulting in assembly disruption (the "Confirmed" column). Note that blank entries in the "Confirmed" column indicate that mutagenesis was not performed to disable those interactions, i.e., it is as yet unknown whether EASAL's predictions are correct.

5.1   *Pentamer Interface with Participating Multimers*

During the formation of larger assembly intermediates two multimers (as opposed to monomers) could assemble across the same interface. We obtained a new pentamer interface atlas for a monomer and a dimer. While the weak-force interactions remain

Table 1 Sensitivity ranking: dimer (top), pentamer (bottom) interfaces

Atom 1	Atom 2	Confirmed
P293	W694, P696	Yes (Bennett, 2012, unpublished manuscript)
R294	E689, E697	Yes (Bennett, 2012, unpublished manuscript) [39]
E689	R298	Yes (Bennett, 2012, unpublished manuscript)
W694	P293, Y397	Yes (Bennett, 2012, unpublished manuscript)
P696	P293	Yes (Bennett, 2012, unpublished manuscript)
Y720	W694	Yes (Bennett, 2012, unpublished manuscript)
N227	Q401	Yes [39]
R389	Y704	
K706	N382	
M402	Q677	Yes (Bennett, 2012, unpublished manuscript)
K706	N382	
N334	T337,Q319	
S292	F397	Yes [39]

the same, the number of hard-sphere sterics increases and changes the interface configuration space significantly. Factoring this into the rankings, we found two other crucial interactions for the pentamer interface: S292-F397 and N227-Q401. Both were confirmed by assembly disruption through mutagenesis, and have been included in the above tables.

Note concerning the trimer interface: We could not obtain useful sensitivity rankings for the trimer interface due to the heavy influence of sterics (caused by interdigitation). This tallied with the fact that mutagenesis of any of the trimer interface interactions could not disrupt assembly. We do not believe that assembly of the AAV2 shell is sensitive to any of the trimer interactions. We conjecture that the assembly proceeds primarily by dimeric and pentameric interface interactions. Trimers interdigitate and contribute to the stability of the capsid after the assembly is complete.

6 Conclusions and Open Questions

We defined the scope of assembly models based on the type of experimental data available for validation. We gave factors influencing assembly, and motivated a multiscale model. We surveyed traditional models and new modes for both the nanoscale and the microscale and highlighted the issues that are still outstanding. We then gave an example of model prediction that could be experimentally validated.

6.1 *Open Questions on Configurational Entropy*

At the moment the exact computation of configurational entropy, i.e., volumes of atlas regions, is done by sampling, which, as mentioned, does not escape the exponential time dependence on dimension. However, for convexified regions, faster methods, for example based on [10], may help with volume computation for atlas regions. Another unresolved issue is that kinetics influence the stucture of equilibrium potential energy basins, which we have not taken into account.

6.2 *Open Questions on Combinatorial Entropy*

The generating function in [7] for counting pathways with the same orbit size does not extend to pathways with a given property, not even those whose intermediate subassemblies are stable. Sacrificing the generating function for an algorithm opens the field to matroid basis-exchange-type algorithms, provided stable subassemblies can be defined appropriately. We gave a randomized counting algorithm [35] based on matroid basis exchange, for counting all assembly trees with stable subassemblies. However, what is needed is to count pathways (i.e., tree orbits) with stable subassemblies. Furthermore, the question is open how to combine microscale kinetics with the above type of orbit counting.

References

1. M. Agbandje-McKenna, A.L. Llamas-Saiz, F. Wang, P. Tattersall, M.G. Rossmann, Functional implications of the structure of the murine parvovirus, minute virus of mice. Structure **6**, 1369–1381 (1998)
2. I. Andricioaei, M. Karplus, On the calculation of entropy from covariance matrices of the atomic fluctuations. J. Chem. Phys. **115**(14), 6289 (2001)
3. B. Berger, P.W. Shor, On the mathematics of virus shell assembly (1994)
4. B. Berger, P.W. Shor, Local rules switching mechanism for viral shell geometry. Technical report, MIT-LCS-TM-527, 1995
5. B. Berger, P. Shor, J. King, D. Muir, R. Schwartz, L. Tucker-Kellogg, Local rule-based theory of virus shell assembly. Proc. Natl. Acad. Sci. U.S.A. **91**, 7732–7736 (1994)
6. M. Bóna, M. Sitharam, The influence of symmetry on the probability of assembly pathways for icosahedral viral shells. Comput. Math. Methods Med. Special Issue on Mathematical Virology **9**(3–4), 295–302 (Hindawi Publishing Corporation, New York, 2008). Stockley and Twarock (Ed.)
7. M. Bóna, M. Sitharam, A. Vince, Enumeration of viral capsid assembly pathways: tree orbits under permutation group action. Bull. Math. Biol. **73**(4), 726–753 (2011). DOI:10.1007/s11538-010-9606-4
8. G.S. Chirikjian, Chapter four – modeling loop entropy, in *Computer Methods, Part C*, ed. by M.L. Johnson, L. Brand. Volume 487 of Methods in Enzymology (Academic, San Diego, 2011), pp. 99–132
9. U. Chittamuru, Sampling configuration space of partial 2-trees in 3D. Master's thesis, University of Florida, 2011
10. M. Dyer, A. Frieze, R. Kannan, A random polynomial-time algorithm for approximating the volume of convex bodies. J. ACM **38**, 1–17 (1991)
11. D. Gfeller, D. Morton, D. Lachapelle, P. De Los Rios, G. Caldarelli, F. Rao, Uncovering the topology of configuration space networks. Phys. Rev. E Stat. Nonlinear Soft Matter Phys. **76**(2 Pt 2), 026113 (2007)

12. M.S. Head, J.A. Given, M.K. Gilson, Mining minima, direct computation of conformational free energy. J. Phys. Chem. A **101**(8), 1609–1618 (1997)

13. U. Hensen, O.F Lange, H. Grubmüller, Estimating absolute configurational entropies of macromolecules: the minimally coupled subspace approach. PLoS ONE **5**(2), 8 (2010)

14. V. Hnizdo, E. Darian, A. Fedorowicz, E. Demchuk, S. Li, H. Singh, Nearest-neighbor nonparametric method for estimating the configurational entropy of complex molecules. J. Comput. Chem. **28**(3), 655–668 (2007)

15. V. Hnizdo, J. Tan, B.J. Killian, M.K. Gilson, Efficient calculation of configurational entropy from molecular simulations by combining the mutual-information expansion and nearest-neighbor methods. J. Comput. Chem. **29**(10), 1605–1614 (2008)

16. W. Im, M. Feig, C.L. Brooks, An implicit membrane generalized Born theory for the study of structure, stability, and interactions of membrane proteins. Biophys. J. **85**(5), 2900–2918 (2003)

17. J.E. Johnson, J.A. Speir, Quasi-equivalent viruses: a paradigm for protein assemblies. J. Mol. Biol. **269**, 665–675 (1997)

18. M. Karplus, J.N. Kushick, Method for estimating the configurational entropy of macromolecules. Macromolecules **14**(2), 325–332 (1981)

19. B.J. Killian, J. Yundenfreund Kravitz, M.K. Gilson, Extraction of configurational entropy from molecular simulations via an expansion approximation. J. Chem. Phys. **127**(2), 024107 (2007)

20. B.M. King, N.W. Silver, B. Tidor, Efficient calculation of molecular configurational entropies using an information theoretic approximation. J. Phys. Chem. B **116**, 2891–2904 (2012)

21. T.-C. Kuo, On Thom-Whitney stratification theory. Math. Ann. **234**, 97–107 (1978). doi:10.1007/BF01420960.

22. Z. Lai, J. Su, W. Chen, C. Wang, Uncovering the properties of energy-weighted conformation space networks with a hydrophobic-hydrophilic model. Int. J. Mol. Sci. **10**(4), 1808–1823 (2009)

23. T. Lazaridis, Effective energy function for proteins in lipid membranes. Proteins **52**(2), 176–192 (2003)

24. T. Lazaridis, M. Karplus, Effective energy function for proteins in solution. Proteins **35**(2), 133–152 (1999)

25. C.J. Marzec, L.A. Day, Pattern formation in icosahedral virus capsids: the papova viruses and nudaurelia capensis β virus. Biophys. J. **65**, 2559–2577 (1993)

26. A. Ozkan, M. Sitharam, EASAL: efficient atlasing and search of assembly landscapes, in *Proceedings of BiCoB*, New Orleans, 2011

27. E. Padron, V. Bowman, N. Kaludov, L. Govindasamy, H. Levy, P. Nick, R. McKenna, N. Muzyczka, J.A. Chiorini, T.S. Baker, M. Agbandje-McKenna, Structure of adeno-associated virus type 4. J. Virol. **79**, 5047–5058 (2005)

28. E. Padron, R. McKenna, N. Muzyczka, N. Kaludov, J.A. Chiorini, M. Agbandje-McKenna, Structurally mapping the diverse phenotype of adeno associatedvirus serotype 4. J. Virol. **80**, 11556–11570 (2006)

29. J. Peters, J. Fan, M. Sitharam, Y. Zhou, Elimination in generically rigid 3D geometric constraint systems, in *Proceedings of Algebraic Geometry and Geometric Modeling*, Nice, 27–29 Sept 2004 (Springer, 2005), pp. 1–16

30. D. Prada-Gracia, J. Gómez-Gardeñes, P. Echenique, F. Falo, Exploring the free energy landscape: from dynamics to networks and back. PLoS Comput. Biol. **5**(6), e1000415 (2009)

31. D. Rapaport, J. Johnson, J. Skolnick, Supramolecular self-assembly: molecular dynamics modeling of polyhedral shell formation. Comput. Phys. Commun. **121**, 231–255 (1998)

32. V.S. Reddy, H.A. Giesing, R.T. Morton, A. Kumar, C.B. Post, C.L. Brooks, J.E. Johnson, Energetics of quasiequivalence: computational analysis of protein-protein interactions in icosahedral viruses. Biophys. J. **74**, 546–558 (1998)

33. R. Schwartz, P.E. Prevelige, B. Berger, Local rules modeling of nucleation-limited virus capsid assembly. Technical report, MIT-LCS-TM-584, 1998

34. R. Schwartz, P.W. Shor, P.E. Prevelige, B. Berger, Local rules simulation of the kinetics of virus capsid self-assembly. Biophys. J. **75**, 2626–2636 (1998)

35. M. Sitharam, M. Agbandje-McKenna, Sampling virus assembly pathway: avoiding dynamics. J. Comput. Biol. **13**(6), 1232–1265 (2006)
36. M. Sitharam, M. Bóna, Combinatorial enumeration of macromolecular assembly pathways, in *Proceedings of the International Conferecnce on Bioinformatics and Applications* (World Scientific, Fort Lauderdale, 2004)
37. M. Sitharam, H. Gao, Characterizing graphs with convex Cayley configuration spaces. Discret. Comput. Geom. **43**, 594–625 (2010)
38. G. Varadhan, Y.J. Kim, S. Krishnan, D. Manocha, Topology preserving approximation of free configuration space, in *ICRA 2006. Proceedings 2006 IEEE International Conference*, 3041–3048 (2006)
39. P. Wu, W. Xiao, T. Conlon, J. Hughes, M. Agbandje-McKenna, T. Ferkol, T. Flotte, N. Muzyczka, Mutational analysis of the adeno-associated virus type 2 (AAV2) capsid gene and construction of AAV2 vectors with altered tropism. J. Virol. **74**(18), 8635–8647 (2000)
40. R. Wu, A. Ozkan, A. Bennett, M. Agbandje-McKenna, M. Sitharam, Robustness measure for AAV2 is correctly predicted by configuration space atlasing using EASAL, in *Proceedings of ACM Bioinformatics and Comptutational Biology*, Orlando, 2012
41. Y. Yao, J. Sun, X. Huang, G.R. Bowman, G. Singh, M. Lesnick, L.J. Guibas, V.S Pande, G. Carlsson, Topological methods for exploring low-density states in biomolecular folding pathways. J. Chem. Phys. **130**(14), 144115 (2009)
42. H.-X. Zhou, M.K Gilson, Theory of free energy and entropy in noncovalent binding. Chem. Rev. **109**(9), 4092–4107 (2009)
43. A. Zlotnick, To build a virus capsid: an equilibrium model of the self assembly of polyhedral protein complexes. J. Mol. Biol. **241**, 59–67 (1994)

The Role of Symmetry in Conformational Changes of Viral Capsids: A Mathematical Approach

Paolo Cermelli, Giuliana Indelicato, and Reidun Twarock

Abstract For many viruses, structural transitions of the viral protein containers, which encapsulate and hence provide protection for the viral genome, form an integral part of their life cycle. We review here two complementary mathematical models for the expansion of an icosahedral viral capsid. The first is based on a geometrical description of the capsid involving a library of point sets obtained by affine extensions of the icosahedral group, and allows us to characterize the space of the possible transition paths between the initial and the final state. In the second approach, the capsid is described as a union of rigid tiles that interact with each other and with the genomic material, placing emphasis on the energetic determinants of the transition event. Both models predict loss of icosahedral symmetry along the transition path, even though the final state is icosahedral.

1 Introduction

Viruses are prime examples of symmetry in biology. In their simplest form, they consist of a protein shell, called a capsid, that envelops and hence protects the genetic material (RNA or DNA). For a large number of viruses, structural

P. Cermelli (✉)
Dipartimento di Matematica, Università di Torino, Via Carlo Alberto 10, 10123 Torino, Italy
e-mail: paolo.cermelli@unito.it

G. Indelicato
Department of Mathematics, York Centre for Complex Systems Analysis, University of York, York, UK
e-mail: giuliana.indelicato@unito.it

R. Twarock
Departments of Mathematics and Biology, York Centre for Complex Systems Analysis, University of York, York, UK
e-mail: reidun.twarock@york.ac.uk

N. Jonoska and M. Saito (eds.), *Discrete and Topological Models in Molecular Biology*, Natural Computing Series, DOI 10.1007/978-3-642-40193-0_10,

rearrangements of the capsid's proteins are an integral part of the life cycle and a prerequisite for the particles becoming infective. For example, in the maturation of the bacteriophage Hong Kong 97 (HK97), the viral head expands from a procapsid to an elongated shell. In cowpea chlorotic mottle virus (CCMV), equine rhinitis A virus (ERAV), tomato bushy stunt virus (TBSV) and red clover necrotic mosaic virus (RCNMV), structural changes result in the opening of pores through which the genomic material is exposed to the environment and eventually released [25, 27, 28, 31].

Here we review two mathematical approaches to the study of the structural transitions of icosahedral viral capsids. The first approach is based on the use of suitable geometric descriptors of the relevant topological features of the capsid. These descriptors are special point arrays that embody the icosahedral symmetry of the capsid [16]. By modelling the non-crystallographic symmetry via projection from lattices in a higher-dimensional space, we can use the theory of crystallographic phase transitions to induce transformations of the point arrays associated with conformational changes of the capsid. In particular, we consider transitions akin to the Bain strain, i.e. a transformation between cubic lattices such that the intermediate lattices maintain a maximal symmetry [3, 23], and demonstrate their relevance in the context of virology. The second approach consists of a coarse-grained model of a viral capsid and focuses on the energetic contributions, rather than geometric (symmetry) arguments.

The question arises as to whether a geometric approach is suitable for characterizing structural transitions in viruses. Viral capsids are complex protein structures, and it is reasonable to assume that their structural changes are driven by chemistry and physics, rather than by six-dimensional geometric features. However, mathematical techniques for explaining virus architecture using symmetry arguments have been highly successful. For example, icosahedral point arrays obtained via affine extensions of the icosahedral group seem to be remarkably good descriptors of capsid geometry [13–16, 32–34], and provide geometric constraints on the full 3D organization of viruses, including genome organization, generalizing an earlier approach of Caspar and Klug [4]. They also provide boundary conditions for tilings that, in turn, could be used to approximate the proteins that make up the capsid. Importantly, the points in the arrays are not necessarily associated with actual atoms in the capsid; rather, they correlate with material boundaries [16]. Indeed, when a configurational change occurs, it may happen that the point sets approximating the initial and final structures have different cardinalities. This fact implies that there is no natural way to construct transition maps from one array to another, simply because there is no natural way of associating points with atoms, or associating points belonging to different arrays. However, the point arrays used to approximate the geometry of the capsid are a union of icosahedral orbits, and crystallographic transformations with maximal intermediate symmetry have the property that they conserve the maximum possible number of points during the transformation. Hence, transformations between point arrays obtained by projecting to 3D crystallographic phase transitions with maximal symmetry have the advantage both of providing a canonical framework for constructing transformations and of conserving the largest possible number of points during the transition.

In this chapter we first review the work presented in [10], in which we applied the above procedure to the structural transition of the cowpea chlorotic mottle virus capsid. We have developed the machinery necessary to determine all pathways with either icosahedral symmetry or a maximal subgroup thereof. The results pave the way to studying transition paths in icosahedral tilings. In this context, we shall also review the work in [11], where we have applied the above machinery to transitions between mathematical quasicrystals. In particular, we were interested in understanding the effect of high-dimensional phase transitions on tilings such as the Penrose tiling. We proved that, for tilings obtained by projection, the transformations of the tiles can be understood in terms of three simple rules: tile flip, tile merger and tile bisection.

The work on CCMV suggests that the symmetry of transition intermediates is not necessarily icosahedral, in contrast to a widespread assumption in the literature. We have therefore constructed a coarse-grained model to investigate the energetic determinants of such structural transitions, and we also review this second approach here. Based on this model, and very general assumptions about the physical forces governing the transition, we have shown that the transition is most likely not icosahedral, especially in the presence of asymmetric components or local fluctuations in the environmental conditions that locally affect the inter-subunit bonds. In particular, we have investigated the stability of viral capsids and constructed explicit transition paths.

The existing approaches are based mainly on shell mechanics (see, for instance, [8]) or biomolecular simulations. For example, one well-developed approach to the prediction of the pathway along which a viral capsid undergoes a conformational change is normal mode analysis. This approach has been applied to capsids in a series of papers [29, 30] with the purpose of identifying 'soft modes', i.e. deformation modes corresponding to valleys or mountain passes on the energy landscape. The analysis, applied to many different ensembles of atoms of the capsid, yields consistently the result that even though the lowest-energy modes involve isotropic expansion of the capsid, there are often other low-energy modes with non-maximal symmetry.

The above is a very important result, but normal mode analysis seems to have the drawback that, since it involves linearized force fields, it can only give information about the initial stages of the conformational changes. Hence, it is natural to enquire whether symmetry loss along the expansion pathway is a general feature, not restricted to small deformations from the closed reference form of the capsid. Our series of papers on the mathematical description of viral capsid transitions suggests that this is indeed the case.

2 Exploring the Role of Symmetry in Conformational Changes of Viral Capsids: The Bain Strain

2.1 3D Point Arrays

Our first step is to construct a library of point sets, also called 3D point arrays, that encode structural information about virus structure. As Crick and Watson observed [6], the protective protein containers (viral capsids) of many viruses are organized with icosahedral symmetry. This means that the proteins are organized into clusters that are positioned following the rotational symmetries of an icosahedron. Caspar and Klug's seminal work recognized the fact that icosahedral symmetry is not the only constraint on virus architecture [4]. These authors introduced quasi-equivalence theory to explain how the proteins may be organized in the fundamental domain of the symmetry group (also called the asymmetric unit in the literature). More recently, we have shown that quasi-equivalence theory is part of a wider set of constraints on virus architecture [13]. In analogy to lattices that encode how crystallographic symmetries occur at different radial levels around their origin, it has been shown in [13] that affine extensions of icosahedral symmetry, a non-crystallographic symmetry, describe how the occurrence of icosahedral symmetry at different radial levels of a structure with long-range order may be coordinated. Via a classification of all possible affine extensions with suitable properties, a library of blueprints has been derived, and it has been demonstrated that a number of viruses, from different families and infecting different hosts, follow these blueprints [16]. These affine symmetries can be visualized as point arrays, which can be generated from a single point in space via an application of the affine extended group [13]. The classification of affine extensions of icosahedral symmetry has thus resulted in a library of point arrays, which we use here as a coarse-grained approximation to a virus, via a description that reduces the complexity of all-atom models and is amenable to analysis.

In order to explain the construction of these point arrays in some detail, we first recall a few notions about the icosahedral group. As an abstract group, the icosahedral group $\mathscr{I}$ is the 60-element group generated by two elements a and b that satisfy the relations $a^2 = b^5 = (ab)^3 = 1$. Its more familiar representation is as the symmetry group of the icosahedron in 3D, denoted $\mathscr{I}_3$, where a is a twofold rotation about an edge of the icosahedron, and b is a fivefold rotation about a vertex. All elements of the icosahedral group are rotations about either twofold, threefold or fivefold axes. We exclude reflections, since the observed capsids lack such symmetry. Notice that by applying the icosahedral group to an arbitrary point in space we obtain orbits with 60 elements, but the orbits of points lying on the twofold, threefold and fivefold axes of the icosahedral symmetry are smaller, and are the vertices of an icosidodecahedron (30 points), a dodecahedron (20 points) and an icosahedron (12 points), respectively. In the following, we will write ICO, DOD and IDD for the icosahedral orbits of the points $(\tau, 1, 0)$, $(1, 1, 1)$ and $(\tau, 0, 0)$, respectively, with $\tau = (1 + \sqrt{5})/2$ (the motivation for such choices of scaling for the polyhedra will be described below). Analogously, by translating the points of a generic icosahedral orbit (60 points) by an arbitrary vector and all of its copies under icosahedral symmetry, we obtain a set of points with 3,600 elements, but there are special translations that yield smaller orbits. Specifically, consider one of the 'small' icosahedral orbits ICO, DOD or IDD. For each of these, there is only a finite number of special translations, parallel to a high-symmetry axis, that map at least one point of the orbit onto a symmetry axis. Such translations are termed *admissible*, and they are similar to the translations that generate lattices from crystallographic point groups. Hence, the admissible translations have the property that they generate a set of points with small cardinality by the following procedure: (i) take a polyhedron from among ICO, DOD and IDD; (ii) choose an admissible translation for that polyhedron, and construct its icosahedral orbit (which is either a rescaled ICO, DOD or IDD); (iii) translate all points of the initial polyhedron by all translations in the orbit of the admissible translation above; and (iv) add the initial polyhedron to the set.

Further, since there is a finite number of admissible translations for each start polyhedron, there is also a finite number of point sets that can be generated in that way: these provide a finite library by which a first approximation of the viral capsid can be attempted. We demonstrate this idea for a two-dimensional example, described in Fig. 1. The group there is the symmetry group D_{10} of the regular pentagon, the starting orbit is the pentagon itself; the admissible translation is indicated by an arrow pointing to a symmetry axis of the pentagon; and the resulting set of points, obtained by applying the admissible translation and its orbit, are indicated. Each of the above point sets is therefore identified by a polyhedron (either ICO, DOD or IDD with suitable scalings; see below) and an admissible translation $\mathbf{t}$. The possible point sets that can thus be obtained are described and labelled in [13]. For instance, the structure labelled 10 in that reference is an icosahedron (ICO) translated along a three-fold axis (DOD) by an amount τ^2, so that $\mathbf{t}_{10} = \tau^2(1, 1, 1)$.

It turns out that the above point sets are too simplistic to provide a useful tool for approximating complex structures such as viral capsids. The next step is to construct a library whose elements are point sets obtained by superimposing two simpler structures from the above classification. In order to keep the feature that the resulting point set is still small enough (relative to the cardinality of generic orbits), a compatibility condition is imposed on the admissible translations of the simpler structures. Specifically, we require that one of the two structures is rescaled so that its admissible translation coincides with the admissible translation of the other structure. This results in an augmented, yet still finite, library of point arrays, which still have considerably fewer points than the superposition of general icosahedral orbits. Each such point set can be labelled by two numbers, each referring to one of the elementary structures used to generate it. For instance, the structure (10-44) is generated by superimposing the structure 10 (ICO, $\mathbf{t}_{10}$), and the structure 44 (IDD, $\mathbf{t}_{44}$), which is an icosidodecahedron with admissible translation $\mathbf{t}_{44} = \frac{1}{2}\tau(1, 1, 1)$, rescaled by a factor 2τ, since $2\tau\mathbf{t}_{44} = \mathbf{t}_{10}$. The classification shows that there are in

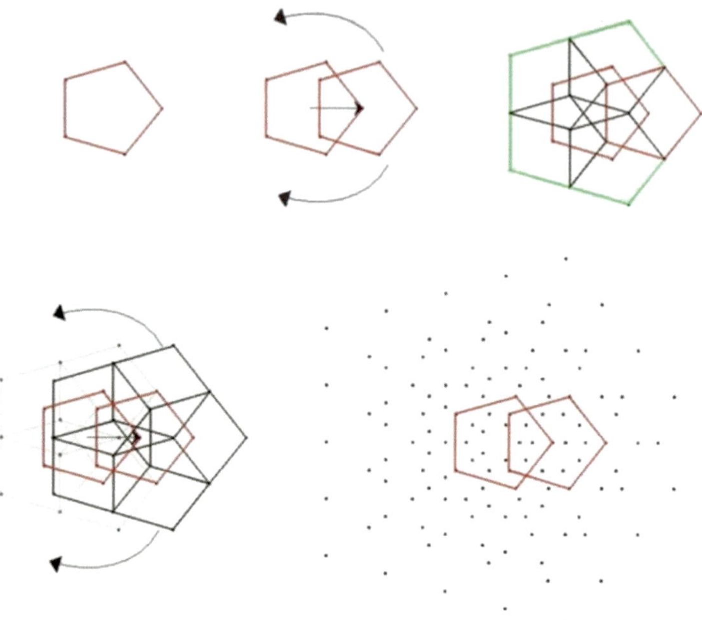

Fig. 1 A two-dimensional example of how affine extensions of non-crystallographic groups yield sets of points with pentagonal symmetry

total 342 such combinations (see [13] for details). Deeper motivations using ideas of affine extensions of Coxeter groups are discussed in [22].

To summarize, we define a point array, or viral configuration, S as a point set obtained by the above procedure. Formally, let $\mathscr{I}_3$ denote the 3D representation of the icosahedral group. As recalled above, if $\mathbf{u}$, $\mathbf{r}$, $\mathbf{t}$ are vectors pointing along either a twofold, threefold, or fivefold axis of $\mathscr{I}_3$, then the icosahedral orbits $\mathscr{I}_3\mathbf{u}$, $\mathscr{I}_3\mathbf{r}$ and $\mathscr{I}_3\mathbf{t}$ correspond to the polyhedra ICO, DOD and IDD with suitable scalings. For every element in the classification of icosahedral point arrays [13], associated to a triple of the form (POLY, POLY′, $\mathbf{t}$), we define a *point array* or *viral configuration* S as the point set

$$S \equiv S(\mathbf{u}, \mathbf{r}, \mathbf{t}) = \mathscr{I}_3\mathbf{u} \cup \mathscr{I}_3\mathbf{r} \cup (\mathscr{I}_3\mathbf{u} + \mathscr{I}_3\mathbf{t}) \cup (\mathscr{I}_3\mathbf{r} + \mathscr{I}_3\mathbf{t}).$$

where $\mathbf{u} \in$ POLY, $\mathbf{r} \in$ POLY′ and $\mathbf{t}$ is a common admissible translation of the two polyhedra. Therefore, every point array is completely determined by a list of 3D vectors $(\mathbf{u}, \mathbf{r}, \mathbf{t})$.

In order to apply the above mathematical concept to virology, we associate a point array with each viral capsid through the algorithm proposed by Keef et al. [16]. Starting from the pdb data available from either the Protein Data Bank or the ViPER website [9], the atoms are first approximated by spheres of radius 1.9 Å

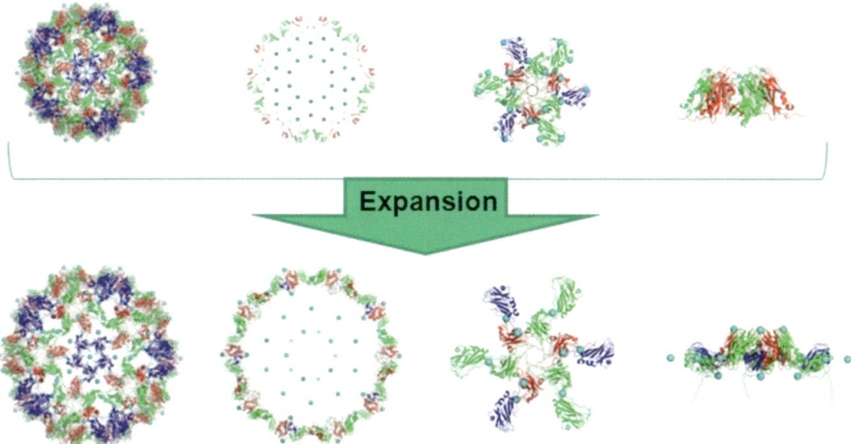

Fig. 2 3D point arrays modelling the layout of the viral capsid and of the pentameric units of CCMV before (*above*) and after (*below*) expansion

around each atomic position, which corresponds to the maximum van der Waals radius of all atoms in the Protein Data Bank file. After aligning each point array with the capsid along their common symmetry axes, each point array is rescaled so that the capsid surface is contained in the convex hull of the points of the array. The goodness of fit of the point array to the capsid is then measured by the sum of two scores: a root-mean-square deviation (RMSD) score and a topography (TOP) score.

The first score is computed as follows: for each point (i, say) in the fundamental domain of the icosahedral group, and for each protein (more precisely, for each quasi-equivalent conformer of the monomer in the fundamental domain), labelled by j, the minimum $R_{i,j}$ of the distances of point i from the van der Waals spheres of the atoms of protein j in the vicinity of point j is computed. The RMSD score is the square root of the mean of the squares of $R_{i,j}$, where only those proteins with distances lower than a cut-off of 2 Å are considered in the sum. In fact, points with distances lower than 2 Å from at least two proteins are likely to lie at the interface between these proteins. The topography score determines which point array best matches the overall surface topography of the virus. In short, the most radially distant 5 % of C_α atoms in the viral capsid are clustered, and the barycentre of each cluster is calculated: the shortest distance of the barycentre from the points of the array is the topography score. The best match for the given capsid virus is the point array with the lowest total score.

Notice that the points of these arrays are located near structurally or functionally important geometric features of the capsid [16], but are not necessarily associated with actual atomic positions. For an example, see the match of points with the capsid of CCMV before and after expansion shown in Fig. 2.

2.2 Lattices and Symmetry

In this section, we review some basic concepts and notation about Bravais lattices that will be useful later. We write $GL(n, \mathbb{Z})$ for the group of $n \times n$ integral matrices with determinant ± 1, $GL(n, \mathbb{R})$, for the group of $n \times n$ invertible real matrices, $O(n)$ and $SO(n)$ for the orthogonal and special orthogonal groups of $\mathbb{R}^n$, and $Sym^+(n, \mathbb{R})$ for the set of $n \times n$ symmetric positive-definite matrices with real coefficients. Given a basis $\{\mathbf{b}_\alpha\}_{\alpha=1,\dots,n}$ in $\mathbb{R}^n$, we denote by $B \in GL(n, \mathbb{R})$ the matrix with columns given by the components of the vectors $\mathbf{b}_\alpha$ in the canonical basis $\{\mathbf{e}_\alpha\}_{\alpha=1,\dots,n}$ of $\mathbb{R}^n$. B is a matrix, but we will often identify it with the linear operator on $\mathbb{R}^n$ of which B is the matrix representation in the canonical basis. Throughout this chapter, we assume that all bases have the same orientation. We denote by $\mathscr{L}(B) = \{\mathbf{x} = \sum_{\alpha=1}^{n} m^\alpha \mathbf{b}_\alpha : m^\alpha \in \mathbb{Z}\}$ the *Bravais lattice* with basis B, i.e. the set of points that are integral linear combinations of the vectors of the *lattice basis* $\{\mathbf{b}_\alpha\}$. All bases of the form MB, with $M \in GL(n, \mathbb{Z})$, generate the same lattice $\mathscr{L}(B)$.

The symmetry of the lattice is described by its *lattice group*

$$\Lambda(B) = \{M \in GL(n, \mathbb{Z}) : \exists Q \in SO(n) \text{ such that } QB = BM\},$$

and its *point group*

$$\mathscr{P}(B) = \{Q \in SO(n) : \exists M \in GL(n, \mathbb{Z}) \text{ such that } QB = BM\}.$$

The point group operations are rotations that map the lattice into itself, and are characterized by the property that they transform lattice bases into lattice bases, whose vectors are therefore integral linear combinations of the original basis vectors. Hence, to each point group operation Q, an integral invertible matrix M representing a change of lattice basis is associated: these matrices form the lattice group. Notice that, in view of the applications to viral capsids, we exclude orthogonal transformations with determinant $= -1$ from the definition of the point and lattice groups, since actual capsids lack these symmetries. A lattice basis is characterized (modulo rotations) by its *lattice metric*

$$C = B^\top B \in Sym^+(n, \mathbb{R}),$$

and, by definition, the lattice group is the subgroup of $GL(n, \mathbb{Z})$ that fixes the metric [23]:

$$M \in GL(n, \mathbb{Z}), \quad M^\top CM = C \qquad \Leftrightarrow \qquad M \in \Lambda(B). \tag{1}$$

Let $\mathscr{L}$ be an n-dimensional lattice with point group $\mathscr{P}$. Consider a subgroup $\mathscr{G}$ of $\mathscr{P}$, and assume that there exists a k-dimensional subspace $E \subset \mathbb{R}^n$ invariant under $\mathscr{G}$. Denote by $E^\perp$ the orthogonal complement of E, so that $\mathbb{R}^n = E \oplus E^\perp$, where $\pi : \mathbb{R}^n \to E$, and $\pi^\perp : \mathbb{R}^n \to E^\perp$ are the corresponding projection operators. If we denote by $\mathscr{G}_E$ the representation of $\mathscr{G}$ on E, then $Q' \in \mathscr{G}_E$ corresponds to $Q \in \mathscr{G}$

in this representation if $Q'\pi\mathbf{v} = \pi Q\mathbf{v}$ for all $\mathbf{v} \in \mathbb{R}^n$, and it follows that $\mathcal{G}$-orbits in $\mathbb{R}^n$ project onto $\mathcal{G}_E$ orbits in E. For a matrix group $\mathcal{H} \subset GL(n, \mathbb{Z})$, we define its centralizer $\mathcal{Z}(\mathcal{H}, \mathbb{R})$ in $GL(n, \mathbb{R})$ as the group

$$\mathcal{Z}(\mathcal{H}, \mathbb{R}) = \left\{ N \in GL(n, \mathbb{R}) : N^{-1}GN = G, \quad \forall G \in \mathcal{H} \right\}.$$

An important property of the elements of the centralizer is that, by definition, for every $\mathbf{v} \in \mathbb{R}^n$, $NG\mathbf{v} = GN\mathbf{v}$, where $N \in \mathcal{Z}(\mathcal{H}, \mathbb{R})$ and $G \in \mathcal{H}$. Hence, $\mathcal{H}$-orbits are mapped into $\mathcal{H}$-orbits by N. Further, if we interpret an element N of the centralizer of $\mathcal{H}$ as a change of basis (or as a linear mapping) of $\mathbb{R}^n$, then the definition above is equivalent to require that every symmetry operation $G \in \mathcal{H}$ has the same matrix representation in the old and in the new basis.

Finally, we recall the definition of the simple cubic (SC), body-centred cubic (BCC) and face-centred cubic (FCC) lattices, defined by

$$\mathcal{L}_{\text{SC}} = \left\{ \mathbf{x} = (x_1, \ldots, x_n) : x_i \in \mathbb{Z}, i = 1, \ldots, n \right\},$$

$$\mathcal{L}_{\text{BCC}} = \left\{ \mathbf{x} = \tfrac{1}{2}(x_1, \ldots, x_n) : x_i \in \mathbb{Z}, x_i = x_j \mod 2, i, j = 1, \ldots, n \right\},$$

$$\mathcal{L}_{\text{FCC}} = \left\{ \mathbf{x} = \tfrac{1}{2}(x_1, \ldots, x_n) : x_i \in \mathbb{Z}, \sum x_j = 0 \mod 2 \right\}.$$

The common point group $\mathcal{P}_{\text{C}}$ of the cubic lattices (of $n!2^{n-1}$-elements) is $\mathcal{P}_C = SO(n) \cap GL(n, \mathbb{Z})$, but their lattice groups are distinct, and not even conjugate in $GL(n, \mathbb{Z})$, since the lattices belong to different Bravais types [23].

2.3 Embedding of the Point Arrays in a 6D Icosahedral Lattice

We show here that point arrays generated by affine extensions of the icosahedral group are projections of subsets of points of a six-dimensional cubic lattice. An analogous assertion has been rigorously proved for affine extensions of Coxeter groups in [22]. It is well known that the smallest dimension in which the icosahedral group is crystallographic is 6. Therefore, let $\mathcal{I}$ be a 6D representation of the icosahedral group [12]: $\mathcal{I}$ leaves all three 6D cubic lattices (SC, FCC and BCC) invariant, and is a subgroup of their common cubic point group. As the three 6D cubic lattices are the only 6D lattices with the above property [19], they are also often referred to as the *icosahedral* 6D lattices. The action of $\mathcal{I}$ on $\mathbb{R}^6$ decomposes into two non-equivalent three-dimensional irreducible representations on two 3D subspaces, which are therefore invariant under icosahedral symmetry and are orthogonal to each other: the *parallel space* and the *orthogonal space*, denoted by E and $E^{\perp}$, respectively. In particular, following [12], we choose as the parallel space E the subspace in which the orthogonal projection of the standard basis of $\mathbb{R}^6$ corresponds to the vectors pointing to the vertices of an icosahedron. We denote by $\mathcal{I}_3$ the 3D representation of the icosahedral group on E.

Recall that icosahedral orbits in $\mathbb{R}^6$ project to icosahedral orbits in $\mathbb{R}^3 = E$, and the same property holds for any subgroup $\mathcal{G} \subset \mathcal{I}$: $\mathcal{G}$-orbits in $\mathbb{R}^6$ project onto $\mathcal{G}_3$-orbits in $\mathbb{R}^3$, where $\mathcal{G}_3$ is the 3D representation of $\mathcal{G}$ in E. In order to associate a point array with a 6D lattice, we use the following facts:

(i) If $\mathcal{L}$ is one of the 6D cubic lattices ($\mathcal{L}_{SC}$, $\mathcal{L}_{FCC}$ and $\mathcal{L}_{BCC}$), the projection $\pi : \mathbb{R}^6 \to E$ is one-to-one onto its image when restricted to $\mathcal{L}$, since E is totally irrational (i.e. $E \cap \mathcal{L} = \{0\}$).

(ii) The icosahedral group commutes with the projection, so that the 6D pre-images of the icosahedral polyhedra are, in turn, icosahedral orbits.

(iii) A dilatation by a factor of τ in 3D corresponds to a symmetry operation of the cubic lattices in 6D, called the quasidilatation.

Consider now a viral configuration $S(\mathbf{u}, \mathbf{r}, \mathbf{t})$ in 3D. Recalling that the icosahedral orbits of $\mathbf{u}$ and $\mathbf{r}$ are a rescaled icosahedron, icosidodecahedron or dodecahedron, as a first step we rescale the standard polyhedra (defined in Sect. 2.1) so that they are the projections of the 6D icosahedral orbits of SC, FCC or BCC lattice vectors, respectively. Then, since the icosahedral orbits of $\mathbf{u}$ and $\mathbf{r}$ are one of these normalized polyhedra at different scalings by τ (i.e. scalings by τ^k with $k \in \mathbb{Z}$; see [13]), we create their 6D counterparts via the action of the quasidilatation. Finally, it turns out that the admissible translation vector $\mathbf{t} \in \mathbb{R}^3$, along which the double-shell structure is translated to generate the point arrays, also belongs to a rescaled polyhedron. Therefore the vector $\mathbf{s} := \pi^{-1}(\mathbf{t})$ and its orbit belong to one of the 6D icosahedral lattices. In this manner, we associate with each 3D point array S a unique set Σ of 6D points in either $\mathcal{L}_{SC}$, $\mathcal{L}_{FCC}$ or $\mathcal{L}_{BCC}$ such that $\pi(\Sigma) = S$. This set is called the *lifted viral configuration* or *lifted point array*. It follows that Σ is a union of icosahedral orbits in $\mathbb{R}^6$ and suitable translates of them:

$$\Sigma = \Sigma(\mathbf{v}, \mathbf{w}, \mathbf{s}) = \mathcal{I}\mathbf{v} \cup \mathcal{I}\mathbf{w} \cup (\mathcal{I}\mathbf{v} + \mathcal{I}\mathbf{s}) \cup (\mathcal{I}\mathbf{w} + \mathcal{I}\mathbf{s}),$$

where $\pi\mathbf{v} = \mathbf{u}$, $\pi\mathbf{w} = \mathbf{r}$, $\pi\mathbf{s} = \mathbf{t}$.

By construction, all points of a given lifted viral configuration Σ are points of some 6D icosahedral lattice, and there exists a unique minimal lattice, which is also icosahedral, that contains any given lifted viral configuration. We say that the lifted point array is *embedded* into this minimal lattice. From the above argument, it follows that since the point arrays can be realized by projection of a 6D lattice onto a completely irrational icosahedrally invariant subspace in 3D, they are subsets of (aperiodic) icosahedral 3D quasilattices (see below).

2.4 Transition Paths for Lattices with Maximal Intermediate Symmetry

In materials science, it is often the case that phase transformations between crystalline substances follow a path along which the intermediate phases have the maximum allowable symmetry. The classical examples are reconstructive phase transformations between cubic phases, such as between SC and BCC or between SC and FCC phases. In these cases, both the parent and the product phase have cubic symmetry, but since they belong to different Bravais types there is no continuous transition between them that maintains cubic symmetry. However, there are transition paths (the Bain strain) that involve small atomic displacements and keep high symmetry (either rhombohedral or tetragonal), and it is a widespread belief that these are minimum-energy paths. We explore here the applicability of this notion in the context of viral transitions: given that the pre- and post-transitional states of the capsid both have icosahedral symmetry, do there exist paths with non-icosahedral but maximal symmetry between them? Or, equivalently, is the transition a simple isotropic expansion of the capsid or given by a more complicated structure?

To answer these questions, we define a lattice transition as a continuous transformation between two lattices $\mathscr{L}_0$ and $\mathscr{L}_1$ along which some symmetry is preserved, described by a common subgroup $\mathscr{H} \subset GL(n, \mathbb{Z})$ of the lattice groups of the intermediate lattices. The definition is best formalized in terms of the lattice group: it $\mathscr{G} \subset \mathscr{P}(\mathscr{L}_0)$ is a subgroup of the point group of $\mathscr{L}_0$, we say that there exists a transition between $\mathscr{L}_0$ and $\mathscr{L}_1$ with intermediate symmetry $\mathscr{G}$ if there exist bases B_0 and B_1 of $\mathscr{L}_0$ and $\mathscr{L}_1$, and a continuous path $B : [0, 1] \to GL(n, \mathbb{R})$, with $B(0) = B_0$ and $B(1) = B_1$, such that the lattice groups of all intermediate lattices $\mathscr{L}(B(t))$ have $\mathscr{H} = B_0^{-1}\mathscr{G}B_0 \subset \Lambda(B_0)$ as a common subgroup, i.e.

$$\mathscr{H} \subset \Lambda(B(t)) \quad \text{for all } t \in [0, 1]. \tag{2}$$

We call the linear mapping

$$T := B_1 B_0^{-1} : \mathscr{L}_0 \to \mathscr{L}_1 \tag{3}$$

the *transition*, and the curve $T(t) = B(t)B_0^{-1}$ is the *transition path*.

As mentioned earlier, we are mostly interested in transitions with maximal symmetry corresponding to a maximal subgroup. The following proposition characterizes lattice transitions (two slightly different proofs of it are given in [10, 11]):

Proposition 1. *Let $\mathscr{L}_0$ and $\mathscr{L}_1$ be two lattices, and let $\mathscr{G} \subset \mathscr{P}(\mathscr{L}_0)$. The following statements are equivalent:*

(i) There exists a transition between $\mathscr{L}_0$ and $\mathscr{L}_1$ with intermediate symmetry $\mathscr{G}$.

(ii) The lattice groups of the initial and the final states have a common non-trivial subgroup: there exist bases B_0 and B_1 of $\mathscr{L}_0$ and $\mathscr{L}_1$ such that for $\mathscr{H} = B_0^{-1}\mathscr{G}B_0$,

$$\mathscr{H} \subset \Lambda(B_0) \cap \Lambda(B_1). \tag{4}$$

(iii) Transitions belong to the centralizer of $\mathscr{G}$. In particular, there exist bases B_0 and B_1 of $\mathscr{L}_0$ and $\mathscr{L}_1$ such that

$$B_1 = RUB_0, \quad \text{with} \quad R \in SO(n) \quad \text{and} \quad U \in \mathscr{Z}(\mathscr{G}, \mathbb{R}) \cap Sym^+(n, \mathbb{R}).$$

$$(5)$$

(iv) *Transition paths belong to the centralizer. In particular, there exist a basis B_0 of $\mathscr{L}_0$ and continuous paths*

$$R : [0, 1] \to SO(n), \quad \text{and} \quad U : [0, 1] \to \mathscr{Z}(\mathscr{G}, \mathbb{R}) \cap Sym^+(n, \mathbb{R}),$$

such that $R(0) = U(0) = I$ and $R(1)U(1)B_0 = B_1$ is a basis of $\mathscr{L}_1$.

(v) *Transitions conserve the lattice metrics: there exist bases B_0 and B_1 of $\mathscr{L}_0$ and $\mathscr{L}_1$ and a continuous path $C : [0, 1] \to Sym^+(n, \mathbb{R})$ such that, letting $C_0 = B_0^\top B_0$, $C_1 = B_1^\top B_1$ and $\mathscr{H} = B_0^{-1}\mathscr{G}B_0$, then $C(0) = C_0$, $C(1) = C_1$ and*

$$M^\top C(t)M = C(t) \quad \text{for all} \quad M \in \mathscr{H} \quad \text{and} \quad t \in [0, 1]. \qquad (6)$$

The following result is an immediate consequence of the above characterization of lattice transitions, and shows that any centralizer of $\mathscr{G}$, not necessarily symmetric, defines a transition with that symmetry (for a proof, see [10]).

Corollary 1. *Any continuous path*

$$T : [0, 1] \to \mathscr{Z}(\mathscr{G}, \mathbb{R}), \qquad T(0) = I, T(1) = B_1 B_0^{-1},$$

where B_0 and B_1 are lattice bases of $\mathscr{L}_0$ and $\mathscr{L}_1$, defines a transition between $\mathscr{L}_0$ and $\mathscr{L}_1$ with intermediate symmetry $\mathscr{G}$.

As an example, consider a transition between the simple cubic and the face-centred lattices in 3D, with intermediate threefold symmetry (Fig. 3). By the above corollary, we can construct such a transition using the centralizers of the group $\mathscr{G} = C_3$ (the threefold rotation group about the main diagonal of the cube) as follows. We define

$$T(t) = \begin{pmatrix} 1 - \frac{1}{2}t & 0 & \frac{1}{2}t \\ \frac{1}{2}t & 1 - \frac{1}{2}t & 0 \\ 0 & \frac{1}{2}t & 1 - \frac{1}{2}t \end{pmatrix}.$$

Then

$$T(0) = B_0 = \begin{pmatrix} 1 & 0 & 0 \\ 0 & 1 & 0 \\ 0 & 0 & 1 \end{pmatrix}, \qquad T(1) = B_1 = \begin{pmatrix} \frac{1}{2} & 0 & \frac{1}{2} \\ \frac{1}{2} & \frac{1}{2} & 0 \\ 0 & \frac{1}{2} & \frac{1}{2} \end{pmatrix},$$

so that T is indeed a transition between the SC and FCC lattices. Moreover, $T(t)$ belongs to the centralizer of C_3 for all t, since

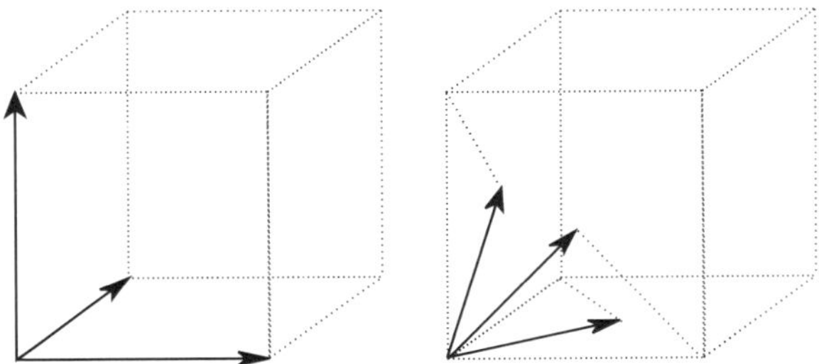

Fig. 3 A Bain transition between a simple cubic lattice and a face-centered cubic lattice with intermediate threefold symmetry

$$GT(t) = T(t)G, \qquad \text{with } G = \begin{pmatrix} 0 & 0 & 1 \\ 1 & 0 & 0 \\ 0 & 1 & 0 \end{pmatrix},$$

where G is the generator of C_3.

2.5 Viral Transitions

Consider a lifted viral configuration $\Sigma = \Sigma(\mathbf{v}, \mathbf{w}, \mathbf{s})$ embedded in a 6D lattice $\mathscr{L}$. A basis B of $\mathscr{L}$ is called admissible for Σ if the $\mathscr{I}$-orbits of the basis vectors and of $\mathbf{v}, \mathbf{w}, \mathbf{s}$ coincide. If B is an admissible basis, we write $\Sigma = \Sigma(B)$. Now let S_0 and S_1 be two 3D viral configurations, with corresponding lifted viral configurations Σ_0 and Σ_1 in 6D, Σ_0 embedded in $\mathscr{L}_0$ and Σ_1 in $\mathscr{L}_1$. We define a *viral transition* between two viral configurations S_0 and S_1 in 3D with intermediate symmetry $\mathscr{G} \subset \mathscr{I}$ as a transition T between the lattices $\mathscr{L}_0$ and $\mathscr{L}_1$ in 6D, such that B_0 and B_1 are admissible for Σ_0 and Σ_1, i.e. $\Sigma_0 = \Sigma(B_0)$ and $\Sigma_1 = \Sigma(B_1)$ (see (3)).

By Corollary 1, given a transition T, the possible transition paths are curves in the centralizer $\mathscr{Z}(\mathscr{G}, \mathbb{R})$ connecting the identity with T. To derive from these paths the information about the actual intermediate structure of a viral capsid, let $T(t) \in \mathscr{Z}(\mathscr{G}, \mathbb{R})$, for $t \in [0, 1]$, be a transition path with intermediate symmetry $\mathscr{G}$. Moreover, let $\mathbf{v}_0$, $\mathbf{w}_0$ and $\mathbf{s}_0$ be three vectors of the basis B_0 such that $\Sigma_0 = \Sigma(\mathbf{v}_0, \mathbf{w}_0, \mathbf{s}_0)$. For $t \in (0, 1)$, we define

$$\mathbf{v}(t) = T(t)\mathbf{v}_0, \quad \mathbf{w}(t) = T(t)\mathbf{w}_0, \quad \mathbf{s}(t) = T(t)\mathbf{s}_0.$$

By definition, $\mathbf{v}(t)$, $\mathbf{w}(t)$ and $\mathbf{s}(t)$ are vectors of the basis $B(t) = T(t)B_0$ of the intermediate lattice. We associate with the transition path $T(t)$, for any $t \in (0, 1)$, a lifted viral configuration $\Sigma(t)$ defined as

$$\Sigma(t) = \mathscr{G}\mathbf{v}(t) \cup \mathscr{G}\mathbf{w}(t) \cup (\mathscr{G}\mathbf{v}(t) + \mathscr{G}\mathbf{s}(t)) \cup (\mathscr{G}\mathbf{w}(t) + \mathscr{G}\mathbf{s}(t)) \subset \mathscr{L}(B(t)).$$

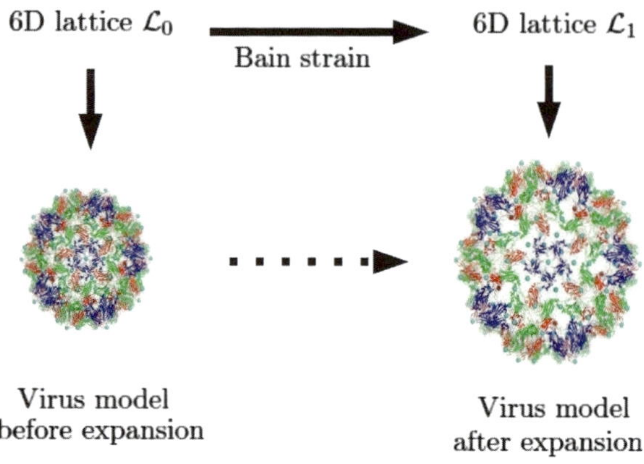

Fig. 4 Viral transitions

The resulting point array, when projected into $\mathbb{R}^3$ via π, yields a family of non-icosahedral point sets $S(t)$ parametrized by t, with constant $\mathscr{G}$-symmetry, that should encode structural boundary conditions on the intermediate viral configurations (see Fig. 4).

The icosahedral group has three maximal subgroups (the tetrahedal group A_4 and the dihedral groups D_{10} and D_6), and therefore we focus on viral transitions with intermediate symmetry corresponding to one of these maximal subgroups. Since viral transitions, by definition, belong to the centralizer of one of these subgroups $\mathscr{G}$, they commute with all its elements and, as a consequence, map $\mathscr{G}$-orbits into $\mathscr{G}$-orbits. Hence, these orbits are conserved along the transition path. Thus, viral transitions are transformations between point arrays that conserve large subsets of points, namely the orbits of the intermediate symmetry group. This property is non-trivial, since in general point arrays corresponding to the initial and final configurations of the capsid have different cardinalities.

2.6 Application to CCMV

We have applied the procedure outlined in the previous sections to transitions in the CCMV capsid during maturation. The point arrays were determined based on the pdb files with the IDs 1cwp for the pre-transition and ccmv_swln_1 for the post-transition configuration. The algorithm of Keef et al. [16] showed that the pre-transition geometry of the CCMV capsid was given by one of two possible viral configurations, and the swollen form of CCMV was best approximated by one of a group of ten other viral configurations.

In [10], we developed a computational algebra technique to determine all possible Bain transitions for a capsid configuration given by a point array S_0 and a post-transition configuration given by an array S_1. The results were as follows:

(i) There exist no Bain transitions with either A_4 or D_{10} symmetry between any of the initial and final configurations for CCMV.

(ii) There exist four Bain transitions with D_6 symmetry, mapping one of the initial configurations corresponding to the point set (10-44) to one of a group of three final configurations.

It is not possible to determine the actual transition path without any information about the physics involved in the process. However, our analysis provides some insights into the likely symmetry of the transition path. In particular, our results for CCMV imply that the configurations of the capsid during the transition will not be icosahedral, unlike the start and end structures, and will have at most D_6 symmetry. D_6 has a representation as the dihedral group of a triangular prism, with one distinguished threefold axis. Assuming that the virus particle will maximize its symmetry throughout the transition, this implies that a threefold axis will play a crucial role during the structural transition. The is result is consistent with a phenomenon observed for transition events in viruses: structural transitions seem to start on a symmetry axis of the spherical particle, for example a threefold axis, and then propagate over the surface of the capsid like a circular wave until the entire particle has undergone the transition. This implies that intermediate configurations presumably preserve one of the symmetry axes, and our method can be used to determine a priori which symmetry axis this is most likely to be.

3 A Case Study of High-Dimensional Symmetry-Preserving Transitions: Transitions of the Penrose Tiling and Icosahedral Tilings of Space

In this section, we review the work presented in [11], where the high-dimensional crystallographic approach was applied to study how quasiperiodic tilings change under symmetry-preserving transformations. The main motivation for that work was to understand the effect of such transformations on the tiles, in order to assess the applicability of this approach to the problem of configurational changes of viral capsids, when proteins are approximated by icosahedral tiles. We focus on tilings associated with cut-and-project quasicrystals (see [26] for an excellent introduction to the subject). Quasicrystals are regular point sets, i.e. sets such that there is a minimum distance between points and a minimum density, that are obtained by projecting a subset of a higher-dimensional lattice onto an irrational subspace. The subset can be chosen to be the intersection of the lattice with a strip between two irrational hyperplanes. If the whole lattice were projected, the resulting set would be dense in the projection subspace, but restricting the set to such a strip guarantees that

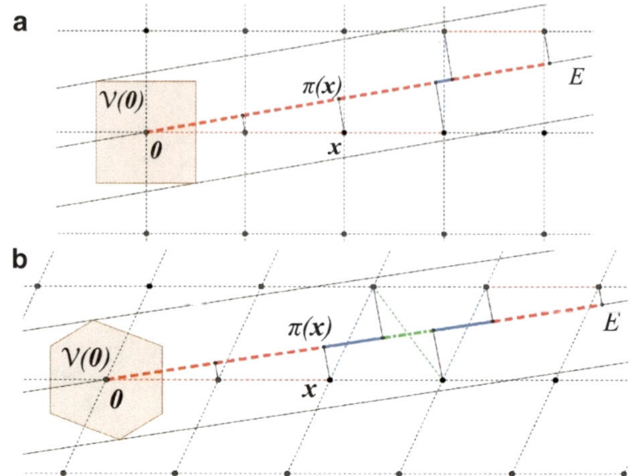

Fig. 5 A 2D square lattice is deformed into a rhombic lattice by an affine deformation. The corresponding 1D quasicrystals are obtained by projecting onto the subspace E the lattice points that lie within a strip whose width is determined by the Voronoi cell $\mathcal{U} = \mathcal{V}(0)$ at the origin (the shaded polygon; see Sect. 3.1). As a result of the lattice deformation, the Voronoi cell changes structure: new facets are created and this, in turn, induces the formation in the projection subspace E of new tiles and changes of shape of the existing ones. (**a**) Square lattice: the Voronoi cell is a square and only two tiles are present (*dashed* and *solid segments*); (**b**) rhombic lattice: the Voronoi cell is a hexagon and there are three different types of tiles (*dashed*, *dotted* and *solid segments*)

the set is regular. Note that the viral configurations defined in the previous sections are subsets of three-dimensional icosahedral quasicrystals.

Aperiodic tilings of the space or the plane can be obtained from a cut-and-project quasicrystal through a general method known as the dualization technique (de Brujin [7]; see also [26] for general references). Dualization provides a general procedure for the construction of aperiodic tilings with non-crystallographic symmetry, such as icosahedral tilings of space. Crystallographic phase transformations of the high-dimensional lattice naturally induce transformations of the quasicrystal, much as in the previous section, and this results in a transformation of the associated tilings. Figure 5 gives a low-dimensional sketch of the procedure: a 2D square lattice is deformed into a rhombic lattice by an affine deformation, the 2D analogue of the Bain strain. The point sets (quasicrystals) change accordingly, and this in turn induces a transformation of the tilings in the projection subspace E through the formation of new tiles and changes of shape of the existing ones. Analogous effects occur in higher dimensions.

Our first application is to the transformations of the Penrose tiling of the plane into the tilings induced by the 5D face-centred cubic and body-centred cubic lattices, with conserved fivefold symmetry in intermediate configurations. The associated quasicrystals are obtained by projection of the 5D lattices onto a plane invariant with respect to a suitable integral representation of the cyclic group C_5. One of the lattice paths in the higher-dimensional space is obtained by compression of the 5D hypercubic cell along a body diagonal, similarly to the rhombohedral strain relating the SC and BCC lattices in 3D (see [23] and Fig. 3). We find that, in projection, the classical Penrose rhombic tiling of the plane transforms into triangle tilings through three basic mechanisms, involving the flipping, bisection and merging of tiles. These mechanisms result from, and indeed correspond to, changes in the geometry of the projection window as a consequence of the deformation of the 5D lattice. We also apply this technique to the study of the transformations of 3D icosahedral quasicrystals and their associated tilings obtained by projection of the SC and FCC lattices in 6D [20, 21], suggesting that the same mechanisms as above occur also in the 3D aperiodic structure transformations that are relevant in the context of virus architecture.

3.1 Cut-and-Project Quasicrystals and Canonical Tilings

Consider an n-dimensional lattice $\mathscr{L}$ with point group $\mathscr{P} = \mathscr{P}(\mathscr{L})$, and a subgroup $\mathscr{H} \subset \mathscr{P}$. We assume that there exists a k-dimensional subspace $E \subset \mathbb{R}^n$ invariant under $\mathscr{H}$, and write, as before, $\pi : \mathbb{R}^n \to E$ and $\pi^\perp : \mathbb{R}^n \to E^\perp$ for the corresponding projection operators. Also, we denote by $\mathscr{U} \subset \mathbb{R}^n$ the Voronoi cell of the lattice at the origin (see e.g. [26]):

$$\mathscr{U} = \{\mathbf{x} \in \mathbb{R}^n : |\mathbf{x} - \mathbf{y}| \geq |\mathbf{x}|, \ \forall \mathbf{y} \in \mathscr{L}\}.$$

$\mathscr{U}$ is invariant under the point group of the lattice [26], and hence also under $\mathscr{H}$. We now fix now a regular shift vector in $\mathbb{R}^n$, i.e. a vector $\mathbf{g}$ (possibly $\mathbf{0}$) such that $(\mathbf{g} + E) \cap \mathscr{F} = \emptyset$ for every d-dimensional facet $\mathscr{F}$ of $\mathscr{U}$, with $d < n - k$. If we define the projection window as

$$\mathscr{W} = \pi^\perp(\mathscr{U}) \subset E^\perp,$$

a cut-and-project quasicrystal is a point set given by

$$(\mathscr{L}, E) := \{\pi(\mathbf{x}) : \pi^\perp(\mathbf{g}) - \pi^\perp(\mathbf{x}) \in \mathscr{W}\} \subset E. \tag{7}$$

When E is totally irrational, i.e., $E \cap \mathscr{L}^* = \{\mathbf{0}\}$, where $\mathscr{L}^*$ is the dual lattice, the set $\pi(\mathscr{L})$ is dense in E; otherwise it is a $\mathbb{Z}$-module, possibly a lattice, in E. There exists a canonical method for constructing aperiodic tilings of the space E using the points of a cut-and-project quasicrystal as vertices [12, 18]. In this method, the Voronoi cells at each point of the lattice define a cell complex that provides a periodic tiling of $\mathbb{R}^n$; its dual complex is also a periodic tiling of $\mathbb{R}^n$, called the Delone tiling. By construction, the vertices of the Delone tiling are lattice points. The Delone tiling of $\mathbb{R}^n$ induces a tiling of E by projection on E of those Delone k-facets dual to the $(n - k)$-facets of the Voronoi tiling that have a non-empty intersection with $\mathbf{g} + E$, where $k = \dim E$. By construction, the tiling of E thus obtained has vertices at the points of the corresponding quasicrystal. If $\mathbf{g}$ is fixed by $\mathscr{H}$, the associated tiling is invariant under the representation of $\mathscr{H}$ on E. In fact, $\mathscr{H}$ commutes with the projection π, so that $\mathscr{H}$-orbits in $\mathbb{R}^n$ project on $\mathscr{H}^{\parallel}$-orbits in E, where $\mathscr{H}^{\parallel}$ is the representation of $\mathscr{H}$ in E.

The techniques summarized above have been extensively applied in the study of quasicrystals and tilings of the plane and space [7, 12, 17, 20, 21, 24].

3.2 Structural Transformations of Cut-and-Project Quasicrystals

Consider two n-dimensional lattices $\mathscr{L}_0$ and $\mathscr{L}_1$, with point groups $\mathscr{P}_0$ and $\mathscr{P}_1$, and two subgroups $\mathscr{H}_0 \subset \mathscr{P}_0$ and $\mathscr{H}_1 \subset \mathscr{P}_1$. Assume that $\mathscr{H}_0$ and $\mathscr{H}_1$ have the same invariant subspaces E, with $\dim E = k$, and consider the cut-and-project quasicrystals $(\mathscr{L}_0, E)$ and $(\mathscr{L}_1, E)$.

Definition 1. We say that there exists a transition between the cut-and-project quasicrystals $(\mathscr{L}_0, E)$ and $(\mathscr{L}_1, E)$ with intermediate symmetry $\mathscr{G} \subset \mathscr{H}_0$ if there exists a transition with intermediate symmetry $\mathscr{G}$ between $\mathscr{L}_0$ and $\mathscr{L}_1$ such that for $T(t) = R(t)U(t)$, with $R(t) \in SO(n)$ and $U(t) \in Sym^+(n, \mathbb{R}) \cap \mathscr{L}(\mathscr{G}, \mathbb{R})$, E is invariant under $\mathscr{G}_t$, i.e.

$$\mathscr{G}_t E = E, \qquad t \in [0, 1], \tag{8}$$

with

$$\mathscr{G}_t = R(t)^{\top} \mathscr{G} R(t) = \mathscr{P}(\mathscr{L}_t), \qquad \mathscr{L}_t = T(t)\mathscr{L}_0. \tag{9}$$

Any such transition defines a family of cut-and-project quasicrystals $(\mathscr{L}_t, E)$ in the same projection space E, all of which have symmetry $\mathscr{G}$.

3.3 Transformations Between Planar Aperiodic Tilings Preserving the Fivefold Symmetry

In this section, we present an example of a transformation of the Penrose tiling that preserves the global fivefold symmetry. We adopt here a five-dimensional approach instead of the usual one based on a 4D minimal embedding [2], because it is simpler to describe the transitions in terms of deformations of the unit cubic cell in $\mathbb{R}^5$. Consider the five-dimensional SC, BCC, and FCC lattices, and the standard basis $(\mathbf{e}_\alpha)_{\alpha=1,...,5}$ in $\mathbb{R}^5$, together with the group $\mathscr{G} = C_5 \subset SO(5)$ of fivefold rotations about the body diagonal $\mathbf{n} = \sum_{\alpha=1}^5 \mathbf{e}_\alpha$ of the unit cube. The group $\mathscr{G}$ which leaves all the three of the above 5D cubic lattices invariant has two mutually orthogonal

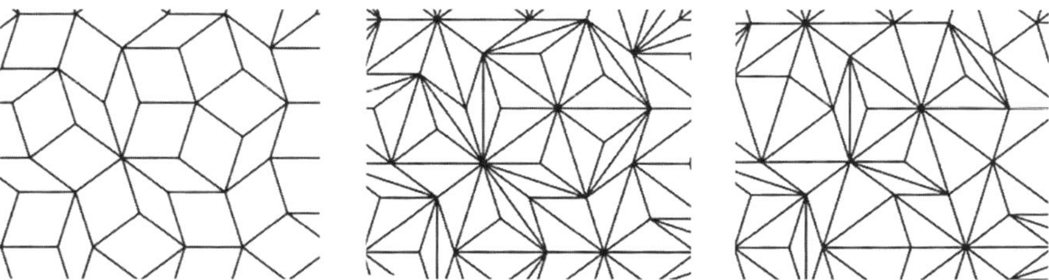

Fig. 6 Three snapshots of a tiling along the transition path (10), showing the splitting of the tiles and their merging

invariant subspaces: the 2D subspace E and its 3D orthogonal complement $E^{\perp}$, with E generated by the vectors

$$\mathbf{v}_1 = \mathbf{e}_1 + \cos(2\pi/5)\mathbf{e}_2 + \cos(4\pi/5)\mathbf{e}_3 + \cos(6\pi/5)\mathbf{e}_4 + \cos(8\pi/5)\mathbf{e}_5,$$

$$\mathbf{v}_2 = \sin(2\pi/5)\mathbf{e}_2 + \sin(4\pi/5)\mathbf{e}_3 + \sin(6\pi/5)\mathbf{e}_4 + \sin(8\pi/5)\mathbf{e}_5.$$

With reference to Sect. 3.1, we choose $\mathbf{g} = \frac{1}{2}\mathbf{n}$; projection of the SC lattice and the related Delone tiling on E produces the well-known Penrose tiling of the plane, whereas projecting the FCC and BCC lattices gives more complex aperiodic planar tilings with global fivefold symmetry about the origin (see also [24]). Corollary 1 guarantees that the C_5-preserving transitions for the associated 5D lattices belong to the centralizer of C_5 in $GL(5, \mathbb{R})$: hence, we consider a specific transition path with fivefold symmetry between the SC and the BCC lattices in 5D (a direct calculation shows that the path does indeed belong to the centralizer of C_5):

$$T(t) = (1-t)I + tB_{\mathrm{BCC}}, \qquad B_{\mathrm{BCC}} = \frac{1}{2}\begin{pmatrix} 1 & 1 & 1 & -1 & -1 \\ -1 & 1 & 1 & 1 & -1 \\ -1 & -1 & 1 & 1 & 1 \\ 1 & -1 & -1 & 1 & 1 \\ 1 & 1 & -1 & -1 & 1 \end{pmatrix}. \tag{10}$$

Since $T(t)\mathbf{n} = (1 - t/2)\mathbf{n}$, this path involves a compression of the unit cube along a body diagonal $\mathbf{n}$. This path also entails, through the dualization technique applied at each step, a transformation between the Penrose and the BCC tilings.

Since the tiles are the projection of the duals of the three-dimensional facets of the Voronoi cell for each intermediate lattice, the transformations of the tilings along these paths can be explained via the changes that occur in these three-dimensional regions, i.e. in terms of rearrangements of the tiles in these regions. In our case, the tiles transform in three possible ways: splitting of a tile into two, tile flips and tile mergers. These correspond to the possible changes of the facets of the Voronoi cell [11]. Figure 6 shows three steps in the transformation from SC to BCC along the path defined by (10): the first step in the transition is a splitting of all tiles along their long diagonal, followed by deletions of vertices and splits/recombinations.

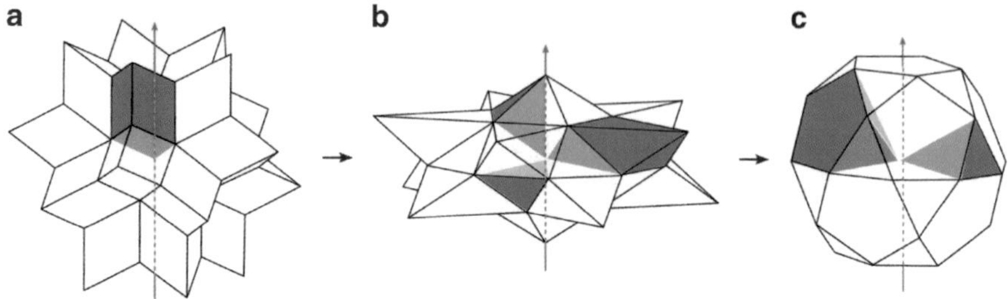

Fig. 7 Three snapshots of the tilings corresponding to the transition path (11)

3.4 Example of a Transformation Between 3D Icosahedral Quasicrystals

We briefly discuss here a transformation with D_{10} symmetry between icosahedral tilings in $\mathbb{R}^3$. As discussed above, such transitions belong to the centralizer of D_{10}. As an example, we consider a path between the six-dimensional SC lattice and the FCC lattice of the form

$$
T(t) = (1-t)I + tB_{\text{FCC}}, \qquad B_{\text{FCC}} = \frac{1}{2}\begin{pmatrix} 1 & 0 & 0 & 0 & 0 & 0 \\ 0 & 1 & 0 & 0 & 0 & 0 \\ 0 & 0 & 1 & 0 & 0 & 0 \\ 0 & 0 & 0 & 1 & 0 & 0 \\ 0 & 0 & 0 & 0 & 1 & 0 \\ 1 & 1 & 1 & 1 & 1 & 2 \end{pmatrix}. \tag{11}
$$

Figure 7 shows three snapshots of a patch of the corresponding 3D tiling around a fixed vertex, i.e. projections on E of a portion of the Delone tiling of the lattices $\mathscr{L}(B(t))$, for $t = 0, 0.233$ and 1. Throughout the transition, the tile arrangements have D_{10} symmetry, and they evolve by mechanisms similar to those discussed for the planar case in Sect. 3.3. Since the point arrays used as descriptors for virus geometry are subsets of 3D icosahedral quasicrystals, icosahedral tilings can be used to obtain additional information about the organization of viral capsids. An example is provided by CCMV; see Fig. 8. The above considerations can hence be used to extend our previous model for structural transitions in CCMV.

4 The Physics of Conformational Changes: A Coarse Grained Model

As mentioned in the introduction, our approach based on point arrays and crystallographic phase transitions is a powerful tool that allows us to canonically construct transformation paths between sets of points with icosahedral symmetry. In turn, this allows us to construct paths between tilings that approximate the arrangement of proteins in a capsid, taking into account the symmetry of the intermediate states.

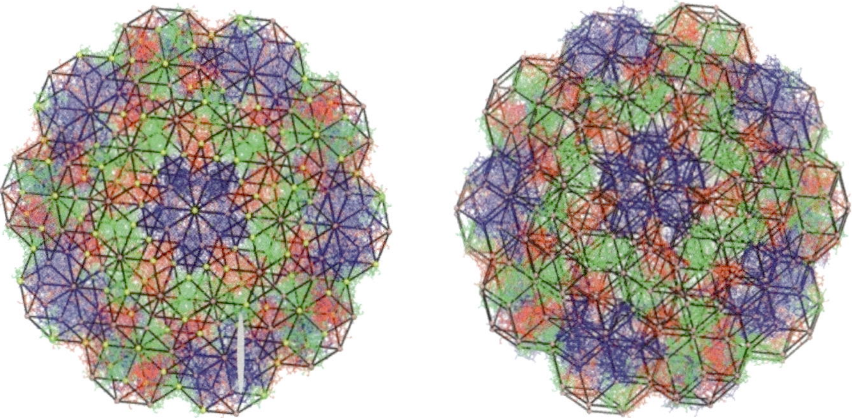

Fig. 8 Two views of an icosahedral tiling of the closed form of the CCMV capsid

The question arises as to how to account for the physics and biochemistry driving the conformational change in this setting. In the quasicrystals community, transitions are described by the so-called phason strain, which represents the orthogonal-space component of a six-dimensional linear transformation, and the energy of a quasicrystal is assumed to be a function of the high-dimensional strain as a whole. We have not explored this path yet, but we have tried a more direct approach, in which we approximate groups of proteins making up the capsid as rigid tiles. Here, their mutual interactions and their interactions with the genomic material inside the capsid are described by suitable potentials or geometrical constraints.

This approach provides a coarse-grained method to explain the main features of the expansion of a large class of viral capsids. Some results, presented in [5], are as follows:

(i) The expansion is a collective rearrangement of the proteins that occurs as a cascade of local events, and is due essentially to the competition between the expansive force of the genomic material acting radially outwards from the inside of the capsid, and the cohesive interaction force between the capsomers. A small local perturbation of the bond strength between two capsomers, due for instance to a change in the chemical environment, can be enough to weaken the total binding force acting on them, so that the capsomers start detaching. This, in turn, decreases the force acting on neighbouring capsomers, and this triggers their detachment, generating an expansion wave over the capsid.

(ii) As a result, the expansion of an icosahedral capsid propagates as a transition wave over the shell: the intermediate states are not icosahedral, and the expansion is not uniform. Eventually, the capsid reaches its final expanded state, which turns out always to be icosahedral.

A fundamental step in the above approach is the identification of the basic building blocks of the capsid, which in this approximation may be regarded as rigid

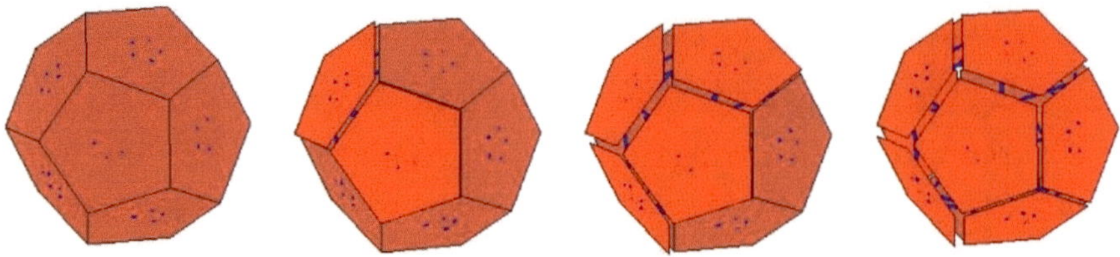

Fig. 9 A few snapshots of the expansion of a dodecahedral capsid according to the model described in Sect. 4

bodies. This can be accomplished by combining a number of different pieces of information about the structure of the capsid, such as knowledge about the assembly and disassembly units, and by using available protein domain decomposition software, such as PiSQRD, developed by Micheletti's group at SISSA in Trieste [1].

We chose as a test case the equine rhynitis A virus, a picornavirus that is believed to release its RNA without disassembling [31]. In this case, both disassembly data and the protein domain decomposition method suggest that the rigid units involved in the capsid expansion are groups of 20 proteins that form a pentagon. We therefore modelled the ERAV capsid as a dodecahedron, and assumed that each pentagonal face can detach by translating along and rotating about its axis. This approximation allows us to reduce the total degrees of freedom to 24. Further, as mentioned above, we assumed that the total energy of the system is the sum of terms accounting for the expansive force of the RNA on the pentagons and a cohesive force acting across their edges. The only parameters of the model are the weights of the pairwise interaction energies, which we call the bond strengths. Finally, we assumed the existence of cross-links, made by peptidic chains, between adjacent capsomers, which maintain the integrity of the capsid as a whole after expansion.

Under very general hypotheses about the energy functions, we were able to prove that it is energetically favourable for the expansion of the capsid to occur through a cascade of local events that propagate over the capsid, which eventually recovers icosahedral symmetry in an open form in which large holes are present. These holes appear to serve the purpose of allowing the release of the RNA into the environment. A few snapshots of the expansion of the ERAV capsid according to our model are shown in Fig. 9.

5   Conclusion

We have presented an overview of two different modelling techniques for structural transitions in viruses. The first is rooted in geometric principles and studies transitions from the point of view of symmetry. The second approach is complementary, and focuses on the energetics of the transition, describing it via a dynamical system and constraints on the stability of the capsid configuration. In particular, it can be used to study how the transition behaviour depends on the distribution of bond strengths across the capsid, and it has highlighted the fact that inhomogeneous solution conditions and the presence of asymmetric components in the capsid result in a wave-like cascade of local expansion events. Both approaches show that the symmetries of the capsid intermediates are likely to be non-icosahedral, in contrast to the pre- and post-transition configurations. These insights shed new light on the mechanisms of capsid transitions, open up new opportunities for biomolecular simulations of the transition events and may potentially lead to the design of new anti-viral strategies that act by blocking these transitions.

Acknowledgements RT and GI thank the Leverhulme Trust for financial support via a Research Leadership Award. PC and GI acknowledge the Italian PRIN 2009 project "Mathematics and Mechanics of Biological Systems and Soft Tissues".

References

1. T. Aleksiev, R. Potestio, F. Pontiggia, S. Cozzini, C. Micheletti, PiSQRD: a web server for decomposing proteins into quasi-rigid dynamical domains. Bioinformatics **25**(20), 2743–2744 (2009)
2. M. Baake, P. Kramer, M. Schlottmann, D. Zeidler, Planar patterns with fivefold symmetry as sections of periodic structures in 4-space. Int. J. Mod. Phys. **B4**, 2217–2268 (1990)
3. E.C. Bain, The nature of martensite. Trans. AIME **70**, 25–46 (1924)
4. D.L.D. Caspar, A. Klug, Physical principles in the construction of regular viruses. Cold Spring Harbor Symp. **27**, 1–24 (1962)
5. P. Cermelli, G. Indelicato, R. Twarock, Non-icosahedral pathways for viral capsid expansion. Phys. Rev. E. **88**, 032710 (2013)
6. F.H.C. Crick, J.D. Watson, The structure of small viruses. Nature **177**, 473–475 (1956)
7. N.G. de Bruijn, Algebraic theory of Penrose's non-periodic tilings of the plane I, II. Nederl. Akad. Wetensch. Indag. Math. **43**(1), 39–52, 53–66 (1981)
8. T. Guérin, R.F. Bruinsma, Theory of conformational transitions of viral shells. Phys. Rev. E **76**, 061911 (2007)
9. http://viperdb.scripps.edu/
10. G. Indelicato, P. Cermelli, D.G. Salthouse, S. Racca, G. Zanzotto, R. Twarock, A crystallographic approach to structural transitions in icosahedral viruses. J. Math. Biol. **64**, 745–773 (2012)
11. G. Indelicato, T. Keef, P. Cermelli, D.G. Salthouse, R. Twarock, G. Zanzotto, Structural transformations in quasicrystals induced by higher dimensional lattice transitions. Proc. R. Soc. **468**, 1452–1471 (2012)
12. A. Katz, Some local properties of the 3-dimensional Penrose tilings, in *Introduction to the Mathematics of Quasicrystals* (Academic Press, Boston 1989), pp. 147–182
13. T. Keef, R. Twarock, Affine extensions of the icosahedral group with applications to the three-dimensional organisation of simple viruses. J. Math. Biol. **59**(3), 287–313 (2009)
14. T. Keef, R. Twarock, Beyond quasi-equivalence: new insights into viral architecture via affine extended symmetry groups, in *Emerging Topics in Physical Virology* (Imperial College Press, London, 2010), pp. 59–83
15. T. Keef, R. Twarock, K.M. Elsawy, Blueprints for viral capsids in the family of Papovaviridae. J. Theor. Biol. **253**, 808–816 (2008)
16. T. Keef, J.P. Wardman, N.A. Ranson, P.G. Stockley, R. Twarock, Structural constraints on the three-dimensional geometry of simple viruses: case studies of a new predictive tool. Acta Cryst. **A69**, 140–150 (2013)

17. P. Kramer, R. Neri, On periodic and non-periodic space fillings of E^m obtained by projection. Acta Cryst. **A40**, 580–587 (1984)
18. P. Kramer, M. Schlottmann, Dualisation of Voronoi domains and Klotz construction: a general method for the generation of proper space fillings. J. Phys. A Math. Gen. **22**, L1097–L1102 (1989)
19. L.S. Levitov, J. Rhyner, Crystallography of quasicrystals; application to icosahedral symmetry. J. Phys. Fr. **49**(11), 1835–1849 (1988)
20. Z. Papadopolos, P. Kramer, D. Zieidler, The F-type icosahedral phase – tilings and vertex models. J. Non-cryst. Solids **153–154**, 215–220 (1993)
21. Z. Papadopolos, R. Klitzing, P. Kramer, Quasiperiodic icosahedral tilings from the six-dimensional BCC lattice. J. Phys. A Math. Gen. **30**, L143–L147 (1997)
22. J. Patera, R. Twarock, Affine extensions of noncrystallographic Coxeter groups and quasicrystals. J. Phys. A Math. Gen. **35**, 1551–1574 (2002)
23. M. Pitteri, G. Zanzotto, *Continuum Models for Phase Transitions and Twinning in Crystals* (CRC/Chapman and Hall, London, 2002)
24. C.A. Reiter, Atlas of quasicrystalline tilings. Chaos Solitons Fract. **14**(7), 937–963 (2002)
25. I.K. Robinson, S.C. Harrison, Structure of the expanded state of tomato bushy stunt virus. Nature **297**, 563–568 (1982)
26. M. Senechal, *Quasicrystals and Geometry* (Cambridge University Press, Cambridge, 1996)
27. M.B. Sherman, H.R. Guenther, F. Tama, T.L. Sit, C.L. Brooks, A.M. Mikhailov, E.V. Orlova, T.S. Baker, S.A. Lommel, Removal of divalent cations induces structural transitions in red clover necrotic mosaic virus, revealing a potential mechanism for RNA release. J. Virol. **80**(21), 10395 (2006)
28. J.A. Speir, S. Munshi, G. Wang, T.S. Baker, J.E. Johnson, Structures of the native and swollen forms of the cowpea chlorotic mottle virus determined by X-ray crystallography and cryo-electron microscopy. Structure **3**, 63–78 (1995)
29. F. Tama, C.L. Brooks III, The mechanism and pathway of pH-induced swelling in cowpea chlorotic mottle virus. J. Mol. Biol. **318**, 733–747 (2002)
30. F. Tama, C.L. Brooks III, Diversity and identity of mechanical properties of icosahedral viral capsids studies with elastic network normal mode analysis. J. Mol. Biol. **345**, 299–314 (2005)
31. T.J. Tuthill, K. Harlos, T.S. Walter, N.J. Knowles, E. Groppelli, D.J. Rowlands, D.I. Stuart, E.E. Fry, Equine rhinitis A virus and its low pH empty particle: clues towards an aphthovirus entry mechanism? PLoS Pathog. **5**(10), e1000620 (2009)
32. R. Twarock, A tiling approach to virus capsid assembly explaining a structural puzzle in virology. J. Theor. Biol. **226**(4), 477–482 (2004)
33. R. Twarock, Mathematical virology: a novel approach to the structure and assembly of viruses. Phil. Trans. R. Soc. **364**, 3357–3373 (2006)
34. R. Twarock, T. Keef, Viruses and geometry: where symmetry meets function. Microbiol. Today **37**, 24–27 (2010)

Minimal Tile and Bond-Edge Types
for Self-Assembling DNA Graphs

Joanna Ellis-Monaghan, Greta Pangborn, Laura Beaudin, David Miller,
Nick Bruno, and Akie Hashimoto

Abstract We employ a model for self-assembling graph-theoretical complexes
using tiles representing branched-junction DNA molecules with free cohesive ends.
We determine the minimum number of tile and bond-edge types necessary to create
a given graph as a self-assembled complex under three different scenarios: (1)
where the incidental creation of complexes of smaller size than the target graph
is acceptable; (2) where the incidental creation of complexes the same size as the
target graph is acceptable, but not smaller complexes; and (3) where no complexes
the same size as or smaller than the target graph are acceptable. In each of these
cases, we find bounds for the minimum number of tile and bond-edge types that
must be designed, and give specific minimum values for common graph classes
(including cycles and trees, as well as complete, bipartite, and regular graphs). For
these classes of graphs, we provide either explicit descriptions of optimal tile sets
or efficient algorithms for generating the desired set.

1 Introduction

The promise of nanotechnology, in particular for drug delivery, biosensors, and
biomolecular computing (see [1, 15, 16, 19, 24]), has driven recent research into
DNA self-assembly of nanoscale geometric constructs. We focus here on the
assembly of graph-theoretical complexes, that is, molecules whose structure may
be modeled as the vertices and edges of a graph. Several different graph-theoretical
complexes have been constructed from self-assembling DNA molecules, including
cubes [3], truncated octahedra [25], rigid octahedra [21], and buckyballs (see the
Molbel announcement in [4]). An essential step in building a self-assembling

J. Ellis-Monaghan (✉) · G. Pangborn · L. Beaudin · D. Miller · N. Bruno · A. Hashimoto
Saint Michael's College, Colchester, VT, USA
e-mail: jellis-monaghan@smcvt.edu; gpangborn@smcvt.edu

N. Jonoska and M. Saito (eds.), *Discrete and Topological Models in Molecular Biology*, 241
Natural Computing Series, DOI 10.1007/978-3-642-40193-0_11,
© Springer-Verlag Berlin Heidelberg 2014

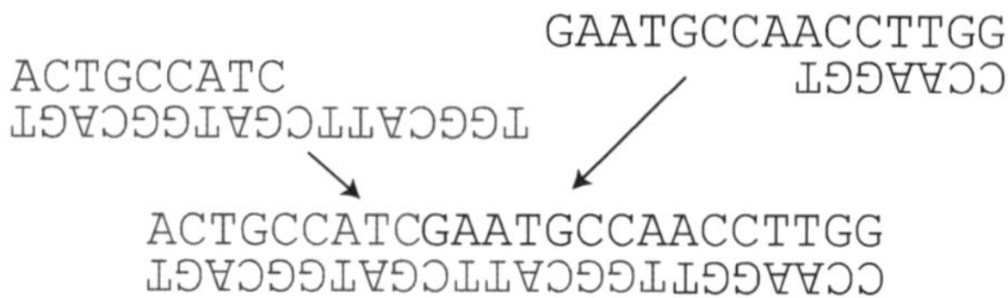

Fig. 1 Two DNA molecules join via bonding of complementary base pairs on cohesive ends

DNA nanoconstruct, whether for biomolecular computing or physical applications, is designing the component molecules. There are several models for DNA self-assembly including linear-strand methods [1, 3, 25], branched-junction molecules [11], and origami foldings [17]. Here we consider the flexible-tile model introduced in [13]. We address approaching this process as efficiently as possible by providing theoretical tools to minimize the number of component molecules that must be designed to create a given nanoconstruct.

In [13], Jonoska et al. introduced a formalism for analyzing the combinatorial properties underlying self-assembly via branched-junction molecules. This involved a combinatorial representation of flexible-armed branched-junction molecules, called *tiles*, and the notion of a *pot*, that is, a set of these tiles. With this, they addressed the questions of guaranteeing that a given pot will produce complexes and also of what complete complexes might be assembled from a given pot. Although there are some difference in the technical details (for example, we use half-edges here, whereas [13] used vertices of degree one), the essential ideas here are based on those of [12–14, 22], particularly the use of a matrix associated with a pot. We emphasize, though, that whereas [12–14, 22] were concerned with the question of what complete complexes might be assembled from a given pot, here we address the inverse problem: given a target graph, what is the smallest-size pot from which a complete complex with the structure of the given graph might be assembled. In a related direction, [8, 9, 11, 18] have demonstrated the utility of this model for biomolecular computing and considered associated computational complexity questions.

Given a connected graph (called the target graph, and allowed to have loops and multiple edges), we want to design a self-assembling DNA complex with this structure. The basic building blocks of our model are k-armed branched-junction molecules, for example those developed by Wang et al. [23], whose arms, in the simplest case, are double strands of DNA with one strand extending beyond the other. This strand forms a 'cohesive end' at the end of the arm that can bond to any other cohesive end with complementary Watson–Crick bases (see Fig. 1). While there are more sophisticated branched-junction molecules, for example those described in [7], the essential property is some form of cohesive attachment at the ends of arms. Like [11–14, 22], we assume that the arms are sufficiently long and flexible enough to achieve the targeted connectivity. We note that although we focus here on DNA assembly, the methods presented may apply to any form of self-assembly, at any scale, that may be modeled by armed components with bonding sites at the ends of the arms.

We seek to construct the target graph from these branched molecules, with a vertex of degree k given by a k-armed branched molecule, and where two vertices in the target graph have an edge between them if and only if their two branched molecules have an arm joined between them via complementary bases on the cohesive ends. Thus, the joined arms form the edges of the target graph. We require that the resulting DNA complex be complete, that is, that it has no unmatched cohesive ends.

We call the combinatorial abstraction of a branched molecule a tile, and the abstraction of a cohesive end together with its complementary cohesive end a bond edge. Tile types and bond-edge types are defined more formally in Sect. 3. We represent the bond-edge types by letters, with the corresponding cohesive ends represented by hatted and unhatted copies of the letter. The tile type is the multiset of letters corresponding to the cohesive end types for the tile. Thus, given a graph, we wish to know the minimum number of tile and bond-edge types that must be designed to construct the complex. We consider this question under three different conditions:

- *Scenario 1.* We allow the possibility that graph-theoretical complexes of smaller size (that is, complexes representing graphs with fewer vertices) than the target graph could be created from the set of tile types used to build the target graph.
- *Scenario 2.* We allow the possibility that graph-theoretical complexes with the same number of vertices as, but not isomorphic to, the target graph could be created from the set of tile or bond-edge types that builds the target graph, but require that no complexes with fewer vertices can be created from the set of tile types used to build the target graph.
- *Scenario 3.* We require that no complexes with a number of vertices less than or equal to that of the target graph can be created from the set of tile types used to build the target graph.

The more restrictive design rules generally decrease the likelihood of extraneous (undesired) complexes being formed in the laboratory, but at the cost of requiring the design of additional component molecules. For flexible tiles, it is impossible to prevent the creation of larger complexes, as we discuss in Sect. 3. For each of these scenarios, we provide general upper and lower bounds for these values and determine specific values for some important classes of graphs. We either provide constructive proofs, that is, we give explicit sets of tile types achieving these bounds, or provide fast algorithms to generate the desired sets.

We let $T_i(G)$, for $i = 1, 2, 3$ in each of the three scenarios, respectively, denote the minimum number of tile types needed to construct a graph G. Likewise, in scenario i, for $i = 1, 2, 3$, we let $B_i(G)$ denote the minimum number of bond-edge types needed to construct G.

Our results for $T_i(G)$ for $i = 1, 2, 3$ on various classes of graphs are summarized in Table 1. Here we write $av(G)$, $ev(G)$, and $ov(G)$ for the numbers of different vertex degrees, different even vertex degrees, and different odd vertex degrees, respectively, that appear in the graph G. By convention, C_n refers to the cycle on n vertices, K_n refers to the complete graph on n vertices, and $K_{m,n}$ refers to the

Table 1 Minimum tile types

Scenario 1	$T_1(G) = $ Minimum number of tile types required if complexes of smaller size than the target graph are allowed
General graph G	$av(G) \le T_1(G) \le ev(G) + 2ov(G)$
Trees	$av(\mathscr{T}) \le T_1(\mathscr{T}) \le av(\mathscr{T}) + 1$
C_n	$T_1(C_n) = 1$
K_n	$T_1(K_n) = 1$ if n is odd, and $T_1(K_n) = 2$ if n is even
$K_{m,n}$	$T_1(K_{m,n}) = 1$ if $n = m$ and is even, and $T_1(K_{m,n}) = 2$ otherwise
K-regular graphs	$T_1(G) = 1$ if n is even, and $T_1(G) = 2$ if n is odd
Scenario 2	$T_2(G) = $ Minimum number of tile types required if complexes of the same size as the target graph but not smaller are allowed
Trees	$T_2(\mathscr{T}) = $ Number of different lesser-size subtree sequences
C_n	$T_2(C_n) = \lceil n/2 \rceil + 1$
K_n	$T_2(K_n) = 2$ if n is even, and $T_2(K_n) = 3$ if n is odd
$K_{m,n}$	$T_2(K_{m,n}) = 2$ if $\gcd(m,n) = 1$, and $T_2(K_{m,n}) = 3$ if $\gcd(m,n) > 1$
Scenario 3	$T_3(G) = $ Minimum number of tile types required if no complexes of the same size as the target graph or smaller are allowed
Trees	$T_3(\mathscr{T}) = $ Cardinality of the maximum subset of induced nonisomorphic trees
C_n	$T_3(C_n) = \lceil n/2 \rceil + 1$
K_n	$T_3(K_n) = n$
$K_{m,n}$	$T_3(K_{m,n}) = m + 1$

complete bipartite graph with partitions of size m and n (where $m \le n$). Lesser-size subtree sequences are defined in Sect. 5, and the maximum subset of induced nonisomorphic trees is defined in Sect. 6. We shall show that $B_1(G) = 1$ for all graphs, and that $B_2(G) + 1 \le T_2(G)$. Although it is not true in general, it is true for the specific classes that we have here, except for trees, that $B_2(G) + 1 = T_2(G)$ and $B_3(G) + 1 = T_3(G)$.

It is difficult at this stage to predict the directions that will be taken in the future by laboratories producing these nanostructures. The graph-theoretical complexes assembled previously have, for the most part, been the skeletons of Platonic or Archimedean solids. The graphs in these two small, finite families may be treated individually, and design strategies for them may be found at [20]. Here we seek general strategies that may be applied to common infinite families of graphs, such as complete or bipartite graphs, and as such may be adapted more readily to emerging design challenges as laboratories begin to build more complex structures.

2 Graph Theory Conventions

The following conventions are used throughout this chapter. Graphs are finite and connected, with loops and multiple edges allowed. The size of a graph is the number of its vertices, and we say that a graph G is smaller than a graph H if G has fewer vertices than H. We typically write n or $|V(G)|$ for the number of vertices of G, and $|E(G)|$ for the number of edges of G. We also consider digraphs, that is, graphs with a direction, usually indicated by an arrow, assigned to each edge. An edge with an assigned direction is also called an *arc*. Because an arm of a branched-junction molecule forms half an edge in the target graph, it is natural to consider half-edges and half-arcs here. In a geometric realization of a graph, for example an embedding of the graph in the plane or in three-space, an edge is simply a curve joining two points, namely the vertices at its endpoints. A half-edge is then simply the subcurve from the midpoint of this curve to one or the other of its endpoints. Half-arcs are similarly defined, with the addition that the curve is directed from one endpoint to the other. In an abstract setting, where the graph G is defined in terms of a vertex set V, and an edge set E consisting of unordered pairs of elements of V, a half-edge may be denoted as, for example, $(v, \{v, u\})$, where $v \in V$ and $\{v, u\} \in E$; here, $(v, \{v, u\})$ is the half-edge of $\{u, v\}$ incident with the endpoint v. A full formalization of the notions of edges, arcs, half-edges, half-arcs, etc. may be found in the introductory chapter, "mixed graphs and their basic parts," of the book by Fleischner [5].

We use two notations for a vertex together with its incident half-edges: v_*, if it is not necessary to specify the half-edges, and $(v; e_1, \ldots, e_k)$, if the half-edges are specified ($e_i = e_j$ for some i, j in the case of a loop).

The applications we consider often involve Eulerian graphs, which are connected graphs wherein the degree of every vertex is even. An Eulerian digraph has directed edges (i.e., arcs) with equal indegree and outdegree at each vertex. A *walk* traverses consecutive edges in a graph (following the direction of the arcs in the case of a digraph), allowing repeated edges and vertices; a *trail* allows repeated vertices but not edges; and a *path* repeats neither. A *circuit* is a closed trail, and a *cycle* is a closed path. Given a connected graph G, an *augmented graph* results from drawing an edge between any two vertices of odd degree and continuing the process until no odd-degree vertices remain (a graph necessarily has an even number of odd-degree vertices). The resulting augmented graph is then Eulerian. A full formalization of these concepts may be found, for example, in the books by Fleischner [5, 6].

3   Combinatorial Formulation of Design Parameters

We formalize the essential combinatorial properties of branched-junction molecules as follows. A *tile* is a graph-theoretical representation of a branched DNA molecule with cohesive ends as a vertex with some half-edges. Cohesive ends are distinguished by cohesive end types (letter labels on the half-edges) such that a cohesive end labeled with an "unhatted" letter can join to a cohesive end labeled with its complementary "hatted" label (e.g., the cohesive end types c and $\hat{c}$ represent complementary strands of bases such as those in Fig. 1, and so could form an edge). The letters used (without regard to being hatted or not) are the *bond-edge types*. Thus, combinatorially, a tile may be thought of as a vertex with labeled half-edges.

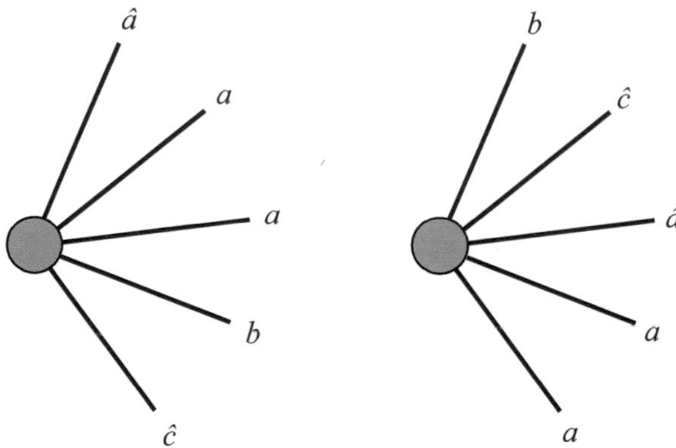

Fig. 2 Two tiles, t_1 and t_2, both representating the same tile type

Because we assume that the branched molecules have flexible arms, we do not distinguish between permutations of cohesive end labels about a vertex, so a tile may be identified with the multiset of its cohesive end labels. In Fig. 2, tiles t_1 and t_2 are of the same tile type, since they correspond to the same multiset, $\{a^2, \hat{a}, b, \hat{c}\}$, where the exponent indicates a repeat in the multiset. A complete complex constructed using tiles t_1 and t_2 would have the bond-edge types a, b, and c. Note that our model differs from the model introduced in [11] in that we allow loops (which may be formed by a tile type containing both a hatted and an unhatted cohesive end corresponding to the same bond-edge type).

A pot P is a set of tile types such that for each cohesive end of type h that appears in any tile $t_i \in P$, there exists a cohesive end of type $\hat{h}$ (its complement) in some tile $t_j \in P$ (possibly $i = j$), and vice versa. We say that a graph G may be constructed from a pot P, or equivalently that the pot P *realizes* G, if there is a map $f : \{v_*\} \to P$ from the set of vertices with half-edges to the tile types with the following properties:

1. If $v_* \mapsto t$, then there is an associated one-to-one correspondence between the cohesive ends of t and the half-edges of v.
2. If $\{u, v\} \in E(G)$, then the two half-edges of $\{u, v\}$ are assigned complementary cohesive ends; that is, if $v_* \mapsto t_1$ with a cohesive end h corresponding to $(v, \{v, u\})$, and $u_* \mapsto t_2$, then t_2 must have a cohesive end $\hat{h}$ corresponding to $(u, \{u, v\})$.

These two conditions taken together ensure that any graph constructed from P is realized by a complete complex; that is, there are no unmatched cohesive ends. Note that any tile type may be used multiple times in the construction of a graph G.

Proposition 1. $B_1(G) \le B_2(G) \le B_3(G)$ *and* $T_1(G) \le T_2(G) \le T_3(G)$.

Proof. It is clear that any pot that realizes G with $B_3(G)$ bond-edge types or $T_3(G)$ tile types, but does not realize any graph smaller than or the same size as but not

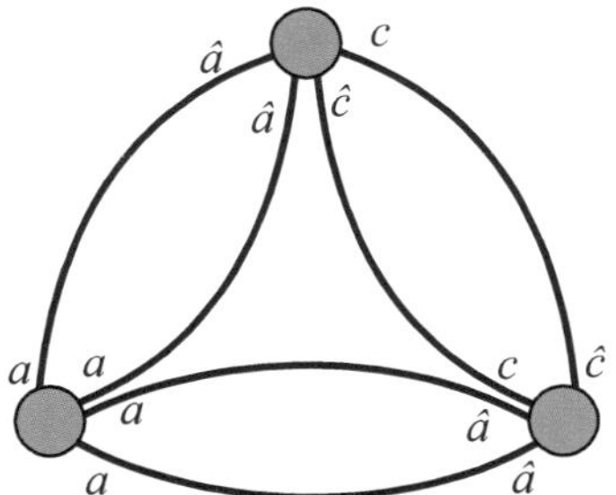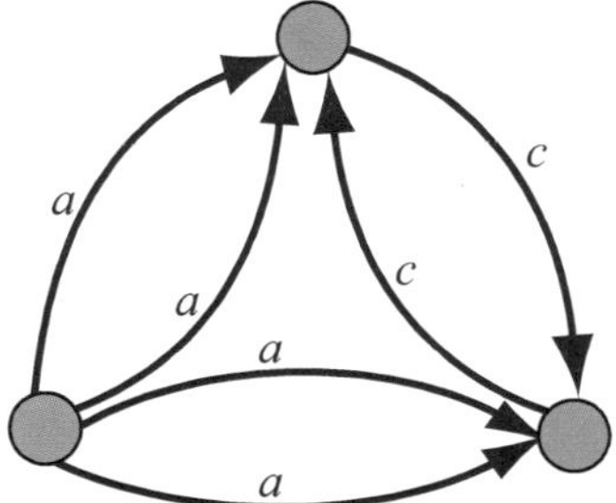

Fig. 3 Two equivalent edge labelings

isomorphic to G, does not realize any graph smaller than G. Hence $B_2(G) \leq B_3(G)$ and $T_2(G) \leq T_3(G)$, and similarly for $B_1(G) \leq B_2(G)$ and $T_1(G) \leq T_2(G)$. $\qquad\square$

Definition 1. Given a pot P, we define $C(P)$ to be the set of graphs that can be constructed from P. The set of graphs of minimum size that may be constructed from P is denoted $C_{\min}(P)$. We write m_P for the size of a smallest graph that may be constructed from P.

Because there typically are arbitrarily many tiles of each type present in the assembly process, a pot that realizes a graph G may realize many other graphs as well. In particular, if a pot realizes a loopless graph G, then it can theoretically realize any covering graph of G (see [10]). A graph H covers a graph G if there is a function from the vertices of H to the vertices of G that induces a surjection on the edges incident to every vertex. For example, by identifying antipodal points, it can be seen that a cube is a covering graph of K_4, so any pot that realizes K_4 will also realize a cube.

Since we require a complex to be complete, we may adopt the convention of orienting edges from unhatted cohesive ends toward hatted cohesive ends. With this convention, a complete complex may be identified with an edge labeling of some orientation of the underlying graph. That is, any complex induces an orientation and an edge labeling of the underlying graph, but what is more important, from any edge labeling of an oriented graph, one may 'read off' a pot from which the corresponding complex may be constructed by listing the multiset of cohesive end types incident with each vertex. For example, the pot read off from either of the labeled graphs in Fig. 3 is $P = \{t_1 = \{\hat{a}^2, \hat{c}, c\}, t_2 = \{a^4\}\}$.

We will frequently use the following propositions and matrix formulation, the essential aspects of which appear in [8, 13, 14]. However, we emphasize that we are addressing a different question here. Whereas prior work began with a pot P and asked what complete complexes could be constructed from that pot, here we begin with a target graph and ask for a minimal pot that realizes it.

Given a pot $P = \{t_1, \ldots, t_p\}$, we define $A_{i,j}$ to be the number of cohesive ends of type a_i on tile t_j, and $\hat{A}_{i,j}$ to be the number of cohesive ends of type $\hat{a}_i$.

The observations below are immediate consequences of requiring complexes to be complete.

Proposition 2. *Let $P = \{t_1, \ldots, t_p\}$ be a pot. Then:*

1. *The total number of hatted cohesive end types must equal the total number of unhatted cohesive end types in a complete complex.*
2. *If a graph G with n vertices may be constructed from the pot P, then there are nonnegative integers R_j for $j = 1, \ldots, p$ (representing the number of each tile of type t_j used in the construction of G) with $\sum_j R_j = n$ and such that $\sum_j R_j(A_{i,j} - \hat{A}_{i,j}) = 0$ for all i. That is, the number of hatted cohesive ends of each type used in the construction of G must equal the number of unhatted cohesive ends of the same type that appear in the construction.*

It is often expedient to encode this information in a matrix.

Definition 2. Let P be a pot with p tile types labeled $t_1, \ldots, t_p$, and let $z_{i,j}$ be the net number of cohesive ends of type a_i on tile t_j, i.e., $z_{i,j} = A_{i,j} - \hat{A}_{i,j}$. Define r_i to be the proportion of tile type t_i used in the assembly process. The following system of equations captures the requirements outlined in Item 2 of Proposition 2:

$$z_{1,1}r_1 + z_{1,2}r_2 + \ldots + z_{1,p}r_p = 0$$

$$\vdots$$

$$z_{m,1}r_1 + z_{m,2}r_2 + \ldots + z_{m,p}r_p = 0$$

$$r_1 + r_2 + \ldots + r_p = 1$$

The *construction matrix of P*, $M(P)$, is the corresponding augmenting matrix:

$$M(P) = \begin{bmatrix} z_{1,1} & z_{1,2} & \cdots & z_{1,p} & 0 \\ \vdots & \vdots & & \vdots & \\ z_{m,1} & z_{m,2} & \cdots & z_{m,p} & 0 \\ 1 & 1 & \cdots & 1 & 1 \end{bmatrix}. \tag{1}$$

Proposition 3. *Let $P = \{t_1, \ldots, t_p\}$ be a pot. Then:*

1. *If a graph G of size n may be constructed from P, using R_j tiles of type T_j, then $(1/n)\langle R_1, \ldots, R_p \rangle$ is a solution of the construction matrix $M(P)$.*
2. *If $\langle r_1, \ldots, r_p \rangle$ is a solution of the construction matrix $M(P)$, and there is a positive integer n such that $nr_j \in \mathbb{Z}_{\geq 0}$ for all j, then there is a graph of size n that may be constructed from P using nr_j tiles of type T_j.*
3. *$m_P = \min\{\text{lcm}\{b_j \mid r_j \neq 0 \text{ and } r_j = a_j/b_j\}$, where $\langle r_1, \ldots, r_p \rangle$ is a solution to $M(P)\}$, and where the minimum is taken over all solutions to $M(P)$ such that $r_j \geq 0$ and a_j/b_j is in reduced form for all j.*

Proof. Item 1 is just a restatement of Item 2 of Proposition 2. Item 2 is a consequence of the use of flexible-armed tiles: provided that there are the same numbers of hatted and unhatted cohesive ends of each bond-edge type summed over all the tiles used, these cohesive ends may join with one another to form a complete complex. Being a solution to $M(P)$ assures this. For Item 3, note that since the entries of $M(P)$ are integers, a solution to $M(P)$ must have rational entries. Thus, the least common multiple of the denominators will be the smallest integer multiplier n that makes all the entries integers, and hence gives the size of the smallest graph corresponding to that solution. We take the minimum of this over all solutions to $M(P)$ with nonnegative entries, i.e., those that correspond to graphs in $C(P)$. $\quad\square$

We make one further observation.

Proposition 4. *If P and P' are pots such that $M(P) \sim M(P')$, then P and P' realize graphs of exactly the same size, so that, in particular, $m_P = m_{P'}$.*

Proof. Suppose G is a graph of size n realized by P. Then, by Proposition 2, the proportions of tiles used, $\langle r_1, \ldots, r_p \rangle$, are a solution to $M(P)$, and $nr_j \in \mathbb{Z}_{\geq 0}$ for all j. Since $M(P) \sim M(P')$, then $\langle r_1, \ldots, r_p \rangle$ is also a solution to $M(P')$ with $nr_j \in \mathbb{Z}_{\geq 0}$ for all j, so, again by Proposition 2, there is a graph G' of size n that is realized by P'. Similarly, if G' is any graph realized by P', then there is a graph G of the same size realized by P. $\quad\square$

Note, however, in Proposition 4 that $M(P) \sim M(P')$ does not necessarily mean that P and P' have the same sets of tiles, nor that they realize the same graphs. In particular, while the number of vertices may be the same for two graphs constructed from proportions $\langle r_1, \ldots, r_p \rangle$ solving both $M(P)$ and $M(P')$, they may have different numbers of edges.

4   Scenario 1

In this scenario, we allow the possibility that graph-theoretical complexes of smaller size could be created from the set of tiles that builds the target graph G. Thus, we seek pots P using the smallest number of bond-edge or tile types so that $G \in C(P)$.

Recall that the vertex degree sequence of a graph G is the sequence, in increasing order, of the degrees of each of the vertices of G. This sequence has repeated values if G has multiple vertices of the same degree. Here, however, we are usually concerned with how many different degrees appear in the graph, so we define the *valency sequence* of G to be the sequence of vertex degrees of G without repeats. The *even-valency sequence* is the sequence of even degrees that appear in G, and the *odd-valency sequence* the sequence of odd degrees. We write $av(G)$, $ev(G)$, and $ov(G)$ for the lengths of the valency, even-valency, and odd-valency sequences, respectively.

We begin with a linear-time algorithm that, given a target graph G, returns at most $ev(G) + 2ov(G)$ tile types from which G may be constructed, and uses exactly one bond-edge type.

Algorithm 1. *Input: A target graph G.*

Output: *At most $ev(G) + 2ov(G)$ tile types from which G may be constructed.*

1. *Create an augmented graph G' from G by adding edges between pairs of odd-degree vertices. G' is then Eulerian.*
2. *Construct an Eulerian circuit for G'.*
3. *Choose a direction to traverse the Eulerian circuit, and then record the orientation of each edge as the circuit is traversed.*
4. *Delete the augmented edges, leaving an orientation $\overrightarrow{G}$ of the original graph G.*
5. *Record a tile type with j cohesive ends of type a and k cohesive ends of type $\hat{a}$ whenever there is a vertex of $\overrightarrow{G}$ with indegree j and outdegree k.*

Correctness: Since the orientation came from an Eulerian circuit in the augmented graph of G, it follows that there is one tile type for each element of the even-valency sequence of G, and it will have equal numbers of cohesive ends of type a and type $\hat{a}$. There will be at most two tile types for each element of the odd-valency sequence of G, and the numbers of cohesive ends of type a and type $\hat{a}$ will differ by 1. Thus, the algorithm outputs at most $ev(G) + 2ov(G)$ tile types.

Running time: The running time is $O(|E(G)|)$, since an Eulerian circuit can be found in linear time and the additional preprocessing and labeling work is also linear.

Corollary 1. $B_1(G) = 1$, *for all G.*

Proof. This follows immediately from Algorithm 1, which uses only one cohesive end letter. $\square$

Theorem 1. $av(G) \leq T_1(G) \leq ev(G) + 2ov(G)$, *and these bounds are tight.*

Proof. The lower bound is clear, since vertices of different degrees require different tile types. The upper bound follows from Algorithm 1. That the bounds are tight follows from observing that, for Eulerian graphs, $av(G) = ev(G) + 2ov(G)$. $\square$

Corollary 2. *For C_n, the cycle on n vertices, $T_1(C_n) = 1$.*

Corollary 3. *If G is a k-regular graph, then $T_1(G) = 1$ if k is even and 2 if k is odd.*

Proof. The even case follows immediately from Theorem 1. The odd case follows from Theorem 1 combined with Proposition 2, which rules out the possibility of constructing the graph from a single tile with an odd number of cohesive ends. $\square$

Corollary 4. *For K_n, the complete graph with n nodes,*
$$T_1(K_n) = \begin{cases} 1 \text{ if } n \text{ is odd,} \\ 2 \text{ if } n \text{ is even.} \end{cases}$$

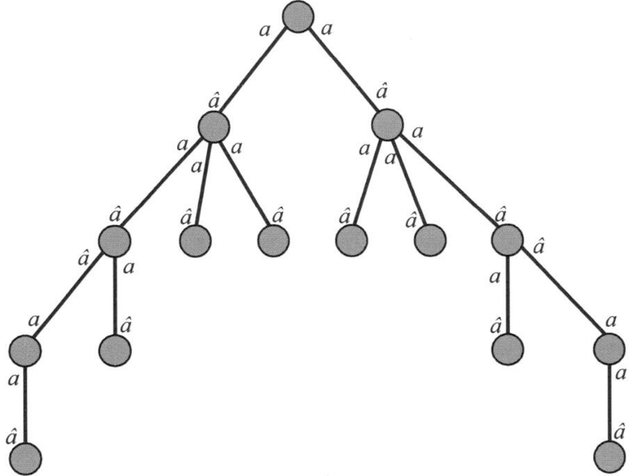

Fig. 4 Tree labeling showing that sufficient conditions are not necessary

Corollary 5. *For $K_{m,n}$, the complete bipartite graph,*

$$T_1(K_{m,n}) = \begin{cases} 1 \text{ if } m = n \text{ and both are even,} \\ 2 \text{ otherwise.} \end{cases}$$

Proposition 5. *If $\mathcal{T}$ is a tree, then $av(\mathcal{T}) \leq T_1(\mathcal{T}) \leq av(\mathcal{T}) + 1$, and these bounds are tight.*

Proof. For the upper bound, choose any vertex of $\mathcal{T}$ as a root, and direct the edges of G away from the root. If the root has degree d, use one tile type with d cohesive ends of type a. For every other element of the valency sequence, create a tile type with one cohesive end of type $\hat{a}$ and the rest of type a. If there are vertices other than the root with degree d, then create an additional tile type with one cohesive end of type $\hat{a}$ and $d - 1$ of type a. This gives at most $av(\mathcal{T}) + 1$ tile types in general, and exactly $av(\mathcal{T})$ tile types if $\mathcal{T}$ has a vertex of unique degree which we then choose as the root. The path on four vertices demonstrates that the upper bound is tight. $\square$

Proposition 6. *If G is a bipartite graph with all vertices of the same degree in the same bipartition, then $T_1(G) = av(G)$.*

Proof. If the bipartition is (X, Y), then all the vertices in X may be represented by a tile type with all cohesive ends of type a, and all vertices in Y by a tile type with all cohesive ends of type $\hat{a}$. $\square$

Proposition 6 gives a second sufficient condition for a tree, other than having a vertex of unique degree, to achieve the lower bound on the minimum number of tile types. However, neither condition is necessary, as illustrated in Fig. 4, which gives a tree that achieves the lower bound while satisfying neither the bipartition condition nor the vertex-of-unique-degree condition.

5 Scenario 2

In this scenario, we allow the possibility that graph-theoretical complexes the same size as, but not isomorphic to, the target graph could be created from the set of tiles that builds the target graph G, but require that no smaller complexes can be built from the set of tiles. Thus, we seek pots P using the smallest number of bond-edge or tile types such that $G \in C_{\min}(P)$.

We begin with a theorem relating the minimum number of bond-edge types and the minimum number of tile types, which will streamline our work.

Theorem 2. *If G is a graph with $n > 2$ vertices, then $B_2(G) + 1 \leq T_2(G)$.*

Proof. Suppose P is a pot with $p = T_2(G)$ tile types and $m \geq B_2(G)$ bond-edge types that realizes G, and consider the construction matrix $M(P)$,

$$\begin{bmatrix} z_{1,1} & z_{1,2} & \cdots & z_{1,p} & 0 \\ \vdots & \vdots & & \vdots & \\ z_{m,1} & z_{m,2} & \cdots & z_{m,p} & 0 \\ 1 & 1 & \cdots & 1 & 1 \end{bmatrix}. \tag{2}$$

Suppose that the rows are not linearly independent, and let row_i refer to the ith row of the construction matrix. Since the bottom row is clearly not in the span of the remaining rows, this means that $\mathrm{row}_m = \sum_{i=1}^{m-1} \alpha_i \, \mathrm{row}_i$, with not all $\alpha_i = 0$. By switching the roles of a_m and $\hat{a}_m$ if necessary (this corresponds to multiplying row_m by -1), we may assume that there is a b with $\alpha_b \notin \{0, -1\}$.

Now let P' be the pot derived from P by first interchanging all the a_m's and $\hat{a}_m$'s in each tile type if necessary so that there is an $\alpha_b \notin \{0, -1\}$ in the equation $\mathrm{row}_m = \sum_{i=1}^{m-1} \alpha_i \, \mathrm{row}_i$, and then replacing every cohesive end of type a_m and $\hat{a}_m$ by a_b and $\hat{a}_b$, respectively, for each tile type.

The construction matrix $M(P')$ is then

$$\begin{bmatrix} z_{1,1} & z_{1,2} & \cdots & z_{1,p} & 0 \\ \vdots & \vdots & & \vdots & \\ z_{b,1} + z_{m,1} & z_{b,2} + z_{m,2} & \cdots & z_{b,p} + z_{m,p} & 0 \\ \vdots & \vdots & & \vdots & \\ 0 & 0 & \cdots & 0 & 0 \\ 1 & 1 & \cdots & 1 & 1 \end{bmatrix}. \tag{3}$$

Note that $M(P') \sim M(P)$ since $M(P')$ is the result of performing elementary row operations on $M(P)$: first multiply row_m by -1 if necessary, then add row_m to row_b, and then add $1/(1 + \alpha_b) \sum_{i=1}^{m-1} \alpha_i \, \mathrm{row}_i$ to row_m. The pot P' clearly constructs G, and, by Proposition 4, it realizes no smaller graphs. However, P' uses one fewer bond-edge type than P.

We may then continue this process of reducing the number of bond-edge types until the rows are no longer linearly dependent, creating a pot P'' with B'' bond-edge types and $T_2(G)$ tile types that realizes G. However, since the rows are linearly independent, then $B'' \leq T_2(G)$. If $B'' = T_2(G)$, then Gauss–Jordan elimination gives the identity matrix, and hence there is no solution to the system of equations, a contradiction, and so $B''(G) + 1 \leq T_2(G)$. Noting that $B_2(G) \leq B''$ completes the proof. $\qquad\square$

Although in the instances of the standard graphs given below it happens that $B_2(G) + 1 = T_2(G)$, in general $T_2(G) - B_2(G)$ may be arbitrarily large, as in the following construction.

Proposition 7. *Given n, there exists a graph of size $n+1$ with $T_2(G) - B_2(G) = n$.*

Proof. Given n, let G be the bipartite graph with one vertex of degree $2^n - 1$ in one partition, and n vertices in the other partition, one of each of the degrees 2^i for $i \in \{0, \ldots, n-1\}$ (G has many multiple edges, with multiplicities 2^i for each i). Since G has $n + 1$ vertices, all with different degrees, $T_2(G) = n + 1$. Moreover, G may be constructed from the pot $P = \{a^i\}_{i=0}^{n-1} \cup \{\hat{a}^{2^n-1}\}$, which uses only one bond-edge type. To see that nothing smaller than G may be constructed from P, note that any graph H constructed from P must use at least one tile of type $\hat{a}^{2^n-1}$. Furthermore, any two copies of a tile of type a^i that are used in the construction of a graph may be replaced by one copy of a tile of type a^{i+1}, resulting in a graph of smaller size. Thus, a graph $H \in C_{\min}(P)$ must have one tile of type $\hat{a}^{2^n-1}$ and may have at most one tile of type a^i for each i. For the number of hatted cohesive ends to equal the number of unhatted cohesive ends, H must use exactly one tile of type a^i for each i. Thus, H is the same size as G, and so $G \in C_{\min}(P)$, and hence $B_2(G) = 1$. $\qquad\square$

Proposition 8. $B_2(C_n) = \lceil n/2 \rceil$.

Proof. If $B_2(C_n) < \lceil n/2 \rceil$, then at least one bond-edge type appears at least three times in the complete complex forming C_n. At least two of the three edges must have the same orientation going around the cycle, when the design strategy is viewed using the labeling convention illustrated on the right-hand side of Fig. 3. These two edges may be detached and rejoined as in Fig. 5 to make two smaller-size complexes. Therefore, $B_2(C_n) \geqslant \lceil n/2 \rceil$.

The following two pots, one for n even and one for n odd, build C_n using $\lceil n/2 \rceil$ bond-edge types, as shown in Fig. 6:

$$P_{\text{even}} = \left\{ t_1 = \{a_1^2\},\ t_i = \{\hat{a}_{i-1}, a_i\} \text{ for } i = 2, \ldots, \left\lceil \frac{n}{2} \right\rceil,\ t_{\lceil n/2 \rceil + 1} = \left\{ \hat{a}^2_{\lceil n/2 \rceil} \right\} \right\}, \tag{4}$$

and

$$P_{\text{odd}} = \left\{ t_1 = \{a_1^2\},\ t_i = \{\hat{a}_{i-1}, a_i\} \text{ for } i = 1, \ldots, \left\lceil \frac{n}{2} \right\rceil - 1,\ t_{\lceil n/2 \rceil} = \left\{ \hat{a}_{\lceil n/2 \rceil - 1}, a_{\lceil n/2 \rceil} \right\}, \right.$$

$$\left. t_{\lceil n/2 \rceil + 1} = \left\{ \hat{a}_{\lceil n/2 \rceil - 1}, \hat{a}_{\lceil n/2 \rceil} \right\} \right\}. \tag{5}$$

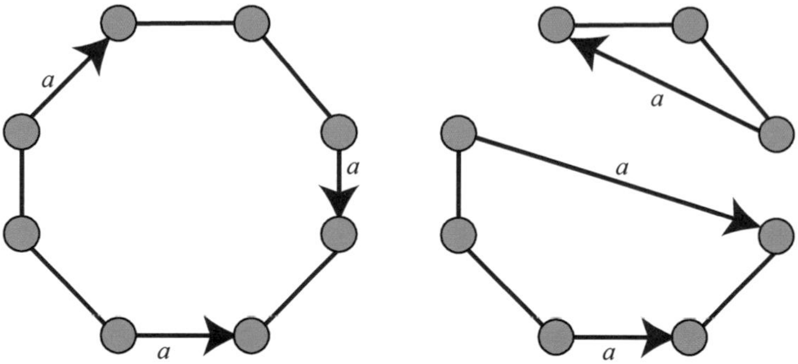

Fig. 5 Smaller graphs formed when there are three edges with the same bond-edge type

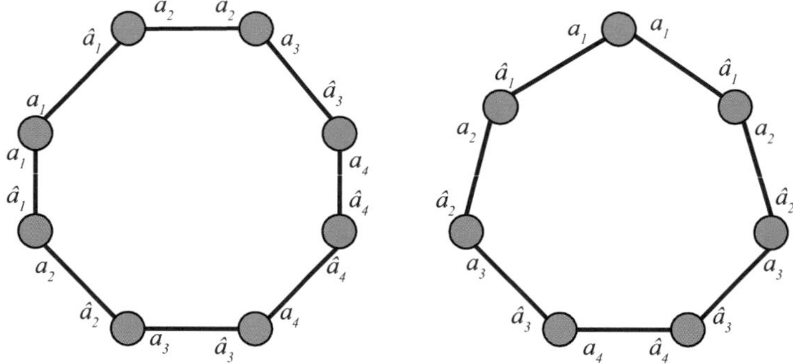

Fig. 6 Cycle constructions

To see that nothing smaller can be made from these pots, note that

$$
M(P_{\text{even}}) =
\begin{bmatrix}
2 & -1 & 0 & 0 & 0 & 0 & 0 & 0 \\
0 & 1 & -1 & 0 & 0 & 0 & 0 & 0 \\
0 & 0 & \ddots & \ddots & 0 & 0 & 0 & 0 \\
0 & 0 & 0 & \ddots & \ddots & 0 & 0 & 0 \\
0 & 0 & 0 & 0 & 1 & -1 & 0 & 0 \\
0 & 0 & 0 & 0 & 0 & 1 & -2 & 0 \\
1 & 1 & 1 & 1 & 1 & 1 & 1 & 1
\end{bmatrix}
\tag{6}
$$

has a unique solution of the form $\langle 1/n, 2/n, \ldots, 2/n, 1/n \rangle$, and

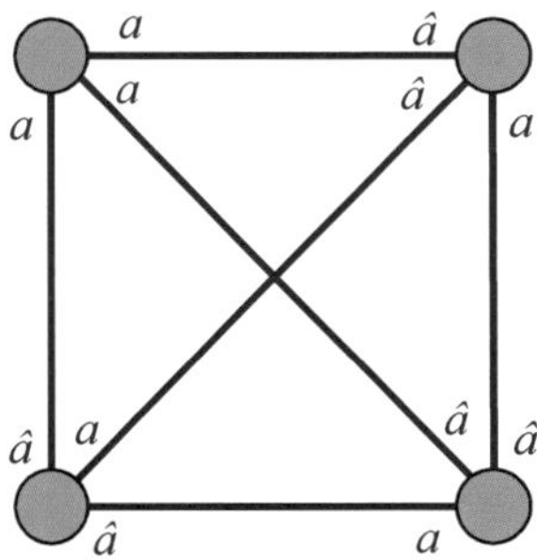
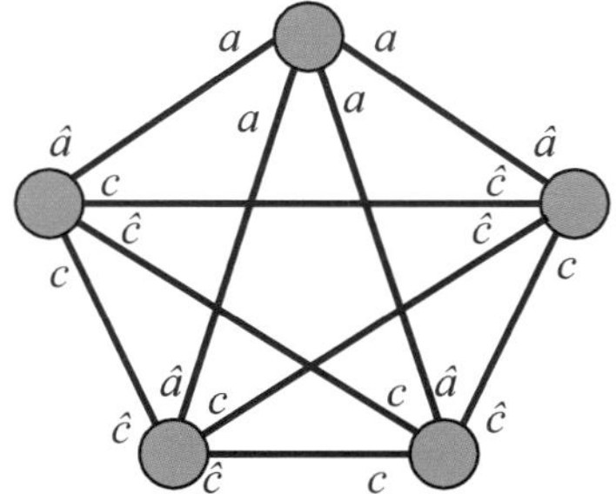

Fig. 7 Constructions of K_4 and K_5

$$M(P_{\text{odd}}) = \begin{bmatrix} 2 & -1 & 0 & 0 & 0 & 0 & 0 \\ 0 & 1 & -1 & 0 & 0 & 0 & 0 \\ 0 & 0 & \ddots & \ddots & 0 & 0 & 0 \\ 0 & 0 & 0 & 1 & -1 & -1 & 0 \\ 0 & 0 & 0 & 0 & -1 & 1 & 0 \\ 1 & 1 & 1 & 1 & 1 & 1 & 1 \end{bmatrix} \tag{7}$$

has a unique solution of the form $\langle 1/n, 2/n, \ldots, 2/n, 1/n, 1/n \rangle$, and apply Proposition 3, Item 3. $\qquad\square$

Corollary 6. $T_2(C_n) = \lceil n/2 \rceil + 1$.

Proof. By Theorem 2, $B_2(C_n) + 1 \le T_2(C_n)$, and the pots given in Proposition 8 achieve this bound. $\qquad\square$

Proposition 9.

$$B_2(K_n) = \begin{cases} 1 & \text{if } n \text{ is even,} \\ 2 & \text{if } n \text{ is odd.} \end{cases}$$

Proof. The following two pots, one for n even and one for n odd, realize K_n, as in Fig. 7:

$$P_{\text{even}} = \left\{ t_1 = \{a^{n-1}\}, \ t_2 = \{\hat{a}^{n/2}, a^{n/2-1}\} \right\}, \tag{8}$$

$$P_{\text{odd}} = \left\{ t_1 = \{a^{n-1}\}, \ t_2 = \{\hat{a}, c^{(n-3)/2}, \hat{c}^{(n-1)/2}\}, \ t_3 = \{\hat{a}, c^{(n-1)/2}, \hat{c}^{(n-3)/2}\} \right\}. \tag{9}$$

For n even, it then suffices to show that P_{even} realizes no smaller-size graph. For n odd, we must show not only that P_{odd} realizes no smaller-size graph, but also that $K_n \notin C_{\min}(P)$ for any pot with just one bond-edge type.

For n even, note that

$$M(P_{\text{even}}) = \begin{bmatrix} n-1 & -1 & 0 \\ 1 & 1 & 1 \end{bmatrix}, \tag{10}$$

which has $\langle 1/n, (n-1)/n \rangle$ as its unique solution, and apply Proposition 3, Item 3. For n odd, note that

$$M(P_{\text{odd}}) = \begin{bmatrix} n-1 & -1 & -1 & 0 \\ 0 & -1 & 1 & 0 \\ 1 & 1 & 1 & 1 \end{bmatrix}, \tag{11}$$

which has $\langle 1/n, (n-1)/2n, (n-1)/2n \rangle$ as its unique solution, where 2 divides $(n-1)$ since n is odd, and apply Proposition 3, Item 3.

Now, by way of contradiction, with n odd, assume there is a pot P with just one bond-edge type with $K_n \in C_{\min}(P)$. The construction matrix of P must have the form

$$M(P) = \begin{bmatrix} z_{1,1} & z_{1,2} & \cdots & z_{1,p} & 0 \\ 1 & 1 & \cdots & 1 & 1 \end{bmatrix}. \tag{12}$$

Since n is odd, each tile type has an even number of cohesive ends, and thus $z_{1,j} \neq 0$ for all j, since a tile with equal numbers of hatted and unhatted cohesive ends could form a graph with one vertex and $(n-1)/2$ loop edges. Again, because each tile type has an even number of cohesive ends, each $z_{1,j}$ must be even. Also, by Proposition 2, Item 1, not all of the $z_{1,j}$'s can have the same sign, so we may assume, by reordering the tile numbers if necessary, that $z_{1,1} > 0$ and $z_{1,2} < 0$. Thus, $M(P)$ is row-equivalent to

$$\begin{bmatrix} 1 & 0 & -\dfrac{-z_{1,3}+z_{1,2}}{z_{1,1}-z_{1,2}} & \cdots & -\dfrac{-z_{1,p}+z_{1,2}}{z_{1,1}-z_{1,2}} & -\dfrac{z_{1,2}}{z_{1,1}-z_{1,2}} \\ 0 & 1 & \dfrac{z_{1,1}-z_{1,3}}{z_{1,1}-z_{1,2}} & \cdots & \dfrac{z_{1,1}-z_{1,p}}{z_{1,1}-z_{1,2}} & \dfrac{z_{1,1}}{z_{1,1}-z_{1,2}} \end{bmatrix}, \tag{13}$$

and so has a solution of the form $\langle -z_{1,2}/(z_{1,1}-z_{1,2}), z_{1,1}/(z_{1,1}-z_{1,2}), 0, \ldots, 0 \rangle$. Since $z_{1,1}$ and $z_{1,2}$ are both even, and both are less than n in absolute value, this solution has the form $\langle a/m, b/m, 0, \ldots, 0 \rangle$, for some integer $m < n$, and so P realizes some graph of size $m < n$ by Proposition 3, Item 3, a contradiction. $\qquad\square$

Corollary 7.

$$T_2(K_n) = \begin{cases} 2 & \text{if } n \text{ is even,} \\ 3 & \text{if } n \text{ is odd.} \end{cases}$$

Proof. By Theorem 2, $B_2(K_n) + 1 \leq T_2(K_n)$, and the pots given in Proposition 9 achieve this bound. $\qquad\square$

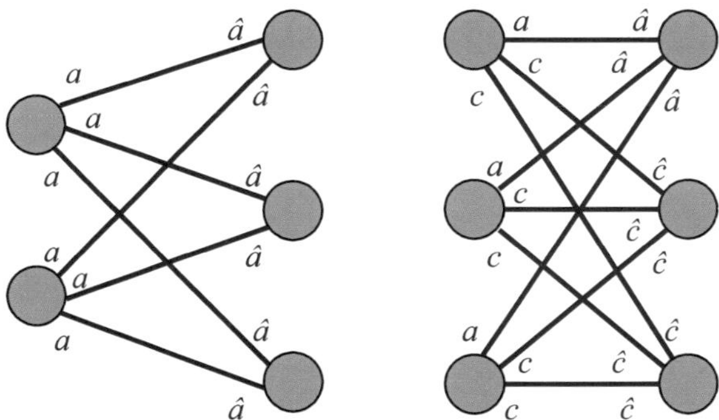

Fig. 8 Constructions of bipartite graphs

Proposition 10.

$$B_2\left(K_{m,n}\right) = \begin{cases} 1 \text{ if } \gcd\left(m,n\right) = 1, \\ 2 \text{ if } \gcd\left(m,n\right) > 1. \end{cases} \tag{14}$$

Proof. This proof follows the same form as that of Proposition 9. The following two pots, P_1 for $\gcd(m,n) = 1$ and P_d for $\gcd(m,n) = d$, realize $K_{m,n}$, as in Fig. 8:

$$P_1 = \{t_1 = \{a^n\},\ t_2 = \{\hat{a}^m\}\}, \tag{15}$$

$$P_d = \{t_1 = \{a, c^{n-1}\},\ t_2 = \{\hat{a}^m\},\ t_3 = \{\hat{c}^m\}\}. \tag{16}$$

It suffices to show that P_1 realizes no smaller-size graph, but we must show that P_d realizes no smaller-size graph and also that if $m = sd$ and $n = td$, then $K_{(sd,td)} \notin C_{\min}(P)$ for any pot with just one bond-edge type.

In the first case, note that

$$M(P_1) = \begin{bmatrix} n & -m & 0 \\ 1 & 1 & 1 \end{bmatrix}, \tag{17}$$

which has $\langle m/(n+m), n/(n+m)\rangle$ as its unique solution, and apply Proposition 3, Item 3. In the second case,

$$M(P_d) = \begin{bmatrix} 1 & -m & 0 & 0 \\ n-1 & 0 & -m & 0 \\ 1 & 1 & 1 & 1 \end{bmatrix}, \tag{18}$$

which has $\langle m/(n+m), 1/(n+m), n-1/(n+m)\rangle$ as its unique solution, so that it follows from Proposition 3, Item 3 that P_d realizes no smaller graph.

Now, with $m = sd$ and $n = td$, by way of contradiction, assume there is a pot P with just one bond-edge type with $K_{m,n} \in C_{min}(P)$. The construction matrix of P must have the form

$$M(P) = \begin{bmatrix} z_{1,1} & z_{1,2} & \cdots & z_{1,p} & 0 \\ 1 & 1 & \cdots & 1 & 1 \end{bmatrix}. \tag{19}$$

Again, $z_{1,j} \neq 0$ for all j, since a tile with equal numbers of hatted and unhatted cohesive ends could form a graph with one vertex and all loop edges, and, by Proposition 2, Item 1, not all of the $z_{1,j}$'s can have the same sign, so we may assume, by reordering the tile numbers and switching the roles of a and $\hat{a}$ if necessary, that t_1 has n cohesive ends with $z_{1,1} > 0$ and t_2 has m cohesive ends with $z_{1,2} < 0$. Now, $M(P)$ is row-equivalent to

$$\begin{bmatrix} 1 & 0 & -\dfrac{-z_{1,3}+z_{1,2}}{z_{1,1}-z_{1,2}} & \cdots & -\dfrac{-z_{1,p}+z_{1,2}}{z_{1,1}-z_{1,2}} & -\dfrac{z_{1,2}}{z_{1,1}-z_{1,2}} \\ 0 & 1 & \dfrac{z_{1,1}-z_{1,3}}{z_{1,1}-z_{1,2}} & \cdots & \dfrac{z_{1,1}-z_{1,p}}{z_{1,1}-z_{1,2}} & \dfrac{z_{1,1}}{z_{1,1}-z_{1,2}} \end{bmatrix}, \tag{20}$$

and so has a solution of the form $\langle -z_{1,2}/(z_{1,1} - z_{1,2}), z_{1,1}/(z_{1,1} - z_{1,2}), 0, \ldots, 0 \rangle$.

If either $z_{1,1} < n$ or $z_{1,2} < m$, then, by Proposition 3, Item 3, this solution implies that P realizes some graph of size less than $m + n$, a contradiction. Thus, $z_{1,1} = n = td$ and $z_{1,2} = m = sd$, and this solution becomes $\langle -sd/(td + sd), td/(td + sd), 0, \ldots, 0 \rangle = \langle -s/(t + s), t/(t + s), 0, \ldots, 0 \rangle$, implying that P realizes some graph of size $s + d < m + n$, again a contradiction. $\square$

Corollary 8.

$$T_2(K_{m,n}) = \begin{cases} 2 & \text{if } \gcd(m, n) = 1, \\ 3 & \text{if } \gcd(m, n) > 1. \end{cases}$$

Proof. By Theorem 2, $B_2(K_{m,n}) + 1 \leq T_2(K_{m,n})$, and the pots given in Proposition 10 achieve this bound. $\square$

We now turn our attention to trees. Let $\mathcal{T}$ be an unrooted tree. Deleting any edge of $\mathcal{T}$ results in two subtrees, one of which has a smaller size than the other (except possibly in the case of one unique edge, which we will call the size-center edge). We call this smaller-size subtree a lesser subtree of $\mathcal{T}$. In the case of a size-center edge, we arbitrarily designate one of the subtrees as the lesser subtree.

Definition 3. If we orient each edge of a tree $\mathcal{T}$ to point toward its lesser subtree, then we say T has a lesser-subtree orientation, or LSO.

See Fig. 9 for a sample tree and its corresponding lesser-subtree orientation. Essentially, creating an LSO corresponds to finding a "most central vertex" (referred to as the size-center vertex), making it the root, and directing all edges away from the root.

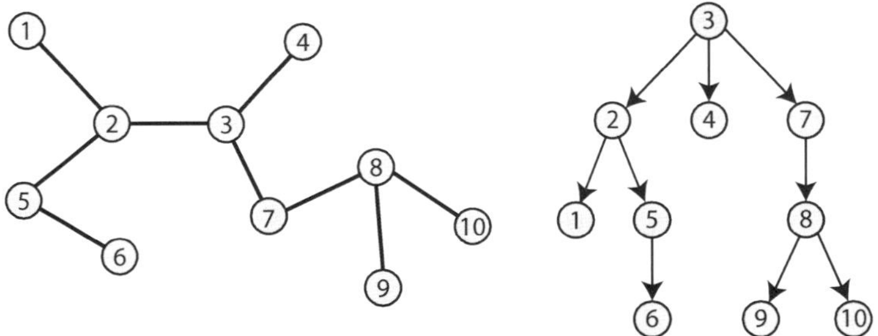

Fig. 9 A tree and its corresponding lesser-subtree orientation

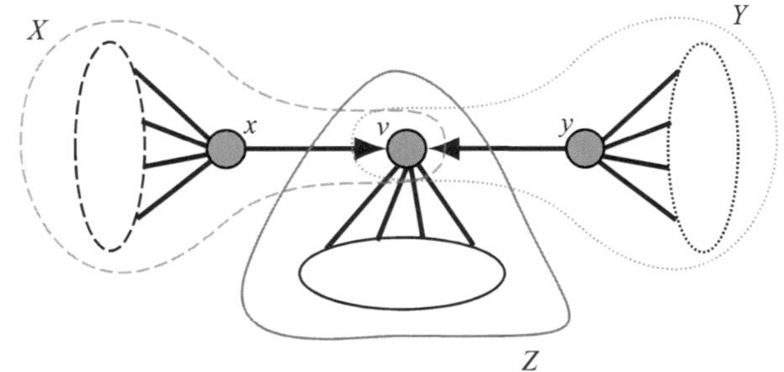

Fig. 10 Three subtrees rooted at v

Lemma 1. *Any vertex in a tree $\mathcal{T}$ with an LSO has at most one incoming edge.*

Proof. Suppose a vertex v has two incoming edges from vertices x and y (see Fig. 10). We may consider three subtrees of the tree rooted at v:

- A tree X, consisting of v, the edge $\{x, v\}$, and the subtree rooted at x;
- A tree Y, consisting of v, the edge $\{y, v\}$, and the subtree rooted at y;
- A tree Z, consisting of v, all edges adjacent to v except $\{v, x\}$ and $\{v, y\}$, and the subtrees rooted at each of the children of v other than x and y.

Let A, B, and C be the sizes of subtrees X, Y, and Z, respectively. Since $v \in X, Y, Z$, then A, B, and C are all of size at least 1. The orientation on $\{x, v\}$ implies that its lesser subtree is $Y \cup Z$ and has size $B + C - 1$, so $B + C - 1 \leq A$. Similarly, the orientation on $\{y, v\}$ implies that its lesser subtree is $X \cup Z$ and has size $A + C - 1$, so $A + C - 1 \leq B$. By the uniqueness of the size-center edge, equality can hold in only one of these cases, so suppose the latter inequality is strict. Thus, $A + C - 1 < B \leq A - C + 1$, and hence $C - 1 < -(C - 1)$, which is impossible, since $1 \leq C$. $\qquad\square$

Corollary 9. *A tree with an LSO has a unique source vertex, which we can call the size-center vertex.*

Proof. A tree with an LSO is a directed acyclic graph and hence has at least one source. Now suppose there are at least two sources, and consider the edge orientations along the unique path between a pair of them. Since the edges incident with the two sources are both directed into this path, there must be some vertex on the path with two incoming edges, a contradiction of Lemma 1.   □

Definition 4. The lesser-subtree sequence $s(v)$ of a vertex v in a tree with an LSO is the nondecreasing sequence of the sizes of the lesser subtrees on its outgoing edges. Thus, $s(v)$ is empty if and only if v is a leaf, it has length $\deg(v)$ if and only if v is the size-center vertex, and it has length $\deg(v) - 1$ otherwise.

Let $Q(\mathcal{T})$ be the number of different sizes of a lesser subtree of a tree $\mathcal{T}$, and let $R(\mathcal{T})$ be the number of distinct lesser-subtree sequences in the LSO of $\mathcal{T}$.

We use the convention that the root is at the top of the tree, with the leaves below.

Algorithm 2. *Input: The target graph, an unrooted tree T.*

Output: A set of $R(\mathcal{T})$ tile types using $Q(\mathcal{T})$ bond-edge types from which $\mathcal{T}$ may be constructed, but nothing smaller.

1. *Create a lesser-subtree orientation for $\mathcal{T}$. Take the size-center vertex to be the root. (Note that all edges will then be directed downward in the tree, from the root toward the leaves.)*
2. *For each directed edge (u, v), label the edge with the bond-edge type b_k, where k is the size of the subtree rooted at v.*
3. *The tile types are given by the labeled orientation of the graph.*

Correctness: The fact that the number of bond-edge types is $Q(\mathcal{T})$ follows directly from our labeling of the edges in step 2. Note that the only vertex that can have no incoming edges is the size-center vertex, and in this case the sum of the indices on the b_i's is $n - 1$, where n is the size of the tree, so the lesser-subtree sequence and tile type for the root will be unique. Every other vertex v has exactly one incoming edge. Note that the label b_k on an edge incoming to v must have $k = 1 + \sum_{i \in s(v)} i$. So, the tile type assigned to another node u will be the same as for v if and only if they have the same lesser-subtree sequence.

The labeling scheme on the edges, corresponding to subtree sizes, demonstrates why no smaller graph can be formed from the pot. Any tile with an incoming edge b_k will force the selection of a tile corresponding to a vertex higher in the original tree: either the root node, or a vertex with incoming edge b_j with $j > k$. Similarly, the outgoing edges will force the selection of tiles corresponding to vertices, with appropriately sized subtrees, that appear lower in the original tree.

Running time: The running time is $O(n^2)$ if, in step 1, each edge is considered in turn, and we compute the connected components created by removing the edge to determine the corresponding lesser subtree. This can be improved to $O(n)$ by choosing an arbitrary root, computing the size of all rooted subtrees using a postorder traversal, and then determining the orientation of each edge by comparing

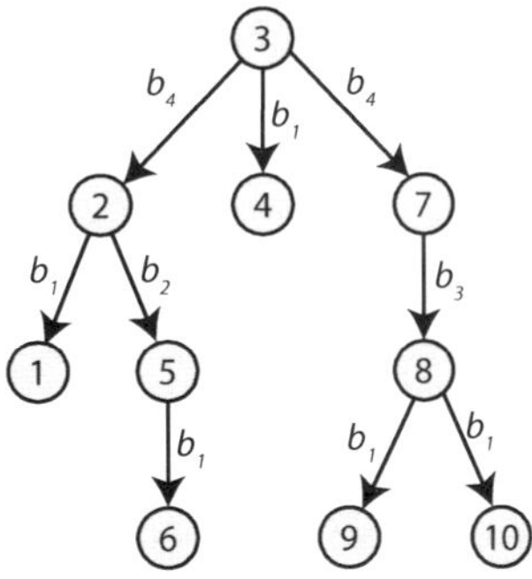

Fig. 11 A tree construction for Scenario 2

the size s of the subtree rooted at its lower endpoint with $n - s$. The labeling work in step 2 is $O(n)$.

See Fig. 11 for the labeling of the tree given in Fig. 9.

Proposition 11. $B_2(\mathcal{T}) = Q(\mathcal{T})$.

Proof. If $B_2(\mathcal{T}) < Q(\mathcal{T})$, then by the pigeonhole principle, there are two edges with the same bond-edge type but different-size lesser subtrees. If these two edges are both oriented either toward or away from their lesser subtrees, then the smaller subtree could replace the larger, resulting in a smaller complex. If one edge is oriented toward its lesser subtree and the other away, then the two lesser subtrees could be joined, again resulting in a smaller complex. Thus $B_2(\mathcal{T}) \geq Q(\mathcal{T})$, but the algorithm returns $Q(\mathcal{T})$ bond-edge types. $\qquad\qquad\square$

Proposition 12. $T_2(\mathcal{T}) = R(\mathcal{T})$.

Proof. Consider a labeling L_1 induced by a complex realizing $T_2(\mathcal{T}) < R(\mathcal{T})$, by labeling each vertex by the index of the tile type and each edge by the index of the bond-edge type used for it in the construction of $\mathcal{T}$. There must be two vertices u, v with the same tile type in the complex realizing $\mathcal{T}$, but with $s(u) \neq s(v)$. Consider also a second labeling L_2, given by Algorithm 2. Since $s(u) \neq s(v)$, the multisets representing the tile types u and v in L_2 must be different. Therefore there must be half-edges adjacent to u and v that have the same label in L_1 but different labels in L_2. Assume that the bond-edge types for these edges in L_2 are b_k and b_j.

If $k \neq j$, then by Algorithm 2, the sizes of the subtrees rooted at the neighbors of u and v are different. Assume, without loss of generality, that the subtree on the edge incident with u is smaller than the corresponding subtree on an edge incident with v. Then, in the case of the L_1 labeling, a copy of the smaller subtree incident with u can replace the larger one incident with v, resulting in a smaller-size graph, a contradiction.

Since bond-edge types in L_2 always have the hatted half-edge toward the smaller subtree, if $k = j$ it must be true that $k = j = n/2$. This means that the bond-edge type b_k corresponds to the unique size-center edge $e = \{u, v\}$. However, this means that in the L_1 labeling, both ends of e have the same cohesive end type, so they need to be complementary. Thus, we have a contradiction and there can be no such labeling L_1. Consequently, $T_2(\mathscr{T}) \geq R(\mathscr{T})$, but Algorithm 2 returns $R(\mathscr{T})$ tile types. $\qquad\square$

6 Scenario 3

In this case, we require that not only may no complexes smaller than the target be created from the given pot, but also that no nonisomorphic complexes of the same size be possible. Thus, we seek pots P using the smallest number of bond-edge or tile types so that $\{G\} = C_{\min}(P)$.

The constraints of Scenario 3 significantly limit the possibilities for tile types and their locations in a realization of a graph.

Lemma 2. *If P is a pot such that $\{G\} = C_{\min}(P)$, and G has no loops, then no tile type $T \in P$ used in the construction of G may have both a hatted and an unhatted cohesive end of the same type.*

Proof. If such a tile, with cohesive ends a and $\hat{a}$, were used in the construction of G to form edges $\{u, v_1\}$ and $\{u, v_2\}$, then P would also realize a graph G' with a loop edge at u and a new edge $\{v_1, v_2\}$, as in Fig. 12. Although G' has the same size as G, it has a loop, and so is not isomorphic to G, contradicting the assertion that $\{G\} = C_{\min}(P)$. $\qquad\square$

Lemma 3. *If P is a pot such that $\{G\} = C_{\min}(P)$, and G has no loops, then no tile type $T \in P$ used in the construction of G may be used for two adjacent vertices in G.*

Proof. Suppose a tile type $T \in P$ was used for adjacent vertices u and v. Since the edge $\{u, v\}$ must be formed from complementary cohesive ends, say a and $\hat{a}$, it follows that T must include both a and $\hat{a}$, contradicting Lemma 2. $\qquad\square$

Recall that the chromatic number $\chi(G)$ of a graph is the minimum number of colors needed to properly color G, i.e., to color the vertices of G so that no two adjacent vertices receive the same color.

Theorem 3. *If G is loopless, then $T_3(G) \geq \chi(G)$.*

Proof. By Lemma 3, $\{G\} = C_{\min}(P)$ implies that no two adjacent vertices are formed from the same tile type. Thus the tile type indices give a proper coloring of G, and hence there must be at least $\chi(G)$ different tile types. $\qquad\square$

Similar constraints hold for bond-edge types in this scenario.

Lemma 4. *If P is a pot such that $\{G\} = C_{\min}(P)$, and two nonadjacent edges $\{u, v\}$ and $\{s, t\}$ of $G = \{V, E\}$ use the same bond-edge type, then G is isomorphic to $G' = \{V, E'\}$, where $E' = E - \{\{u, v\}, \{s, t\}\} \cup \{\{u, t\}, \{s, v\}\}$.*

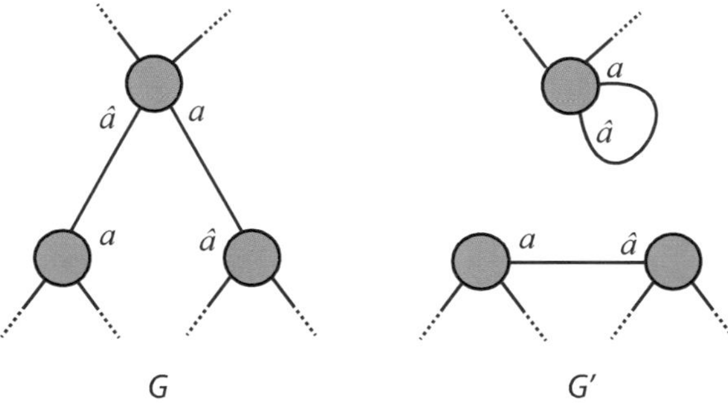

G   G′

Fig. 12 Resulting nonisomorphic graphs. Note that G' must be connected, since otherwise its components would have smaller size than G, contradicting the assertion that $G = C_{\min}(P)$

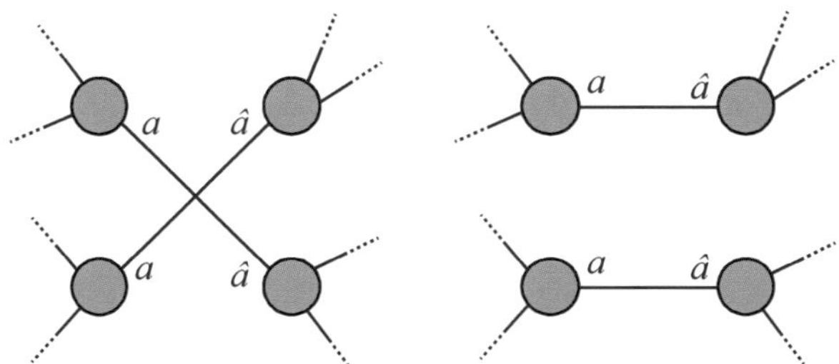

Fig. 13 Swapping nonadjacent edges

Proof. If P realizes G with $\{u, v\}$ and $\{s, t\}$ receiving the same bond-edge type, then P also realizes G', as in Fig. 13. Since G' is the same size as G, and $\{G\} = C_{\min}(P)$, this means that G is isomorphic to G'. □

Proposition 13. $T_3(C_n) = \lceil n/2 \rceil + 1$ *and* $B_3(C_n) = \lceil n/2 \rceil$.

Proof. The fact that $T_3(C_n) \geq \lceil n/2 \rceil + 1$ and $B_3(C_n) \geq \lceil n/2 \rceil$ follows from Proposition 8, Corollary 6, and Proposition 1. Equality holds by noting that if P is the pot given in Proposition 8, then $\{C_n\} = C_{\min}(P)$. □

Proposition 14. $T_3(K_n) = n$.

Proof. This follows immediately from Theorem 3. □

Proposition 15. $B_3(K_n) = n - 1$.

Proof. We proceed by induction on n, first noting that $K_3 = C_3$ and $B_3(C_3) = 2$ by Proposition 13. Now assume that n is the minimum value such that $B_3(K_n) < n-1$, and let v be a vertex of K_n. Since there are fewer than $n - 1$ bond-edge types, one of them must be repeated in the tile type used for v, and, by Lemma 2, the cohesive ends must be either both unhatted or both hatted. Without loss of generality, we assume that both are unhatted and that the bond-edge type is a. We call the two

Fig. 14 Construction of K_4

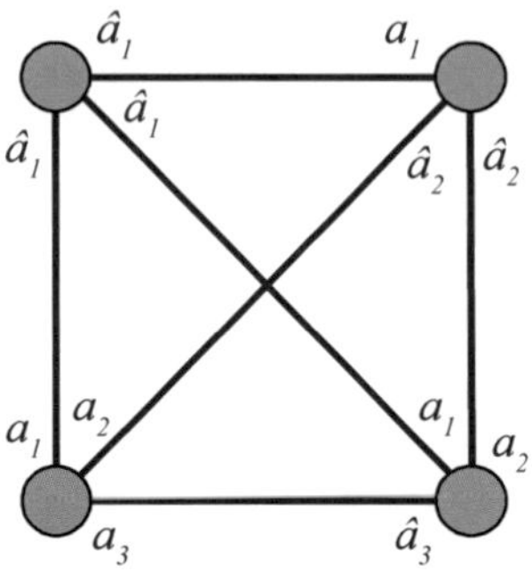

edges involved e and f. Since, by induction, the complete graph on the vertices $V(K_n) - v$ requires $n - 2$ bond-edge types, there is an edge g using bond-edge type a in this subgraph. Since K_n has no multiple edges, and multiple edges would be created by exchanging the cohesive ends of any two nonadjacent edges of K_n, as in Lemma 4, then by Lemma 4, the edge g must be adjacent to both e and f. However, this forms a triangle using just one bond-edge type, and thus one of the vertices of the triangle must have both an a and an $\hat{a}$ cohesive end, contradicting Lemma 2. Thus, $B_3(K_n) \geq n - 1$.

For the equality, we claim that the pot $P = \{t_1 = \{\hat{a}_1^{n-1}, t_i = \{a_1, \ldots, a_{i-1}, \hat{a}_i^{n-i}\}\}$, for i from 2 to $n - 1, t_n = \{a_1, \ldots, a_{n-1}\}\}$, using $n - 1$ bond-edge types, has $\{K_n\} = C_{\min}(P)$. That this pot realizes K_n may be seen from Fig. 14. No graph of smaller size can be constructed from this pot, since the construction matrix

$$M(P) = \begin{bmatrix} -(n-1) & 1 & 1 & 1 & 1 & 1 & 0 \\ 0 & -(n-2) & 1 & 1 & 1 & 1 & 0 \\ 0 & \cdots & \ddots & \ddots & \vdots & \vdots & 0 \\ 0 & \cdots & \cdots & \ddots & \ddots & \vdots & 0 \\ 0 & 0 & 0 & 0 & -1 & 1 & 0 \\ 1 & 1 & 1 & 1 & 1 & 1 & 1 \end{bmatrix} \tag{21}$$

has a unique solution, $\langle 1/n, \ldots, 1/n \rangle$, so $m_P = n$ and $K_n \in C_{\min}(P)$. To see that equality holds, note that any graph $G' \in C_{\min}(P) - \{K_n\}$ must use exactly one of each tile, and must have loops or multiple edges. However, no loops may be realized from this tile set, and for a multiple edge to be formed, there must be two tiles with two sets of complementary cohesive ends, and there is no such pair. $\square$

Next we consider complete bipartite graphs. For $K_{m,n}$, we take $m \leq n$, write X for the partition of size m and Y for the partition of size n, and observe in the following that Lemma 4 places considerable restrictions on the construction of $K_{m,n}$.

Fig. 15 Construction of $K_{2,3}$

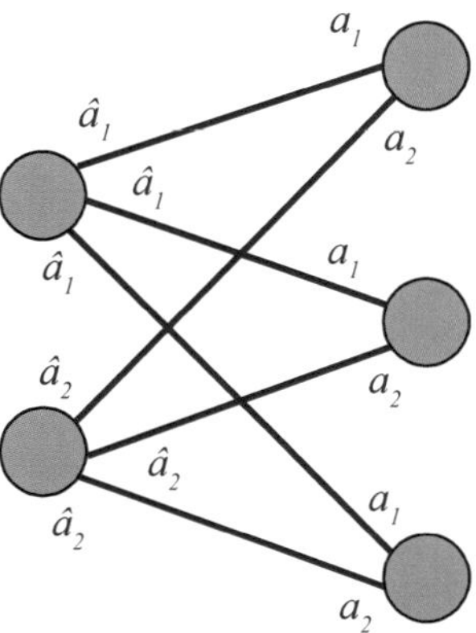

Lemma 5. *In any pot P such that $\{K_{m,n}\} = C_{\min}(P)$, the following must be true:*

1. *Any two edges formed from the same bond-edge type must be incident with the same vertex, and, moreover, that vertex must have the two corresponding cohesive ends either both hatted or both unhatted.*
2. *Two tiles corresponding to vertices in the same partition of $K_{m,n}$ (whether of the same or different tile types) cannot both have two cohesive ends of the same kind.*

Proof. Item 1 follows from Lemma 4 in the case of a bipartite graph by observing that the reconfiguring of the edges results in multiple edges (if the two hatted cohesive ends are on tiles used for vertices in the same partition) or edges in the same partitions (if the two hatted cohesive ends are on tiles used for vertices in different partitions). That the two cohesive ends must both be hatted or unhatted follows from Lemma 2, since $K_{m,n}$ has no loops.

Item 2 follows from Item 1, which implies that all four of the edges involved must be pairwise adjacent. This cannot occur in a complete bipartite graph which has no multiple edges. ∎

Proposition 16. $B_3(K_{m,n}) = m.$

Proof. By Lemma 5, Item 1, if a bond edge of type a appears, all edges of that type must be incident with the same vertex. Thus, there can be at most n edges with each bond-edge type. But $K_{m,n}$ has mn edges, so there must be at least m bond-edge types.

The following pot achieves this bound:

$$P = \{\{t_i = \{\hat{a}_i^n\}\ \text{for}\ i\ \text{from}\ 1\ \text{to}\ m\},\ t_{m+1} = \{a_1 \ldots a_m\}\}. \tag{22}$$

This P realizes $K_{m,n}$ when there is one copy each of the tiles $t_1, \ldots, t_m$ in one partition and n copies of tile t_{m+1} in the other partition (see Fig. 15). That P does not realize any graph of smaller size than $K_{m,n}$ follows from the construction matrix,

$$M(P) = \begin{bmatrix} 1 & 1 & 1 & 1 & 1 \\ -n & 0 & 0 & 1 & 0 \\ 0 & \ddots & \vdots & \vdots & 0 \\ 0 & 0 & -n & 1 & 0 \end{bmatrix}, \tag{23}$$

which has the unique solutions $\langle 1/(m+n), \ldots, 1/(m+n), n/(m+n) \rangle$. To see that no graph of size $m+n$ not isomorphic to $K_{m,n}$ may be realized by this pot, note that any such graph $G' \in C_{\min}(P) - \{K_{m,n}\}$ must use exactly the same numbers of each tile type as $K_{m,n}$. Since the tiles form a bipartition in which tiles with all cohesive ends hatted are on one side and tiles with all cohesive ends unhatted are on the other, G' must also be bipartite. Thus, if G' is not isomorphic to $K_{m,n}$, then it must have loops or multiple edges. However, no loops may be realized from this tile set, and for a multiple edge to be formed, there must be two tiles with two sets of complementary cohesive ends, but there is no such pair. $\square$

Proposition 17. $T_3(K_{m,n}) = m + 1$.

Proof. Clearly, $T_3(K_{1,n}) = 2$ for all n, so we may assume $m \geq 2$. By way of contradiction, suppose $T_3(K_{m,n}) \leq m$, and let P be a pot using m tiles with $\{K_{m,n}\} = C_{\min}(P)$.

Then, since by Lemma 3 no tile type can represent a vertex in both X and Y, the vertices in each of X and Y must be represented by $m - 1$ or fewer tiles. Thus, there are two vertices, say x_1 and x_2, in X using the same tile type, and two vertices y_1 and y_2 in Y using the same tile type. Suppose the edge from x_1 to y_1 uses bond-edge type a. Then the tile type at x_2 also has a cohesive end labeled a, and by Lemma 5, Item 1, the edge using bond-edge type a must also be incident with y_1. Thus, the tile type at y_1 has two cohesive ends with the same label. But y_2 uses the same tile type, so there are two tile types forming two vertices, y_1 and y_2, in the same partition of $K_{m,n}$, so that both have two of the same kind of cohesive end. This contradicts Lemma 5, Item 2, and so $T_3(K_{m,n}) \geq m$.

The pot in Proposition 16 achieves $T_3(K_{m,n}) = m + 1$. $\square$

Finally, we consider trees. For a rooted tree, we call the height the number of vertices in the longest path from the root to any leaf node, so a single vertex has height 1, and the tree consisting of two vertices joined by an edge has height 2. The level of a node v is the number of vertices in the path from the root to v. For a tree $\mathscr{T}$ with a lesser-subtree orientation we consider the set of trees induced by considering the subtree rooted at each vertex (i.e., the tree consisting of the vertex and all of its descendants in the tree). Let $I(\mathscr{T})$ denote the size of the maximum subset of nonisomorphic trees in this set.

Algorithm 3. *Input: The target graph, a tree $\mathscr{T}$.*

Output: A set of $I(\mathscr{T})$ tiles using $I(\mathscr{T}) - 1$ bond-edge types from which $\mathscr{T}$ may be constructed but no smaller or same-size graphs.

Fig. 16 A tree construction for Scenario 3

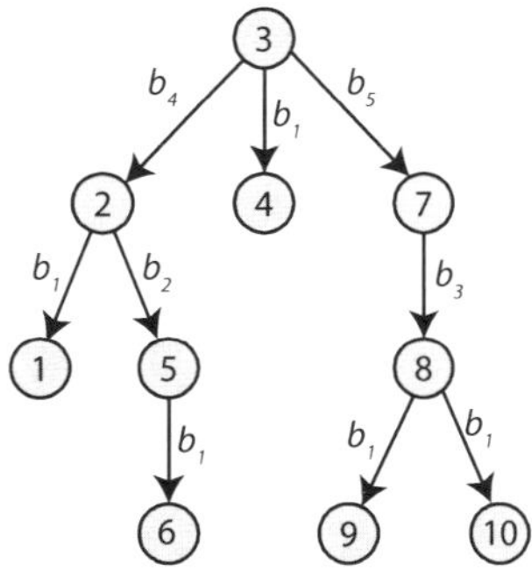

1. *Create a lesser-subtree orientation of the graph, and choose the unique size-center vertex to be the root. (Note that all edges will be directed downward in the tree.) Let h be the height of this rooted tree.*
2. *Label all edges incident to leaf nodes b_1.*
3. *Set $j = 1$.*
4. *For i from h to 1, examine each edge (u, v) from a node at level $i - 1$ to a node at level i successively.*

   - *If the edge is already labeled b_1, continue and consider the next edge.*
   - *Examine the subtree rooted at v.*
   - *If it is isomorphic to the subtree below edge e for any edge e that has already been labeled, then give (u, v) the same label as edge e. Otherwise, let $j = j + 1$, and label (u, v) by b_j.*

5. *The tile types are given by the labeled orientation of the graph.*

Correctness: The fact that the number of bond-edge types is $I(\mathcal{T}) - 1$ follows directly from our labeling of the edges. Note that the only vertex that can have no incoming edges is the size-center vertex, and in this case the set of subtrees rooted at that vertex and the corresponding tile type for the root will be unique. Every other vertex v has exactly one incoming edge. The tile type assigned to another node u will be the same as for v if and only if the subtrees rooted at those nodes are isomorphic.

The labeling scheme for the edges, corresponding to subtree isomorphisms, demonstrates why no nonisomorphic complex the same size as or smaller than the target graph can be formed from the pot. Any tile with an incoming edge b_k will force the selection of a tile corresponding to a vertex higher in the original tree: either the root node or a vertex with an incoming edge b_j, where j corresponds to the label of a subtree isomorphism that has a subtree corresponding to the label k. Similarly, the outgoing edges will force the selection of tiles corresponding to subtree isomorphisms that appear lower in the original tree.

Running time: The running time of this algorithm is $O(n^3)$, since each tree isomorphism may be checked in linear time [2].

See Fig. 16 for the labeling of the tree given in Fig. 9.

Proposition 18. $B_3(\mathcal{T}) = I(\mathcal{T}) - 1.$

Proof. If $B_3(\mathcal{T}) < I(\mathcal{T}) - 1$, then by the pigeonhole principle, there are two edges with the same bond-edge type but with nonisomorphic lesser subtrees. If these two edges are both oriented either toward or away from their lesser subtrees, then the smaller (or same-size) subtree could replace the larger. If the subtree is smaller, the result is clearly nonisomorphic. If the subtree is the same size, the fact that the result is nonisomorphic follows from the fact that the lesser-subtree orientation will be unaffected. If one edge is oriented toward its lesser subtree and the other away, then the two lesser subtrees could be joined, again resulting in a nonisomorphic complex no larger than the target graph. Thus $B_3(\mathcal{T}) \geq I(\mathcal{T}) - 1$, but the algorithm returns $I(\mathcal{T}) - 1$ bond-edge types. $\square$

Proposition 19. $T_3(\mathcal{T}) = I(\mathcal{T}).$

Proof. Consider a labeling L_1 induced by a complex realizing $T_2(\mathcal{T}) < I(\mathcal{T})$. There must be two vertices u, v with the same tile type in the complex realizing T, but where the subtree rooted at u is not isomorphic to the subtree rooted at v. Consider also a second labeling L_2, given by Algorithm 2. Since the corresponding subtrees are nonisomorphic, the multisets representing the tile types u and v in L_2 must be different. Therefore there must be half-edges adjacent to u and v that have the same label in L_1 but different labels in L_2. Assume, without loss of generality, that the subtree on the edge incident with u is smaller than (or the same size as) the corresponding one on an edge incident with v. Then, in the case of the L_1 labeling, a copy of the smaller (or same-size) subtree incident with u can replace the one incident with v, resulting in a nonisomorphic complex no larger than the target graph. $\square$

7 Conclusion

Although we have found optimal design strategies for some common infinite families of graphs, there are many other graph-theoretical constructs to consider, as well as other assembly models. Recent further work in this area may be found at [20]. Also, several open questions immediately arise. These include the following:

1. In Scenario 1, there is a need to characterize trees that need $av(\mathcal{T})$ tile types for their assembly and those that need $av(\mathcal{T}) + 1$.
2. Although for each of the examples given here we were able to find a pot that simultaneously achieved both the minimum number of bond-edge types and the minimum number of tile types, it is not clear that this is always possible. There is a need to determine whether or not it is alway possible to find a pot that simultaneously achieves both minimums, and if it is not always possible, to determinine conditions under which it is possible.
3. There is a need to characterize graphs for which $B_2(G) + 1 = T_2(G)$.

4. There is a need to find a bounding relation between $B_3(G)$ and $T_3(G)$ similar to that for Scenario 2, if one exists, and again to determine conditions for when equality holds.

Acknowledgements We thank Dan Archdeacon, Natasha Jonoska, Bruce Sagan, Ned Seeman, and Anna Staninska for a number of informative conversations. We also thank Saint Michael's College students Brian Hopper and Paul Jarvis, who considered some closely related problems.

Support was provided by the National Science Foundation through award 1001408, by the National Security Agency, and by the Vermont Genetics Network through Grant Number P20 RR16462 from the INBRE Program of the National Center for Research Resources (NCRR), a component of the National Institutes of Health (NIH). The contents of this chapter are solely the responsibility of the authors and do not necessarily represent the official views of the NSF, NCRR, or NIH.

References

1. L. Adleman, Molecular computation of solutions to combinatorial problems. Science **266**, 1021–1024 (1994)
2. A. Aho, J. Hopcroft, J. Ullman, *The Design and Analysis of Computer Algorithms* (Addison-Wesley, Reading, MA 1974)
3. J. Chen, N. Seeman, Synthesis from DNA of a molecule with the connectivity of a cube. Nature **350**, 631–633 (1991)
4. Cornell University, Self-assembled DNA Buckyballs For Drug Delivery. ScienceDaily (2005, August 31). Retrieved 23 Oct 2013, from http://www.sciencedaily.com/releases/2005/08/050829074441.htm
5. H. Fleischner, *Eulerian Graphs and Related Topics. Part 1. Vol. 1.* Volume 45 of Annals of Discrete Mathematics (North-Holland, Amsterdam, 1990)
6. H. Fleischner, *Eulerian Graphs and Related Topics. Part 1. Vol. 2.* Volume 50 of Annals of Discrete Mathematics (North-Holland, Amsterdam, 1991)
7. H. Gu, J. Chao, S. Xiao, N. Seeman, Dynamic patterning programmed by DNA tiles captured on a DNA origami substrate. Nat. Nanotechnol. **4**(4), 245–248 (2009)
8. N. Jonoska, G. McColm, Complexity classes for self-assembling flexible tiles. Biosystems **410**(4–5), 332–346 (2009)
9. N. Jonoska, N. Seeman, Computing by molecular self-assembly. Interface Focus **2**, 504–511 (2012)
10. N. Jonoska, S. Karl, M. Saito, Creating 3-dimensional graph structures with DNA, in *Proceedings of the First Annual Meeting, DIMACS Series in Discrete Mathematics and Theoretical Computer Science*, vol. 44 (American Mathematical Society, Providence, 1998), pp. 123–136
11. N. Jonoska, S. Karl, M. Saito, Three dimensional DNA structures in computing. Biosystems **52**, 143–153 (1999)
12. N. Jonoska, G. McColm, A. Staninska, Expectation and variance of self-assembled graph structures, in *Proceedings of the 11th International Conference on DNA Computing*, DNA'05 (Springer, Berlin/Heidelberg, 2006), pp. 144–157
13. N. Jonoska, G. McColm, A. Staninska, Spectrum of a pot for DNA complexes, in *DNA Computing*, vol 4287, Lecture Notes in Computer Science ed. by C. Mao, T. Yokomori (Springer, Berlin/Heidelberg, 2006), pp. 83–94. ISBN:978-3-540-49024-1, DOI:10.1007/11925903_7, URL:http://dx.doi.org/10.1007/11925903_7
14. N. Jonoska, G. McColm, A. Staninska, On stoichiometry for the assembly of flexible tile DNA complexes. Nat. Comput. **10**(3), 1121–1141 (2011)

15. T. LaBean, H. Li, Constructing novel materials with DNA. Nanotoday **2**(2), 26–35 (2007)
16. Y. Roh, R. Ruiz, S. Peng, J. Lee, D. Luo, Engineering DNA-based functional materials. Chem. Soc. Rev. **40**, 5730–5744 (2011)
17. P. Rothemund, Folding DNA to create nanoscale shapes and patterns. Nature **440**, 297–302 (2006)
18. P. Sa-Ardyen, N. Jonoska, N. Seeman, Self-assembling DNA graphs. Nat. Comput. **2**(4), 427–438 (2003)
19. N. Seeman, Nanomaterials based on DNA. Annu. Rev. Biochem. **79**, 65–87 (2007)
20. Self-assembly design strategies, http://sites.google.com/site/nanoselfassembly/ (2013)
21. W. Shih, J. Quispe, G. Joyce, A 1.7 kilobase single-stranded DNA that folds into a nanoscale octahedron. Nature **427**, 618–621 (2004)
22. A. Staninska, The graph of a pot with DNA molecules, in *Proceedings of the 3rd Annual Conference on Foundations of Nanoscience (FNANO'06)*, Snowbird, Apr 2006, pp. 222–226
23. Y. Wang, J. Mueller, B. Kemper, N. Seeman, Assembly and characterization of five-arm and six-arm DNA branched junctions. Biochemistry **30**(23), 5667–5674 (1991)
24. H. Yan, S. Park, G. Finkelstein, J. Reif, T. LaBean, DNA-templated self-assembly of protein arrays and highly conductive nanowires. Science **301**, 1882–1884 (2003)
25. Y. Zhang, N. Seeman, Construction of a DNA-truncated octahedron. J. Am. Chem. Soc. **116**, 1661–1669 (1994)

Part III
Gene Rearrangements

Programmed Genome Processing in Ciliates

Aaron David Goldman, Elizabeth M. Stein, John R. Bracht,
and Laura F. Landweber

Abstract The ciliates are a group of protists distinguished by the hair-like cilia on
their cell surfaces. Ciliates also possess two types of nuclei, a germline micronucleus
and a somatic macronucleus. The micronuclear genome contains segmented genes
divided by spacer sequences of DNA that are removed to generate the macronuclear
genome during development. For some species, certain micronuclear gene segments
can be reordered and/or inverted with respect to their final gene sequence in the
macronucleus. This chapter explores the similarities of and differences between
micronuclear genomes and the processes of macronuclear development across
different ciliate species.

1 Sex and Nuclear Dimorphism in Ciliates

The ciliates (phylum Ciliophora) are a highly diverse, monophyletic clade of
unicellular protists composed of more than 8,000 species that diverged from a
common ancestor with alveolates like *Plasmodium* about 1.25 billion years ago
[1]. The name "ciliate" describes the most obvious feature that unites members
of this group, the presence of hair-like cilia on their cell surfaces that function in
motility, feeding, and sensation. But within the cell, ciliates also share the otherwise

A.D. Goldman (✉)
Department of Biology, Oberlin College, Oberlin, OH 44074, USA
e-mail: adg@princeton.edu

J.R. Bracht • L.F. Landweber
Department of Ecology and Evolutionary Biology, Princeton University, Princeton,
NJ 08544, USA
e-mail: jbracht@princeton.edu; lfl@princeton.edu

E.M. Stein
Georgetown University, Washington, DC 20057, USA
e-mail: ems264@georgetown.edu

N. Jonoska and M. Saito (eds.), *Discrete and Topological Models in Molecular Biology*,
Natural Computing Series, DOI 10.1007/978-3-642-40193-0_12,
© Springer-Verlag Berlin Heidelberg 2014

Fig. 1 Images of three model ciliate species superimposed onto an evolutionary tree. Cells were stained to reveal DNA (in *cyan*) and tubulin (in *green*). Locations of micronuclei are indicated by "MIC" and macronuclei by "MAC". The guide tree is based on Doak et al. [46] (Image credits: *Tetrahymena thermophila* (credit Kensuke Kataoka), *Paramecium tetraurelia* (credit Janine Beisson), and *Oxytricha trifallax* (credit Wenwen Fang))

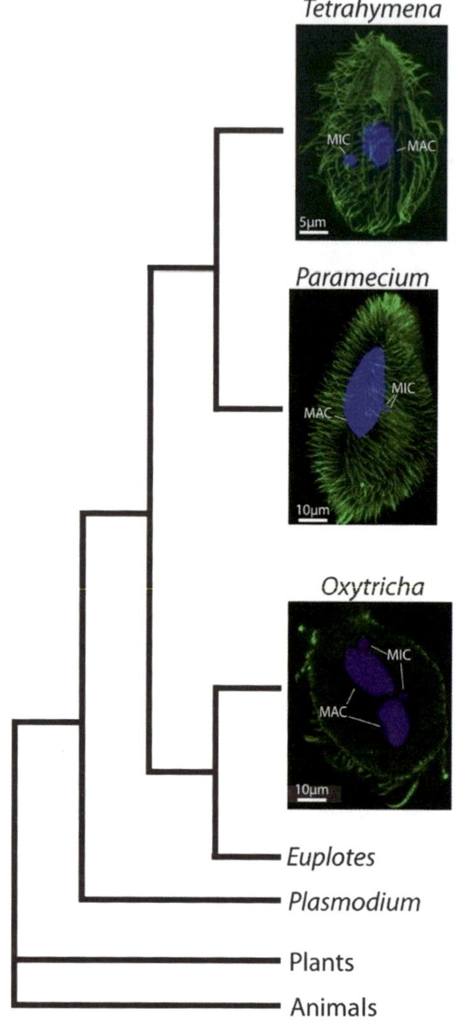

uncommon trait of nuclear dimorphism, wherein each cell has two distinct kinds of nuclei, a micronucleus and a macronucleus, that contain related but different genomes (Fig. 1).

Ciliates engage in nonreproductive sex in which a haploid version of each genome is transferred between mating partners, transforming the mating cells into progeny (Fig. 2). The micronucleus serves the role of sexual exchange, much like sperm and egg cells in animals. The micronuclear genome is diploid and composed of long eukaryotic chromosomes. It becomes haploid through meiosis, again like the genomes of developing sperm and egg cells, and is exchanged between mating partners. However, the micronuclear genomes of ciliates are segmented such that

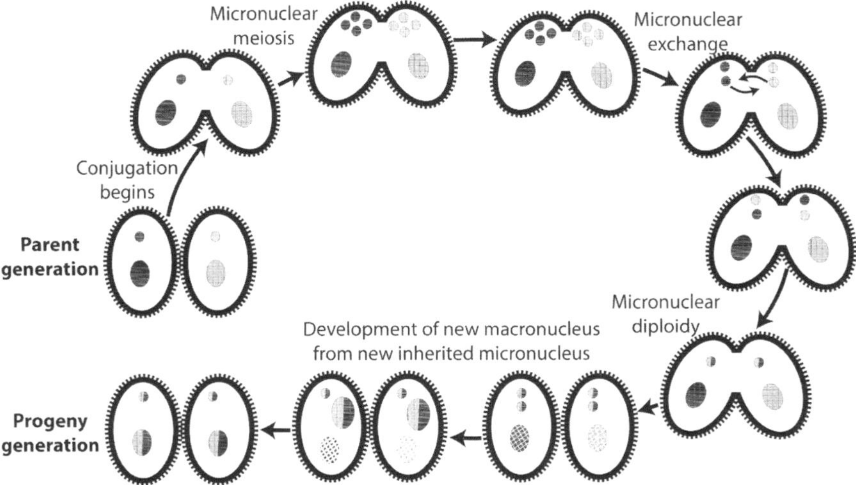

Fig. 2 The ciliate sexual cycle is sex without reproduction. During conjugation, the micronuclei of both cells undergo meiosis. One haploid micronucleus is exchanged between each cell and combined with one haploid micronucleus retained by the cell to create a new diploid micronucleus. The new micronucleus undergoes mitosis, and one of the progeny micronuclei differentiates into a new macronucleus. The old macronucleus degrades. The result is two sexual progeny, each possessing a new micronucleus and macronucleus with alleles from both conjugating cells

individual pieces of genes and other informational sequences are in a nonfunctional form. These genes are interrupted by spacer DNA that is eliminated following sexual exchange. In some species, segments of the same gene can be inverted and out of order with respect to one another. In these cases, the micronuclear genes are said to be "scrambled".

The macronucleus serves the somatic function, with genes in their functional form. Macronuclear genomes are not exchanged during sex. After sexual exchange, the parent macronuclear genome is dismantled and a new macronuclear genome is built from micronuclear DNA (Fig. 2). This process is referred to as "development", and the developing macronucleus is referred to as the "anlagen", a phrase borrowed from animal embryology that refers to an early stage in animal development. Macronuclear development requires that micronuclear segments are faithfully joined to create a functioning genome with open reading frames and appropriate regulatory elements. Unlike micronuclear DNA, macronuclear DNA typically exists as very short chromosomes that can be as small as a few hundred base pairs. In some ciliate species, these small chromosomes often encode only a single gene. Macronuclear chromosomes usually exist at high copy number, ranging from tens to thousands of copies in a single macronucleus.

Since the parental macronucleus is not exchanged between mating partners, there must be some mechanism capable of developing a macronuclear genome from micronuclear DNA. Simple mitotic replication of each parent cell's macronucleus would remove the benefit of sexual recombination between micronuclei. Several mechanisms for macronuclear development have been proposed which usually rely on noncoding RNAs to provide information about the content and/or order of micronuclear segments in the macronuclear genome. Recent evidence suggests that alternative models appear to be true for several different groups within the ciliates. This chapter is dedicated to reviewing our understanding of macronuclear development and DNA differentiation in different groups of ciliates.

2   Genome Structure and Macronuclear Development in Oligohymenophorea

The Oligohymenophorea are a class of ciliates that includes two well-studied genera, *Tetrahymena* and *Paramecium*. Within these genera, *T. thermophila*, *P. tetraurelia*, and *P. primaurelia* represent the most commonly used model organisms for elucidating the molecular processes that underlie macronuclear development. *Paramecium* and *Tetrahymena* both contain a single ovoid macronucleus. However, *Tetrahymena* species have only one micronucleus, while *Paramecium* species have at least two micronuclei [2]. The sequence of events that occurs following meiosis and conjugation differs between *Paramecium* and *Tetrahymena*. Following meiosis and conjugation in *P. tetraurelia*, for example, the diploid zygotic micronucleus divides mitotically in two rounds, producing four daughter nuclei. Two of the four remain as micronuclei, and the other two become macronuclei. In *T. thermophila*, however, only one mitotic division of the zygote occurs, and one of the resulting micronuclei becomes a macronucleus (for reviews, see [2, 3]).

In addition to these slight differences in the overall life cycles of *Paramecium* and *Tetrahymena*, the DNA content of the two organisms also differs. The *Paramecium* macronucleus contains chromosomes at around 800× copy number, and the chromosome size ranges from 50 to 1,000 kb [4]. In *Tetrahymena* species, however, macronuclear chromosomes are present at an average of 45× copy number, although the length of each chromosome is similar to the lengths of those of *Paramecium tetraurelia,* 700 kbp (Orias, http://www.lifesci.ucsb.edu/~genome/Tetrahymena/). Each *Tetrahymena* haploid micronuclear genome contains an estimated 6,000 eliminated sequences [5], comprising between 10 % and 20 % of the genome and ranging in size from approximately 500 to 20,000 bp (Yao et al. [6]; for a review, see Yao et al. [7]).

Following conjugation, in both *Tetrahymena* and *Paramecium*, several rounds of DNA replication amplify the micronuclear DNA in the developing macronucleus. Around the same time, early in the DNA amplification period, spacer DNA in between macronuclear segments is precisely excised. In the developing macronucleus of *Tetrahymena*, approximately 30 % of the micronuclear DNA sequence is eliminated (http://www.broadinstitute.org/annotation/genome/Tetrahymena/). Most of the deleted DNA in *Tetrahymena* consists of moderately repetitious sequences with copy numbers of around 200 that may be distantly related to transposable elements. Some unique sequences are also found among the eliminated DNA (for reviews, see [2, 3]).

Each *Paramecium* haploid micronuclear genome contains approximately 45,000 eliminated sequences [8]. Some of these eliminated sequences are found in non-coding regions, but most interrupt open reading frames, making their precise excision necessary for the developing ciliate [9]. In addition, the spacer sequences eliminated from *Paramecium* micronuclear DNA are short. In *P. tetraurelia*, these eliminated sequences range from 26 to 5,316 bp. Seventy-six percent of them are less than 100 bp, and almost one-third are between 26 and 30 bp ([9]; supported by [8]). Furthermore, all *Paramecium* eliminated sequences are flanked by 5′-TA-3′ dinucleotide repeats. After these sequences are deleted, one dinucleotide repeat is retained on the new somatic chromosome [9]. An 8-bp consensus sequence for these repeats, which is a larger version of the typical 5′-TA-3′ repeat, has been proposed – 5′-TA(C/T)AG(C/T)N(A/G)-3′ [10]. The similarity of this consensus sequence to the ends of the Tc1-related transposons suggests that these eliminated sequences may have evolved from ancestral transposons [10, 11].

Following the excision of eliminated sequences, telomeres are added to the ends of the new macronuclear chromosomes. In *Tetrahymena*, telomeric sequences consistently end in single-stranded overhangs of 16 to 20 repeated bases of the form 3′-dG$_4$T$_2$-5′ [12]. In *Paramecium*, however, telomeres consist of a mixture of 3′-dG$_4$T$_2$-5′ and 3′-dG$_3$T$_3$-5′ sequences [13]. Following the addition of these telomeres in both organisms, the chromosomes replicate continually until a final copy number is reached. After this replication, macronuclear development is complete, and vegetative proliferation follows.

The epigenetic control of macronuclear development by maternal cells has been studied substantially in ciliates. In *Paramecium*, some of these effects appear to result from small RNA molecules of approximately 22–23 nucleotides that accumulate in the parental macronucleus prior to development of the new macronucleus, following sexual exchange [14]. Several experiments in *Paramecium* have revealed that the silencing of certain genes through the RNA interference pathway (RNAi) correlates with the accumulation of these small RNAs ([15–17]; Lepère et al. 2008). Small RNAs may be involved in homology-dependent maternal silencing of genes in *Tetrahymena* as well, and can be produced at a few loci in the organism [18]. Although these small RNAs may be involved in shutting off maternally controlled genes, they cannot account for other epigenetic phenomena, such as maternal inheritance of alternately arranged genes, that are not the result of the action of RNAi on specific, maternally controlled genes [19]. An alternate model, the "scan RNA model", has been proposed to account for macronuclear development in oligohymenophoreans [20, 21].

The scan RNA model for programmed DNA elimination in ciliates (Fig. 3), originally proposed by Mochizuki et al. [20] and modified by Lepère et al. [17], suggests that portions of the micronuclear DNA are bidirectionally transcribed at the start of conjugation to form double-stranded RNA molecules. These double-stranded RNA molecules are processed by a dicer-like enzyme to form scan RNAs [22]. Although the model is somewhat vague about the mechanism by which these

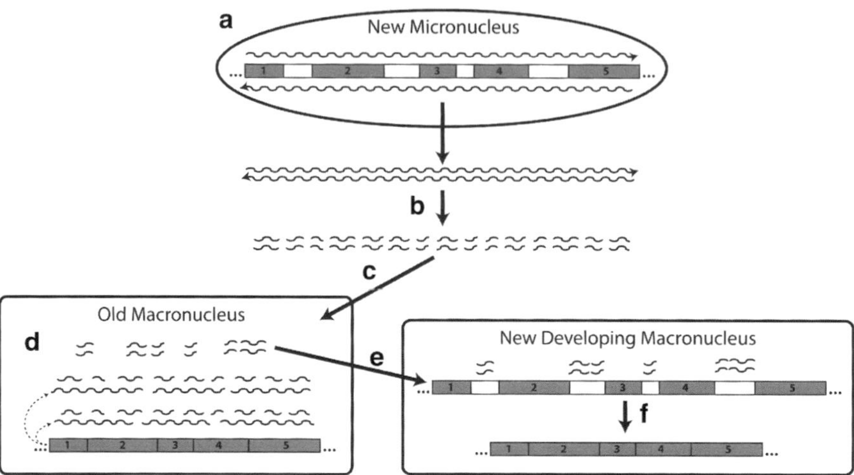

Fig. 3 The scan RNA model of macronuclear development in the Oligohymenophorea. (**a**) After sexual exchange, long bidirectional RNA transcripts are produced from the new micronuclear DNA. (**b**) The RNAs are cut into smaller RNAs and (**c**) enter the old macronucleus. (**d**) Those RNA segments with a macronuclear sequence are retained by long RNAs transcribed from the old macronuclear genome. (**e**) RNAs that are not homologous to the old macronuclear sequence are transferred to the developing macronucleus. (**f**) These RNAs correctly signal the micronuclear sequence for deletion, allowing the new macronuclear genome to form

scan RNAs function, it has been suggested that they base-pair to longer RNA transcripts of the maternal macronuclear genome [15, 23]. Micronuclear scan RNAs that paired to homologous sequences would be degraded, while those that did not would migrate to the newly developing macronucleus, and target regions of the developing macronuclear genome for deletion [20].

Mochizuki and Gorovsky [24] conducted early tests of the scan RNA model in *Tetrahymena thermophila*. Two hours following conjugation, the small RNAs hybridized to DNA found only in the micronuclear genome three times as often as they did to macronuclear DNA. Eight hours following conjugation, the small RNAs hybridized to DNA found only in the micronuclear genome twenty times as often as to macronuclear DNA. This progression was recently confirmed by deep sequencing of RNAs [21]. Scan RNA transcription was observed to originate from the micronuclear DNA and the majority of the transcripts match eliminated sequences. Those that match retained macronuclear sequences are depleted from the set of small RNA sequences.

A protein in the Piwi family called Twi1p is required for both scan RNA stability and proper DNA elimination [20]. Twi1p concentrations were previously observed to mirror the presence of scan RNAs from the parental macronucleus to the developing macronucleus, indicating that Twi1p may be associated with the scan RNAs during conjugation [24]. A similar phenomenon is likely to control *Paramecium* macronuclear development [17].

In *Paramecium,* mutation of the TA repeat marking the ends of eliminated DNA can inhibit elimination [25]. This observation implies that some aspects of DNA excision may not simply be mediated by scan RNAs alone, and may require some sequence recognition. Indeed, eliminated DNA segments in *Paramecium* are marked by an 8 bp inverted repeat motif (containing the TA repeat) similar to that found in *Tc1/*mariner transposons (Klobutcher and Herrick [10]; confirmed by Arnaiz et al. [8]); this motif is important for targeting the PiggyBac transposase that mediates DNA cleavage events [26, 27].

A modified scan RNA model has been proposed, where the scanning occurs by base-pairing across maternal RNA transcripts of the macronuclear genome rather than the maternal macronuclear DNA itself [17]. However, this modification does not completely account for the specificity of information transfer achieved during macronuclear development and programmed DNA deletion, since RNA can tolerate multiple noncanonical base pairs [19]. This conflicting observation suggests that further examination is required for a full understanding of oligohymenophorean genome development and the role of scan RNAs in this process.

3 Genome Structure and Macronuclear Development in the Stichotrichia

The stichotrichs are a subclass of the spirotrichous ciliates and include the genera *Oxytricha* and *Stylonychia*. *O. trifallax* and *S. lemnae* are representative model organisms for stichotrich genome rearrangement. Genome rearrangement in these organisms is much more complex at the molecular level than it is in the oligo-hymenophoreans *Paramecium* and *Tetrahymena*. In many stichotrichs, more than 95 % of the micronuclear genome content is eliminated during macronuclear development [28, 29]. Perhaps the most important feature of stichotrich macronuclear development is that the macronuclear segments within the micronuclear genome can be scrambled, with the order of gene pieces inverted and permuted with respect to one another. Because the scan RNA model explains only the removal of eliminated segments, the rearrangement of a scrambled micronuclear genome requires a different mechanism than that of macronuclear development in the Oligohymenophorea.

The micronuclear chromosomes of stichotrichs are similar in size and structure to those of the Oligohymenophorea. Stichotrich macronuclear genomes, however, are composed of very short chromosomes called "nanochromosomes", which typically contain only one gene ([30–32]; Swart et al. 2013). The *O. trifallax* macronuclear genome, for example, is composed of over 16,000 nanochromosomes that have an average length of 3.2 kbp and are each represented at around 1,000× copy number [32]. The *Oxytricha* and *Stylonichia* 3′ telomeres have G-rich single-stranded tails that indicate an unusual "G-quartet" DNA structure via

Hoogsteen base-pairing of single stranded G-rich regions [33]. *S. lemnae* telomeres were observed to form these structures in vivo [34].

Thousands of genes are composed of segments that are scrambled in the micronuclear genome. Macronuclear-destined segments in the micronuclear genome typically contain short sequences that are repeated on both sides of the segment, though only one copy is retained in the final macronuclear version. These "pointer" sequences can be very short, usually 2 bp or more, but possibly a single base pair, and sometimes do not provide a perfect match with their counterpart. It is not possible that information from these pointers alone can guide genome rearrangement.

Macronuclear development in stichotrichs begins with the formation of polytene chromosomes of micronuclear DNA. These giant chromosomes form as a result of extensive genome replication without mitosis [28, 35]. Polytene micronuclear DNA provides the source of DNA for the developing macronuclear genome. Following the formation of polytene chromosomes and the elimination of spacer DNA and transposable elements, between 20 % and 30 % of the remaining macronuclear segments are unscrambled by inversion and permutation (reviewed in [19]). The scrambling of the macronuclear sequence in the micronucleus can be very complex. One well-described gene encodes DNA polymerase α in *O. trifallax*, which appears in the micronuclear genome as ~50 permuted macronuclear segments [36, 37].

Homologous recombination at pointers appears to be necessary for the rearrangement of some macronuclear chromosomes. In early estimates, the pointers that exist between scrambled segments were approximately 9 bp in average length, whereas those that exist between nonscrambled segments were 4 bp in average length [38]. No consensus sequence for pointers has been identified, but some motifs may be enriched [37]. Prescott et al. [39] proposed that the eliminated sequences would form a loop, allowing the pointers to align, at which point a recombinase would cut the sequences at either end and they would undergo recombination, excising the eliminated DNA in the form of a closed circle. However, although pointers may be necessary for unscrambling to occur, their short lengths suggest that they are probably not capable of doing so alone.

A major feature of the scan RNA model is that the primary function of scan RNAs is to tag portions of the germline genome for deletion. Scan RNAs, by themselves, are probably too short to direct the permutation and inversion process involved in genome unscrambling in these organisms. In addition, some eliminated regions between macronuclear segments in *Oxytricha* and *Stylonychia* are smaller than the scan RNAs. Prescott et al. [39] proposed an alternative model to the scan RNA model, in which a template derived from the parent macronucleus facilitates correct rearrangement in the developing macronucleus (Fig. 4).

Nowacki et al. [40] tested the hypothesis that these templates are composed of RNAs with sequences derived from the parental macronuclear genome. RNA interference that targeted predicted templates resulted in aberrant rearrangements and incomplete elimination of micronuclear spacer DNA. In a more direct experiment, Nowacki et al. [40] injected developing *O. trifallax* with artificial nanochromosomes (in the form of double-stranded DNA) or templates (in the form of single- or

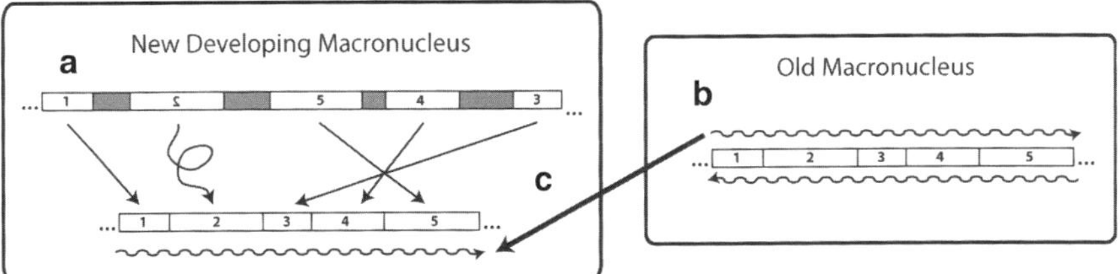

Fig. 4 The template model of macronuclear development in the Stichotrichia. We refer the reader to Bracht et al. [3] for a more comprehensive representation of the model. (**a**) Segments of micronuclear DNA must be retained to form the developing macronuclear genome. (**b**) RNA transcripts of complete nanochromosomes are generated from the parental macronucleus and delivered to the new, developing macronucleus. (**c**) Retained segments of micronuclear DNA are guided by the RNA template and rejoin to form the new macronuclear genome

double- stranded RNA), both of which contained macronuclear segments that were out of order. In both sets of experiments, the resulting macronuclear genome contained nanochromosomes with the misordered segments. These artificially ordered nanochromosomes appeared to be stable through multiple sexual generations after their original introduction, indicating that the misordered nanochromosome itself must have been producing templates capable of guiding rearrangements.

Several other recent observations can be added to the template model of genome rearrangement. First, point substitutions artificially introduced into the template sometimes appear in the new macronuclear genome if they are close to a boundary between retained macronuclear segments. One explanation for this observation is that the micronuclear DNA is imprecisely cut somewhere near the retained sequence and that any missing macronuclear sequence is filled in from the template by an RNA-dependent DNA polymerase [40]. Some of this cutting may be performed by a class of transposons called "telomere-bearing elements", or TBEs, which appear to play an important role in macronuclear assembly [41].

It was recently discovered that an additional class of RNAs related to piRNAs (small RNAs that interact with the protein Piwi) are required for identifying the micronuclear sequences that are retained in the macronuclear genome [42, 43]. This phenomenon represents an evolutionary "sign change" in the following sense. In the Oligohymenophorea, small RNAs (scan RNAs) mark micronuclear DNA for removal; but in *Oxytricha*, small RNAs (piRNAs) mark DNA for retention. In both cases, small RNAs mark the minority fraction of the micronuclear genome. Another recent study suggests that destruction of repetitive DNA and the parental macronuclear genome in *Oxytricha* requires methylation and/or hydroxymethylation of cytosine nucleotides in the eliminated DNA [44]. Similar methylation was previously observed in transposon-like sequences of developing macronuclear DNA in *S. lemnae* [45].

Even with the recent discoveries of template RNA and piRNAs in *Oxytricha*, our understanding of genome rearrangement is limited. It is not known whether there are templates for every gene, or just for a subset that require rearrangement; it is not known whether macronuclear nanochromosomes are rearranged simultaneously or at times depending on the role of their gene products with respect to the developmental process; and, aside from TBEs, very little is known about the proteins that orchestrate the developmental process. Stichotrichous ciliates exhibit the most complex genome rearrangements yet described, and these outstanding issues suggest that the mechanisms underlying the genome rearrangement process are accordingly complex.

4   Genome Structure and Macronuclear Development in Other Ciliates

Most of what we know about genome rearrangement in ciliates comes from the study of four genera – *Paramecium, Tetrahymena, Oxytricha*, and *Stylonychia*. Yet, these species represents only a small subset of ciliate diversity. Noteworthy research on the ciliate genera *Euplotes, Chilodonella*, and *Nyctotherus* has yielded a partial understanding of macronuclear genome development and diversity in these distantly related groups. Surveying macronuclear development across the full diversity of ciliates is an indispensible step toward resolving the evolutionary history of nuclear dimorphism.

Euplotes is a ciliate of the class Spirotrichea, which also includes *Oxytricha* and *Stylonychia*, but *Euplotes* does not belong to the subclass Stichotrichea, to which *Oxytricha* and *Stylonychia* belong. The macronuclear segments in the *Euplotes* micronuclear genome are thought to be in order, rather than scrambled as in *Oxytricha* and *Stylonichia*; however, this has not been explored at a genome level. Nonetheless, *Euplotes* is more closely related to *Oxytricha* and *Stylonychia* than to the oligohymenophoreans *Paramecium* and *Tetrahymena* [46]. Like the stichotrichs, *Euplotes* deletes up to 95 % of its micronuclear genome throughout the course of macronuclear development. In *Euplotes*, as in other ciliates, genome rearrangement follows the general pattern of elimination, chromosomal fragmentation, and amplification. First, as in *Oxytricha* and *Stylonychia*, the beginning of macronuclear development in Euplotes coincides with the formation of gigantic polytene chromosomes and the subsequent degradation of these chromosomes into gene-sized molecules. Within *Euplotes, E. crassus* is the best-studied species, and a draft of its macronuclear genome sequence is available [47].

In macronuclear development in *E. crassus*, the first elements deleted are the transposon-like Tec elements that are present within gene segments, of which approximately 15,000 copies are dispersed throughout the micronuclear genome [48]. *Tec* elements that interrupt the macronuclear sequence are deleted precisely. The sequence 5′-TTdGdAdA-3′, which flanks each *Tec* transposon, appears to be necessary for this deletion [49]. Approximately one-third of the *Tec* elements interrupt genes [50]. On the other hand, *Tec* elements that do not interrupt macronuclear DNA may or may not be specifically excised, indicating that *Tec* elements are eliminated at different times and, potentially, by different mechanisms based on their location within the micronuclear genome [48].

The Chromatin structure in *E. crassus* likely plays a role in marking sequences for elimination [51]. Three types of chromatin structure appear to be involved in differentiating between excised elements and the macronuclear sequence. First, the macronuclear sequence in the micronuclear genome has a typical nucleosome structure. A second type of chromatin structure, which exhibits atypically long sequence stretches between nucleosomes, was observed for micronuclear sequences containing *Tec* elements and telomeres [52]. Finally, a third, highly compact chromatin structure was observed for the circular forms of the excised *Tec* elements.

Chilodonella is a member of class Phyllopharyngea, and includes four known species: *C. cyprini*, *C. fluviatilis*, *C. hexasticha*, and *C. uncinata*. *C. uncinata* is the best-studied species of *Chilodonella*. Although *Tetrahymena*, *Paramecium*, and *Chilodonella* were once classified as holotrichous ciliates, a subclass of the phylum Ciliophora, *Chilodonella* processes its macronuclear genome much more extensively than do *Tetrahymena* and *Paramecium* [53].

The micronuclear genome architecture of *Chilodonella* has been minimally explored, with descriptions of a small number of genes, such as those encoding α-tubulin and β-tubulin for *C. uncinata* [54, 55]. The extensively fragmented macronuclear genome contains nanochromosomes that are typically less than 15 kbp in length and are probably composed of only a single gene [53]. Katz and Kovner [56] recently discovered that the micronuclear genome of *Chilodonella* contains scrambled genes. For example, the first gene segment of actin, including the start codon and the 5′untranslated region (5′-UTR), is inverted relative to adjacent sequences [56].

Nyctotherus is a genus of the order Clevelandellida. *N. ovalis* is the best-studied species within *Nyctotherus*. One difficulty that arises when attempting to study the genome architecture of *Nyctotherus* is that cells must be obtained by hand from the hindguts of cockroaches, where they reside [57]. In spite of this difficulty, studies have indicated that the *Nyctotherus* macronuclear genome contains tiny gene-sized chromosomes similar to those in other ciliates such as *Oxytricha* [58], and a pilot macronuclear genome sequence is available [57]. The fact that *Nyctotherus* macronuclear genomes are composed of gene-sized chromosomes suggests that either this phenomenon appeared early in ciliate evolution or it arose in parallel at multiple points during ciliate evolution [57].

Only a handful of ciliate species have been studied at the molecular level. These species represent a small subset of the full diversity of the ciliate phylum on our planet. The two best studied groups, the Oligohymenophorea and the Stichotrichia, do not share key features of their macronuclear and micronuclear genome architectures and have strikingly different developmental processes. The further investigation of ciliates outside of these two well-studied groups can help resolve the question of the ancestral state and subsequent evolution of nuclear dimorphism.

5 Conclusion: Why Nuclear Dimorphism?

The existence of nuclear dimorphism and the extensive genomic processing that takes place during macronuclear development appear, at first glance, to defy an evolutionary explanation. However, traits such as polytene chromosomes, gene-sized macronuclear chromosomes, and even possibly scrambled genes could be adaptive and contribute to selection efficiency [56, 59, 60]. Gene scrambling appears linked to the presence of polytene chromosome formation and gene-sized macronuclear chromosomes, and together these phenomena could enable enhanced generation of diverse proteins.

These extensive processing events also destroy gene linkages, allowing selection to operate more directly on individual genes. Zufall et al. [61] suggested that extensive processing allows heterogeneous rates of protein evolution. Meanwhile, purifying selection maintains conserved macronuclear sequences within rapidly evolving micronucleus-limited sequences. For example, in *C. uncinata*, despite the fact that the eliminated micronuclear sequences are highly divergent, the macronuclear sequences for the genes encoding α-tubulin and actin have been maintained throughout the evolution of the species [56].

The molecular mechanisms underlying ciliate macronuclear development are some of the most complex and intriguing examples of genetically and epigenetically programmed genome rearrangement that have been observed. Macronuclear processing and genome rearrangement in ciliates is an extreme case of the type of genome plasticity that exists in the development of some metazoa [62, 63], the establishment of some cancers [64, 65], and the evolution of genomes [66]. Examining the unique phenomena associated with ciliate nuclear dimorphism will contribute to our understanding of the genomic and epigenetic limits of cellular life [67].

References

1. L.W. Parfrey, D.J. Lahr, A.H. Knoll, L.A. Katz, Estimating the timing of early eukaryotic diversification with multigene molecular clocks. Proc. Natl. Acad. Sci. U.S.A. **108**, 13624–13629 (2011)
2. D.M. Prescott, The DNA of ciliated protozoa. Microbiol. Rev. **58**(2), 233–267 (1994)
3. J.R. Bracht, W. Fang, A.D. Goldman, E. Dolzhenko, E.M. Stein, L.F. Landweber, Genomes on the edge: programmed genome instability in ciliates. Cell **152**, 406–416 (2013)
4. E. Dubois, J. Bischerour, A. Marmignon, N. Mathy, V. Régnier, M. Bétermier, Transposon invasion of the *Paramecium* germline genome countered by a domesticated PiggyBac Transposase and the NHEJ pathway. Int. J. Evol. Biol. **2012**, 436196 (2012)
5. J.N. Fass, N.A. Joshi, M.T. Couvillion, J. Bowen, M.A. Gorovsky, E.P. Hamilton, E. Orias, K. Hong, R.S. Coyne, J.A. Eisen, D.L. Chalker, D. Lin, K. Collins, Genome-scale analysis of programmed DNA elimination sites in *Tetrahymena thermophila*. G3 (Bethesda) **1**, 515–522 (2011)
6. M.C. Yao, J. Choi, S. Yokoyama, C.F. Austerberry, C.H. Yao, DNA elimination in *Tetrahymena*: a developmental process involving extensive breakage and rejoining of DNA at defined sites. Cell **36**, 433–440 (1984)

7. M.C. Yao, S. Duharcourt, D.L. Chalker, Genome-wide rearrangements of DNA in ciliates, in *Mobile DNA, vol II*, ed. by N.L. Craig et al. (ASM Press, Washington, DC, 2002), pp. 730–758

8. O. Arnaiz, N. Mathy, C. Baudry, S. Malinsky, J.M. Aury, C.D. Wilkes, O. Garnier, K. Labadie, B.E. Lauderdale, A. Le Mouël, A. Marmignon, M. Nowacki, J. Poulain, M. Prajer, P. Wincker, E. Meyer, S. Duharcourt, L. Duret, M. Bétermier, L. Sperling, The *Paramecium* germline genome provides a niche for intragenic parasitic DNA: evolutionary dynamics of internal eliminated sequences. PLoS Genet. **8**, e1002984 (2012)

9. A. Gratias, M. Bétermier, Developmentally programmed excision of internal DNA sequences in *Paramecium aurelia*. Biochimie **83**, 1009–1022 (2001)

10. L.A. Klobutcher, G. Herrick, Consensus inverted terminal repeat sequence of *Paramecium* IESs: resemblance to termini of Tc1-related and *Euplotes* Tec transposons. Nucleic Acids Res. **23**(11), 2006–2013 (1995)

11. L.A. Klobutcher, G. Herrick, Developmental genome reorganization in ciliated protozoa: the transposon link. Prog. Nucleic Acid Res. Mol. Biol. **56**, 1–62 (1997)

12. E.R. Henderson, E.H. Blackburn, An overhanging 3′ terminus is a conserved feature of telomeres. Mol. Cell. Biol. **9**, 345–348 (1989)

13. J.D. Forney, E.H. Blackburn, Developmentally controlled telomere addition in wild-type and mutant paramecia. Mol. Cell. Biol. **8**, 251–258 (1988)

14. G. Lepère, M. Nowacki, V. Serrano, J.F. Gout, G. Guglielmi, S. Duharcourt, E. Meyer, Silencing-associated and meiosis-specific small RNA pathways in *Paramecium tetraurelia*. Nucleic Acids Res. **37**(3), 903–915 (2009)

15. O. Garnier, V. Serrano, S. Duharcourt, E. Meyer, RNA-mediated programming of developmental genome rearrangements in *Paramecium tetraurelia*. Mol. Cell. Biol. **24**, 7370–7379 (2004)

16. M. Nowacki, W. Zagorski-Ostoja, E. Meyer, Nowa1p and Nowa2p: novel putative RNA binding proteins involved in trans-nuclear crosstalk in *Paramecium tetraurelia*. Curr. Biol. **15**, 1616–1628 (2005)

17. G. Lepere, M. Betermier, E. Meyer, S. Duharcourt, Maternal noncoding transcripts antagonize the targeting of DNA elimination by scanRNAs in *Paramecium tetraurelia*. Genes Dev. **22**, 1501–1512 (2008)

18. S.R. Lee, K. Collins, Two classes of endogenous small RNAs in *Tetrahymena thermophila*. Genes Dev. **20**, 28–33 (2006)

19. M. Nowacki, K. Shetty, L.F. Landweber, RNA-mediated epigenetic programming of genome rearrangements. Annu. Rev. Genomics Hum. Genet. **12**, 367–389 (2011)

20. K. Mochizuki, N.A. Fine, T. Fujisawa, M.A. Gorovsky, Analysis of a *Piwi*-related gene implicates small RNAs in genome rearrangement in *Tetrahymena*. Cell **110**(6), 689–699 (2002)

21. U.E. Schoeberl, H.M. Kurth, T. Noto, K. Mochizuki, Biased transcription and selective degradation of small RNAs shape the pattern of DNA elimination in *Tetrahymena*. Genes Dev. **26**, 1729–1742 (2012)

22. K. Mochizuki, M.A. Gorovsky, A Dicer-like protein in *Tetrahymena* has distinct functions in genome rearrangement, chromosome segregation, and meiotic prophase. Genes Dev. **19**, 77–89 (2005)

23. L. Aronica, J. Bednenko, T. Noto, L.V. DeSouza, K.W. Siu, J. Loidl, R.E. Pearlman, M.A. Gorovsky, K. Mochizuki, Study of an RNA helicase implicates small RNA-noncoding RNA interactions in programmed DNA elimination in *Tetrahymena*. Genes Dev. **22**, 2228–2241 (2008)

24. K. Mochizuki, M.A. Gorovsky, Conjugation-specific small RNAs in *Tetrahymena* have predicted properties of scan (scn) RNAs involved in genome rearrangement. Genes Dev. **18**, 2068–2073 (2004)

25. A. Gratias, G. Lepère, O. Garnier, S. Rosa, S. Duharcourt, S. Malinsky, E. Meyer, M. Bétermier, Developmentally programmed DNA splicing in *Paramecium* reveals short-distance crosstalk between DNA cleavage sites. Nucleic Acids Res. **36**(10), 3244–3251 (2008)

26. C. Baudry, S. Malinsky, M. Restituito, A. Kapusta, S. Rosa, E. Meyer, M. Betermier, Piggy-Mac, a domesticated PiggyBac transposase involved in programmed genome rearrangements in the ciliate *Paramecium tetraurelia*. Genes Dev. **23**, 2478–2483 (2009)

27. C.Y. Cheng, A. Vogt, K. Mochizuki, M.C. Yao, A domesticated PiggyBac transposase plays key roles in heterochromatin dynamics and DNA cleavage during programmed DNA deletion in *Tetrahymena thermophila*. Mol. Biol. Cell **21**, 1753–1762 (2010)
28. D. Ammermann, G. Steinbrück, L. von Berger, W. Hennig, The development of the macronucleus in the ciliated protozoan *Stylonychia mytilus*. Chromosoma **45**, 401–429 (1974)
29. M.R. Lauth, B.B. Spear, J. Heumann, D.M. Prescott, DNA of ciliated protozoa: DNA sequence diminution during macronuclear development of *Oxytricha*. Cell **7**, 67–74 (1976)
30. D.C. Hoffman, R.C. Anderson, M.L. Dubois, D.M. Prescott, Macronuclear gene-sized molecules of hypotrichs. Nucleic Acids Res. **23**, 1279–1283 (1995)
31. M.T. Swanton, J.M. Heumann, D.M. Prescott, Gene-sized DNA molecules of the macronuclei in three species of hypotrichs: size distributions and absence of nicks. Chromosoma **77**, 217–227 (1980)
32. E. Swart, J.R. Bracht, V. Magrini, P. Minx, X. Chen, Y. Zhou, J.S. Khurana, A.D. Goldman, M. Nowacki, K. Schotanus, S. Jung, R.S. Fulton, A. Ly et al., The *Oxytricha trifallax* macronuclear genome: a complex eukaryotic genome with 16,000 tiny chromosomes. PLoS Biol. **11**, e1001473 (2013)
33. J.R. Williamson, M.K. Raghuraman, T.R. Cech, Monovalent cation-induced structure of telomeric DNA: the G-quartet model. Cell **59**, 871–880 (1989)
34. C. Schaffitzel, I. Berger, J. Postberg, J. Hanes, H.J. Lipps, A. Pluckthun, In vitro generated antibodies specific for telomeric guanine-quadruplex DNA react with *Stylonychia lemnae* macronuclei. Proc. Natl. Acad. Sci. U.S.A. **98**, 8572–8577 (2001)
35. P. Alonso, J. Perez-Silva, Giant chromosomes in protozoa. Nature **205**, 313–314 (1965)
36. D.C. Hoffman, D.M. Prescott, Evolution of internal eliminated segments and scrambling in the micronuclear gene encoding DNA polymerase alpha in two *Oxytricha* species. Nucleic Acids Res. **25**, 1883–1889 (1997)
37. L.F. Landweber, T.C. Kuo, E.A. Curtis, Evolution and assembly of an extremely scrambled gene. Proc. Natl. Acad. Sci. U.S.A. **97**, 3298–3303 (2000)
38. D.M. Prescott, M.L. DuBois, Internal eliminated segments (IESs) of Oxytrichidae. J. Eukaryot. Microbiol. **43**, 432–441 (1996)
39. D.M. Prescott, A. Ehrenfeucht, G. Rozenberg, Template-guided recombination for IES elimination and unscrambling of genes in stichotrichous ciliates. J. Theor. Biol. **222**, 323–330 (2003)
40. M. Nowacki, V. Vijayan, Y. Zhou, K. Schotanus, T.G. Doak, L.F. Landweber, RNA-mediated epigenetic programming of a genome-rearrangement pathway. Nature **451**, 153–158 (2008)
41. M. Nowacki, B.P. Higgins, G.M. Maquilan, E.C. Swart, T.G. Doak, L.F. Landweber, A functional role for transposases in a large eukaryotic genome. Science **324**, 935–938 (2009)
42. W. Fang, X. Wang, J.R. Bracht, M. Nowacki, L.F. Landweber, Piwi-interacting RNAs protect DNA against loss during oxytricha genome rearrangement. Cell **151**, 1243–1255 (2012)
43. A.M. Zahler, Z.T. Neeb, A. Lin, S. Katzman, Mating of the Stichotrichous ciliate *Oxytricha trifallax* induces production of a class of 27 nt small RNAs derived from the parental macronucleus. PLoS One **7**(8), e42371 (2012)
44. J.R. Bracht, D.H. Perlman, L.F. Landweber, Cytosine methylation and hydroxymethylation mark DNA for elimination in *Oxytricha trifallax*. Genome Biol. **13**, R99 (2012)
45. S. Juranek, H.J. Wieden, H.J. Lipps, De novo cytosine methylation in the differentiating macronucleus of the stichotrichous ciliate *Stylonychia lemnae*. Nucleic Acids Res. **31**, 1387–1391 (2003)
46. T.G. Doak, A.R.O. Cavalcanti, N.A. Stover, D.M. Dunn, R. Weiss, G. Herrick, L.F. Landweber, Sequencing the *Oxytricha trifallax* macronuclear genome: a pilot project. Trends Genet. **19**, 603–607 (2003)
47. D.V. Vinogradov, O.V. Tsoĭ, A.V. Zaika, A.V. Lobanov, A.A. Turanov, V.N. Gladyshev, M.S. Gel'fand, Draft macronuclear genome of a ciliate *Euplotes crassus*. Mol. Biol. (Mosk.) **46**, 361–366 (2012)
48. C.L. Jahn, Differentiation of chromatin during DNA elimination in *Euplotes crassus*. Mol. Biol. Cell **10**, 4217–4230 (1999)

49. J.W. Jaraczewski, C.L. Jahn, Elimination of Tec elements involves a novel excision process. Genes Dev. **7**, 95–105 (1993)
50. C.L. Jahn, L.A. Nilles, M.F. Krikau, Organization of the *Euplotes crassus* micronuclear genome. J. Protozool. **35**, 590–601 (1988)
51. D.S. Gross, W.T. Garrard, Nuclease hypersensitive sites in chromatin. Annu. Rev. Biochem. **57**, 159–197 (1988)
52. C.L. Jahn, Z. Ling, C.M. Tebeau, L.A. Klobutcher, An unusual histone H3 specific for early macronuclear development in *Euplotes crassus*. Proc. Natl. Acad. Sci. U.S.A. **94**, 1332–1337 (1997)
53. J.L. Riley, L.A. Katz, Widespread distribution of extensive chromosomal fragmentation in ciliates. Mol. Biol. Evol. **18**, 1372–1377 (2001)
54. L.A. Katz, E. Lasek-Nesselquist, O.L.O. Snoeyenbos-West, Structure of the micronuclear α-tubulin gene in the phyllopharyngean ciliate *Chilodonella uncinata*: implications for the evolution of chromosomal processing. Gene **315**, 15–19 (2003)
55. R.A. Zufall, L.A. Katz, Micronuclear and macronuclear forms of β-tubulin genes in the ciliate *Chilodonella uncinata* reveal insights into genome processing and protein evolution. J. Eukaryot. Microbiol. **54**(3), 275–282 (2007)
56. L.A. Katz, A.M. Kovner, Alternative processing of scrambled genes generates protein diversity in the ciliate *Chilodonella uncinata*. J. Exp. Zool. **314B**, 480–488 (2010)
57. G. Ricard, R.M. de Graaf, B.E. Dutilh, I. Duarte, T.A. van Alen, H.A.M. van Hoek, B. Boxma, G.W.M. Van der Staay, S.Y. Moon-van der Staay, W.J. Chang, L.F. Landweber, J.H.P. Hackstein, M.A. Huynen, Macronuclear genome structure of the ciliate *Nyctotherus ovalis*: single-gene chromosomes and tiny introns. BMC Genomics **9**, 587 (2008)
58. C. McGrath, R.A. Zufall, L.A. Katz, Variation in macronuclear genome content of three ciliates with extensive chromosomal fragmentation: a preliminary analysis. J. Eukaryot. Microbiol. **54**, 242–46 (2007)
59. L.F. Landweber, Why genomes in pieces? Science **318**, 405–407 (2007)
60. L.F. Landweber, Making sense of scrambled genomes. Science **319**, 901–902 (2008)
61. R.A. Zufall, C.L. McGrath, S.V. Muse, L.A. Katz, Genome architecture drives protein evolution in ciliates. Mol. Biol. Evol. **23**(9), 1681–1687 (2006)
62. J.J. Smith, F. Antonacci, E.E. Eichler, C.T. Amemiya, Programmed loss of millions of base pairs from a vertebrate genome. Proc. Natl. Acad. Sci. U.S.A. **106**, 11212–11217 (2009)
63. J.J. Smith, C. Baker, E.E. Eichler, C.T. Amemiya, Genetic consequences of programmed genome rearrangement. Curr. Biol. **22**, 1524–1529 (2012)
64. H. Li, J.L. Wang, G. Mor, J. Sklar, A neoplastic gene fusion mimics trans-splicing of RNAs in normal human cells. Science **321**, 1357–1361 (2008)
65. J.D. Rowley, T. Blumenthal, The cart before the horse. Science **321**, 1302–1304 (2008)
66. D. Sankoff, J.H. Nadeau, Chromosome rearrangements in evolution: from gene order to genome sequence and back. Proc. Natl. Acad. Sci. U.S.A. **100**(20), 11188–11189 (2003)
67. A.D. Goldman, L.F. Landweber, Oxytricha as a modern analog of ancient genome evolution. Trends Genet. **28**, 382–388 (2012)

The Algebra of Gene Assembly in Ciliates

Robert Brijder and Hendrik Jan Hoogeboom

Abstract The formal theory of intramolecular gene assembly in ciliates is fitted into the well-established theories of Euler circuits in 4-regular graphs, principal pivot transformations, and delta-matroids.

1 Introduction

Gene assembly is an intricate process that occurs in the class of unicellular organisms called ciliates. During this process a nucleus, called the *micronucleus*, is transformed into a functionally and structurally different nucleus, called the *macronucleus*. This is accomplished using complicated DNA splicing and recombination operations. Gene assembly has been studied formally on the level of individual genes (e.g., [16, 23]).

The theory of Euler circuits in 4-regular graphs was initiated in a seminal paper by Kotzig [31]. Bouchet developed the theory further by relating it to delta-matroids [6] and isotropic systems [5, 7]. In [6], Bouchet used a matrix transformation that turns out to be "almost" a principal pivot transform (PPT) [42]. PPT, delta-matroids, and isotropic systems have many interesting properties which have direct consequences for the theory of Euler circuits in 4-regular graphs.

Although, at first glance, the formal theory of gene assembly seems to be related to the theory of Euler circuits in 4-regular graphs (this will become clear when we

R. Brijder
Hasselt University, Diepenbeek, Belgium

Transnational University of Limburg, Diepenbeek, Belgium
e-mail: robert.brijder@uhasselt.be

H.J. Hoogeboom (✉)
Leiden Institute of Advanced Computer Science, Leiden University, Leiden, The Netherlands
e-mail: hoogeboom@liacs.nl

N. Jonoska and M. Saito (eds.), *Discrete and Topological Models in Molecular Biology*,   289
Natural Computing Series, DOI 10.1007/978-3-642-40193-0_13,
© Springer-Verlag Berlin Heidelberg 2014

consider Fig. 5 in Sect. 3), there have been few attempts to fit the former theory into the latter. In this chapter, we do exactly this. We show that the formal model of gene assembly can be defined quite efficiently in terms of 4-regular graphs. In a survey-style fashion, we discuss consequences of known results in the theory of 4-regular graphs (including, for example, results related to PPT and delta-matroids) for the theory of gene assembly.

This chapter is organized as follows. In Sect. 2 we briefly recall the biology of gene assembly in ciliates and its string model [21, 22] (see also [23]). In Sect. 3, we view gene assembly in terms of Euler circuit transformations (or, more generally, circuit partition transformations) in 4-regular graphs, and in terms of the corresponding local and edge complementation transformations on looped circle graphs. These operations of local and edge complementation turn out to be special cases of principal pivot transform defined on arbitrary square matrices (see Sect. 4). In Sect. 5, we find that we can view local and edge complementation in terms of a very elementary operation, called pivot, on set systems. Finally, in Sect. 6, we combine pivot on set systems with another operation, called loop complementation, on set systems. Together, the two operations turn out to form a group, which enables us to replace intricate graph operations by a simple algebra on set systems.

2   Gene Assembly in Ciliates

Ciliates contain two different kinds of nuclei, which differ both functionally and structurally. The relatively large *macronucleus* (MAC for short) has many copies of short chromosomes, each containing only a single gene or just a few genes. The *micronucleus* (MIC for short) contains a much smaller number of chromosomes, each containing numerous genes (as is usual for chromosomes in general). The germ-line MIC is used only for reproduction, while the somatic MAC is used for general cell regulation. During sexual reproduction, a newly formed MIC is transformed into a MAC. This process is called *gene assembly* and is accomplished using extensive DNA splicing and recombination operations.

The genetic material in the MIC is scrambled: the genes are broken up into segments, called *macronuclear destined sequences* (or MDSs for short), which are reordered and possibly inverted with respect to the corresponding MAC genes. Moreover, the MDSs in the MIC genes are separated by *internal eliminated sequences* (IESs for short), which are not part of the genes. For example, the MIC form of the Actin I gene of the ciliate *Sterkiella nova* is depicted in Fig. 1 and can be described as the string $I_0\, M_3\, I_1\, M_4\, I_2\, M_6\, I_3\, M_5\, I_4\, M_7\, I_5\, M_9\, I_6\, \overline{M}_2\, I_7\, M_1\, I_8\, M_8\, I_9$ [38], where the M_i's are MDSs and the I_i's are IESs. Note that the inversion of the MDS M_2 is indicated by a bar. The MDSs $M_1, \ldots, M_9$ are oriented and numbered according to the order in which they occur in the corresponding "unscrambled" MAC gene; see Fig. 2. Note that consecutive MDSs overlap (the gray segments in Fig. 2). These segments are called *pointers* in the MIC gene, as they indicate the

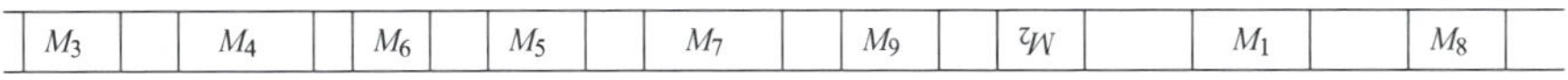

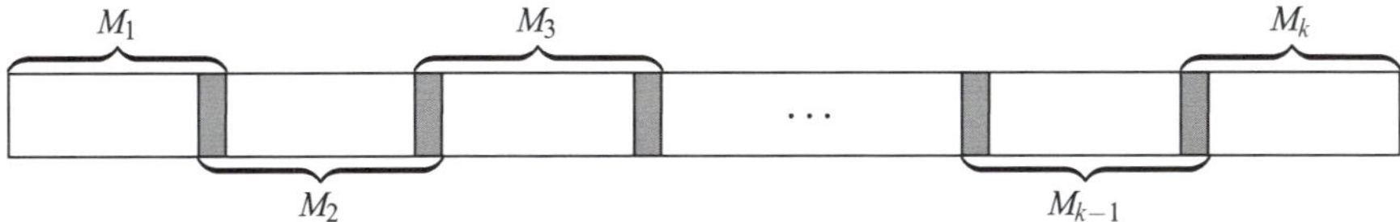

Fig. 1 The structure of the MIC gene encoding for the Actin I protein in *Sterkiella nova*

Fig. 2 The structure of a MAC gene consisting of κ MDSs

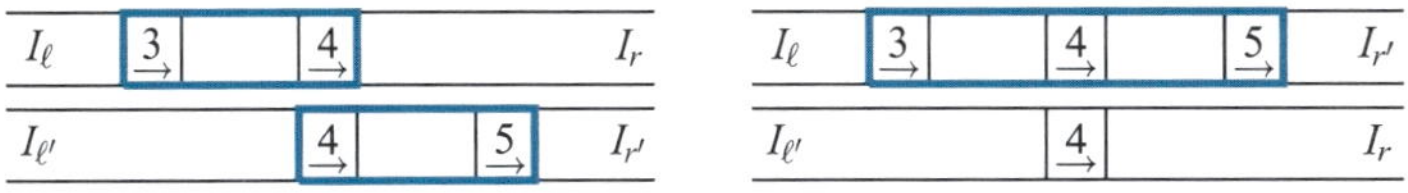

Fig. 3 Recombination on pointer 4 joins MDSs M_3 and M_4. The left and right pointers of M_3 are denoted by 3 and 4, respectively (and similarly for M_4)

complex recombination schema that is to be performed to obtain the corresponding MAC gene.

The generic form of recombination aligns two MDSs on their common pointer and then performs a crossover operation at that pointer; see Fig. 3. In that way, the two segments are joined into a larger MDS segment. Note that the order (or the level of parallelism) in which recombination operations are applied has no influence on the outcome. This general *confluency* property of recombination ensures that the MAC gene is uniquely obtained from the MIC gene by performing recombination on each pointer pair (regardless of the order in which recombination takes place) [19]. Biologically, the pointer pair no longer exists after recombination, as it cannot be used for another recombination operation. Mathematically, it turns out to be worthwhile to leave the pointer pair for further consideration.

We consider the intramolecular model for gene assembly presented in [23, 39]. In this model, three specific types of recombination operations are distinguished; see Fig. 4. (a) Two consecutive MDSs (i.e., a single IES separates the two MDSs) having the same orientation can be recombined by *loop excision*. In that process, a circular molecule is removed from the segment containing an IES. (b) Two MDSs in opposite orientations can be recombined by *hairpin recombination*. This operation inverts the segment that was originally between the two MDSs. This segment may contain other MDSs. (c) Two interleaved pairs of consecutive MDSs, where the MDSs in each pair are in the same orientation, can be recombined by *double loop recombination*. During this operation, two segments between the MDSs are swapped. A sequence φ of recombination operations (of these types) is called *successful* for a given MIC gene g if (i) φ is applicable to (defined on) g and

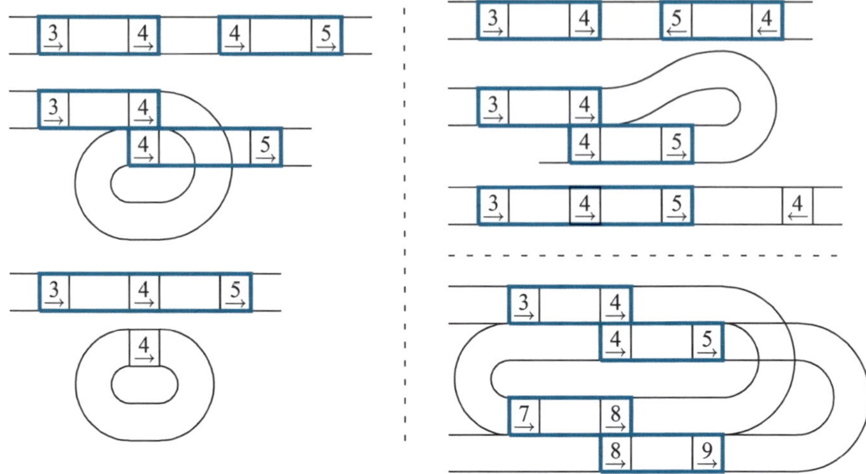

Fig. 4 Three operations: loop excision (*left-hand side*), hairpin recombination (*upper-right corner*), and double loop recombination (intermediate stage only, *lower-right corner*)

(ii) applying φ on g yields the MAC gene corresponding to g. Because of the above-mentioned confluence property of recombination in general, we have the result that φ is successful for g iff φ is applicable to g and each pointer pair is used exactly once in φ.

The above recombination operations are formalized on strings as follows. We fix a positive integer κ. We denote pointers and their orientation by the alphabet $\Pi = \{1, 2, \ldots, \kappa\} \cup \{\bar{1}, \bar{2}, \ldots, \bar{\kappa}\}$. The inversion of the string $w = w_1 w_2 \ldots w_n \in \Pi^*$ is the string $\bar{w} = \bar{w}_n \ldots \bar{w}_2 \bar{w}_1$, where we let $\bar{\bar{p}} = p$ for each $p \in \Pi$.

A *directed double-occurrence string*, or *doc-string* for short (called a legal string in [23]), is a string w over Π that contains each pointer of w exactly twice, in either orientation (barred or unbarred). The MIC gene is then encoded by concatenating the pointers (including their orientations) in the same order as they appear in the MIC gene. Hence, if there are κ MDSs, then MDS M_i (for $i \in \{2, \ldots, \kappa - 1\}$) corresponds to $i\,(i+1)$, its inversion $\overline{M_i}$ corresponds to $\overline{(i+1)\,i}$, MDSs M_1 and M_κ correspond to 2 and κ, respectively, and their inversions correspond to $\bar{2}$ and $\bar{\kappa}$, respectively (recall that M_1 and M_κ have only one neighboring MDS). Note that there is no pointer 1. Thus the MIC form of the Actin I gene of *Sterkiella nova* mentioned above is written as $34\ 45\ 67\ 56\ 78\ 9\ \overline{3\bar{2}}\ 2\ 89$, with spaces added for clarity.

The three recombination operations can be described using doc-strings in the following straightforward manner [21, 22]. First we define the following three mappings on doc-strings. Let $u_1, \ldots, u_5 \in \Pi^*$, and let $p, q \in \{1, \ldots, \kappa\}$. Then, $u \setminus p$ deletes occurrences of p and $\bar{p}$ in u; if $u = u_1\,p\,u_2\,\bar{p}\,u_3$, then $u * p = u_1\,p\,\bar{u}_2\,\bar{p}\,u_3$; and if $u = u_1\,p\,u_2\,q\,u_3\,p\,u_4\,q\,u_5$, then $u * \{p, q\} = u_1\,p\,u_4\,q\,u_3\,p\,u_2\,q\,u_5$. In a similar way, we define $u * p$ in the case $u = u_1\,\bar{p}\,u_2\,p\,u_3$ (i.e., the bars on the two occurrences of p are swapped), and $u * \{p, q\}$ in the case where the positions of p and q are interchanged and/or in the case where the two occurrences in the p-pair or q-pair are barred.

Then (a) $u \setminus p$ models loop excision, provided u contains pp (or $\bar{p}\bar{p}$) as consecutive pointers; (b) $u * p \setminus p$ models hairpin recombination; and (c) $u * \{p, q\} \setminus p \setminus q$ models double loop recombination.

Given a doc-string, any sequence of these three operations that reduces this string to the empty string is called a *successful reduction*. Note that if a doc-string w represents a MIC gene g, then successful reductions of w correspond precisely to successful reductions of g. It is easily verified that every doc-string has a successful reduction [20]. This reduction is usually not unique.

3 Graph Models

It is not surprising that graph-theoretical concepts are important tools in modeling and understanding the process of gene assembly in ciliates. Consider for instance the diagram of the Actin I gene of *Sterkiella nova* as depicted by Prescott [37]. A simplified representation is given in Fig. 5 (see Example 1 below for details). The structure of the genetic material is given as a "bicolored" graph, with pointers as vertices, and MDS and IES segments as edges. We can read both the original MIC sequence and the target MAC sequence from the graph. If we follow the IES and MDS edges in an alternating fashion, then we obtain the MIC, and if we follow the edges according their colors, then we obtain the MAC (with flanking IESs).

Example 1. In the MDS–IES description of the MIC form of Actin I of *Sterkiella nova* (see Sect. 2), we can explicitly add the pointers flanking the MDSs to obtain the sequence $\pi = I_0\, p_3\, M_3\, p_4\, I_1\, p_4\, M_4\, p_5\, I_2\, p_6\, M_6\, p_7\, I_3\, p_5\, M_5\, p_6\, I_4\, p_7\, M_7\, p_8\, I_5\, p_9\, M_9\, I_6\, \overline{p}_3\, \overline{M}_2\, \overline{p}_2\, I_7\, M_1\, p_2\, I_8\, p_8\, M_8\, p_9\, I_9$.

One may view π as an Eulerian path in a multigraph G: the pointers p_i are the vertices of G, and the strings between the vertices are the (labeled) edges of G. In this way, π induces Fig. 5. Apart from the MDS edges M_i (between p_i and p_{i+1}) and the IES edges I_i, there are also "mixed" edges, such as the loop $I_7\, M_1$ on p_2, caused by the fact that M_1 has no initial pointer. Note that we may have parallel edges; for example, there are two edges from p_5 to p_6.

The MAC form is obtained by recombining MDSs at each pointer. For example, at p_5 in π we have both $M_4\, p_5\, I_2$ and $I_3\, p_5\, M_5$, whereas in the MAC form M_4 and M_5 are joined at vertex (pointer) p_5, and we have both $M_4\, p_5\, M_5$ and $I_3\, p_5\, I_2$. Recall from Fig. 3 that when MDSs are recombined at a pointer, IESs are at the same time joined together at that pointer.

As MDS M_2 is inverted in the MIC form, it has to be read "backwards" in the MAC form. Thus, in the MAC form we follow edge $p_3\, \overline{M}_2\, p_2$ in the opposite direction. Also, when recombining M_1 and M_2 at p_2, IESs I_7 and I_8 are at the same time joined together at p_2.

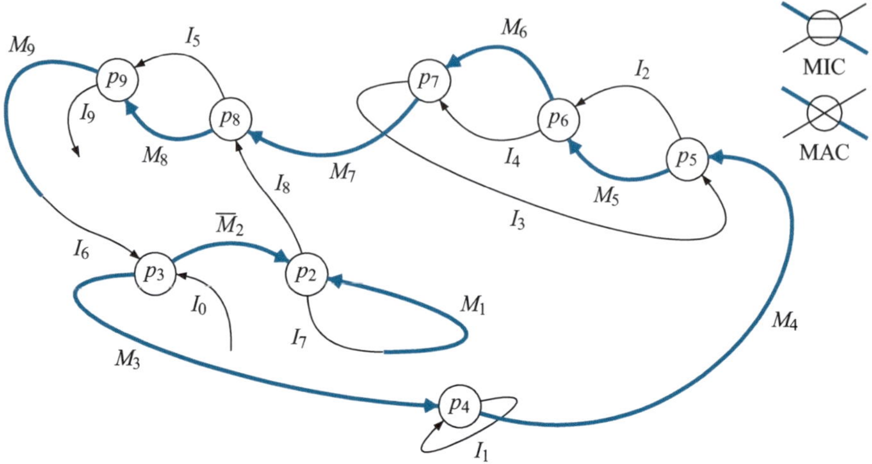

Fig. 5 Actin I gene of *Sterkiella nova*. Schematic diagram, based on [37]

In this way, the MAC form of the gene consists of three molecules. The string (where the pointers are omitted) $\overline{I}_9 \overline{I}_5 \overline{I}_8 \overline{I}_7 M_1 M_2 \ldots M_8 M_9 I_6 \overline{I}_0$ represents the strand consisting of the recombined MDSs and flanking IESs. The MAC form also has two circular IES molecules (that are excised), I_1 and $I_2 I_4 I_3$ (again the pointers are omitted).   □

In the manner described above, every (unbarred) doc-string defines a 4-regular multigraph (every vertex has degree 4; we allow loops and parallel edges) together with an Euler cycle (which visits every edge of the graph exactly once). We start by representing every pointer pair by a vertex. We then follow the string, adding an edge as we step from pointer to pointer. We treat the string as if it is circular, and connect the last pointer to the first. Obviously, the multigraph is 4-regular, and the string traces an Euler cycle through the multigraph. The result is very similar to the representation of the gene in Fig. 5, if we merge the initial and final "edges" I_0 and I_9. Conversely, every 4-regular multigraph with an Euler cycle induces a (unbarred) doc-string w (in fact, a set of "equivalent" doc-strings that are obtained from w by conjugation).

We now briefly describe the theory of operations on cycles in 4-regular multigraphs as initiated by Kotzig [31] and continued by Bouchet. Given a 4-regular multigraph, we obtain a set of cycles by "joining" pairwise the edges at each vertex. These pairings can be unambiguously described using a fixed Euler circuit as an anchor; see Fig. 6a–c. The pairings may follow the Euler circuit; otherwise, they may reconnect in a way that may or may not agree with the orientation of the Euler circuit. The pairings in Fig. 6a–c are called smoothings in [3] and transitions in [6].

Example 2. (1) Consider the (unbarred) doc-string $w = 126134563245$. Then w defines a 4-regular graph G along with an Euler circuit E_w in G; see Fig. 7a. For

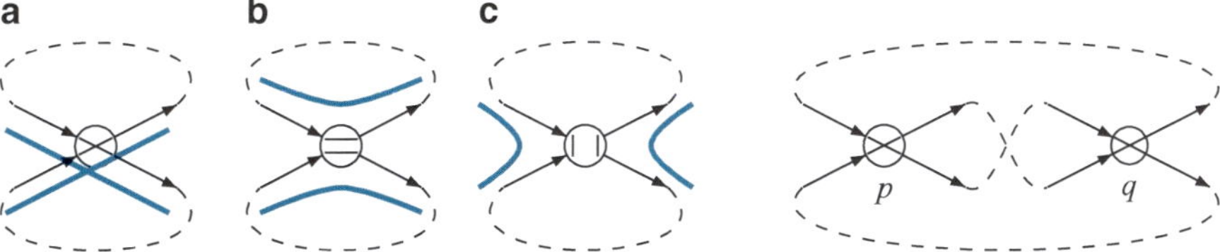

Fig. 6 Three ways to connect pairs of edges in a 4-regular graph relative to an Eulerian cycle; (**a**) following the cycle; (**b**) orientation consistent; (**c**) orientation inconsistent. Two interleaved vertices in the cycle (*rightmost diagram*)

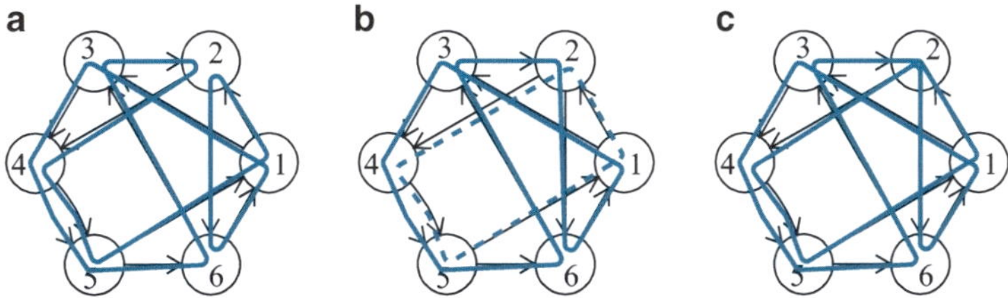

Fig. 7 Recombining edges of an Euler cycle (see Example 2)

convenience, the edges in G are directed according to w. (2) If we take the pairing at vertex 2 that is different from E_w in an orientation-consistent way, and if we take the same pairings as in E_w at the other vertices, then we obtain two disjoint cycles, $w_a = 1245$ and $w_b = 26134563$; see Fig. 7b. (3) If we take the pairing at vertex 2 that is different from E_w in an orientation-inconsistent way, and if we take the same pairings as in E_w at the other vertices, then we obtain the Euler circuit described by $w' = 123654316245$; see Fig. 7c.   □

Note how an orientation-inconsistent transition induces a "reversal" of part of the original Euler circuit, which agrees with hairpin recombination. Note also that an orientation-consistent transition breaks the Euler circuit into two disconnected parts. Now consider two vertices p and q that are interleaved in the Euler circuit, occurring in the order $\cdots p \cdots q \cdots p \cdots q \cdots$: see Fig. 6 (right). Then a synchronized orientation-consistent transition at both p and q again yields an Euler circuit. This Euler circuit is obtained from the original one by swapping two segments in exactly the same way as in double loop recombination.

Then, to correctly model the gene assembly process, we have to keep track of which pointers (vertices) we can apply successive transitions to while maintaining an Euler circuit. Both the orientation and the interleavings may change during the process. The tool that we use is a *circle graph*, which represents the intersections of the chords in a circle: each chord is represented by a vertex, and two vertices are adjacent iff the corresponding chords intersect. A (barred) doc-string w defines a circle graph $C(w)$ in a natural way if we write w in a circular way and connect the pointer pairs. Additionally, we encode the relative orientation of the pointers of each

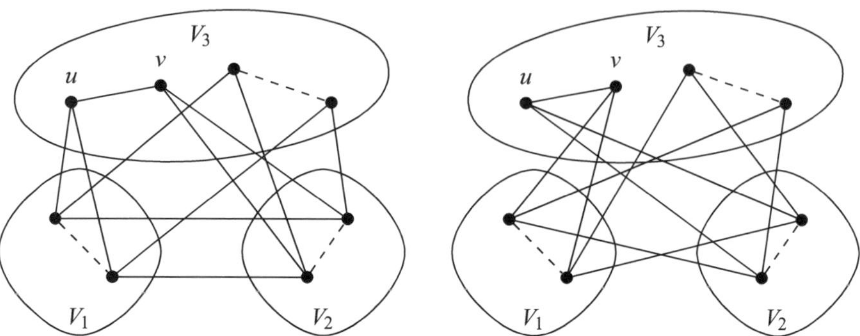

Fig. 8 A pivot on an edge $\{u, v\}$ in a graph. The adjacency between two vertices x and y is toggled iff $x \in V_i$ and $y \in V_j$ with $i \neq j$. Note that u and v are adjacent to all vertices in V_3 – these edges are omitted in the diagram. The operation does not affect edges adjacent to vertices outside the sets V_1, V_2, V_3

pointer pair by adding a loop to a vertex when the two pointers of the pointer pair have different orientations (i.e., one is with a bar, and the other is without a bar). In [23], $\pm$ signs are used instead of loops, and the corresponding graph, equivalent to a (looped) circle graph, is called a *signed graph*. The advantage of using loops instead of $\pm$ signs will become clear in Sect. 4. The (looped) circle graph $C(w)$ of the doc-string $w = 1\,2\,6\,\bar{1}\,3\,4\,\bar{5}\,6\,3\,\bar{2}\,4\,5$ is given in Fig. 9 (middle, bottom row).

From now on, by a *graph* we mean an undirected graph G where loops are allowed, but parallel edges are not allowed. More precisely, $G = (V, E)$, where V is a finite set of *vertices* and $E \subseteq \{\{x, y\} \mid x, y \in V\}$ is a set of *edges* (we have $\{x\} \in E$ iff x is a looped vertex). If the graph G is clear from the context, then we shall simply denote its vertex set by V. For $X \subseteq V$, we denote the subgraph of G induced by X by $G[X]$.

We now define the basic operations of local and edge complementation on graphs [7, 31]. If G is a graph with a looped vertex u, then the *local complement* of G on u, denoted by $G * u$, is obtained from G by complementing the edges in the neighborhood $N_G(u) = \{v \in V \mid \{u, v\} \in E, u \neq v\}$ of u in G. Thus, for $v, w \in N_G(u)$, $e = \{v, w\}$ is an edge of $G * u$ iff e is not an edge of $G * u$ (we allow $v = w$, i.e., e is a loop). All other edges remain the same in G and $G * u$.

For an edge $\{u, v\}$ of G where u and v are distinct unlooped vertices, we define the *edge complement* of G on $\{u, v\}$, denoted by $G * \{u, v\}$, as follows. The closed neighborhood of vertex w, denoted by $N_G[w]$, equals $N_G(w) \cup \{w\}$. The neighbors of u and v can be partitioned into the three sets $N_G[u] \setminus N_G[v]$, $N_G[v] \setminus N_G[u]$, and $N_G[u] \cap N_G[v]$. The graph $G * \{u, v\}$ is obtained from G by complementing all pairs $\{x, y\}$ such that x and y are each neighbors of u or v, but not in the same partition; see Fig. 8. This will not change any adjacencies to vertices not adjacent to u and v, nor will it change any loops.

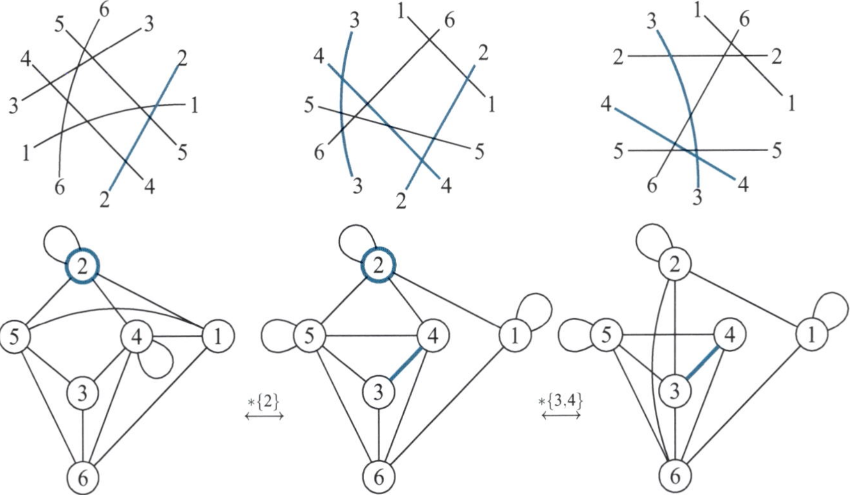

Fig. 9 Local complement on looped vertex 2 (*left-hand side*) and edge complement on unlooped edge $\{3, 4\}$ (*right-hand side*). The *top row* indicates how the pointer segments overlap in the underlying doc-strings; the *bottom row* contains the circle graphs

Theorem 1 ([31]). *Let w be a doc-string and let $p, q \in \{1, \ldots, \kappa\}$. If $w * p$ is defined (i.e., w contains both p and $\bar{p}$), then $C(w) * p = C(w * p)$. If $w * \{p, q\}$ is defined, then $C(w) * \{p, q\} = C(w * \{p, q\})$.*

Of course, Theorem 1 may be reformulated using Euler cycles in 4-regular multigraphs instead of doc-strings.

Example 3. Consider the doc-string $w = 1\,2\,6\,\bar{1}\,3\,4\,\bar{5}\,6\,3\,\bar{2}\,4\,5$. This defines the circle graph $C(w)$ in Fig. 9 (middle). If we complement the neighborhood $N_{C(w)}(2) = \{1, 4, 5\}$ of vertex 2 in $C(w)$, we obtain the graph $C(w) * \{2\} = C(1\,2\,\bar{3}\,6\,5\,\bar{4}\,\bar{3}\,1\,\bar{6}\,\bar{2}\,4\,5)$; see Fig. 9 (left). The edge complement on the unlooped edge $\{3, 4\}$ in $C(w)$ yields $C(w) * \{3, 4\} = 1\,2\,6\,\bar{1}\,3\,\bar{2}\,4\,5\,6\,3\,4\,5$, depicted in Fig. 9 (right). $\qquad\square$

Theorem 1 suggests a generalization of the three recombination operations on doc-strings that model loop excision, hairpin recombination, and double loop recombination (defined in Sect. 2). We have that (a) removing an isolated unlooped vertex corresponds to loop excision, (b) local complementation followed by the deletion of the vertex involved corresponds to hairpin recombination, and (c) edge complementation followed by the deletion of the vertices involved corresponds to double loop recombination.

A *reduction* of a graph G is a sequence of these three operations, and a *successful reduction* of G is a reduction of G to the empty graph. Every graph has a successful reduction: we can apply local complement reductions until there are no more loops, then apply edge complement reductions until the graph contains only isolated unlooped vertices, and finally remove these isolated unlooped vertices.

If a doc-string w can be successfully reduced by a sequence of operations, then the associated circle graph $C(w)$ can be rewritten by the corresponding sequence of graph operations. A similar result holds for the converse, except that we may have to reorder the loop excision operations [21,22]. For the string $w = 2332$, for example, there is a unique sequence of two loop excisions ("inside out"), whereas its circle graph $C(w)$ consists of two isolated unlooped vertices which can be removed in any order.

Since not every graph is a circle graph $C(w)$ for some doc-string w, the operations of local complementation $* p$ and edge complementation $*\{p, q\}$ are generalizations of the corresponding operations $* p$ and $* \{p, q\}$, respectively, for doc-strings.

We remark that a polynomial called the *Martin polynomial* [32] (its multivariate variant is called the *transition polynomial* [30]) has been defined with respect to Euler circuits C in 4-regular multigraphs. This polynomial records the number of circuits $c_T(C)$ obtained when one performs on C a set T of transitions of the form described as in Fig. 6. In a similar way to that described in this section, the Martin polynomial corresponds to a graph polynomial called the *interlace polynomial* [4] (or *Tutte–Martin polynomial* [8]), in which local and edge complementation play a central role. Interestingly, the well-known *Tutte polynomial* on the diagonal coincides with the interlace polynomial when consideration is restricted to bipartite graphs [2]. A number of variations of the Martin and interlace polynomials have been studied in the literature, as we may restrict or loosen the allowed types of transitions T. Among them is the *Penrose polynomial* [1, 35] and the *bracket polynomial* for graphs [41]. We refer the reader to [24, 25] for a detailed survey of these polynomials.

4   Matrices

With the definitions of local complementation $* u$ and edge complementation $*\{u, v\}$ in place, we are now interested in *sequences* of these operations (and, in particular, successful reductions). Since the definition of edge complementation is already complicated in itself, it seems even more difficult to reason about the effect of sequences such as $*\{u, v\}*\{v, w\}$. Fortunately, it turns out that sequences of local and edge complementations correspond to (a special case of) the so-called principal pivot transform operation on square matrices. This leads to a different perspective in which sequences of local and edge complementations are much easier to study. Let us first recall the principal pivot transform operation.

Let V be a finite set, and let A be a $V \times V$ matrix, i.e., a matrix where the columns and rows are indexed by V. For a set $X \subseteq V$ we use $A[X]$ to denote the principal submatrix induced by X (i.e., the rows and columns are indexed by X). Moreover, we define $A \setminus X = A[V \setminus X]$. Let A be a $V \times V$ matrix (over an arbitrary field), and let $X \subseteq V$ be such that $A[X]$ is nonsingular, i.e., $\det A[X] \neq 0$. The *principal* *pivot transform* (*PPT* or *pivot* for short) of A on X, denoted by $A * X$, is defined as follows (see [43]). If

$$A = \begin{array}{c} \\ X \\ V \setminus X \end{array} \begin{pmatrix} \overset{X}{P} & \overset{V \setminus X}{Q} \\ R & S \end{pmatrix},$$

then

$$A * X = \begin{array}{c} \\ X \\ V \setminus X \end{array} \begin{pmatrix} \overset{X}{P^{-1}} & \overset{V \setminus X}{-P^{-1}Q} \\ RP^{-1} & S - RP^{-1}Q \end{pmatrix}.$$

Hence, $A * X$ is defined iff $A[X]$ is nonsingular. The matrix $(A * X) \setminus X = S - RP^{-1}Q$ is called the *Schur complement* of X in A.

Pivot is sometimes considered a partial inverse, since A and $A * X$ are related as follows, where the vectors x_1 and x_2 correspond to the elements of X. In fact, the following relation defines $A * X$ given A and X [42]:

$$A \begin{pmatrix} x_1 \\ y_1 \end{pmatrix} = \begin{pmatrix} x_2 \\ y_2 \end{pmatrix} \text{ iff } A * X \begin{pmatrix} x_2 \\ y_1 \end{pmatrix} = \begin{pmatrix} x_1 \\ y_2 \end{pmatrix}. \tag{1}$$

Note that if $\det A \neq 0$, then $A * V = A^{-1}$. By Eq. (1), we see that a pivot operation is an involution (i.e., an operation of order 2) and, more generally, if $(A * X) * Y$ is defined, then $A * (X \Delta Y)$ is defined (applying the symmetric difference of X and Y) and the resulting matrices are equal. Note that in order to apply the pivot $* X$ to matrix A, it is required that $A[X]$ is nonsingular.

We may apply pivot to graphs through its adjacency matrix representation. The adjacency matrix $A(G)$ of a graph $G = (V, E)$ is a $V \times V$ matrix $(a_{u,v})$ over $\mathbb{F}_2$ (the binary field) with $a_{u,v} = 1$ iff $\{u, v\} \in E$. Obviously, for $X \subseteq V$, $A(G[X]) = A(G)[X]$. In this chapter, we make no distinction between G and $A(G)$ and so we write, for example, $\det G$ to denote $\det A(G)$, the determinant of $A(G)$ computed over $\mathbb{F}_2$. In this way, graphs correspond precisely to symmetric $V \times V$ matrices over $\mathbb{F}_2$. By convention, the determinant of the empty matrix (or graph) is 1.

For a graph G and nonempty $X \subseteq V$, X is called *elementary in G* if $G[X]$ is nonsingular and, for all nonempty $Y \subsetneq X$, $G[Y]$ is singular. Hence, if X is elementary in G, then $G * X$ is defined, but $G * Y$ is not defined for any nonempty proper subset of X. It is easy to see that if X is elementary in G, either (1) $X = \{u\} \in E(G)$ (i.e., X is a loop) or (2) $X = \{u, v\} \in E(G)$ and $\{u\}, \{v\} \notin E(G)$ (i.e., X is an edge on unlooped vertices). Geelen [26] observed that a pivot in case (1) is precisely a local complementation and a pivot in case (2) is precisely an edge complementation.

Indeed, if vertex u has a loop in G, then the matrix $G[\{u\}]$ is equal to the 1×1 identity matrix:

$$u \quad u \begin{pmatrix} 1 \end{pmatrix}.$$

Hence, $* \{u\}$ is indeed applicable to G, and

$$G * \{u\} = \begin{array}{c} u \\ V \setminus \{u\} \end{array} \begin{pmatrix} \overset{u}{1} & \overset{V \setminus \{u\}}{\chi_u^T} \\ \chi_u & G[V - u] - \chi_u \chi_u^T \end{pmatrix},$$

where χ_u is the column vector belonging to u, without the element at the position (u, u). One may easily verify that $G * \{u\}$ is indeed the graph obtained from G by applying local complementation on u.

Turning to edge complementation, if $\{u, v\}$ is an edge in G and u and v are unlooped vertices, then the matrix $G[\{u, v\}]$ is equal to

$$\begin{array}{c} u \\ v \end{array} \begin{pmatrix} \overset{u}{0} & \overset{v}{1} \\ 1 & 0 \end{pmatrix}.$$

Hence, $* \{u, v\}$ is indeed applicable to G, and

$$G * \{u, v\} = \begin{array}{c} u \\ v \\ V \setminus \{u, v\} \end{array} \begin{pmatrix} \overset{u}{0} & \overset{v}{1} & \overset{V \setminus \{u, v\}}{\chi_v^T} \\ 1 & 0 & \chi_u^T \\ \chi_v & \chi_u & G[V - u - v] - (\chi_v \chi_u^T + \chi_u \chi_v^T) \end{pmatrix},$$

where χ_u is the column vector of G belonging to u without the elements at positions (u, u) and (v, u) (and similarly for χ_v). One may again verify that $G * \{u, v\}$ is indeed the graph obtained from G by applying edge complementation on $\{u, v\}$.

Having thus characterized local and edge complementation in terms of pivot, we are ready to study sequences of these operations. For example, let u, v, and w be mutually distinct unlooped vertices of G. If $(G * \{u, v\}) * \{v, w\}$ is defined (i.e., if $\{u, v\}$ is an edge of G and $\{v, w\}$ is an edge of $G * \{u, v\}$), then we immediately find that $(G * \{u, v\}) * \{v, w\} = G * (\{u, v\} \Delta \{v, w\}) = G * \{u, w\}$ (and that $\{u, w\}$ is an edge of G, since $G * \{u, w\}$ is defined and u and w are unlooped vertices). In general, we have the following *confluence* result.

Let $\varphi = * X_1 * X_2 \cdots * X_n$ be a sequence of pivot operations (we assume left-associativity of the pivot operation). The *support* of φ, denoted by $\sup(\varphi)$, is defined as $\Delta_i X_i$, i.e., the set of vertices that occur an odd number of times in φ. If φ is applicable to a graph G, then, by the above, $G\varphi = G * (X_1 \Delta X_2 \Delta \ldots \Delta X_n) = G * \sup(\varphi)$. This observation may be seen as a highly generalized version of the confluence property of DNA recombination described in Sect. 2. If we specialize this observation to the case where the $* X_i$'s are elementary (i.e., local or edge complementations), then we obtain the following.

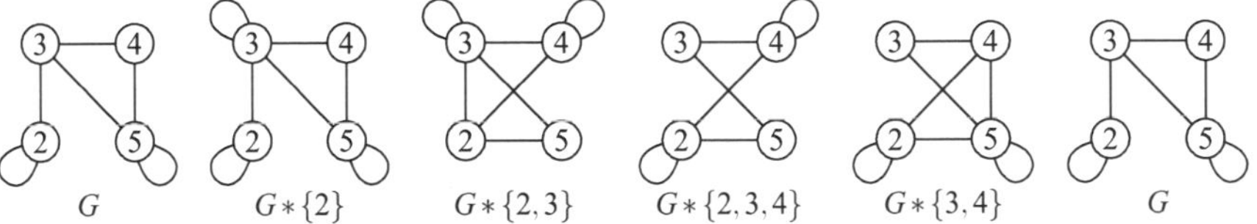

Fig. 10 Circle graph G for $4534\bar{5}23\bar{2}$ (*left* and *right*). $G * \{2, 3, 4\}$ is computed twice, as $G *$ $\{2\} * \{3\} * \{4\}$ and as $G * \{3, 4\} * \{2\}$ (reading from *right* to *left*)

Theorem 2 ([17]). *If φ and φ' are applicable sequences of local and edge complementations for a graph G, then $\sup(\varphi) = \sup(\varphi')$ implies $G\varphi = G\varphi'$.*

Special cases of this result are mentioned in the literature. The *triangle equality* stated above, $* \{u, v\} * \{v, w\} = *\{u, w\}$ for a graph with an induced loopless triangle $\{u, v, w\}$, can be found as [4, Lemma 10], [27, Proposition 1.3.5], and [33, Proposition 2.5]. A "classical" proof typically involves keeping track of numerous neighboring edges. Also, commutativity of edge complementation has been obtained in the context of gene assembly by Harju et al. [29]: if two disjoint edge complementations are applicable in either order, then the two results are identical. In short, $* \{u, v\} * \{w, x\} = * \{w, x\} * \{u, v\}$.

Example 4. Consider the MDS sequence $M_4\, M_3\, \bar{M}_5\, M_2\, \bar{M}_1$. This defines the pointer sequence $4534\bar{5}23\bar{2}$, which in turn has the circle graph G depicted in Fig. 10. The figure illustrates that $G * \{2\} * \{3\} * \{4\} = G * \{3, 4\} * \{2\}$. $\square$

Note also that G is nonsingular iff there is a sequence φ of local and edge complementations with $\sup(\varphi) = V$ such that $G\varphi$ is defined. Moreover, if this is the case, then we may choose φ in such a way that each vertex of V appears exactly once. In the context of gene assembly, we thus find that there is a sequence of hairpin and double loop recombinations that transforms a MIC gene into the corresponding MAC gene iff the circle graph G corresponding to the MIC gene is nonsingular. Moreover, if this is the case, then the circle graph corresponding to the MAC gene is $G * V = G^{-1}$, the inverse matrix of the adjacency matrix of G! Thus, from this point of view, we have the curious fact that the construction of the MAC gene entails inverting a matrix. The intermediate products obtained during the transformation of a MIC gene into its MAC gene, using only hairpin and double loop recombinations, correspond in this way to partially inverted matrices.

For each possible set S of operation types (loop excision, hairpin recombination, and double loop recombination), there is a characterization of the existence of a sequence of recombination operations that transforms a MIC gene into the corresponding MAC gene, where each recombination operation is of a type from S; see [23, Sect. 13.3] when S contains loop excision, and [15, 17] for the remaining cases where only hairpin recombination and/or double loop recombination are allowed.

We remark that an extension of the interlace polynomial from graphs to arbitrary matrices (over some field), using PPT instead of local and edge complementation on graphs, has been studied in [13, 28].

5   Set Systems

In this section, we provide yet another perspective on local and edge complementation. It turns out that we may define local and edge complementation (and pivot for graphs in general) in terms of a very elementary operation on set systems, essentially involving only the symmetric difference.

First we recall a fundamental result on PPT due to Tucker [43] (see also [18, Theorem 4.1.1] and [34]). This result allows one to formulate the applicability of the pivot $* Y$ to the resulting matrix $A * X$ in terms of the applicability of the pivot $*(X \Delta Y)$ to the original matrix A.

Proposition 1 ([43]). *Let A be a $V \times V$ matrix, and let $X \subseteq V$ be such that $A[X]$ is nonsingular. Then, for all $Y \subseteq V$, $\det(A * X)[Y] = \det A[X \Delta Y] / \det A[X]$.*

We remark here that Proposition 1 for the case $Y = V \setminus X$ is called the Schur determinant formula, $\det((A * X) \setminus X) = \det A / \det A[X]$, and was shown as early as 1917 by Issai Schur [40].

A *set system* (over V) is an ordered pair $M = (V, D)$ with V a finite set and D a family of subsets of V. We write simply $Y \in M$ to denote $Y \in D$. Let M be a set system over V. We define, for $X \subseteq V$, the pivot (often called *twist* in the literature; see, e.g., [26]) $M * X = (V, D * X)$, where $D * X = \{Y \Delta X \mid Y \in D\}$.

For a $V \times V$ matrix A, we let $\mathcal{M}_A = (V, D_A)$ be the set system with $D_A = \{X \subseteq V \mid \det A[X] \neq 0\}$. As observed in [6], we have, by Proposition 1, $Z \in \mathcal{M}_{A*X}$ iff $\det((A * X)[Z]) \neq 0$ iff $\det(A[X \Delta Z]) \neq 0$ iff $X \Delta Z \in \mathcal{M}_A$ iff $Z \in \mathcal{M}_A * X$. Hence $\mathcal{M}_{A*X} = \mathcal{M}_A * X$.

Using the adjacency matrix representation of graphs, we may carry the notion of $\mathcal{M}_A$ for matrices over to graphs. Let G be a graph. Given only the set system $\mathcal{M}_G = (V, D_G)$, one can (re)construct the graph G: $\{u\}$ is a loop in G iff $\{u\} \in D_G$, and $\{u, v\}$ is an edge in G iff $(\{u, v\} \in D_G) \oplus ((\{u\} \in D_G) \wedge (\{v\} \in D_G))$ (where $\oplus$ denotes the exclusive or); see [9, Property 3.1]. Hence the function $\mathcal{M}_{(\cdot)}$ which assigns to each graph G its set system $\mathcal{M}_G$ is injective. In this way, the family of graphs (with a set V of vertices) can be considered as a subset of the family of set systems (over the set V).

As $\mathcal{M}_{G*X} = \mathcal{M}_G * X$, the pivot operation for graphs coincides with the pivot operation for set systems. Therefore, pivot on set systems forms an alternative definition of pivot on graphs. Note that while for a set system M over V, $M * X$ is defined for all $X \subseteq V$, for a graph G, $G * X$ is defined precisely when $\det G[X] = 1$, or, equivalently, when $X \in D_G$, which in turn is equivalent to $\emptyset \in D_G * X$. Thus, for example, whereas on a graph $* \{u\} * \{v\}$ and $* \{u, v\}$ cannot both be defined, they are both defined on set systems, where they have the same outcome.

Example 5. Consider the circle graph G in Example 4 (see Fig. 10). The corresponding set system equals $\mathscr{M}_G = (V, \{\varnothing, 2, 5, 34, 23, 25, 45, 35, 234, 245, 345\})$, where $V = \{2, 3, 4, 5\}$. Note that we have abbreviated sets in this example, so, for example, 245 denotes $\{2, 4, 5\}$. Then $\mathscr{M}_G * \{3\} = (V, \{3, 23, 35, 4, 2, 235, 345, 5, 24, 2, 345, 45\})$. This does not represent a graph (as $\varnothing$ is not a set of $\mathscr{M}_G * \{3\}$). $\square$

It turns out that $\mathscr{M}_G$ for graphs G has a special structure, that of a *delta-matroid* [6]. A consequence of the fact that $\mathscr{M}_G$ is a delta-matroid is that the maximal sets of $\mathscr{M}_G$ with respect to inclusion, denoted by $\max(\mathscr{M}_G)$, are all of cardinality equal to the rank $r(A(G))$ of the matrix $A(G)$. Thus, if G is the circle graph corresponding to a MIC gene, then the nullity $n(A(G)) = |V| - r(A(G))$ of $A(G)$ (i.e., the dimension of the null space of $A(G)$) is equal to the number of loop recombinations in *every* successful transformation of that gene to its MAC form. Equivalently, the number of loops created during the transformation of a MIC gene to its MAC gene is equal to the nullity of the adjacency matrix of the circle graph corresponding to that MIC gene. In fact, a set $S_l \subseteq V$ is the support of the loop excision part of a successful reduction of G iff $V \setminus S_l \in \max(\mathscr{M}_G)$.

Example 6. Consider the circle graph G in Example 4 and the corresponding set system $\mathscr{M}_G$ (see Example 5). As G has nullity 1, the maximal sets of $\mathscr{M}_G$ are of cardinality 3. Considering the maximal sets $\{2, 3, 4\}, \{2, 4, 5\}, \{3, 4, 5\}$ of $\mathscr{M}_G$, we see that loop recombination can be performed on every pointer except 4. $\square$

6 Loop Complementation

The concept of local complementation defined in this chapter is defined only on looped vertices, and thus cannot be applied to simple graphs. A related concept, which is also called local complementation, is defined for each simple graph G and vertex u of G: local complementation of G on u complements the neighborhood of u (without introducing loops). By abuse of notation, we denote local complementation for simple graphs G by $G * \{u\}$ also. The operation of edge complementation can be defined as for graphs with loops. In this context of simple graphs, we have the "curious" identity $* \{u, v\} = * \{u\} * \{v\} * \{u\} = * \{v\} * \{u\} * \{v\}$ [7, Corollary 8.2], which is not valid for graphs with loops. In fact, in that case, the left and right sides of the equation do not have the same support.

The operation loop of complementation is useful for dealing with loops. For a graph G and a set X of vertices of G, the *loop complementation* of G on X, denoted by $G + X$, is the graph obtained from G by toggling the loops on the vertices in X. This operation can be faithfully represented on set systems. For a set system $M = (V, D)$ and an element $u \in V$, the *loop complementation* of M on u, denoted by $M + u$, is the set system (V, D'), where $D' = D\Delta\{X \cup \{u\} \mid X \in D, u \notin X\}$. As the operation is commutative, we can extend it to $M + X$ for a set $X \subseteq V$ by performing the $+ u$'s, $u \in X$, in any order. We have $\mathscr{M}_{G+X} = \mathscr{M}_G + X$ for any set $X \subseteq V$.

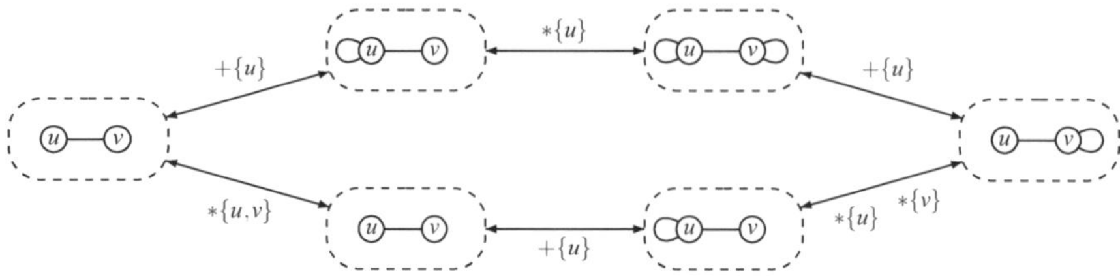

Fig. 11 Verification of applicability of $*\{u, v\} + \{u\} * \{u\} * \{v\} + \{u\} * \{u\} + \{u\}$ to any graph F having an edge $\{u, v\}$, where both u and v are nonloop vertices

Pivot $* X$ and loop complementation $+ X$ for set systems together form an interesting algebra. On a single common element u, the operations $* u$ and $+ u$ are involutions (i.e., of order 2) generating a group isomorphic to the group S_3 of permutations on three elements [12]. In particular, we have $+u*u+u = *u+u*u$, which is the third involution (in addition to pivot and loop complementation). On different elements $u \neq v$, the operations commute; thus, $* u+v = +v*u$, $+u+v = +v + u$, and $*u * v = *v * u$.

This algebra makes it possible to understand the relation between edge complementation and local complementation for simple graphs mentioned above. First we note that the sequence of operations $\varphi = *\{u, v\} + \{u\} * \{u\} * \{v\} + \{u\} * \{u\} + \{u\}$ is applicable to any graph with an edge $\{u, v\}$, where u and v are unlooped vertices, by checking the existence of loops on u and v in successive stages; see Fig. 11. Then we observe that φ is the identity on set systems using the group structure (using the fact that, for set systems, we have $*\{u, v\} = *\{u\} * \{v\}$). This makes φ the identity on any graph where it is applicable (without having to consider the involved graph operations). We can project φ from graphs to simple graphs by skipping the loop complementation operations, and obtain the equality $*\{u, v\} = *\{u\} * \{v\} * \{u\}$.

Inspired by and motivated by the context of gene assembly, Brijder and Hoogeboom [10] implicitly studied the interplay of loop complementation and pivot (but only for the case of doc-strings). This interplay led to an extension of the interlace polynomial (including the related bracket polynomial for graphs) and an extension of the Penrose polynomial from graphs and matroids to delta-matroids [11, 14].

7   Discussion

We have fitted the theory of gene assembly in ciliates into the theory of 4-regular graphs and have carried over results from the latter theory to the former. Interestingly, operations on Euler circuits in 4-regular graphs (see Fig. 6) also occur (often implicitly) in the context of other topics in computational molecular biology. For instance, the monograph [36] by Pevzner has three chapters where the operations of Fig. 6 are used: Chap. 2, on restriction mapping, refers to them under the names of "order exchange" and "order reflexion"; Chap. 5, on sequencing by hybridization, features rearrangements of Eulerian cycles; and Chap. 10, on genome rearrangements, studies reversal in the so-called breakpoint graph. Hence we expect that these topics (and others) may benefit from a similar approach as is done in this chapter; to carry over the general theory of 4-regular graphs to these topics.

References

1. M. Aigner, The Penrose polynomial of graphs and matroids, in *Surveys in Combinatorics*, vol 288, ed. by J.W.P. Hirschfeld. London Mathematical Society Lecture Note Series (Cambridge University Press, Cambridge, 2001), pp. 11–46. doi:10.1017/CBO9780511721328.004
2. M. Aigner, H. van der Holst, Interlace polynomials. Linear Algebra Appl. **377**, 11–30 (2004). doi:10.1016/j.laa.2003.06.010
3. A. Angeleska, N. Jonoska, M. Saito, DNA recombination through assembly graphs. Discret. Appl. Math. **157**(14), 3020–3037 (2009). doi:10.1016/j.dam.2009.06.011
4. R. Arratia, B. Bollobás, G. Sorkin, The interlace polynomial of a graph. J. Comb. Theory B **92**(2), 199–233 (2004). doi:10.1016/j.jctb.2004.03.003
5. A. Bouchet, Isotropic systems. Eur. J. Comb. **8**, 231–244 (1987). doi:10.1016/S0195-6698(87)80027-6
6. A. Bouchet, Representability of Δ-matroids, in *Proceedings of the 6th Hungarian Colloquium of Combinatorics, Colloquia Mathematica Societatis János Bolyai*, Eger, vol. 52 (North-Holland, 1987), pp. 167–182
7. A. Bouchet, Graphic presentations of isotropic systems. J. Comb. Theory B **45**(1), 58–76 (1988). doi:10.1016/0095-8956(88)90055-X
8. A. Bouchet, Tutte-Martin polynomials and orienting vectors of isotropic systems. Graphs Comb. **7**(3), 235–252 (1991). doi:10.1007/BF01787630
9. A. Bouchet, A. Duchamp, Representability of Δ-matroids over $GF(2)$. Linear Algebra Appl. **146**, 67–78 (1991). doi:10.1016/0024-3795(91)90020-W
10. R. Brijder, H. Hoogeboom, The fibers and range of reduction graphs in ciliates. Acta Inform. **45**, 383–402 (2008). doi:10.1007/s00236-008-0074-3
11. R. Brijder, H. Hoogeboom, Interlace polynomials for delta-matroids (2010). [arXiv:1010.4678]
12. R. Brijder, H. Hoogeboom, The group structure of pivot and loop complementation on graphs and set systems. Eur. J. Comb. **32**, 1353–1367 (2011). doi:10.1016/j.ejc.2011.03.002
13. R. Brijder, H. Hoogeboom, Nullity invariance for pivot and the interlace polynomial. Linear Algebra Appl. **435**, 277–288 (2011). doi:10.1016/j.laa.2011.01.024
14. R. Brijder, H. Hoogeboom, Bicycle matroids and the Penrose polynomial for delta-matroids (2012). [arXiv:1210.7718]
15. R. Brijder, H. Hoogeboom, Binary symmetric matrix inversion through local complementation. Fundam. Inform. **116**(1–4), 15–23 (2012). doi:10.3233/FI-2012-664
16. R. Brijder, M. Daley, T. Harju, N. Jonoska, I. Petre, G. Rozenberg, Computational nature of gene assembly in ciliates, in *Handbook of Natural Computing*, ed. by G. Rozenberg, T. Bäck, J. Kok, vol. 3 (Springer, Berlin/London, 2012), pp. 1233–1280. doi:10.1007/978-3-540-92910-9_37
17. R. Brijder, T. Harju, H. Hoogeboom, Pivots, determinants, and perfect matchings of graphs. Theor. Comput. Sci. **454**, 64–71 (2012). doi:10.1016/j.tcs.2012.02.031
18. R. Cottle, J.S. Pang, R. Stone, *The Linear Complementarity Problem* (Academic, San Diego, 1992)

19. A. Ehrenfeucht, I. Petre, D. Prescott, G. Rozenberg, Circularity and other invariants of gene assembly in ciliates, in *Words, Semigroups, and Transductions*, ed. by M. Ito et al. (World Scientific, Singapore, 2001), pp. 81–97. doi:10.1142/9789812810908_0007
20. A. Ehrenfeucht, T. Harju, I. Petre, G. Rozenberg, Characterizing the micronuclear gene patterns in ciliates. Theory Comput. Syst. **35**, 501–519 (2002). doi:10.1007/s00224-002-1043-9
21. A. Ehrenfeucht, I. Petre, D. Prescott, G. Rozenberg, String and graph reduction systems for gene assembly in ciliates. Math. Struct. Comput. Sci. **12**, 113–134 (2002). doi:10.1017/S0960129501003516
22. A. Ehrenfeucht, T. Harju, I. Petre, D. Prescott, G. Rozenberg, Formal systems for gene assembly in ciliates. Theor. Comput. Sci. **292**, 199–219 (2003). doi:10.1016/S0304-3975(01)00223-7
23. A. Ehrenfeucht, T. Harju, I. Petre, D. Prescott, G. Rozenberg, *Computation in Living Cells – Gene Assembly in Ciliates* (Springer, Berlin/New York, 2004)
24. J. Ellis-Monaghan, C. Merino, Graph polynomials and their applications I: the Tutte polynomial, in *Structural Analysis of Complex Networks*, ed. by M. Dehmer (Birkhäuser, Boston, 2011), pp. 219–255. doi:10.1007/978-0-8176-4789-6_9
25. J. Ellis-Monaghan, C. Merino, Graph polynomials and their applications II: Interrelations and interpretations, in *Structural Analysis of Complex Networks*, ed. by M. Dehmer (Birkhäuser, Boston, 2011), pp. 257–292. doi:10.1007/978-0-8176-4789-6_10
26. J. Geelen, A generalization of Tutte's characterization of totally unimodular matrices. J. Comb. Theory B **70**, 101–117 (1997). doi:10.1006/jctb.1997.1751
27. F. Genest, Graphes eulériens et complémentarité locale. Ph.D. thesis, Université de Montréal, 2002. Available online: `arXiv:math/0701421v1`
28. R. Glantz, M. Pelillo, Graph polynomials from principal pivoting. Discret. Math. **306**(24), 3253–3266 (2006). doi:10.1016/j.disc.2006.06.003
29. T. Harju, C. Li, I. Petre, G. Rozenberg, Parallelism in gene assembly. Nat. Comput. **5**(2), 203–223 (2006). doi:10.1007/s11047-005-4462-0
30. F. Jaeger, On transition polynomials of 4-regular graphs, in *Cycles and Rays*, ed. by G. Hahn, G. Sabidussi, R. Woodrow. NATO ASI Series, vol. 301 (Kluwer, Dordrecht, 1990), pp. 123–150. doi:10.1007/978-94-009-0517-7_12
31. A. Kotzig, Eulerian lines in finite 4-valent graphs and their transformations, in *Theory of Graphs, Proceedings of the Colloquium*, Tihany, 1966 (Academic, New York, 1968), pp. 219–230
32. P. Martin, Enumérations eulériennes dans les multigraphes et invariants de Tutte-Grothendieck. Ph.D. thesis, Institut d'Informatique et de Mathématiques Appliquées de Grenoble (IMAG), 1977. Available online: http://tel.archives-ouvertes.fr/tel-00287330_v1/
33. S. Oum, Rank-width and vertex-minors. J. Comb. Theory B **95**(1), 79–100 (2005). doi:10.1016/j.jctb.2005.03.003
34. T. Parsons, Applications of principal pivoting, in *Proceedings of the Princeton Symposium on Mathematical Programming*, ed. by H. Kuhn (Princeton University Press, Princeton, 1970), pp. 567–581
35. R. Penrose, Applications of negative dimensional tensors, in *Combinatorial Mathematics and Its Applications*, Oxford, ed. by D. Welsh (Academic, 1971), pp. 211–244
36. P. Pevzner, *Computational Molecular Biology: An Algorithmic Approach* (The MIT Press, Cambridge, MA/ London, 2000)
37. D. Prescott, Genome gymnastics: unique modes of DNA evolution and processing in ciliates. Nat. Rev. **1**, 191–199 (2000). doi:10.1038/35042057
38. D. Prescott, A. Greslin, Scrambled Actin I gene in the micronucleus of Oxytricha nova. Dev. Genet. **13**, 66–74 (1992). doi:10.1002/dvg.1020130111
39. D. Prescott, A. Ehrenfeucht, G. Rozenberg, Molecular operations for DNA processing in hypotrichous ciliates. Eur. J. Protistol. **37**, 241–260 (2001). doi:10.1078/0932-4739-00807
40. J. Schur, Über Potenzreihen, die im Innern des Einheitskreises beschränkt sind. Journal für die reine und angewandte Mathematik **147**, 205–232 (1917). http://resolver.sub.uni-goettingen.de/purl?PPN243919689_0147

41. L. Traldi, L. Zulli, A bracket polynomial for graphs, I. J. Knot Theory Ramif. **18**(12), 1681–
    1709 (2009). doi:10.1142/S021821650900766X
42. M. Tsatsomeros, Principal pivot transforms: properties and applications. Linear Algebra Appl.
    307(1–3), 151–165 (2000). doi:10.1016/S0024-3795(99)00281-5
43. A. Tucker, A combinatorial equivalence of matrices, in *Combinatorial Analysis, Proceedings of
    Symposia in Applied Mathematics*, vol. X, Columbia University, 24–26 April 1958 (American
    Mathematical Society, 1960), pp. 129–140. doi:10.1090/psapm/010

Invariants of Graphs Modeling Nucleotide Rearrangements

Egor Dolzhenko and Karin Valencia

Abstract Nucleotide rearrangements occur in many biological systems. These genome reorganisations are especially widespread in ciliates, making these protozoans an attractive model system for experimental, computational, and theoretical studies. Rearrangements of ciliate chromosomes are modeled by the so-called assembly graphs. Edges of these graphs represent double-stranded DNA molecules, while vertices correspond to DNA recombination sites. This work is an expository article in which we discuss topological and combinatorial invariants of assembly graphs. The topological invariant, called the genus range, gives information about possible spatial arrangement of the corresponding DNA molecule. The combinatorial invariant, called the assembly polynomial, is closely related to the possible products of the rearrangement modeled by the assembly graph.

1 Introduction

Complex genome rearrangements are common in many biological contexts. These include trans-splicing of RNA transcripts [4,13], somatic recombination (e.g., V(D)J recombination in jawed vertebrates and VLR recombination in jawless fish [1]), and the development of the somatic genome from its precursor germline genome in binucleate ciliates. Furthermore, complex genome rearrangements have been recently associated with 2–3 % of human cancers [16].

E. Dolzhenko (✉)
Department of Mathematics and Statistics, University of South Florida, Tampa, FL 33620, USA
e-mail: egor.dolzhenko@gmail.com

K. Valencia
Department of Mathematics, Imperial College London, South Kensington Campus, Office: 640, London, SW7 2AZ, UK
e-mail: karin.valencia06@imperial.ac.uk

N. Jonoska and M. Saito (eds.), *Discrete and Topological Models in Molecular Biology*,
Natural Computing Series, DOI 10.1007/978-3-642-40193-0_14,
© Springer-Verlag Berlin Heidelberg 2014

Developmental genome rearrangements in ciliates are particularly striking due to their complexity and scale. Computational techniques for analysis of rearrangements in ciliates [8] perform well in other biological contexts, indicating that ciliates are also suitable model organisms for computational studies.

Theoretical models are, for the most part, focused on elucidating the biological mechanisms behind genome-wide DNA rearrangements, both in ciliates and other contexts. The chief aim of the work presented herein is to contribute to this effort by studying topological and polynomial invariants of assembly graphs, since these graphs have been specifically designed to model nucleotide rearrangement in ciliates and other biological systems [3]. This article is a short summary of results presented in [5] and [6].

1.1 Genome-Wide Rearrangements in Ciliates and Assembly Graphs

A more comprehensive exposition of the biology of genome-wide rearrangements in ciliates can be found in the chapter by Goldman et al. of this book. Here we present a transition from the biology to mathematical ideas and questions.

Ciliates constitute a large group of unicellular protozoans that possess two functionally distinct types of nuclei: the germline micronucleus (MIC) – a transcriptionally silent nucleus whose purpose is to pass on genetic information to sexual progeny – and the somatic macronucleus (MAC) – a transcriptionally active nucleus made up of short, gene-sized chromosomes.

After conjugation, old macronuclei disintegrate, while some of the new micronuclei transform into the new macronuclei in a process that involves thousands of deletions of the non-coding DNA and rearrangement of the remaining pieces, to create new somatic chromosomes. This process of developing MAC from MIC is called differentiation. Such DNA transactions are achieved by performing homologous DNA recombination at specific places in the MIC DNA sequences [15]. In this article, a mathematical model of the intricate process of differentiation is presented and used to further investigate features of this system.

Portions of MIC chromosomes involved in the rearrangement consist of three types of subsequences: macronuclear destined sequences (MDSs), internal eliminated sequences (IESs) and pointer sequences. MDSs are subsequences of MAC chromosomes that are sometimes present in a permuted or inverted order in the MIC, relative to their order in MAC chromosomes. IESs are segments of noncoding DNA positioned between MDS. Pointer sequences are short segments of DNA flanking MDSs that appear in pairs. If the MDSs are numbered by integers according to their order in MAC chromosomes, the upstream pointer of MDS $i - 1$ would be the same as the downstream pointer of MDS i. Formation of transcriptionally competent genes in the macronucleus requires that the non-coding IESs and one partner of each pair of pointer sequences are excised, and that the MDSs are spliced together in order. This is achieved through a series of homologous recombination events that happen at the pairs of pointer sequences (Fig. 1).

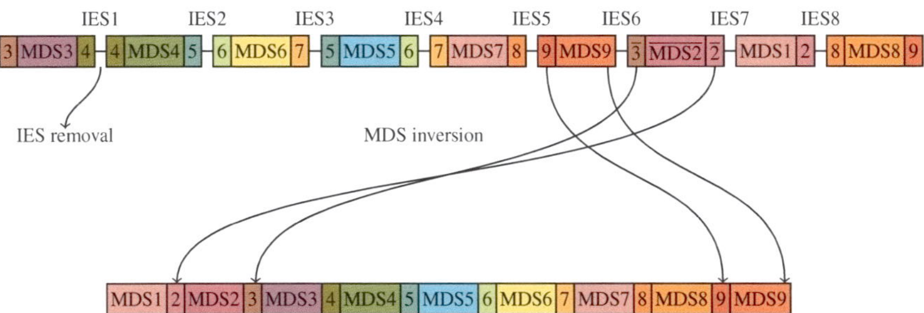

Fig. 1 A schematic representation of assembly of a MAC chromosome (*bottom*) of *Oxytricha trifallax* containing an Actin I gene from pieces (MDSs) of its MIC counterpart (*top*) (The diagram is adapted from [3])

1.2 Mathematical Model of Differentiation

In [3] the authors developed a graph-theoretic model of differentiation. In summary, in this model, the pointer sequences in a MIC chromosome inherit numbering from the MDSs as explained above. Reading the chromosome from left to right, and taking note of the numbered pointer sequences only, yields a sequence of numbers with each integer appearing either 0 or 2 times, called a double-occurrence word. Given a double-occurrence word, a graph is constructed as follows: place a crossing on the plane for each integer in the double-occurrence word, and label them accordingly. Pick a point (base point) in the plane and, following a chosen direction, connect the crossings in the order of appearance in the double-occurrence word making sure that consecutive edges are not neighbours (see section 1.3). When all of the crossings are exhausted, connect the path back to the base point. There may be instances when the edges self-cross, or cross other edges, but their crossings are not new vertices, and to avoid confusion an under crossing is drawn instead (this is intuitively illustrated in Fig. 2). Specifically, the graph constructed is called a "simple assembly graph" (see below for mathematical definitions). In this model, edges of the graph represent double stranded DNA in the MIC, and vertices correspond to the alignment of the DNA recombination sites. Smoothing (removal) of vertices from these graphs corresponds to homologous recombination.

In this chapter, this simple assembly graph model of differentiation is adopted, and results from [5] and [6] are summarized.

1.3 Mathematical Preliminaries

A vertex v of a graph G is said to be rigid if it has a cyclic order of the adjacent edges assigned to it. We draw rigid vertices in such a way that taking a regular disc neighborhood of the vertex and traveling around its boundary, the edges appear in

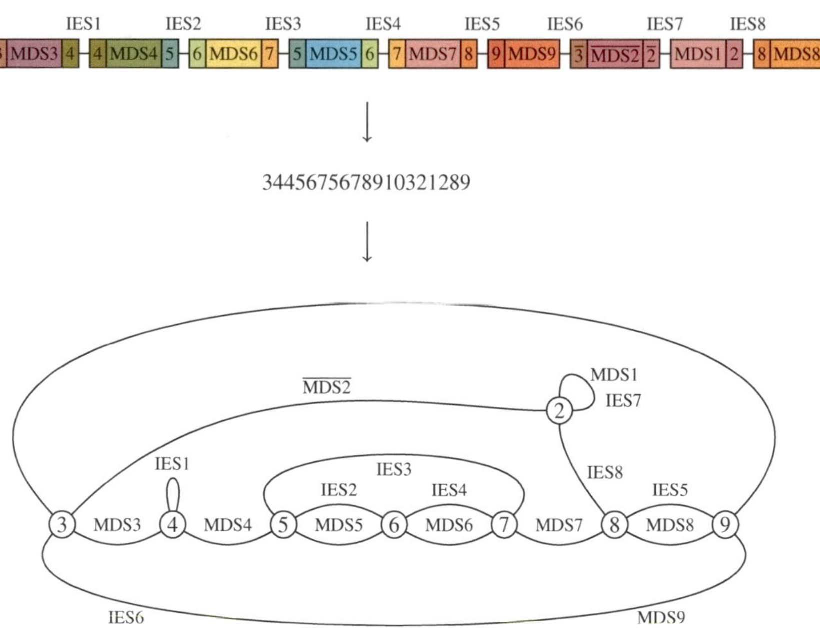

Fig. 2 Annotated MIC sequence (*top*) that gives rise to a double-occurrence word (*middle*) and an assembly graph (*bottom*) (Image modified from [3])

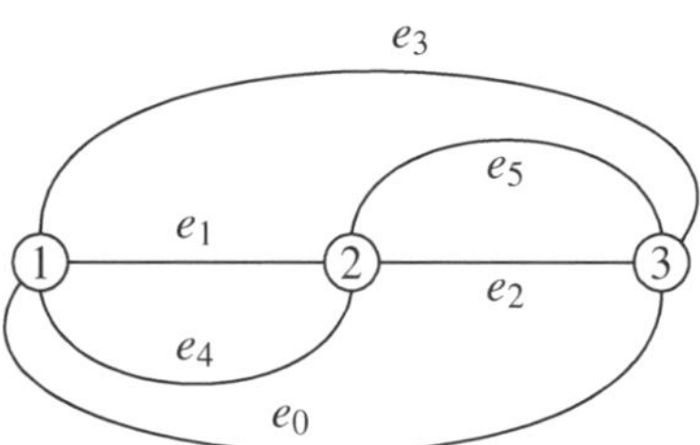

Fig. 3 Example of an assembly graph

the specified cyclic order. If v is a rigid vertex of G, e and e' are said to be neighbors (with respect to v) if one of these edges is the immediate successor of the other when traveling around the vertex in some direction. For example, in Fig. 3, e_1, e_4 are neighbors with respect to vertex 1. An assembly graph $\mathcal{A}$ is a connected 4-regular graph whose vertices are rigid. Assembly graphs may have loops (which are counted twice when calculating the degree of a vertex). The size of an assembly graph $\mathcal{A}$ is the number of vertices of the graph, and is denoted $|\mathcal{A}|$. Two assembly graphs are isomorphic if they are isomorphic as graphs and the graph isomorphism preserves the cyclic order of edges associated with every vertex. A path in an assembly graph is called a transverse path (or a transversal) if consecutive edges of the path are never neighbors with respect to their common incident vertex. Transversals are denoted by

α. An assembly graph that has an Eulerian (visiting all edges) transversal is called a simple assembly graph. Note that the mathematical model of differentiation of a single molecule developed in [3] and outlined above, yields graphs that are simple assembly graphs. Only these assembly graphs are considered in this chapter and henceforth are called assembly graphs.

Let $\mathcal{A}$ be an assembly graph and $\alpha = v_0, e_0, \ldots, v_n, e_n, v_0$ be its transversal path. Let $w = v_0 \cdots v_n$ be the subsequence of α consisting of the vertices only; each vertex appears twice in w. Call w a double-occurrence word (DOW) corresponding to the assembly graph $\mathcal{A}$. Conversely, an assembly graph is associated to a DOW w as described in Sect. 1.1. It was proved in [3] that isomorphic classes of assembly graphs are in one-to-one correspondence with equivalence classes of DOW. Therefore an assembly graph with DOW w is denoted $\mathcal{A}(w)$.

An embedding of a graph in a surface is called cellular if each component of the complement of the graph in the surface is an open disc. For a graph G, the minimum orientable genus of G, denoted $g_{min}(G)$, is the smallest non-negative integer g such that G admits an embedding in a closed (compact, empty boundary) orientable surface F of genus g. The maximum orientable genus of G, denoted $g_{max}(G)$, is the largest non-negative integer g such that G admits a cellular embedding in a closed orientable surface F of genus g. The *genus range $gr(G)$* of a graph G is the set of values of genera over all surfaces into which G can be embedded cellularly.

2 Genus Ranges of Assembly Graphs

In this section, the genus ranges of assembly graphs is discussed. Intuitively, an embedding of a graph into a surface is a drawing of the graph on the surface in such a way that its edges may intersect only at their endpoints. The genus is a topological measure of the spatial complexity of graphs and hence provides information about the spatial organization of the DNA at the moment of rearrangement. Only the orientable genus is considered. The work presented here is a summary of [5].

Specifically, the following two problems are of interest:

Problem 1.

(a) Characterise the sets of integers that appear as genus ranges of assembly graphs with n 4-valent vertices for each positive integer n.

(b) Characterise the assembly graphs with a given set of genus ranges.

The two questions in Problem 1 were tackled in [5] as follows. First, computer calculations were used to find genus ranges of assembly graphs with up to seven vertices (Fig. 4). This information was then used to make various observations about genus ranges and the following were proved in the general case.

- A genus range of an assembly graph is always a set of consecutive integers (Lemma 2.10 in [5]).
- Every set $\{m, m+1, \ldots, m'\}$ for $0 \leq m < m' \leq n$ appears as a genus range of some assembly graph with $2n$ vertices (Theorem 6.1 in [5]).

$$n = 2 \quad \{0\}, \{\mathbf{1}\}$$
$$n = 3 \quad \{0\}, \{0,1\}, \{1\}, \{\mathbf{1,2}\}$$
$$n = 4 \quad \{0\}, \{0,1\}, \{1\}, \{0,1,2\}, \{\mathbf{1,2}\}$$
$$n = 5 \quad \{0\}, \{0,1\}, \{1\}, \{0,1,2\}, \{1,2\}, \{2\}, \{1,2,3\}, \{\mathbf{2,3}\}$$
$$n = 6 \quad \{0\}, \{0,1\}, \{1\}, \{0,1,2\}, \{1,2\}, \{2\}, \{0,1,2,3\}, \{1,2,3\}, \{\mathbf{2,3}\}, \{3\}$$
$$n = 7 \quad \{0\}, \{0,1\}, \{1\}, \{0,1,2\}, \{1,2\}, \{2\}, \{0,1,2,3\}, \{1,2,3\}, \{2,3\}, \{3\},$$
$$\{1,2,3,4\}, \{2,3,4\}, \{\mathbf{3,4}\}$$

Fig. 4 Genus ranges of assembly graphs with up to seven vertices. The bolded genus ranges correspond to the tangled cord (Sect. 2.1)

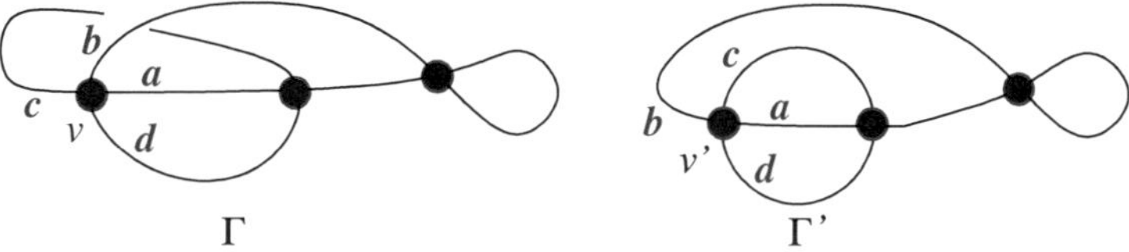

Fig. 5 The graphs that are isomorphic as non-rigid vertex graphs but not as rigid vertex graphs and have different genus ranges

- Every set $\{m, m+1, \ldots, m'\}$ for $0 \leq m < m' \leq n$ without the set $\{0, 1, \ldots, n\}$ appears as a genus range of some assembly graph with $2n - 1$ vertices.
- No assembly graph with $2n - 1$ vertices has genus range $\{0, \ldots, n\}$ (Lemma 3.9) nor genus range $\{n\}$ (Lemma 3.10 in [5]).
- Families of graphs that achieve certain sets of genus ranges are constructed, including a family of graphs with $2n$ vertices that have genus range $\{0, 1, \ldots, n\}$ (Proposition 4.5 in [5]).
- The genus range of the special subfamily of assembly graphs, called tangled cords, is characterized (Theorem 5.5 in [5] and Theorem 1 below).

Remark 1. There exist equivalent results about the genus ranges of classic (non-rigid vertex) graphs [9,14]. However, a priori these do not necessarily extend to rigid vertex graphs: two graphs that are isomorphic as non-rigid vertex graphs may not necessarily be equivalent as rigid vertex graphs. Figure 5 shows such an example where the genus range of these graphs (without considering the rigidity of the vertices) is $\{0, 1, 2\}$, however, the genus range of the graph with rigid vertices to the left is $\{1\}$.

2.1 Genus of the Tangled Cord

In this section the genus range of a special family of assembly graphs is found in the general case. Preliminary data suggests that these graphs may appear as subgraphs of graphs modeling ciliate rearrangements. Furthermore, odd-sized tangled cords

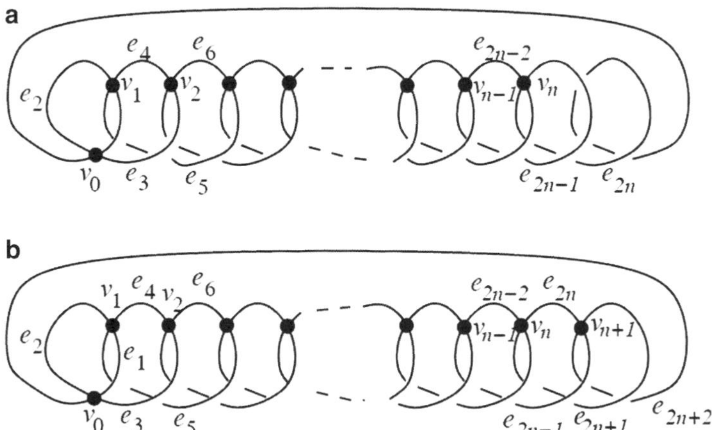

Fig. 6 Tangled cords (**a**) $\mathcal{T}_n$ and (**b**) $\mathcal{T}_{n+1}$

have been shown to maximize (see Fig. 4) the genus range over all assembly graphs with the same number of vertices (the same is conjectured for even-sized graphs) [5]. This implies that tangled cords represent the most complex DNA-recombination intermediates that appear during ciliate nuclei differentiation.

Angeleska et al. [2, 3] studied paths in assembly graphs that correspond to fully assembled MAC chromosomes. These so-called Hamiltonian polygonal paths visit every vertex exactly once and every pair of edges in the path that are adjacent to a vertex v are v-neighbors. It turns out that the tangled cord has the largest number of Hamiltonian polygonal paths among all odd-sized assembly graphs with a fixed number of vertices [6]. Hence tangled cords can "facilitate" the largest number of distinct MAC chromosomes.

Definition 1. The tangle cord, denoted $\mathcal{T}_n$, is an assembly graph corresponding to the unsigned DOW

$$1213243\cdots(n-1)(n-2)n(n-1)n.$$

For example, $\mathcal{T}_1 = 11$, $\mathcal{T}_2 = 1212$, $\mathcal{T}_3 = 121323$, etc. A tangle cord with n vertices (and $2n$ edges) is illustrated in Fig. 6. Specifically, for $\mathcal{T}_n$, the adjacent edges to each vertex are listed below, in the (rigid) order that they are encountered, clockwise around the vertex, up to cyclic permutation:

$$v_1 : e_1, e_3, e_{2n}, e_2$$
$$v_2 : e_1, e_5, e_2, e_4$$
$$v_i : e_{2(i-1)}, e_{2i}, e_{2(i-2)+1}, e_{2i+1} \text{ for } i \neq 1, 2, n-1, n$$
$$v_{n-1} : e_{2(n-2)}, e_{2n-2}, e_{2(n-3)+1}, e_{2n-1}$$
$$v_n : e_{2n}, e_{2n-2}, e_{2n-1}, e_{2n-3}$$

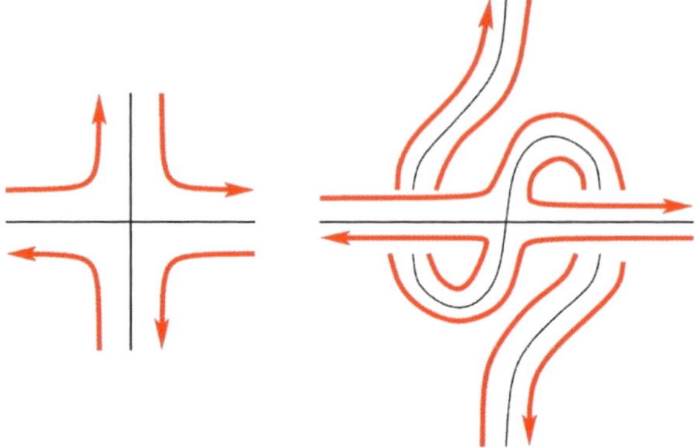

Fig. 7 Possible connectivity types of the boundary components of a ribbon surface, locally at a vertex

The main result of this section is the following.

Theorem 1. *Let $\mathcal{T}_n$ be the tangled cord with n vertices. Then*

$$gr(\mathcal{T}_n) = \begin{cases} \{\frac{n-2}{2}, \frac{n}{2}\} & \text{if } n \text{ is even,} \\ \{\frac{n-1}{2}, \frac{n+1}{2}\} & \text{if } n \text{ is odd.} \end{cases}$$

Two approaches for proving this result are summarized below. These differ in that while one works directly with the assembly graphs and rigid vertices, the other uses standard results for non-rigid vertex graphs in the final steps of finding the genus. The proofs overlap in two key places: the use of a regular surface neighborhood of the graph and formulas of the genus as a function of the number of boundary components of this surface, derived from the Euler characteristic.

In [5] it was shown that the genus range of assembly graphs consists of consecutive integers.

We construct a regular surface neighborhood $\mathcal{F}$ of $\mathcal{T}_n$, called a ribbon surface obtained by thickening the edges of the graph as ribbons (see [11, 12]). This is a closed, connected, orientable surface with circle boundary components. Due to rigidity, locally, at any vertex the boundary components of $\mathcal{F}$ have one of the two possible connectivity types illustrated in Fig. 7. Therefore there are 2^n (possibly not distinct) ribbon surfaces for a $\mathcal{T}_n$.

The formula $g(\mathcal{F}) = (1/2)(n - b(\mathcal{F}) + 2)$ (a consequence of the Euler characteristic of a surface, where $g(\mathcal{F})$, $b(\mathcal{F})$ and n denote the genus of $\mathcal{F}$, the number of boundary components of $\mathcal{F}$ and the number of vertices of $\mathcal{T}_n$, respectively [11, 19]) implies that to find the genus range of $\mathcal{T}_n$ it is enough to find the numbers of boundary components of all the ribbon surfaces $\mathcal{F}$ of $\mathcal{T}_n$. In the approach in [5], it is shown by induction, and by checking all possible connectivity types, that if n is odd (resp. even) the number of boundary components of any $\mathcal{F}$ of $\mathcal{T}_n$ is either 1 or 3 (resp. 2 or 4) and each of these situations occur.

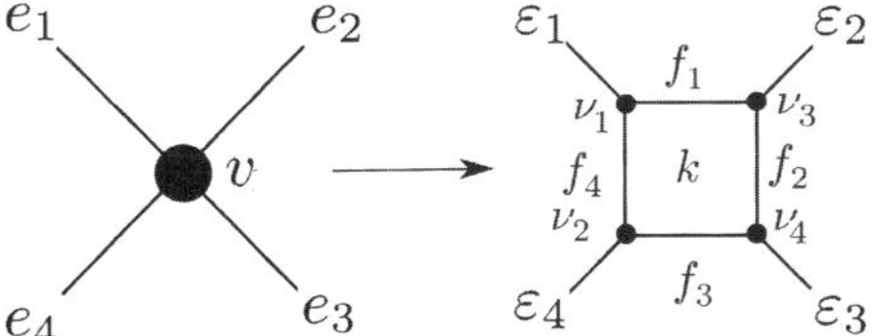

Fig. 8 Squared perturbation at a rigid 4-valent vertex yields four new non-rigid 3-valent vertices $v_{4k-3}, v_{4k-2}, v_{4k-1}, v_{4k}$ and eight new edges $f_1, f_2, f_3, f_4, \varepsilon_1, \varepsilon_2, \varepsilon_3, \varepsilon_4$

In the other approach [18], from a tangled cord, one can construct a classic (non-rigid vertex) regular 3-valent graph by replacing each rigid vertex of $\mathcal{T}_n$ by a squared perturbation as indicated in Fig. 8. Each squared perturbation at each rigid vertex is explicitly chosen to be unknotted and unlinked; in particular they bound a planar disc. This preserves the rigidity structure of the vertices of $\mathcal{T}_n$. Denote the squared perturbation of a tangled cord by $\mathcal{T}_{sq_n}$.

As above, one constructs a ribbon surface $\mathcal{F}$ of $\mathcal{T}_{sq_n}$. Again, there are two types of connectivity of the boundary components locally at each squared perturbation.[1] These are illustrated in Fig. 9a, and for simplicity they are denoted $\odot$ and $\otimes$.

Standard results of the genus of classic graphs are used to find the genus $\mathcal{T}_{sq_n}$. As above, the genus range of a graph consists of consecutive numbers. In addition, the genus formula from the Euler characteristic gives the maximal and minimal number of boundary components over all ribbon surfaces of $\mathcal{T}_{sq_n}$, respectively. The result follows from a lemma equivalent to the observations when rigid vertices were considered. That is, if n is odd (resp. even) the number of boundary components of any $\mathcal{F}$ of $\mathcal{T}_{sq_n}$ is either $n + 1$ or $n + 3$ (resp. $n + 2$ or $n + 4$) and each of these situations occur.

3 Assembly Polynomials

This section discusses a polynomial invariant of assembly graphs called assembly polynomial (AP). These polynomials were first proposed by Burns at al. [6] as a way to study connectedness properties of the assembly graphs, and are a natural extension of Jones and Tutte polynomials [10, 17]. The AP's definition relies on two operations called p- and n-smoothings, modeling the excision of DNA during

---

[1] Locally at each non-rigid vertex there are two possible connectivities of the boundary components of a ribbon surface [figure] and [figure] . However, since each squared perturbation represents a rigid vertex of $\mathcal{T}_n$, only the cases where all four vertices of a squared perturbation have the same connectivity of the boundary components are considered.

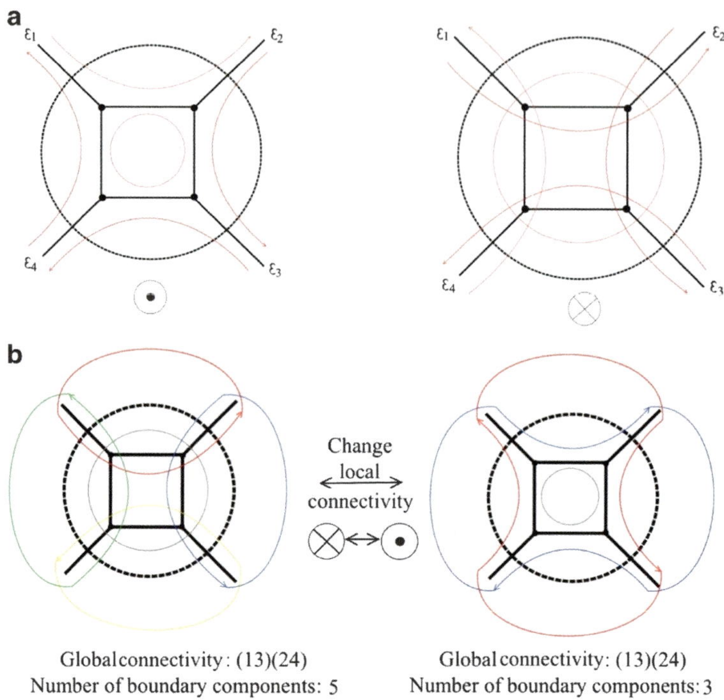

Fig. 9 (**a**) Ribbon surfaces at a squared perturbation (compare with Fig. 7). (**b**) Example of the global connectivity of the boundary components at a vertex with local connectivities $\otimes$ and $\odot$. The number of boundary components change if the local connectivity is swapped between $\otimes$ and $\odot$

the rearrangement. The AP counts the number of smoothing operations it takes to separate a given graph into a given number of connected components. We start with a few definitions.

Let $\mathcal{A} = \mathcal{A}(w)$ be an assembly graph with n vertices and let $\alpha = v_0 e_1 v_1 \cdots e_n v_{2n}$ (with $v_n = v_0$) be $\mathcal{A}$'s transversal. Consider $2n$ vertices on a unit circle labeled by $v_0, \ldots, v_{2n-1}$ (in this order) and $2n$ arcs between these vertices that we identify with $\mathcal{A}$'s edges, so that the arc between v_i and v_{i+1} is identified with e_{i+1}. Note that each vertex of $\mathcal{A}$ appears exactly twice on this list. If each pair of vertices with the same label is connected by a chord, then the result is the chord diagram $\mathcal{C} = \mathcal{C}(\mathcal{A})$ of $\mathcal{A}$ (see Fig. 10a, b). To ensure that $\mathcal{C}(\mathcal{A})$ is well-defined, we identify chord diagrams under rotations and reflections.

Each assembly graph defines a circle graph via its chord diagram. Given a chord diagram $\mathcal{C} = \mathcal{C}(\mathcal{A})$ of an assembly graph $\mathcal{A}$, consider a graph whose vertices are the distinct labels of $\mathcal{C}$. Two vertices are connected by an edge if the corresponding chords intersect. The resulting graph is called the circle graph of $\mathcal{A}$ and is denoted by $\mathcal{G}(\mathcal{A})$ (see Fig. 10b, c).

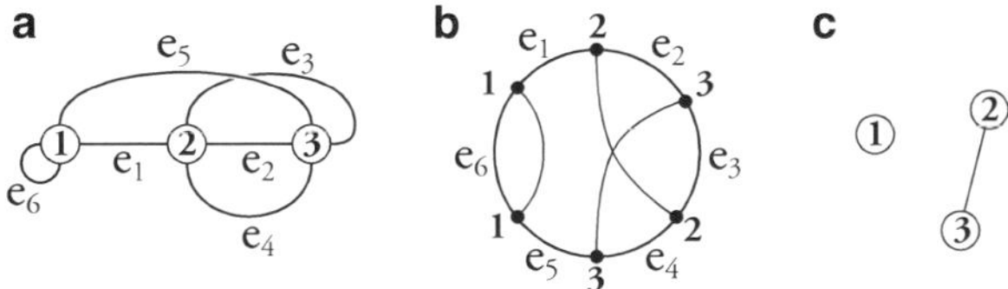

Fig. 10 (**a**) Assembly graph $\mathcal{A} = \mathcal{A}(123231)$, (**b**) chord diagram $\mathcal{C} = \mathcal{C}(\mathcal{A})$, and (**c**) circle graph $\mathcal{G}(\mathcal{C})$

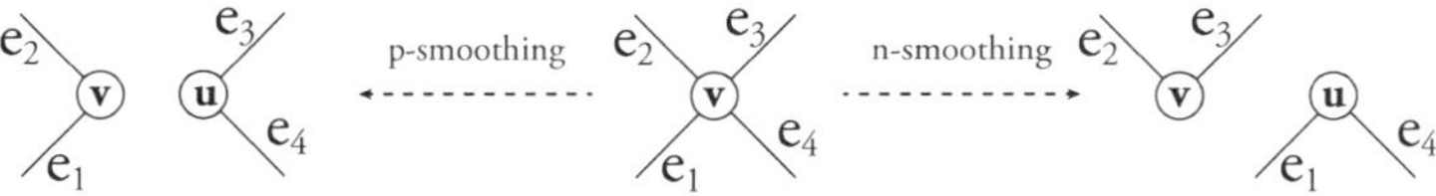

Fig. 11 p- and n-smoothings of a rigid vertex v

We consider two operations on assembly graphs with rigid vertices called p- and n-smoothings. Let v be a vertex of an assembly graph $\mathcal{A}$ with transversal α, and let e_1, e_3, e_2, and e_4 be the edges incident to v in order of their appearance in α. Consider a graph obtained from $\mathcal{A}$ by replacing a vertex v with vertices v' and v'' so that (1) v' is the new endpoint of edges e_1 and e_2 and (2) the vertex v'' is the new endpoint of e_3 and e_4. We say that the resulting graph is obtained by p-smoothing of a vertex v and denote it by $p(\mathcal{A}, \alpha, v)$. The result of the n-smoothing of a vertex v is the graph $n(\mathcal{A}, \alpha, v)$ which is defined similarly except that v' is the new endpoint of e_2 and e_3, and v is the new endpoint of e_1 and e_4 (see Fig. 11). Observe that in both cases, the new graph has one rigid vertex fewer than the original.

The graphs obtained by p- and n-smoothings of vertices in assembly graphs are called intermediates to emphasize that they represent partially rearranged DNA molecules. We are particularly interested in intermediates obtained by smoothing all of the rigid vertices in an assembly graph because they represent possible products of the rearrangement. For an assembly graph with k vertices and an a priori chosen transversal, these intermediates are specified by k-tuples from $\{p, n\}^k$ once the order of vertices has been fixed. We call these tuples smoothings of a graph. That is (p, n, p)-smoothing of a graph with rigid vertices v_1, v_2, and v_3 defines an intermediate obtained by p-smoothing of vertices v_1 and v_3, and n-smoothing of vertex v_2.

For a smoothing s of an assembly graph $\mathcal{A}$ with transversal α, define $\pi(\mathcal{A}, \alpha, s)$ to be the number of symbols p in s, and $\mu(\mathcal{A}, \alpha, s)$ to be the number of connected components in an intermediate corresponding to s-smoothing of $\mathcal{A}$. When the graph and a transversal are clear from the context, we simply write $\pi(s)$ and $\mu(s)$.

Definition 2. An assembly polynomial of an assembly graph $\mathcal{A}$ with vertices $v_1, \ldots, v_k$, corresponding to the transversal α is

$$P(\mathcal{A}, \alpha) = \sum_{s \in \{p, n\}^k} p^{\pi(s)} t^{\mu(s) - 1}$$

Table 1 Assembly words with three distinct symbols and their assembly polynomials

Assembly word	Assembly polynomial
112233	$1 + 3pt + 3p^2t^2 + p^3t^3$
121233	$t + 2p + p^2 + pt^2 + 2tp^2 + tp^3$
122133	$1 + 3pt + 3p^2t^2 + p^3t^3$
122313	$t + 2p + p^2 + pt^2 + 2tp^2 + tp^3$
122331	$1 + 3pt + 3p^2t^2 + p^3t^3$
112323	$t + 2p + p^2 + pt^2 + 2tp^2 + tp^3$
121323	$1 + 2p + pt + 2p^2 + tp^2 + tp^3$
123123	$3pt + t^2 + 3p^2 + tp^3$
123213	$1 + 2p + pt + 2p^2 + tp^2 + tp^3$
123231	$t + 2p + p^2 + pt^2 + 2tp^2 + tp^3$
112332	$1 + 3pt + 3p^2t^2 + p^3t^3$
121332	$t + 2p + p^2 + pt^2 + 2tp^2 + tp^3$
123132	$1 + 2p + pt + 2p^2 + tp^2 + tp^3$
123312	$t + 2p + p^2 + pt^2 + 2tp^2 + tp^3$
123321	$1 + 3pt + 3p^2t^2 + p^3t^3$

where the sum is taken over all possible smoothings of $\mathcal{A}$. For an assembly word w, we define $P(w) = P(\mathcal{A}(w))$.

Proposition 1. *If α and α' are two transversals of an assembly graph $\mathcal{A}(w)$ then*

$$P(w, \alpha) = P(w, \alpha').$$

The above proposition shows that the assembly polynomial is the same for any choice of the transversal. Thus we can refer to the assembly polynomial of a word w by $P(w)$.

Table 1 contains a list of assembly words with three distinct symbols and their corresponding assembly polynomials.

An assembly word is *strongly irreducible* if it does not contain a proper subword that is a double-occurrence word. For example, 123123 is irreducible, but 1221 is not (because it contains a subword 22).

Proposition 2 (Lemma 6.5 in [6]). *If w is an assembly word that is not strongly irreducible, then $P(w) = P(u)P(v)$ for some assembly words u and v.*

We show that assembly graphs that define isomorphic circle graphs also define the same assembly polynomials. We also show that the converse is false, i.e., there are assembly graphs with identical assembly polynomials and non-isomorphic circle graphs. The proof of this result requires a few additional definitions.

A *share* of a chord diagram consists of two disjoint closed arcs that contain both endpoints of each chord they intersect [7]. We assume (without loss of generality) that no endpoint of an arc in a share is also an endpoint of a chord. Two shares are called *complementary* if they are disjoint and contain the endpoints of all of the chords of the diagram. Because the segments of the chord diagram's unit circle are identified with the edges of the assembly graph, we can denote the endpoints of

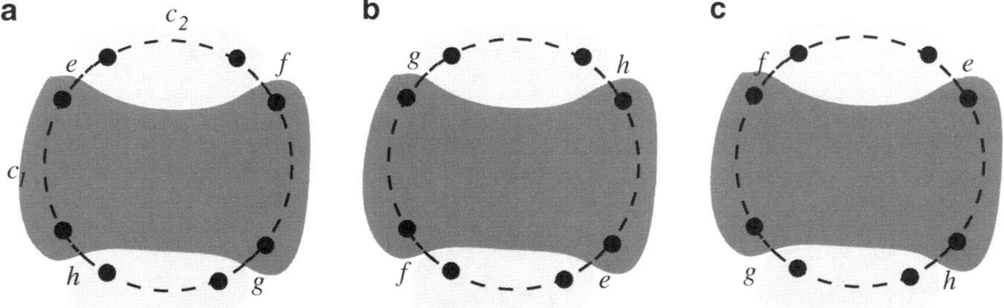

Fig. 12 A chord diagram $\mathcal{C}(\mathcal{A})$ (*left*) with two complementary shares c_1 and c_2. The *middle* graph (**b**) is obtained by rotating share c_1 and the graph to the *right* is obtained by reflecting the same share along a vertical axis

the arcs in a share by the corresponding edges of the assembly graph. We call these edges the *border edges* of the share; e.g., $\{e, f, g, h\}$ is the set of border edges of share c_1 (and also of the complement share c_2) for the example depicted in Fig. 12a.

A *mutation* of a chord diagram is obtained by separating the chord diagram into two complementary shares and recombining the shares into a new chord diagram through a rotation or a reflection. In Fig. 12b, c two mutations of the chord diagram in (a) are depicted, in (b) by rotation of c_1 and in (c) by reflection of c_1 about the vertical axis. One can describe these operations as pairwise swapping of the border edges $\{e, f, g, h\}$. Hence the rotation (Fig. 12b) can be described by transposing edges e and g (as well as h and f), while the reflection can be described as a transposition of e and f (as well as g and h). In this way, we can specify a rotation or reflection of a share by a permutation of the border edges (here we assume that symbols for all border edges are distinct even if some of them represent the same edge). We refer to these permutations as *insertion* maps. There are four insertion maps for each pair of complementary shares corresponding to the three ways of pairwise swapping the edges and the identity.

Two chord diagrams are said to be *mutant* if they are related by a sequence of mutations. S. Chumutov and S. Lando showed that chord diagrams define the same circle graphs if and only if they are mutant [7]. We use the same terminology for assembly graphs, i.e., two assembly graphs are mutant if their chord diagrams are. In the proof of the following proposition we show that mutant assembly graphs, and hence the graphs that define isomorphic circle graphs, have the same assembly polynomials. The result and the proof are similar to those for Jones polynomials and mutant knots [10].

Proposition 3. *Let $\mathcal{A}$ and $\mathcal{A}'$ be two assembly graphs. If $\mathcal{C}(\mathcal{A})$ is isomorphic to $\mathcal{C}(\mathcal{A}')$ then $P(\mathcal{A}) = P(\mathcal{A}')$.*

Proof. Two assembly graphs define the same circle graph if their chord diagrams are related by a sequence of mutations [7]. Thus the claim is proven by showing that a single mutation preserves the assembly polynomial.

Let $\mathcal{A}$ be an assembly graph and consider the corresponding chord diagram $\mathcal{C} = \mathcal{C}(\mathcal{A})$. Consider a pair of complementary shares c_1 and c_2 of $\mathcal{C}$ depicted in Fig. 12. Note that c_1 and c_2 have the same border edges and we denote the set of these edges by $\mathcal{B} = \{e, f, g, h\}$.

There are exactly four ways to form a new chord diagram by rotating and reflecting the share c_1 corresponding to four assembly graphs $\mathcal{A}_i$, $1 \le i \le 4$, one of which is the original assembly graph $\mathcal{A} = \mathcal{A}_1$. For each $i = 2, 3, 4$, $\mathcal{A}_i$ is obtained from $\mathcal{A}_1$ by one of the insertion maps: $(ef)(gh)$ (reflection vertically), $(eh)(fg)$ (reflection horizontally) and $(eg)(fh)$ (rotation) all of which belong to $\mathrm{Sym}(\mathcal{B})$. We can show that for every $1 \le i, j \le 4$, a given smoothing separates both $\mathcal{A}_i$ and $\mathcal{A}_j$ into the same number of connected components (and hence the equality of $P(\mathcal{A}_i)$ and $P(\mathcal{A}_j)$ follows).

Consider a smoothing s. For every $1 \le i \le 4$, let $\mathcal{K}_i$ be the collection of connected components resulting from s-smoothing of $\mathcal{A}_i$. Define $\mathcal{K}_i^1$ (resp. $\mathcal{K}_i^2$) to be the portion of $\mathcal{K}_i$ obtained by restricting the components in $\mathcal{K}_i$ to the edges of c_1 (resp. c_2) including the portions of the border edges. After smoothing, the border edges will be shared between $\mathcal{K}_i^1$ and $\mathcal{K}_i^2$. This means that after smoothing s, each border edge belongs to a component of $\mathcal{K}_i$ whose one portion is in $\mathcal{K}_i^1$ and the other in $\mathcal{K}_i^2$. Hence, the border edges either all belong to the same component in $\mathcal{K}_i$ or there are two components in $\mathcal{K}_i$ each containing a pair of edges.

Let $a, b, c, d \in \mathcal{B}$. After a smoothing s, each pair of the edges a, b, c, d can be either connected within a component of $\mathcal{K}_i$ or not. For each $i = 1, 2, 3, 4$, consider $\phi_i \in \mathrm{Sym}(\mathcal{B})$ such that $\phi_i = (ab)(cd)$ if the pair of edges a, b belongs to one component and the pair c, d belongs to another component in $\mathcal{K}_i$. We take $\phi_i = id \in \mathrm{Sym}(\mathcal{B})$ the identity map to represent the case when all four border edges remain in the same component in $\mathcal{K}_i$. Since the components that do not contain the border edges are the same in every $\mathcal{K}_i$, this implies that $|\mathcal{K}_i| = |\mathcal{K}_j|$ if and only if $\phi_i = \phi_j$.

Let $K = \{(ab)(cd) \mid (ab)(cd) \in \mathrm{Sym}(\mathcal{B})\}$ be a Klein 4-subgroup of $\mathrm{Sym}(\mathcal{B})$. Let $\rho_{i,j} \in \mathrm{Sym}(\mathcal{B})$ be the insertion map transforming $\mathcal{A}_i$ to $\mathcal{A}_j$ as described above. Observe that $\rho_{i,j} \in K$ and because $\rho_{i,j}$ is its own inverse, $\rho_{i,j} = \rho_{j,i}$. For every $1 \le i \le 4$, both ϕ_i and ϕ_j belong to K and $\rho_{i,j}\phi_i\rho_{i,j} = \phi_j$. However, because K is abelian, and $\rho_{i,j} = \rho_{i,j}^{-1}$, $\phi_i = \rho_{i,j}\phi_i\rho_{i,j} = \phi_j$ implying $|\mathcal{K}_i| = |\mathcal{K}_j|$ as needed.

The converse of the above proposition is false. The assembly polynomial of both 12345156246737 and 12345261564737 is

$$P = 1 + 4p + 9p^2 + 16p^3 + 17p^4 + 12p^5 + 4p^6 + 3pt + 10p^2t + 14p^3t$$
$$+ 15p^4t + 8p^5t + 3p^6t + p^7t + 2p^2t^2 + 5p^3t^2 + 3p^4t^2 + p^5t^2.$$

However the circle graph of 12345156246737 has two vertices of degree 1 and the circle graph of 12345261564737 has only one such vertex.

Acknowledgements We wish to thank F. Din-Houn Lau and Kylash Rajendran, Erica Flapan, Mauro Mauricio and Julian Gibbons for insightful discussions. We are greatful to Natasha Jonoska and Masahico Saito for many useful suggestions and help with preparing the manuscript. ED has been supported by the NSF grants DMS-0900671, KV is supported by EP/G0395851.

References

1. N.M. Alder, B.I. Rogozin, M.L. Iyer, V.G. Glazko, D.M. Cooper, Z. Pancer, Diversity and function of adaptive immune receptors in a jawless vertebrate. Science **23**, 1970–1973 (2005)
2. A. Angeleska, N. Jonoska, M. Saito, L.F. Landweber, RNA-guided DNA assembly. J. Theor. Biol. **248**, 706–720 (2007)
3. A. Angeleska, N. Jonoska, M. Saito, DNA recombination through assembly graphs. Discret. Appl. Math. **157**, 3020–3037 (2009)
4. L. Bonen, Trans-splicing of pre-mRNA in plants, animals, and protists. FASEB J. **7**, 40–46 (1993)
5. D. Buck, E. Dolzhenko, N. Jonoska, M. Saito, K. Valencia, Genus ranges of 4-regular rigid vertex graphs (2012). (arXiv:1211.4939)
6. J. Burns, E. Dolzhenko, N. Jonoska, T. Muche, M. Saito, Four-regular graphs with rigid vertices associated to DNA recombination. Discret. Appl. Math. **161**, 1378–1394 (2013). http://dx.doi.org/10.1016/j.dam.2013.01.003
7. S.V. Chmutov, S.K. Lando, Mutant knots and intersection graphs. Algebra Geome Topol. **7**, 1579–1598 (2007)
8. E. Dolzhenko, E.C. Swart, A.D. Goldman, L.F. Landweber, STAGR: software to annotate genome rearrangement, extended abstract, ISMB/ECCB, 2011. http://www.iscb.org/uploaded/css/86/20246.pdf
9. R.A. Duke, The genus, regional number, and Betti number of a graph. Can. J. Math. **18**, 817–822 (1966)
10. V.F.R. Jones, A polynomial invariant for knots via von Neumann algebra. Bull. Am. Math. Soc.(N.S.) **12**, 103–111 (1985)
11. N. Jonoska, M. Saito, Boundary components of thickened graphs. Lect. Notes Comput. Sci. **2340**, 70–81 (2002)
12. N. Jonoska, N.C. Seeman, G. Wu, On existence of reporter strands in DNA-based graph structures. Theor. Comput. Sci. **410**, 1448–1460 (2002)
13. W. Marande, G. Burger, Mitochondrial DNA as a genomic jigsaw puzzle. Science **19**, 415 (2007)
14. B. Mohar, C. Thomassen, *Graphs on Surfaces* (The Johns Hopkins University Press, Baltimore, 2001)
15. D.M. Prescott, Genome gymnastics: unique modes of DNA evolution and processing in ciliates. Nat. Rev. Genet. **1**, 191–198 (2000)
16. J.P. Stephens, D.C. Greenman, B. Fu, F. Yang, G.R. Bignell, Massive genomic rearrangement acquired in a single catastrophic event during cancer development. Cell **144**, 27–40 (2011)
17. W.T. Tutte, Graph-polynomials. Adv. Appl. Math. **32**(1–2), 5–9 (2004)
18. K. Valencia, Topological investigation of bacterial site-specific recombination and genome differentiation in ciliates. Ph.D. Dissertation, Department of Mathematics, Imperial College London, 2013
19. J.W.T. Youngs, Minimal imbeddings and the genus of a graph. J. Math. Mech. **12**(2), 303–315 (1963)

Part IV
Topological Models and Spatial DNA Embeddings

Introduction to DNA Topology

Isabel K. Darcy, Stephen D. Levene, and Robert G. Scharein

Abstract In this expository chapter we give an elementary introduction to DNA and to proteins that can knot and link circular DNA, with a special focus on recombination. We also describe the Ernst and Sumners tangle model of the action of proteins on circular DNA.

1 Introduction to DNA

In 1953, Rosalind Franklin and her student R. G. Gosling published their crystal structure of DNA upon which our current models of DNA are based. As discussed in her 1953 paper [36], "DNA is a helical structure" with "two co-axial molecules." The "co-axial molecules," shown as blue ribbons in Fig. 1a, refer to a chain of phosphate groups connected via sugar groups. The sugar groups are shown in red in Fig. 2. Each sugar group is connected to a base: A, T, G, or C. These bases spell out our genetic information, i.e., they form our DNA sequence. Using Franklin's data without her knowledge, Watson and Crick came to similar conclusions regarding the structure of DNA and published their model in the same issue of Nature [90].

I.K. Darcy (✉)
Department of Mathematics and Program in Applied Mathematical and Computational Sciences, University of Iowa, 14 MLH, Iowa City, IA 52242, USA
e-mail: idarcymath@gmail.com

S.D. Levene
Departments of Bioengineering, Molecular and Cell Biology, and Physics, University of Texas at Dallas, 800 West Campbell Road, Richardson, TX 75080, USA
e-mail: Stephen.Levene@utdallas.edu

R.G. Scharein
Hypnagogic Software, Vancouver, BC, Canada
e-mail: rob@knotplot.com

N. Jonoska and M. Saito (eds.), *Discrete and Topological Models in Molecular Biology*, Natural Computing Series, DOI 10.1007/978-3-642-40193-0_15,
© Springer-Verlag Berlin Heidelberg 2014

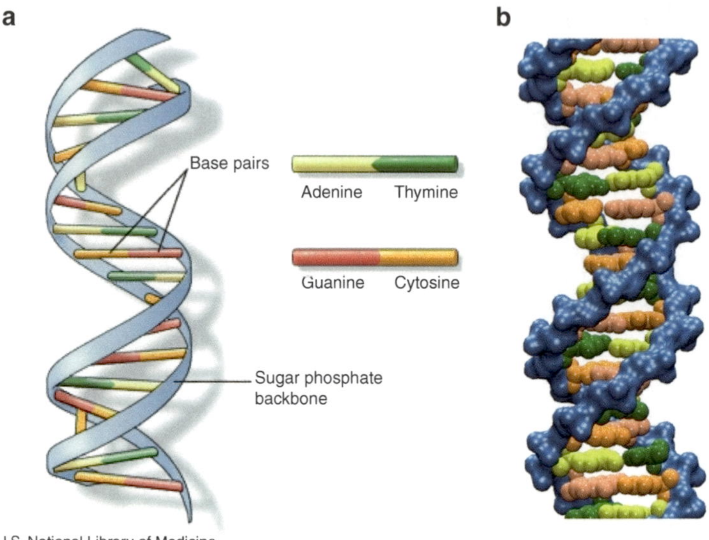

Fig. 1 (**a**) Structure of DNA. Per Franklin and Gosling, the "period is 34 Å" and "one repeating unit contains ten nucleotides on each of two … co-axial molecules." They conclude "The phosphate groups lie on the outside of the structural unit, on a helix of diameter about 20 Å" and "the sugar and base groups must accordingly be turned inwards towards the helical axis." [36]. The four bases are called adenine (A), thymine (T), cytosine(C), and guanine (G). Figure courtesy of the National Library of Medicine (NLM). (**b**) Computer model of a DNA molecule

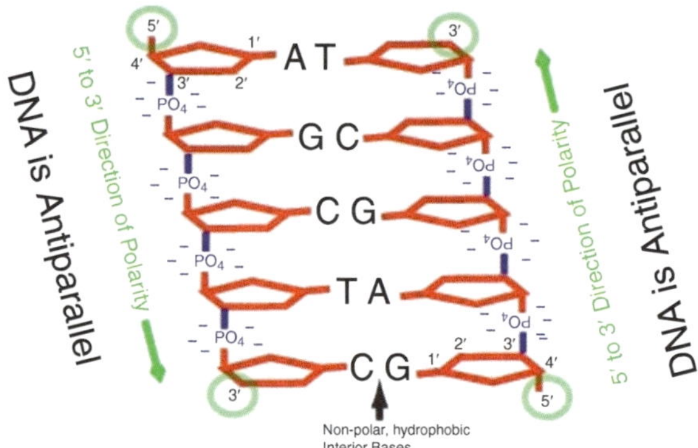

Fig. 2 Chemical structure of DNA. The sugars are shown in red. Note that the pairing of backbone strands is antiparallel (Figure from [50])

In most cases, the base A pairs with the base T while the base G pairs with the base C [90]. Thus, knowing the sequence of one strand of double-stranded DNA (dsDNA) means knowing the sequence of both strands. However, the two strands making up dsDNA are read in opposite directions. Rosalind Franklin noted that the two sugar–phosphate backbones are antiparallel [35]. A sugar residue contains five

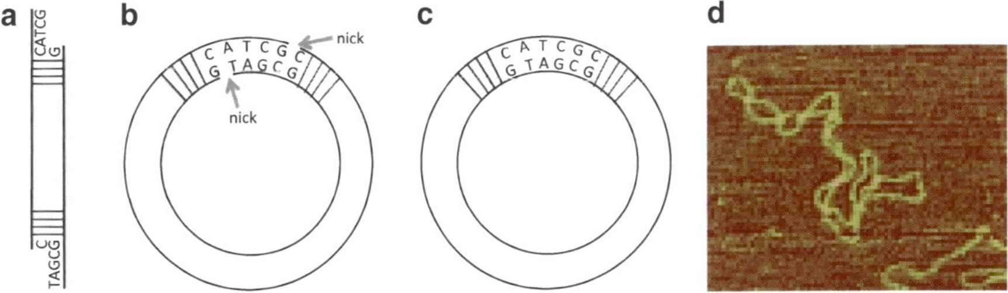

Fig. 3 (**a**) Linear DNA with sticky ends. (**b**) Nicked circular DNA. (**c**) Closed circular DNA. (**d**) Atomic-force microscopy image of supercoiled DNA (Unpublished data courtesy of Dr. Alexandre Vetcher)

carbons that are numbered from $1'$ to $5'$, as shown in Fig. 2 for the sugar in the upper left corner, which is connected to the base A. The $5'$ carbon of this sugar is circled in green. Its $3'$ carbon is connected via a phosphate bond to the $5'$ carbon of the next sugar (which is connected to the base G). Hence the chemistry of this connection can be used to assign an orientation to a DNA strand. Sequences are read from $5'$ to $3'$. Thus the strand on the left is read from top to bottom, and thus its sequence is AGCTC. The direction of the strand on the right goes from bottom to top. Hence its sequence is read GAGCT. Thus both AGCTC and GAGCT refer to exactly the same double-stranded DNA sequence.

In the laboratory, molecular biologists often work with linear DNA that has sticky ends. The end of the DNA is *sticky* if the portion at the end is single-stranded, as shown in Fig. 3a. This means that there are unpaired bases. If linear DNA contains two sticky ends that have complementary sequences, then if the DNA is sufficiently long, the linear DNA will circularize to form nicked circular DNA as shown in Fig. 3b. The DNA is called *nicked* because the phosphate backbone is not closed. A protein called ligase is needed to create a phosphodiester bond to close the nicks to form *closed circular DNA* (Fig. 3c). This closed DNA can be modeled by an annulus. Since the phosphate backbones are antiparallel, DNA cannot form a Möbius band (under normal circumstances). DNA has a preferred twist of about 10.5 base pairs per turn [36, 68, 85]. Closed circular DNA is called *relaxed* if it is as close as possible to its preferred twist. In nature, DNA is usually underwound, and hence it supercoils negatively (Fig. 3d). Since the DNA is underwound, the two strands are easier to pull apart for replication or transcription. For an elementary introduction to DNA, see [12]. For more on DNA topology, see [4].

2   DNA Knots and Topoisomerase

There are many beautiful knot tables in the literature and online. For an excellent introduction to knot theory, see [1]. Knots were first tabulated by Tait in the late 1800s [81]. The knot/link table shown in Fig. 4 was created by KnotPlot [72] based

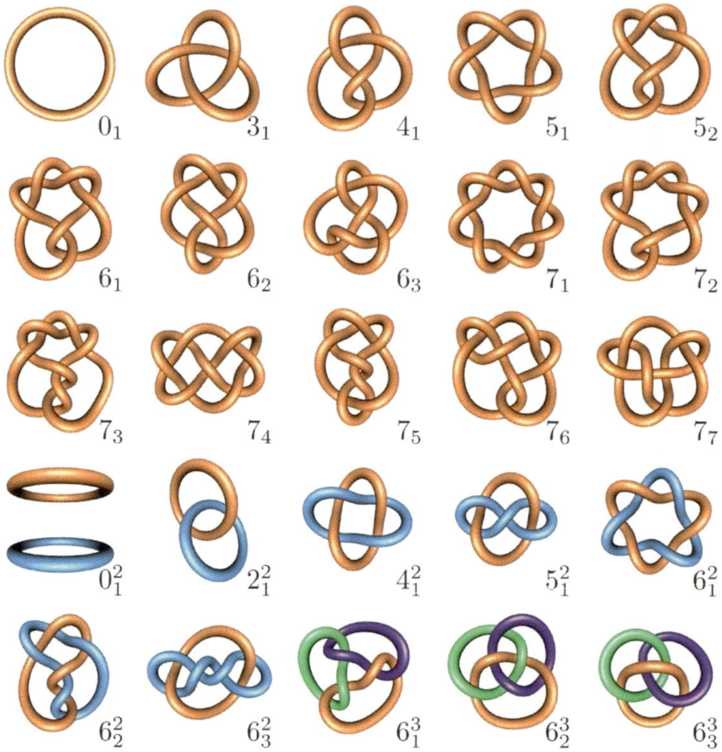

Fig. 4 Knot/link table containing prime knots up to seven crossings and two- and three-component prime links up to six crossing. Twist knots include 3_1 (the trefoil knot), 4_1 (the Fig. 8 knot), 5_2, 6_1, and 7_2. Torus knots include 3_1, 5_1, and 7_1. The unknot, 0_1, can also be considered to be a twist knot and a torus knot

on data provided by Dale Rolfsen. The knot n_k refers to the kth knot in the list of knots containing n crossings in their minimal crossing diagram. The superscript in the link table refers to the number of components. Mathematicians sometimes use the term "link" to include knots (and in rare cases, a link may be referred to as a knot). Molecular biologists normally use the term *catenane* when referring to links with at least two components. Most tables, including the one in Fig. 4, only contain *prime knots*. These are knots that cannot be subdivided into two or more simpler nontrivial knots. Knots that are not prime are called *composite*. Composite knots correspond to the operation of tying two separate knots sequentially in a piece of rope and closing the ends, as shown in Fig. 5a, where the individual prime knots are colored differently.

A knot is called *chiral* if it cannot be smoothly deformed into its own mirror image. The mirror image of a knot K is denoted K^*. The simplest example of a chiral knot is 3_1; it is shown together with its mirror image, 3_1^*, in Fig. 5b. Most

Fig. 5 (**a**) The composite knot $3_1\#4_1$. (**b**) Chiral pair of trefoil knots, 3_1 (in *gold*) and 3_1^* (in *blue*)

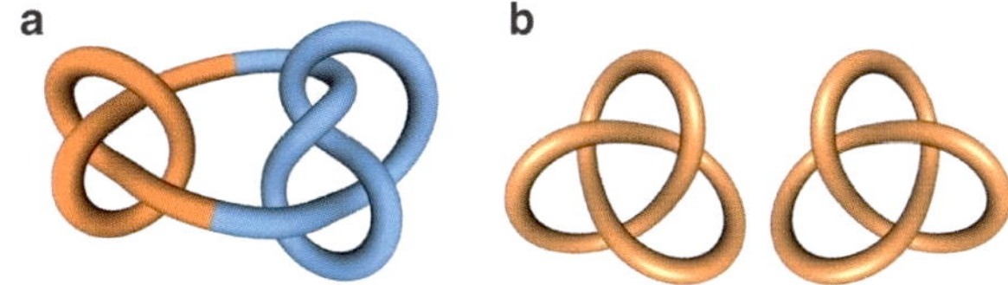

knot tables like the one in Fig. 4 list only one enantiomer of a chiral pair. Knots that are not chiral are *achiral*. The simplest three achiral knots are 0_1, 4_1, and 6_3.

One of the most beautiful knot tables is the one created by topoisomerase I. This protein acting on nicked circular DNA, was able to create all different types of knots up to six crossings, 10 of the 16 possible seven-crossing knots, and a few eight- and nine- crossing knots [30]. There are two main types of topoisomerase. Type I topoisomerase will break and reconnect one strand of DNA, while type II topoisomerases will break and reconnect both strands of dsDNA. Thus type I topoisomerases can knot circular single-stranded DNA (ssDNA) as well as nicked DNA. Type II topoisomerases can knot dsDNA [89]. The normal function of topoisomerases is to keep DNA unknotted, unlinked, and properly supercoiled. For more on topoisomerase, see [87, 88].

3 Recombinases

The genome of any organism must possess two key characteristics. It must be stable enough to pass accurate information through inheritance, yet remain sufficiently dynamic to respond to selective environmental pressures. These requirements create a tension between genome integrity and flexibility. In all organisms, recombination systems are the principal mechanism that regulates genome stability. Chromosomal breakages and mutations stemming from problems in DNA replication or environmental stress can be repaired by recombination. Most organisms have multiple recombination pathways by which damage can be repaired, underscoring the importance of this process.

We focus here on several examples of site-specific recombination mechanisms. These processes involve interactions among specialized DNA-sequence elements that also contain specific binding sites for recombination proteins. The requirements for sequence specificity and specialized proteins distinguish site-specific recombination from general or homologous recombination, which can occur with arbitrary DNA sequences that share very high levels of sequence identity (see [58, 92] for reviews). Site-specific-recombination target sequences form the point of genetic exchange and usually are present in few copies in the genome. Often these sites are present in pairs; in the case of the bacteriophage-lambda integration site, only one copy is present in the *E. coli* genome. This extraordinary degree of specificity leads to precisely defined genetic rearrangements. In the examples considered here, the rearrangements that occur are essentially uniquely defined.

Another important attribute of a site-specific recombination locus is the polarity of the recombination site. These loci are frequently nonpalindromic and therefore have an intrinsic polarity (Fig. 6). Recombination normally occurs only when a pair of recombination sites has been juxtaposed in a particular spatial alignment, thereby imparting both positional and orientational specificity to these systems [37,73]. This specificity has important biological consequences; moreover, the site-orientation specificity leads to the formation of specific DNA topologies in the recombination products. If two recombination sites are oriented in opposite polarities on a circular DNA molecule as shown in Fig. 6a, then the sites are said to be *inversely repeated*. Recombination on inversely repeated sites is called an *inversion* because it results in the inversion of one of the DNA segments between the two recombination sites with respect to the other DNA segment. In Fig. 6b, the two recombination loci are oriented in the same direction and are thus called *directly repeated*. Recombination on directly repeated sites results in the *deletion* (also called *excision*) of a DNA segment, changing the number of components of the substrate. The reverse reaction is called *integration*.

When supercoiled DNA substrates are used in reconstituted in-vitro (i.e., in the test tube) recombination reactions, it is possible to examine the topological changes that take place during recombination. For intramolecular recombination reactions, supercoiled plasmid substrates bearing inversely oriented sites generate knotted recombination products, whereas supercoiled substrates containing directly repeated sites generate topologically linked circles called catenanes (Fig. 6). The knots and catenanes that are formed during recombination are never random; instead, recombination generates a highly restricted subset of all the possible knotted or catenated structures that can be formed. For example, all of the knots with up to 13 crossings are known – there are over 12,000 topologically distinct knots. Integrative recombination on a circular substrate with inverted sites yields only seven of the possible knots containing up to 13 crossings, each containing an odd number of crossings. Among all possible recombination mechanisms that can lead to the formation of a knotted DNA product, the formation of this particular set of observed products can be ascribed uniquely to a particular mechanism. The topological specificity of site-specific recombination systems has been exploited to great effect in unraveling the mechanisms of many site-specific recombinases (see, e.g., [10, 21, 23, 34, 41, 79, 80, 82–84, 95]).

The complementarity of DNA strands normally plays a very limited role in site-specific recombination, more as a feature of specific recombinase–DNA interactions than a necessity for homologous pairing or strand exchange. Unlike other modes of recombination, site-specific recombination is conservative in that no DNA is gained or lost during the recombination reaction. This aspect of site-specific recombination applies both at the level of genetic information (recombination products are merely permutations of the original parental DNA) and at the level of actual DNA nucleotides (no DNA synthesis or nucleolytic degradation is involved).

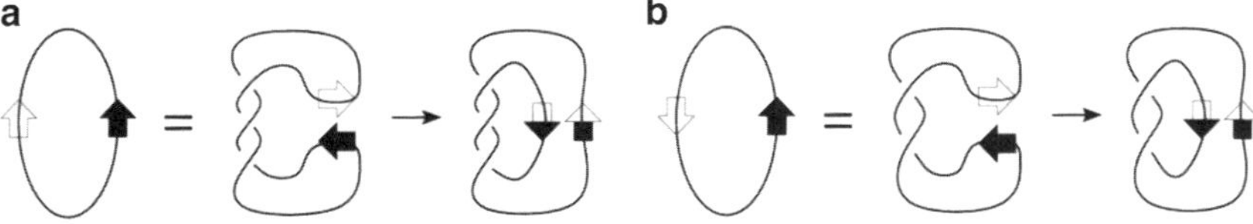

Fig. 6 (**a**) Recombination on inversely repeated sites on a circular DNA molecule can result in knotted DNA. (**b**) Recombination on directly repeated sites changes the number of components of the DNA substrate

All site-specific recombination systems that have been investigated to date fall into two superfamilies: the lambda-integrase and resolvase/invertase families. Particular examples from the lambda-integrase family are discussed below. The products of the lambda-integrase recombination reaction depend on the orientation and disposition of recombination sites; this variability permits systems such as lambda-integrative recombination to carry out excisive as well as integrative recombination in a highly regulated fashion. The two superfamilies are also distinct in terms of the intermediate structure of the DNA segments undergoing recombination; whereas lambda-integrase-type mechanisms proceed through a four-stranded DNA intermediate called a Holliday junction, the resolvase/invertase mechanisms do not. Site-specific recombination systems participate in a wide range of biological processes in both prokaryotes and eukaryotes: viral integration, antigenic variation, gene duplication and copy-number control, and the integration of antibiotic resistance cassettes. For a very nice review of site-specific recombination, including resolvase/invertase recombination, see [42].

3.1 Holliday Junctions

In 1964, Robin Holliday proposed that recombination could be mediated by a hypothetical DNA structure consisting of four polynucleotide strands associated by a single-stranded crossover (Fig. 7) [48]. This structure, later to be named the Holliday junction, has played a central conceptual role in models of recombination. A large body of evidence has accumulated in the intervening decades that substantiates the role of these junctions in both general recombination mechanisms [46, 93, 94] and those belonging to the lambda-integrase superfamily of site-specific recombinases [3, 21].

Although the existence of this intermediate structure is no longer questioned, the details of Holliday junction geometry remain controversial. Since the early 1990s, a wide range of biochemical and biophysical tools have been used to characterize the conformation of these recombination intermediates, both as complexes with recombination proteins [6, 15, 39, 43, 49, 84] and as free DNA molecules [13, 14, 16, 18–20]. Many studies of protein-free junctions were focused on the structure and dynamics of immobile four-way DNA junctions, in which four synthetic DNA

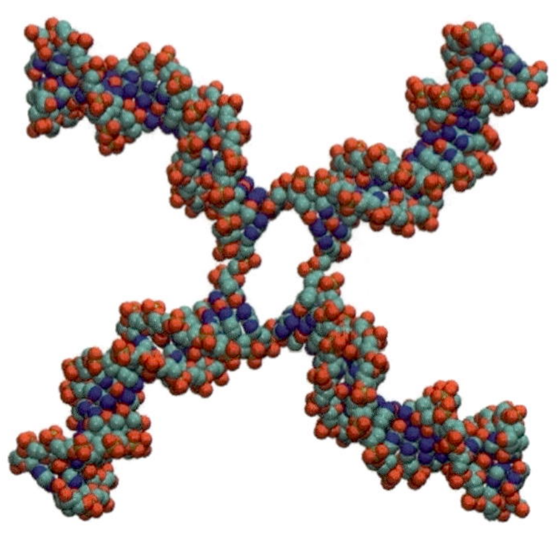

Fig. 7 Molecular model of a Holliday-like four-way DNA junction. The DNA sequences of the four strands in this structure lack the symmetry of a true Holliday junction, thereby inhibiting migration of the junction's branch point

strands designed with specific patterns of homology have been annealed together (Fig. 7) [51]. The limited homology among the DNA strands fixes the branch point of the junction; thus, such structures lack the ability to undergo branch migration, an essential isomerization step in general recombination. The extent to which the behavior of such immobile analogs actually mimics that of mobile junctions is an interesting issue that has remained largely unaddressed. However, it is clear that even immobile four-way junctions are conformationally quite flexible, a feature that is likely to be, if anything, more pronounced in fully mobile junctions [77]. Several groups have succeeded in obtaining high-resolution X-ray structures of four-way junctions [31, 61, 62]. These high-resolution structures exemplify many features that are consistent with those of immobile junctions based on studies in solution.

3.2 λ-Int: Integration and Excision of Phage Genomes

The λ-integrase (λ-Int) system is vital to the lysogenic stage of the life cycle of bacteriophage λ and is one of the most intensively studied site-specific recombination systems. A notable feature of this system is the nonsymmetrical nature of the integrative and excisive recombination reactions: although the strand exchange activities are identical for both integration and excision of the phage-λ genome, each reaction has distinct requirements for specific DNA sequences at the recombining loci and the subsets of protein cofactors involved in recombination.

Integration of phage λ occurs at a unique 25-bp site, termed *attB*, on the 4.6-Mbp *E. coli* chromosome. The catalytic activity for strand exchange resides in the λ-encoded integrase protein (Int), which functions in concert with a number of DNA-binding accessory proteins: the integration host factor (IHF) and factor for inversion stimulation (Fis) proteins of *E. coli*, and the λ-excisionase (Xis), which is phage-encoded. In contrast to the *attB* site, which by itself has negligible affinity for the recombination proteins, the recombination locus on the phage genome, *attP*, is about 250 bp in size and has multiple binding sites for Int and the accessory factors. Integrative recombination most likely involves assembly of Int and IHF proteins to form an organized nucleoprotein structure called the intosome [5], which subsequently captures a protein-free *attB* site during synapsis [67]. The products of the integrative recombination reaction are a functionally distinct pair of new recombination sites called *attL* and *attR* that are no longer competent to participate in subsequent rounds of integrative recombination. Instead, these sites are substrates for excisive recombination, a reaction that requires Fis and Xis in addition to Int and IHF. By coupling recombination to the intracellular levels of specific protein factors, tight regulation of the phage-λ life cycle can be achieved in vivo. The topology of λ-Int recombination is discussed in [21].

3.3   Cre and XerC/D: Excision and Resolution of DNA Dimers

The genome of bacteriophage P1 is a 90-kbp circular molecule; as with all circular genomes, daughter molecules must be decatenated after replication [91]. This process is facilitated by a protein called Cre recombinase, a phage-encoded member of the λ-Int superfamily. The Cre mechanism acts on specific sites, denoted *loxP* in a multistep reaction scheme that involves fusion followed by resolution (Fig. 8a). A common feature of the λ-Int superfamily is phosphoryl transfer based on a catalytic tyrosine residue. The enzymatic reaction progresses in two distinct stages (Fig. 8b): an initial round of strand cleavage followed by DNA strand exchange to form a stable recombinase-bound Holliday junction. The junction is resolved by a second set of tyrosine-catalyzed cleavage and strand-exchange steps that lead to recombinant products.

The wild-type *loxP* target site for Cre is a 34-bp DNA sequence that consists of two 13-bp inverted repeats flanking an asymmetric 8-bp core region [38]. The core sequence confers an overall directionality on the *lox*P site. Recombination of directly repeated *loxP* sites leads to the exclusive formation of deletion products, whereas recombination of inversely repeated *loxP* sites results in an inversion of the intervening DNA sequence with respect to the parental substrate [84].

Normal replication of the *E. coli* chromosome yields intermediate forms consisting of multiply linked circular DNA molecules. The linked intermediates are resolved to unlinked monomers by the action of type II topoisomerases, most notably topo IV [78]. However, homologous recombination during replication generates concatenated dimers at a significant frequency; such structures cannot be resolved by topoisomerases. These dimers are instead resolved by the XerC/D system, which also belongs to the λ-Int superfamily. The activity of Xer is tightly coupled to that of FtsK, a molecular machine that controls the transport of DNA across the intercellular septum during cell division [66]. Cells lacking functional XerC or XerD genes cannot properly segregate daughter chromosomes. These

                                                                I.K. Darcy et al.

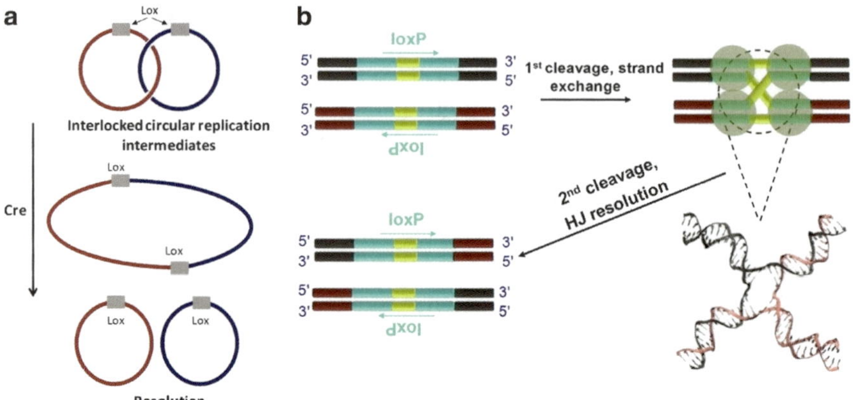

Fig. 8 (**a**) Unlinking by Cre recombinase: Cre activity initially generates circular concatamers from linked circular substrates. This fused intermediate is subsequently unlinked via an excision reaction that is also mediated by Cre. (**b**) Mechanism of Cre acting on a pair of *loxP* target sequences. The *loxP* site consists of two inversely repeated, 13-bp Cre-binding sequence elements (*cyan*) flanking an 8-bp spacer region (*yellow*). Recombination takes place via ordered and reversible strand cleavage, exchange, and resolution reactions. The central intermediate is a Holliday junction, shown in an open, square planar conformation similar to that in Fig. 7

cells develop an anomalous filamentous-growth phenotype, in which cells elongate without dividing [59].

The target site for Xer recombination on the *E. coli* chromosome is a 28-bp sequence called *dif*. This sequence is located opposite the chromosomal replication origin and consists of a pair of 11-bp inverted repeats that flank a central 6-bp spacer region. Unlike the *loxP* site of Cre, the *dif* repeats are targeted by the Xer C and D subunits. In other respects, however, the similarities between Xer and Cre are more striking than their differences. Like the Cre–*loxP* mechanism, Xer recombination proceeds via a Holliday-junction intermediate [7].

Plasmids in *E. coli* can also become dimerized during replication. Resolution of these concatameric forms occurs via Xer activity at plasmid sequences such as *cer* and *psi*. These sequences contain the 28-bp core sequence from the chromosomal *dif* element in addition to flanking sequences that bind the accessory proteins PepA and either ArgR (in *cer*) [74] or ArcA (in *psi*) [8]. These proteins are required for recombination and play a role in organizing the active synaptic complex of proteins and DNA sequences needed for site pairing and strand exchange. This also ensures that recombination occurs exclusively via an intermolecular pathway involving directly repeated target sites. A similar mode of synaptic-complex organization occurs in gamma/delta resolvase recombination (based on a serine recombinase) and accounts for exclusivity of deletion in that system [71].

4 The Tangle Model for Protein Action

In this section, we start by giving some mathematical background on tangles and then describe how tangles are applied to study protein action. An N-*string tangle* is a collection of N disjoint arcs properly embedded in a three-dimensional ball (3-ball) that have their endpoints fixed on the 2-sphere boundary of the 3-ball. Examples of 2-string tangles are shown in Fig. 9. A tangle is *rational* if it can be formed from a zero crossing tangle by moving the endpoints in an arbitrary fashion with the constraint that they remain on the boundary of the 2-sphere and the arcs are confined to stay within the 3-ball. It is common practice and sufficient to consider only 180° rotations about the horizontal and vertical axes. Conway introduced rational tangles in a paper that was concerned with enumerating prime knots and links [17]. He discovered a curious relationship between rational tangles and the set of extended rational numbers[1] (hence the name *rational tangle*). He showed that two rational tangles are equivalent (in the sense that one may be converted into the other, keeping the boundary of the 3-ball fixed) if and only if the continued fractions corresponding to the individual tangles are equal to the same extended rational number. The details of this proof are beyond the scope of this chapter, however, a few examples will be illustrative. First of all, let us consider how to associate a tangle with a continued fraction. We denote a tangle by a list of integers $(c_1, c_2, \ldots, c_n)$, n odd, and create this tangle by starting with the 0 tangle and rotating about the horizontal axis by $c_1 \times 180°$, followed by a rotation about the vertical axis by $c_2 \times 180°$, alternating the axes in this way until we reach the end of the list. Since n is odd, we always end in horizontal twists. For the tangle $(c_1, c_2, \ldots, c_n)$, we assign the rational number

$$r = c_n + \cfrac{1}{c_{n-1} + \cfrac{1}{\ldots + \frac{1}{c_1}}}.$$

Figure 10 shows several rational tangles and one nonrational tangle. We can see that tangle $(2, 3, 4)$ and tangle $(-1, -1, -4, 1, 3)$ have the same rational number $30/7$, and therefore must be equivalent tangles from Conway's theorem.

Tangles that cannot be formed from the operations described above are known as *nonrational tangles*. Informally, these are tangles whose construction would require one of the endpoints of an arc to leave the 2-sphere, and pass through the 3-ball and around one of the arcs inside. An example is shown in Fig. 10e. Nonrational tangles are not uncommon in DNA; however, their analysis is considerably more complicated than the rational case, and we will not discuss them further in this chapter.

Ernst and Sumners [34] were the first to apply tangles to DNA biology. In their model, a protein complex binding N segments of DNA is represented by

---

[1] The rational numbers plus infinity.

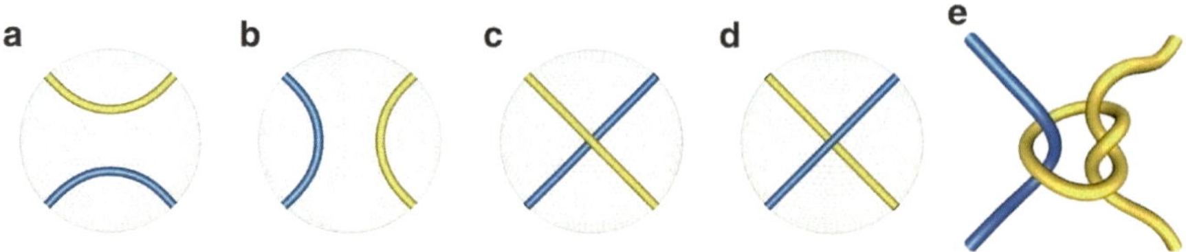

Fig. 9 Various tangles. (**a**) The 0 tangle. (**b**) The infinity tangle. (**c**) the $+1$ tangle. (**d**) The -1 tangle. (**e**) The $\frac{1}{2} + -\frac{1}{3}$ tangle

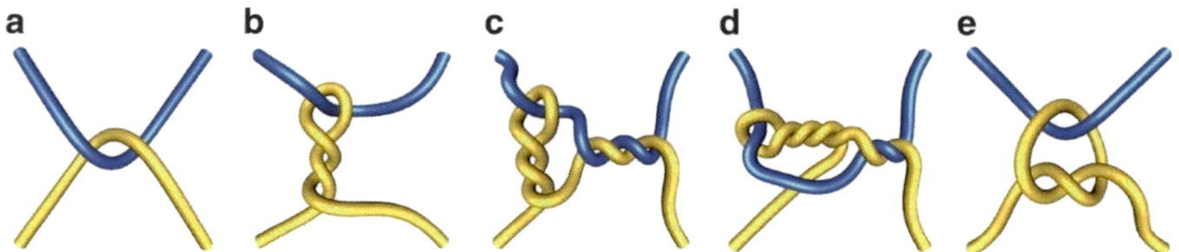

Fig. 10 Four rational tangles with their associated rational numbers and one nonrational tangle. (**a**) The (2) tangle. (**b**) The $(2, 3, 0)$ tangle $= \frac{2}{7}$ tangle. (**c**) The $(2, 3, 4)$ tangle $= \frac{30}{7}$ tangle. (**d**) The $(-1, -1, -4, 1, 3)$ tangle $= \frac{30}{7}$ tangle. Note that the signs of the integers in $(-1, -1, -4, 1, 3)$ determine the handedness of the crossings. (This tangle can be created in KnotPlot by using the command: tangle 1z1z4z13o.) (**e**) A nonrational tangle

a tangle ball, and the DNA itself by the disjoint arcs. Of course, this is a highly simplified model of the binding of proteins with DNA. A sphere is a very rough approximation to a protein complex, and the DNA is likely to exist in more complicated conformations than the arcs seen in the above illustrations. For example, the DNA likely winds around the tangle ball rather than being embedded within the 3-ball.

If at least two DNA segments are bound in a protein–DNA complex, then this complex is referred to as a synaptic complex. The protein complex together with the segments of DNA bound by protein is called a synaptosome. In many cases it is possible to prove that a tangle modeling a synaptosome is rational (e.g., [22, 23, 32–34, 47]), but there are also several biological reasons why rational tangles are the most likely models of synaptosomes. Although an upper bound for the number of DNA crossings that can be bound in a synaptosome has not yet been determined, it is believed that synaptosomes cannot be overly complicated. The simplest nonrational two-string tangles are the five-crossing tangles shown in Figs. 9e and 10e. Moreover, as protein complexes bind supercoiled DNA (Fig. 3d), rational tangles are likely models of synaptosomes, since such tangles are formed by adding twists. Lastly, a tangle is rational if and only if one can push the strings to lie on the boundary of the 3-ball so that the strings do not cross themselves on the 3-ball. Thus if DNA wraps around a protein complex without crossing itself and if the protein complex can be modeled by a topological sphere, the tangle modeling the synaptosome must be rational. Since DNA is negatively charged, it is unlikely to cross itself on the

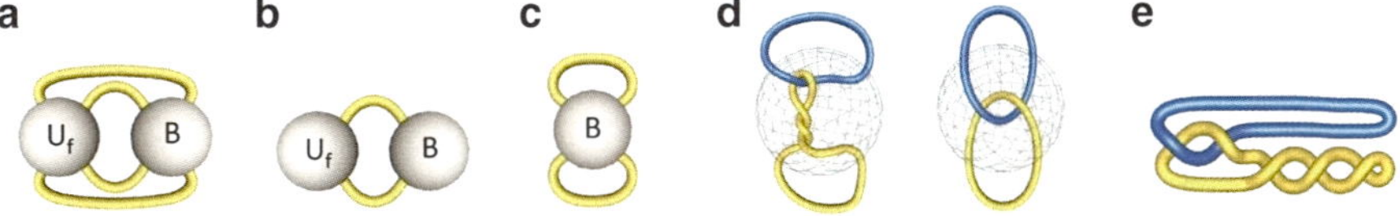

Fig. 11 (**a**) $N(U_f + B)$. (**b**) Tangle addition, $U_f + B$. (**c**) Numerator closure of a tangle $N(B)$. (**d**) $N(\frac{2}{7}) = N(2)$. (**e**) Same link as in (**d**), but shown in 2-bridge = 4-plat form

boundary of a protein. There are, however, protein complexes that can be modeled by higher-genus objects such as tori; but in all known cases, the protein–DNA complex can still be modeled by a spherical tangle.

Let K represent knotted circular DNA. For those who like to think of this circular DNA as living in S^3, then the tangle model of Ernst and Sumners [34] divides S^3 into two tangles (Fig. 11a). One tangle, B, models the synaptosome (i.e., the protein complex together with the portion of DNA bound by protein), whereas the unbound DNA is in the complementary tangle, U_f. For simplicity, this is written as the tangle equation $N(U_f + B)$. Tangle addition, $U_f + B$, corresponds to the operation shown in Fig. 11b. The numerator closure operation is shown in Fig. 11c. The numerator closure of a rational tangle is a rational knot/link (Fig. 11d). By rotating the vertical crossings in a rational tangle so that they appear horizontal, it is easy to see that rational knots/links are equivalent to 2-bridge knots/links (also known as called 4-plats); see the example in Fig. 11e. If $ac \geq 0$, then two rational knots $N(a/b) = N(c/d)$ are equivalent if $a = c$ and $bd^{-1} = 1 \bmod a$.

Let the tangle B represent the synaptosome before protein action, and let the tangle E represent it after protein action. Recall that the tangle U_f represents the DNA not bound by protein. A protein action that changes the knot K_1 into the knot K_2 is represented by the system of two tangle equations (Fig. 12):

$$N(U_f + B) = K_1, \tag{1}$$

$$N(U_f + E) = K_2. \tag{2}$$

The starting conformation of the DNA, K_1, is called the *substrate*, while K_2 is called the *product*. Different recombinases have different topological mechanisms. A tangle model of Cre recombination, is shown in Fig. 12b, and tangle model of Xer acting on *psi* sites, is shown in Fig. 12c. For a tangle model of Xer acting on Ftsk, see [76]. The software packages TangleSolve [70], TopoICE-R [25], and TopoICE-X [27] can be used to solve certain types of tangle equations and visualize the solutions.

In the original tangle model, the tangle B is divided into a sum of two tangles $B = U_b + P$, where the tangle P represents the local action of the protein, and the tangle U_b represents protein-bound DNA whose conformation is unchanged by protein action. Often proteins act on very short segments of DNA, and thus one can often assume that P is a zero-crossing tangle. Protein action is represented by

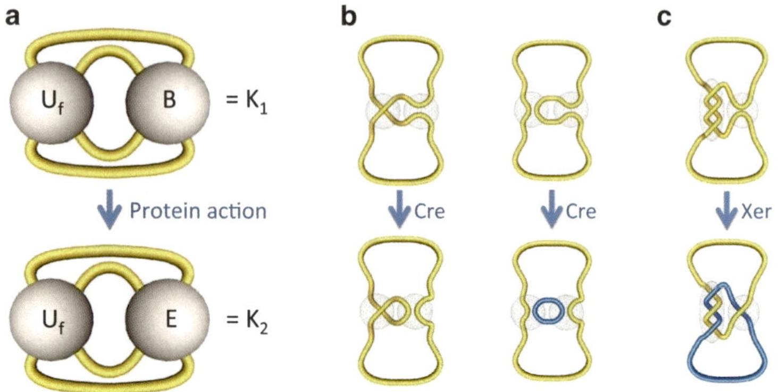

Fig. 12 (a) Protein action is represented by the tangle equations $N(U_f + B) = K_1$, $N(U_f + E) = K_2$. (b) A tangle model for Cre recombination. (c) Tangle model for Xer acting on *psi* sites

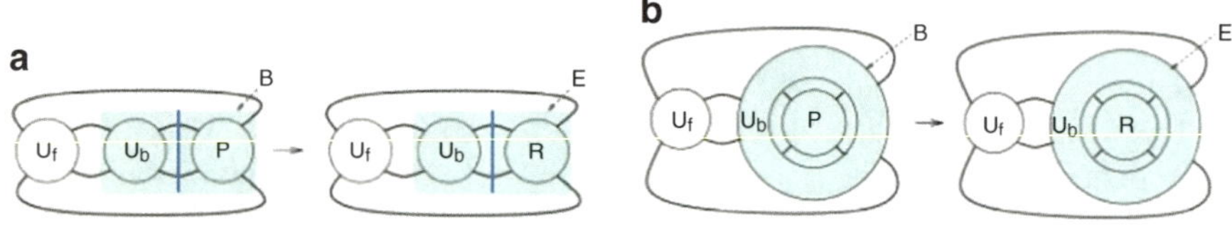

Fig. 13 (a) The Ernst and Sumners model of protein action is represented by the tangle equations $N(U_f + U_b + P) = K_1$, $N(U_f + U_b + R) = K_2$. (b) A more general tangle model

replacing the tangle P by the tangle R, as shown in Fig. 13 and modeled by the equations $N(U_f + U_b + P) = K_1$, $N(U_f + U_b + R) = K_2$. However, this model assumes that the local protein action can be separated from the remaining protein-bound DNA by a disk that intersects the strings of tangle B in exactly two points. However, this eliminates potential biologically relevant models. A more general tangle model is shown in Fig. 13b [23]. Note that the tangle B is transformed into the tangle E via replacing the subtangle P with the subtangle R. For more on rational subtangle replacement, see [9].

5   Concluding Remarks

We have only touched the surface regarding the topological modeling of protein–DNA complexes. In addition to modeling the action of proteins that can knot circular DNA, tangles can also be applied to probe the structure of multiple DNA segments in any stable protein–DNA complex regardless of the action (or inaction) of the protein via an experimental technique called difference topology [2, 26, 28, 40, 45, 52–57, 63–65]. This technique uses a recombinase or topoisomerase to trap crossings bound by the protein under study. This requires knowledge regarding how the recombinase or topoisomerase acts. However, there are still unsolved problems regarding how these proteins act. With respect to recombination, we can only determine all solutions to the tangle equations modeling these reactions in special cases, even when the substrate and product are both rational knots/links (e.g. [10,24,29,47]). Although we can solve all tangle equations modeling topoisomerase action when the substrate and product are rational knots/links [24], there are still questions regarding preferred pathways [11,44,60,69,75,86].

Acknowledgements This research was supported in part by a grant from the joint Division of Mathematical Sciences (DMS) and National Institute of General Medical Sciences (NIGMS) Initiative to Support Research in the Area of Mathematical Biology (grant number NSF 0800285 to I.K.D. and NSF DMS-0800929 to S.D.L). We wish to thank Massa Shoura for providing Figs. 1b, 8, and 9.

References

1. C. Adams, *The Knot Book: An Elementary Introduction to the Mathematical Theory of Knots* (American Mathematical Society, Providence, RI, 2004)
2. A. Akopian, S. Gourlay, H. James, S.D. Colloms, Communication between accessory factors and the Cre recombinase at hybrid psi-loxP sites. J. Mol. Biol. **357**(5), 1394–1408 (2006)
3. P. Argos, A. Landy, K. Abremski, J.B. Egan, E. Haggard-Ljungquist, R.H. Hoess, M.L. Kahn, B. Kalionis, S.V.L. Narayana, L.S. Pierson, N. Sternberg, J.M. Leong, The integrase family of site-specific recombinases: regional similarities and global diversity. EMBO J. **5**(2), 433–440 (1986)
4. A.D. Bates, A. Maxwell, *DNA Topology*. In Focus (IRL Press at Oxford University Press, Oxford/New York, 1993)
5. M. Better, C. Lu, R.C. Williams, H. Echols, Site-specific DNA condensation and pairing mediated by the int protein of bacteriophage lambda. Proc. Natl. Acad. Sci. U S A **79**(19), 5837–5841 (1982)
6. T. Biswas, H. Aihara, M. Radman-Livaja, D. Filman, A. Landy, T. Ellenberger, A structural basis for allosteric control of DNA recombination by lambda integrase. Nature **435**(7045), 1059–1066 (2005)
7. G.W. Blakely, A.O. Davidson, D.J. Sherratt, Sequential strand exchange by XerC and XerD during site-specific recombination at dif. J. Biol. Chem. **275**(14), 9930–9936 (2000)
8. M. Bregu, D.J. Sherratt, S.D. Colloms, Accessory factors determine the order of strand exchange in Xer recombination at psi. EMBO J. **21**(14), 3888–3897 (2002)
9. D. Buck, K. Baker, The classification of rational subtangle replacements between rational tangles. Algebra. Geome Topol. **13**, 1413–1463 (2013)
10. D. Buck, M. Mauricio, Connect sum of lens spaces surgeries: application to Hin recombination. Math. Proc. Camb. Philos. Soc. **150**(3), 505–525 (2011)
11. G.R. Buck, E.L. Zechiedrich, DNA disentangling by type-2 topoisomerases. J. Mol. Biol. **340**(5), 933–939 (2004)
12. C. Calladine, H. Drew, B. Luisi, A. Travers, *Understanding DNA: The Molecule and How it Works* (Elsevier, San Diego, CA, 2004)
13. S.M. Chen, F. Heffron, W.J. Chazin, Two-dimensional 1H NMR studies of 32-base-pair synthetic immobile Holliday junctions: complete assignments of the labile protons and identification of the base-pairing scheme. Biochemistry **32**(1), 319–326 (1993)
14. S.M. Chen, W.J. Chazin, Two-dimensional 1H NMR studies of immobile holliday junctions: nonlabile proton assignments and identification of crossover isomers. Biochemistry **33**(38), 11453–11459 (1994)

15. Y. Chen, U. Narendra, E.L. Iype, M.M. Cox, A.P. Rice, Crystal structure of a Flp recombinase-Holliday junction complex: assembly of an active oligomer by helix swapping. Mol. Cell **6**(4), 885–897 (2000)
16. R.M. Clegg, A.I. Murchie, D.M. Lilley, The solution structure of the four-way DNA junction at low-salt conditions: a fluorescence resonance energy transfer analysis. Biophys. J. **66**(1), 99–109 (1994)
17. J.H. Conway, An enumeration of knots and links, and some of their algebraic properties, in *Computational Problems in Abstract Algebra (Proceedings of a Conference Held at Oxford, 1967)* (Pergamon, Oxford, 1970), pp. 329–358
18. J.P. Cooper, P.J. Hagerman, Gel electrophoretic analysis of the geometry of a DNA four-way junction. J. Mol. Biol. **198**(4), 711–719 (1987)
19. J.P. Cooper, P.J. Hagerman, Geometry of a branched DNA structure in solution. Proc. Natl. Acad. Sci. U S A **86**(19), 7336–7340 (1989)
20. J.P. Cooper, P.J. Hagerman, Analysis of fluorescence energy transfer in duplex and branched DNA molecules. Biochemistry **29**(39), 9261–9268 (1990)
21. N.J. Crisona, R.L. Weinberg, B.J. Peter, D.W. Sumners, N.R. Cozzarelli, The topological mechanism of phage lambda integrase. J. Mol. Biol. **289**(4), 747–775 (1999)
22. M. Culler, C.M. Gordon, J. Luecke, P.B. Shalen, Dehn surgery on knots. Bull. Am. Math. Soc. (N.S.) **13**(1), 43–45 (1985)
23. I.K. Darcy, Biological distances on DNA knots and links: applications to Xer recombination. J. Knot Theory Ramif. **10**(2), 269–294 (2001). Knots in Hellas '98, Vol. 2 (Delphi)
24. I.K. Darcy, Solving unoriented tangle equations involving 4-plats. J. Knot Theory Ramifi. **14**(8), 993–1005 (2005)
25. I.K. Darcy, R.G. Scharein, TopoICE-R: 3D visualization modeling the topology of DNA recombination. Bioinformatics **22**(14), 1790–1791 (2006)
26. I.K. Darcy, A. Bhutra, J. Chang, N. Druivenga, C. McKinney, R.K. Medikonduri, S. Mills, J. Navarra Madsen, A. Ponnusamy, J. Sweet, T. Thompson, Coloring the Mu transpososome. BMC Bioinform. **7**, 435 (2006)
27. I.K. Darcy, R.G. Scharein, A. Stasiak, 3D visualization software to analyze topological outcomes of topoisomerase reactions. Nucleic Acids Res. **36**(11), 3515–3521 (2008)
28. I.K. Darcy, J. Luecke, M. Vazquez, Tangle analysis of difference topology experiments: applications to a Mu protein-DNA complex. Algebra. Geome Topol. **9**(4), 2247–2309 (2009)
29. I.K. Darcy, K. Ishihara, R.K. Medikonduri, K. Shimokawa, Rational tangle surgery and Xer recombination on catenanes. Algebra. Geome Topol. **12**(2), 1183–1210 (2012)
30. F.B. Dean, A. Stasiak, T. Koller, N.R. Cozzarelli, Duplex DNA knots produced by Escherichia coli topoisomerase I. Structure and requirements for formation. J. Biol. Chem. **260**(8), 4975–4983 (1985)
31. B.F. Eichman, M. Ortiz-Lombardia, J. Aymami, M. Coll, P.S. Ho, The inherent properties of DNA four-way junctions: comparing the crystal structures of Holliday junctions. J. Mol. Biol. **320**(5), 1037–1051 (2002)
32. C. Ernst, Tangle equations. J. Knot Theory Ramifi. **5**(2), 145–159 (1996)
33. C. Ernst, Tangle equations. II. J. Knot Theory Ramifi. **6**(1), 1–11 (1997)
34. C. Ernst, D.W. Sumners, A calculus for rational tangles: applications to DNA recombination. Math. Proc. Camb. Philos. Soc. **108**(3), 489–515 (1990)
35. R.E. Franklin, R.G. Gosling, Evidence for 2-chain helix in crystalline structure of sodium deoxyribonucleate. Nature **172**(4369), 156–157 (1953)
36. R.E. Franklin, R.G. Gosling, Molecular configuration in sodium thymonucleate. Nature **171**(4356), 740–741 (1953)
37. M. Gellert, H. Nash, Communication between segments of DNA during site-specific recombination. Nature **325**(6103), 401–404 (1987)
38. K. Ghosh, G.D. Van Duyne, Cre-loxP biochemistry. Methods **28**(3), 374–383 (2002)
39. D.N. Gopaul, F. Guo, G.D. Van Duyne, Structure of the Holliday junction intermediate in Cre-loxP site-specific recombination. EMBO J. **17**(14), 4175–4187 (1998)

40. S.C. Gourlay, S.D. Colloms, Control of Cre recombination by regulatory elements from Xer recombination systems. Mol. Microbiol. **52**(1), 53–65 (2004)
41. I. Grainge, M. Bregu, M. Vazquez, V. Sivanathan, S.C. Ip, D.J. Sherratt, Unlinking chromosome catenanes in vivo by site-specific recombination. EMBO J. **26**(19), 4228–4238 (2007)
42. N.D. Grindley, K.L. Whiteson, P.A. Rice, Mechanisms of site-specific recombination. Ann. Rev. Biochem. **75**, 567–605 (2006)
43. F. Guo, D.N. Gopaul, G.D. van Duyne, Structure of Cre recombinase complexed with DNA in a site-specific recombination synapse. Nature **389**(6646), 40–46 (1997)
44. A.H. Hardin, S.K. Sarkar, Y. Seol, G.F. Liou, N. Osheroff, K.C. Neuman, Direct measurement of DNA bending by type IIA topoisomerases: implications for non-equilibrium topology simplification. Nucleic Acids Res. **39**(13), 5729–5743 (2011)
45. R.M. Harshey, M. Jayaram, The mu transpososome through a topological lens. Crit. Rev. Biochem. Mol. Biol. **41**(6), 387–405 (2006)
46. W.D. Heyer, Biochemistry of eukaryotic homologous recombination, in *Molecular Genetics of Recombination*, ed. by A. Aguilera, R. Rothstein (Springer, Heidelberg, 2007), pp. 95–133
47. M. Hirasawa, K. Shimokawa, Dehn surgeries on strongly invertible knots which yield lens spaces. Proc. Am. Math. Soc. **128**(11), 3445–3451 (2000)
48. R. Holliday, A mechanism for gene conversion in fungi. Genet. Res. Camb. **5**, 282–304 (1964)
49. K.E. Huffman, S.D. Levene, DNA-sequence asymmetry directs the alignment of recombination sites in the Flp synaptic complex. J. Mol. Biol. **286**, 1–13 (1999)
50. A. Hughes, Primary DNA molecular structure (2005). Connexions web site, Available at: http://cnx.org/content/m11411/1.8/
51. N.R. Kallenbach, R.I. Ma, N.C. Seeman, An immobile nucleic acid junction constructed from oligonucleotides. Nature **305**, 829–831 (1983)
52. E. Kilbride, M.R. Boocock, W.M. Stark, Topological selectivity of a hybrid site-specific recombination system with elements from Tn3 res/resolvase and bacteriophage P1 loxP/Cre. J. Mol. Biol. **289**(5), 1219–1230 (1999)
53. E.A. Kilbride, M.E. Burke, M.R. Boocock, W.M. Stark, Determinants of product topology in a hybrid Cre-Tn3 resolvase site-specific recombination system. J. Mol. Biol. **355**(2), 185–195 (2006)
54. S. Kim, A 4-string tangle analysis of DNA-protein complexes based on difference topology. ProQuest LLC, Ann Arbor, 2009. Ph.D. thesis, The University of Iowa
55. S. Kim, A generalized 4-string solution tangle of DNA-protein complexes. J. Korean Soc. Ind. Appl. Math. **15**(3), 161–175 (2011)
56. S. Kim, I.K. Darcy, Topological analysis of DNA-protein complexes, in *Mathematics of DNA Structure, Function and Interactions*. IMA Volumes in Mathematics and its Applications, vol. 150 (Springer, New York, 2009), pp. 177–194
57. K. Kimura, V.V. Rybenkov, N.J. Crisona, T. Hirano, N.R. Cozzarelli, 13S condensin actively reconfigures DNA by introducing global positive writhe: implications for chromosome condensation. Cell **98**(2), 239–248 (1999)
58. S.C. Kowalczykowski, D.A. Dixon, A.K. Eggleston, S.D. Lauder, W.M. Rehrauer, Biochemistry of homologous recombination in Escherichia coli. Microbiol. Rev. **58**(3), 401–465 (1994)
59. C. Lesterlin, C. Pages, N. Dubarry, S. Dasgupta, F. Cornet, Asymmetry of chromosome Replichores renders the DNA translocase activity of FtsK essential for cell division and cell shape maintenance in Escherichia coli. PLoS Genet. **4**(12), e1000288 (2008)
60. Z. Liu, L. Zechiedrich, H.S. Chan, Action at hooked or twisted hooked DNA juxtapositions rationalizes unlinking preference of type-2 topoisomerases. J. Mol. Biol. **400**(5), 963–982 (2010)
61. P.K. Mandal, S. Venkadesh, N. Gautham, Structure of d(cgggtacccg)4 as a four-way Holliday junction. Acta. Crystallogr. Sect. F Struct. Biol. Cryst. Commun. **67**(Pt 12), 1506–1510 (2011)
62. M. Ortiz-Lombardia, A. Gonzalez, R. Eritja, J. Aymami, F. Azorin, M. Coll, Crystal structure of a DNA holliday junction. Nat. Struct. Biol. **6**(10), 913–917 (1999)
63. S. Pathania, M. Jayaram, R. Harshey, Path of DNA within the Mu transpososome: transposase interaction bridging two Mu ends and the enhancer trap five DNA supercoils. Cell **109**, 425–436 (2002)

64. Z.M. Petrushenko, C.H. Lai, R. Rai, V.V. Rybenkov, DNA reshaping by MukB. Right-handed knotting, left-handed supercoiling. J. Biol. Chem. **281**(8), 4606–4615 (2006)
65. C. Price, A biological application for the oriented skein relation. ProQuest LLC, Ann Arbor, 2012. Ph.D. thesis, The University of Iowa
66. G.D. Recchia, M. Aroyo, D. Wolf, G. Blakely, D. Sherratt, FtsK-dependent and -independent pathways of Xer site-specific recombination. EMBO J. **18**(20), 5724–5734 (1999)
67. E. Richet, P. Abcarian, H.A. Nash, Synapsis of attachment sites during lambda integrative recombination involves capture of a naked DNA by a protein-DNA complex. Cell **52**(1), 9–17 (1988)
68. E.D. Ross, R.B. Den, P.R. Hardwidge, L.J. Marter, Improved quantitation of DNA curvature using ligation ladders. Nucleic Acids Res. **27**(21), 4135–4142 (1999)
69. V.V. Rybenkov, C. Ullsperger, A.V. Vologodskii, N.R. Cozzarelli, Simplification of DNA topology below equilibrium values by type II topoisomerases. Science **277**(5326), 690–693 (1997)
70. Y. Saka, M. Vazquez, Tanglesolve: topological analysis of site-specific recombination. Bioinformatics **18**(7), 1011–1012 (2002)
71. G.J. Sarkis, L.L. Murley, A.E. Leschziner, M.R. Boocock, W.M. Stark, N.D. Grindley, A model for the gamma delta resolvase synaptic complex. Mol. Cell **8**(3), 623–631 (2001)
72. R.G. Scharein, Interactive topological drawing. ProQuest LLC, Ann Arbor, 1998. Ph.D. thesis, The University of British Columbia
73. A.M. Segall, H.A. Nash, Architectural flexibility in lambda site-specific recombination: three alternate conformations channel the attL site into three distinct pathways. Genes Cells **1**(5), 453–463 (1996)
74. H. Senechal, J. Delesques, G. Szatmari, Escherichia coli ArgR mutants defective in cer/Xer recombination, but not in DNA binding. FEMS Microbiol. Lett. **305**(2), 162–169 (2010)
75. Y. Seol, A.H. Hardin, M.P. Strub, G. Charvin, K.C. Neuman, Comparison of DNA decatenation by Escherichia coli topoisomerase IV and topoisomerase III: implications for non-equilibrium topology simplification. Nucleic Acids Res. **41**, 4640–4649 (2013)
76. K. Shimokawa, K. Ishihara, M. Vazquez, Tangle analysis of DNA unlinking by the Xer/FtsK system. Bussei Kenkyu **92**, 89–92 (2009)
77. L.S. Shlyakhtenko, V.N. Potaman, R.R. Sinden, Y.L. Lyubchenko, Structure and dynamics of supercoil-stabilized DNA cruciforms. J. Mol. Biol. **280**(1), 61–72 (1998)
78. C. Sissi, M. Palumbo, In front of and behind the replication fork: bacterial type iia topoisomerases. Cell Mol. Life Sci. **67**(12), 2001–2024 (2010)
79. W.M. Stark, D.J. Sherratt, M.R. Boocock, Site-specific recombination by Tn3 resolvase: topological changes in the forward and reverse reactions. Cell **58**(4), 779–790 (1989)
80. D.W. Sumners, C. Ernst, S.J. Spengler, N.R. Cozzarelli, Analysis of the mechanism of DNA recombination using tangles. Q. Rev. Biophys. **28**(3), 253–313 (1995)
81. P. Tait, On knots. Trans. R. Soc. Edinb. **28**, 145–190 (1877)
82. M. Vazquez, D.W. Sumners, Tangle analysis of Gin site-specific recombination. Math. Proc. Camb. Philos. Soc. **136**(3), 565–582 (2004)
83. M. Vazquez, S.D. Colloms, D.W. Sumners, Tangle analysis of Xer recombination reveals only three solutions, all consistent with a single three-dimensional topological pathway. J. Mol. Biol. **346**(2), 493–504 (2005)
84. A.A. Vetcher, A.Y. Lushnikov, J. Navarra-Madsen, R.G. Scharein, Y.L. Lyubchenko, I.K. Darcy, S.D. Levenc, DNA topology and geometry in Flp and Cre recombination. J. Mol. Biol. **357**(4), 1089–1104 (2006)
85. A.A. Vetcher, A.E. McEwen, R. Abujarour, A. Hanke, S.D. Levene, Gel mobilities of linking-number topoisomers and their dependence on DNA helical repeat and elasticity. Biophys. Chem. **148**(1–3), 104–111 (2010)
86. A. Vologodskii, Theoretical models of DNA topology simplification by type IIA DNA topoisomerases. Nucleic Acids Res. **37**(10), 3125–3133 (2009)
87. J.C. Wang, Cellular roles of DNA topoisomerases: a molecular perspective. Nat. Rev. Mol. Cell Biol. **3**(6), 430–440 (2002)

88. J.C. Wang, *Untangling the Double Helix: DNA Entanglement and the Action of the DNA Topoisomerases* (Cold Spring Harbor Laboratory Press, Cold Spring Harbor, NY, 2009)
89. S.A. Wasserman, N.R. Cozzarelli, Supercoiled DNA-directed knotting by T4 topoisomerase. J. Biol. Chem. **266**(30), 20567–20573 (1991)
90. J.D. Watson, F.H. Crick, Molecular structure of nucleic acids; a structure for deoxyribose nucleic acid. Nature **171**(4356), 737–738 (1953)
91. G. Witz, A. Stasiak, DNA supercoiling and its role in DNA decatenation and unknotting. Nucleic Acids Res. **38**(7), 2119–2133 (2010)
92. C. Wyman, R. Kanaar, Homologous recombination: down to the wire. Curr. Biol. **14**(15), R629–R631 (2004)
93. C. Wyman, D. Ristic, R. Kanaar, Homologous recombination-mediated double-strand break repair. DNA Repair **3**(8–9), 827–833 (2004)
94. K. Yamada, M. Ariyoshi, K. Morikawa, Three-dimensional structural views of branch migration and resolution in DNA homologous recombination. Curr. Opin. Struct. Biol. **14**(2), 130–137 (2004)
95. W. Zheng, C. Galloy, B. Hallet, M. Vazquez, The tangle model for site-specific recombination: a computer interface and the TnpI-IRS recombination system, in *Knot Theory for Scientific Objects-Proceedings of the International Workshop on Knot Theory for Scientific Objects held in Osaka (Japan), March 8–10, 2006*, vol 1 ed. by Akio Kawauchi. OCAMI Studies (Osaka Municipal universities Press, Sakai, 2007), pp. 251–271, http://www.omup.jp/modules/tinyd1/index.php?id=25&easiestml_lang=en

Reactions Mediated by Topoisomerases and Other Enzymes: Modelling Localised DNA Transformations

Dorothy Buck

Abstract Many proteins cleave and reseal DNA molecules in precisely orchestrated ways. Modelling these reactions has often relied on the axis of the DNA double helix being circular, so these cut-and-seal mechanisms can be tracked by corresponding changes in the knot type of the DNA axis. However, when the DNA molecule is linear, or the protein action does not manifest itself as a change in knot type, or the knot types are not 4-plats, these knot-theoretic models are less germane. We thus give a taxonomy of local DNA axis configurations. More precisely, we characterise all rational tangles obtained from a given rational tangle via a rational-subtangle replacement. This classification is then endowed biologically with a distance that determines how many enzyme-mediated reactions of a particular type are needed to proceed from one local DNA conformation to another, or indeed if it is even possible. We conclude by discussing a variety of biological applications of this categorisation, including reactions mediated by type II topoisomerase, site-specific recombinase and transposase.

1 Introduction

The two strands of the DNA double helix wrap around an imaginary axis. How this axis is contorted in space (topologically and geometrically) affects many cellular processes, including replication, recombination and transcription. This chapter considers the topological and geometric conformations of this axis in space. We model the localised transformations mediated by type II topoisomerases, as well as recombinases, transposases and other enzymes, that make transient breaks in DNA, followed by rearrangement and ligation.

D. Buck (✉)
Department of Mathematics, Imperial College London, London, UK
e-mail: d.buck@imperial.ac.uk

N. Jonoska and M. Saito (eds.), *Discrete and Topological Models in Molecular Biology*, 347
Natural Computing Series, DOI 10.1007/978-3-642-40193-0_16,
© Springer-Verlag Berlin Heidelberg 2014

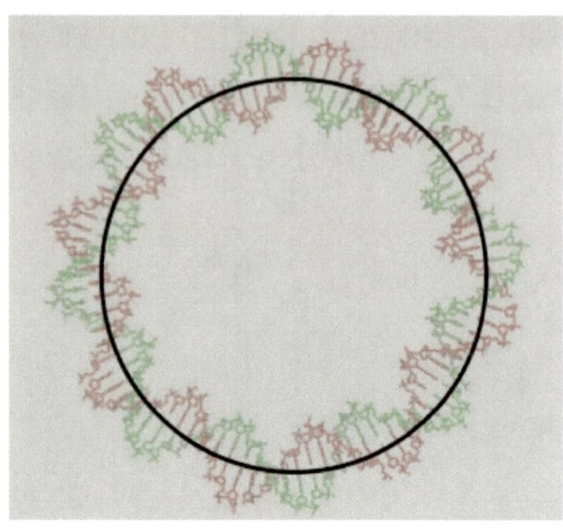

Fig. 1 The axis of DNA can be circular

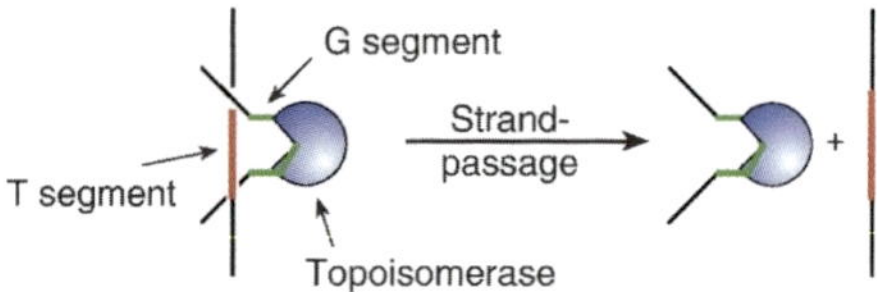

Fig. 2 Type II topoisomerases unlink DNA molecules by performing local crossing changes

The axis of the DNA molecule can exhibit many non-trivial conformations. For example, mitochondrial DNA, chloroplast DNA, some viral DNA and bacterial genomic DNA are all circular (see Fig. 1). In the laboratory, experiments are typically conducted with plasmid DNA: small circular molecules most commonly of around 4,000 bp (although they range from 1 to 10 kbp in length). Although human genomic DNA is not circular, in cells DNA wraps around a core histone to form a nucleosome fibre, which undergoes many further folding process to form higher-order, large chromatin fibre loops. As each loop is attached to a protein scaffolding, the DNA inside each loop can be viewed as topologically constrained similarly to covalently closed circular DNA.

Moreover, this circular DNA can become knotted or linked as a result of a variety of cellular processes. For example, in *E. coli*, newly replicated daughter DNA molecules are non-trivially linked, with the linking number proportional to the number of base pairs of the original circular DNA molecule. The unlinking of these daughter molecules is one of the first steps in proliferation, and is mediated by type II topoisomerases [24, 30, 33, 34] (see Fig. 2).

1.1 Localised DNA Transformations

Most proteins, such as the topoisomerases mentioned above, that change the DNA topology or geometry act locally – that is, the binding, rearrangement and

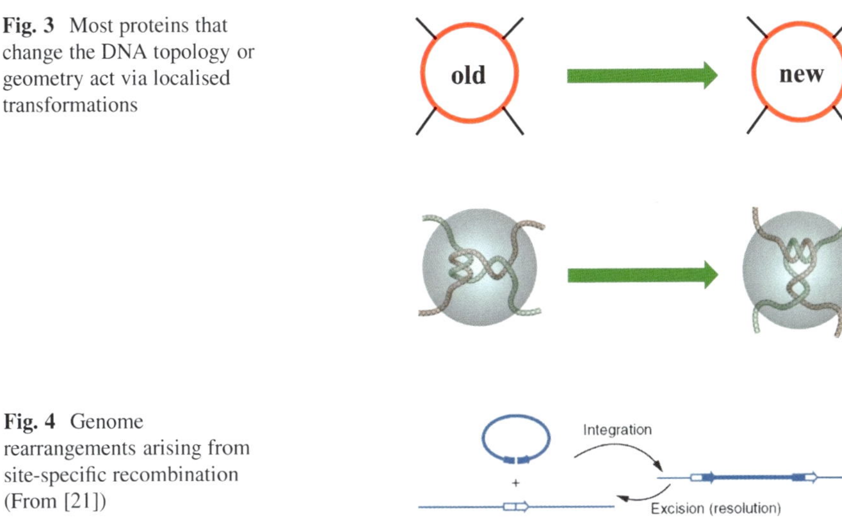

Fig. 3 Most proteins that change the DNA topology or geometry act via localised transformations

Fig. 4 Genome rearrangements arising from site-specific recombination (From [21])

subsequent ligation all occur within a small, well-defined spatial region. Type II topoisomerases bind to a segment of double-stranded DNA, *the gate segment*, which is then transiently cleaved. Subsequently, a second DNA segment, the *transport segment*, is passed through the 'open' gate segment, and the gate segment is then ligated. (The crossing change effected then changes the linking number of the underlying molecule by 2, hence the terminology of 'type II topoisomerases'.) As this crossing change occurs within a restricted region, leaving the rest of the DNA molecule unchanged, the unknotting/unlinking reactions mediated by type II topoisomerases are examples of localised DNA transformations (see Fig. 3).

Site-specific recombination is the reshuffling of the DNA sequence, and is mediated by a protein, a site-specific recombinase [21]. Site-specific recombinases fall into two families, the serine recombinases and the tyrosine recombinases, classified according to their active nucleophile.

The result of site-specific recombination can be the excision, insertion or inversion of a sequence, as shown in Fig. 4. In the simplest case, as with some serine recombinases, a tetramer of recombinases binds to two short identical DNA sequences (the crossover sites), makes double-stranded breaks in both, interchanges the ends via subunit rotation and then ligates before releasing. Most tyrosine recombinases perform this in two steps, where they cleave single backbones, interchange them and then repeat before ligating (see Fig. 5). In some cases, such as recombination mediated by tyrosine recombinases (defined below) which includes branch migration within an intermediary Holliday junction, there are larger-scale effects, yet even here, the recombinases form phosphotyrosol intermediates at well-defined short DNA sequences.

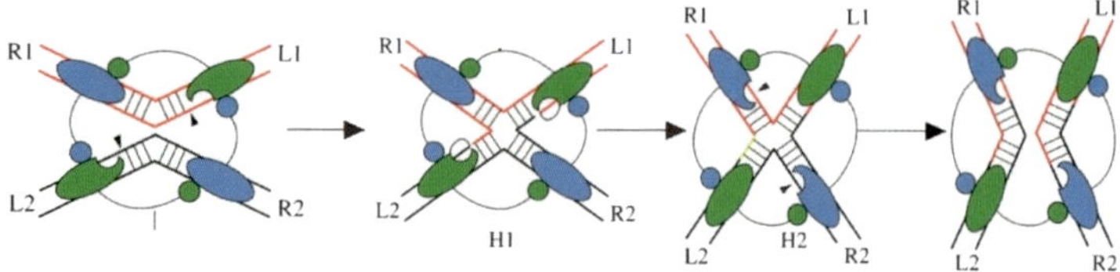

Fig. 5 Steps of site-specific recombination mediated by a tyrosine recombinase. The enzymes bind, cleave and ligate short, defined DNA sequences. The intermediary stage (images 2 and 3) is a Holliday junction

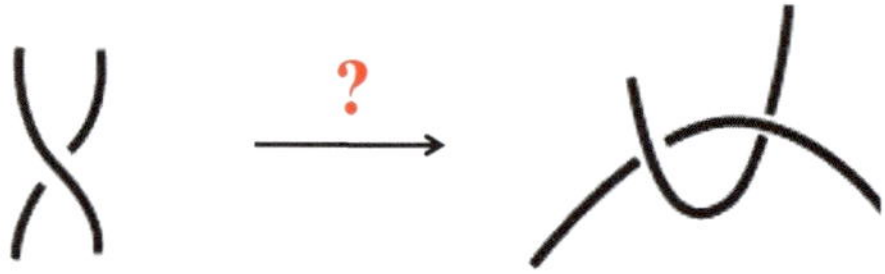

Fig. 6 Subquestion 1: What is the enzyme choreography? The arc represents the axis of the double helix

2   Guiding Questions

In this arena, the motivating question is to understand how these unknotting/ unlinking or recombination reactions proceed. Ideally, a dynamic, all-atom understanding of these localised transformations would be available. However, despite studies of the crystal structures of various recombinase–DNA co-complexes and single-molecule experiments (as well as numerous simulations, including Metropolis Monte Carlo and molecular dynamics techniques), this is still unfortunately a distant fantasy. So the main issue is how to unveil the salient features of this process.

This breaks down into two subquestions about these protein-mediated reactions, the first querying the verb (the mechanism) and the second a noun (the initial or final DNA conformation). First, what is the enzyme mechanism or choreography? (See Fig. 6.) That is, given the substrate and product conformations, what is the process that takes the former to the latter? We are primarily concerned with the movement of the DNA segments during this process, rather than, for example, the exact biochemical bonds.

An example of this is the question of what happens during recombination mediated by the recombinase PhiC31. This is a member of the important but poorly understood family of large serine recombinases. Initially there were two competing models for how recombination proceeded: either via a *subunit rotation mechanism* (that is, the two sites are simultaneously cleaved, one cleaved side is then rotated right-handed by 180° relative to the other and then the DNA is resealed) or a *domain-swapping mechanism*. A combination of biochemical experiments, including yoking PhiC31 to Tn3 resolvase, a well-understood recombinase, together with a topological analysis of the reaction using coherent band surgery, enabled Buck and colleagues to showed that PhiC31 acts via the subunit rotation mechanism [28] (Fig. 7).

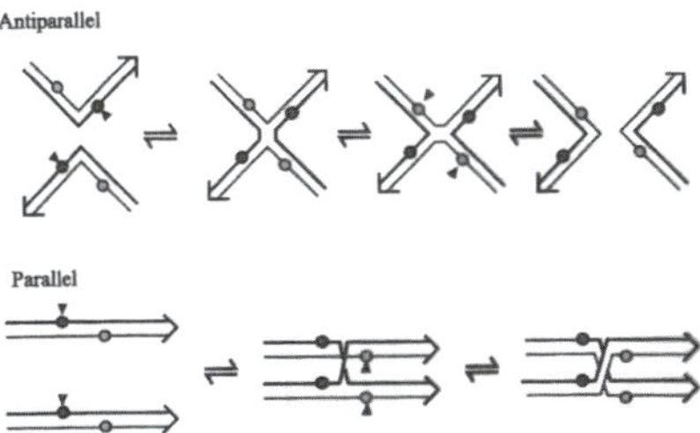

Fig. 7 Are the sites aligned in an anti-parallel or parallel orientation during site-specific recombination mediated by the Flp recombinase? Note in particular the difference in the Holliday junctions. The small dots denote the four recombinase proteins, and the base pairs are omitted for clarity

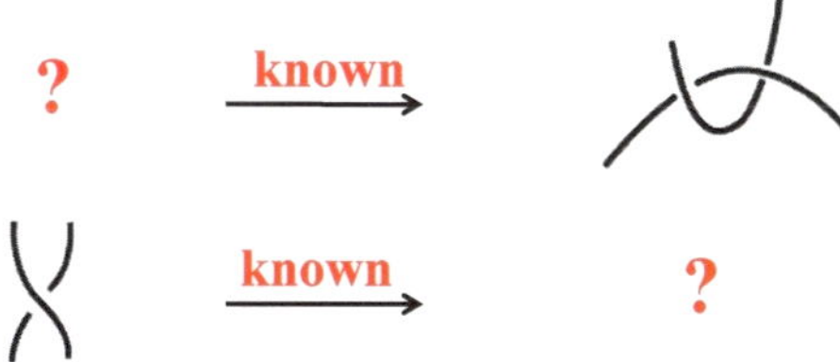

Fig. 8 Subquestion 2: what is the structure of the pre or post-recombinant local DNA conformations?

The second subquestion assumes knowledge of this enzymatic pathway and either the initial or the final DNA conformation, and queries what is the other DNA conformation (see Fig. 8).

One setting for this is during site-specific recombination when the initial DNA conformation, the *synaptic complex*, is understood, for example from a crystal structure, and the recombinase mechanism is well understood. Then the second subquestion asks if we can predict the final DNA conformation, the *post-recombinant complex*. The resolvase subfamily of the serine recombinases has been shown to utilise a subunit rotation mechanism, and there are crystal structures of the synaptic complex for several members (see, e.g., [21]). We then hope to predict, or at least constrain the possibilities for, the post-recombinant complex.

3 Earlier Treatments

Previous work has been able to answer these questions *if* this localised action yields a change in DNA knot type.

One particularly fruitful avenue has been topological modelling of the action of these proteins. This field was initiated by the Ernst–Sumners tangle model, which helped determine both structural and mechanistic information about particular site-specific recombination systems [17]. (Tangles are defined below.) In this model, the DNA is represented as a sum of two two-strand tangles where the recombination replaces one of the summands with another tangle. In the original applications, the substrate and product DNA were each taken to be two-bridge knots or links. Then Ernst and Sumners proved that certain summands were rational, and so could apply the cyclic surgery theorem [14] to their corresponding double branched covers (as two-bridge knots and links lift to lens spaces and rational tangles lift to solid tori) to determine the exact tangles.

Modified versions of the original tangle model have been helpful in understanding several DNA–protein interactions. For example, site-specific recombination mediated by the Flp recombinase, a member of the tyrosine family of recombinases, yields a spectrum of $(2, p)$-torus knots or links, depending on the orientation of the initial sites. Using a combination of biochemical experiments, including utilising a mutant version of Flp which binds but does not cleave, together with new topological proofs in a three-summand setting of the tangle model, Buck, Grainge and Jayaram showed that the recombination sites must be oriented in an anti-parallel alignment [19].

There have been several other useful developments and generalisations of this model in the 20 years since its introduction (see [4] and references therein, as well as [10, 11]). More recently, Cabrera-Ibarra and Lizárraga-Navarro developed a model that allows potential tangles to be three-strand braids [12, 13], and Buck and Mauricio developed a model that allows substrates and products to include composite knots [8].

In another vein, one can also make reasonable assumptions about the enzyme mechanism to *predict* the knots and links arising from site-specific recombination. In some cases, this has been experimentally possible, by using electron microscoy or atomic force microscopy to determine the precise knot or link. However, these techniques are limited both by the difficulty involved in resolving the crossings and the paucity of experimentalists with the necessary training. An alternate experimental method has been to perform gel electrophoresis, which will resolve knots by the minimal crossing number. Unfortunately, there are many knots which share the same minimal crossing number, so alternative methods are needed.

A topological model has thus been successfully developed for unknot, unlink and torus knot/link substrates [3, 7] and twist knot/link substrates [9, 32]. This model, under very mild assumptions, can dramatically restrict the putative knot/link products. For example, it predicts that recombination mediated by a serine recombinase on an unlink substrate can only yield an unknot, an unlink or a Hopf link.

4   The Limitations

This range of topological models has dramatically increased our understanding of localised DNA transformations in a variety of settings. However, they thus far have all relied on one basic assumption: the localised DNA transformation effects a corresponding global change in the topology of the DNA. (For example, it is assumed that the type II topoisomerase-mediated crossing change transforms a DNA trefoil into an unknot, rather than the crossing change being of a trivial Reidemeister type I removal of writhe.)

However, this is not always the case. For example, the topology of the DNA axis may not change (for example, the most common product of recombination of an unknot mediated by a tyrosine recombinase is another unknot). Or the resulting DNA product may be a topology type not yet covered by the model (e.g. an algebraic, non-Montesinos knot). Or, most dramatically, the DNA axis may be linear (and unconstrained) throughout the reaction. (Note: although the study of open knots, partly spurred on by the recent discovery of 'knotted' proteins, is an exciting and emerging area, it has yet to be considered in this context, as most of the time the linear DNA is not knotted in this 'open' sense.)

5   The Solution: Think Locally

To overcome these obstacles, we model these localised DNA transformations exclusively locally. That is, we model the reaction without placing restrictions on the global topology of the DNA axis. This enables us to consider a wider variety of reactions, including those where the cleavage/ligation occurs within a subregion of a larger complex. Perhaps unsurprisingly though, the topological methods used to model these are different as well.

One similarity with the tangle model described above is the sensible topological description of localised DNA conformations. As visualised by electron microscopy and atomic force microscopy in situ (at physiological conditions), DNA in vivo and in vitro is (negatively) plectonemically supercoiled and, typically, branched. That is, the DNA axis naturally forms rows of twists, broken by branch points, where additional rows of twists can emanate in another direction. Thus, as with the Ernst–Sumners method above, a natural model for two nearby segments of DNA is a (two-string) rational tangle.

A *tangle* is an ordered pair (B^3, t), where B^3 is a 3-ball and t is a pair of properly embedded arcs, i.e. two arcs whose four endpoints are on the boundary of B^3, typically parameterised as NE, SE, SW, NW, considered up to strong equivalence. A *rational* tangle is composed of alternating horizontal and vertical rows of twists, and via a continued fraction expansion can be written as a rational number (or ∞). Each rational tangle is endowed with a meridional disc whose boundary is on the boundary of the 3-ball and which separates the two tangle arcs. Additionally, each tangle is endowed with a core arc that joins the midpoints of the two tangle segments, and thus the tangle can be thought of as the neighbourhood of this core arc. (See Fig. 9, and [4] for a more detailed introduction to rational tangles.)

We then model the enzymatic reaction as a *rational tangle replacement*: the removal of one rational tangle from a 3-manifold with a properly embedded 1-manifold and replacement with another, leaving the rest of the (linear or circular)

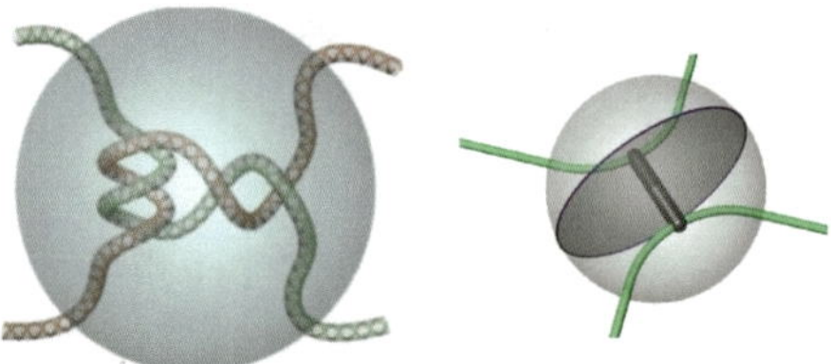

Fig. 9 *Left:* Two segments of (double-stranded) DNA, one *red* and one *green*, are modelled as a rational tangle. *Right:* A rational tangle, its meridional disc and the core arc (Figure courtesy of K. Baker)

DNA molecule(s) unchanged. For example, the crossing change initiated by a type II topoisomerase is modelled as replacing a (-1)-tangle with a $(+1)$-tangle (or vice versa). While this is a very simple replacement, we can consider more intricate replacements as well. One measure of this intricacy is via the *distance* Δ between two rational tangles, defined as half the number of intersections of the boundaries of their respective meridional discs. In this notation, the distance between the type II topoisomerase-mediated crossing change is 2. (Note, however, that this 'distance' is not a metric; it parallels the notion of distance between Dehn surgeries. See [5] for more details of this correspondence.)

We emphasise that in this model there are many options for the ambient space: for example, it could be S^3 containing a circular or linear segment, or itself a 3-ball with properly embedded arcs (i.e. a tangle itself). Thus we can also readily model more complicated protein–DNA systems, such as the resolvases, which utilise additional resolvase subunits to trap a fixed number of supercoils within a larger complex (itself a rational tangle) and perform the cleavage/rearrangement/ligation reaction within a smaller complex (a rational *sub*tangle).

6 Results

Within this paradigm, the two questions above about these protein-mediated reactions – what is the verb (the mechanism) and what is one of the nouns (the initial or final DNA configuration) – can be rephrased. The topological answers to these questions and the related proofs were found determined in joint work with Ken Baker [6].

The first question asks *which* tangle replacements can occur, i.e. which rational tangles are at a distance Δ from a given rational tangle by a full rational-tangle replacement.

Theorem 1 ([6, 17]). *If a (full) rational-tangle replacement of distance d takes the p/q-tangle to the u/v-tangle, then the following relation must hold: $pv-qu = \pm d$.*

This can be generalised to consider the much larger and more difficult question of which tangles are related via a rational-*sub*tangle replacement, or RSR.

Theorem 2 ([6]). *If a rational-subtangle replacement of distance d takes the p/q-tangle to the u/v-tangle, then u/v belongs to one of the following families of rational numbers (depending on p/q and d):*

$$I.\ (d \geq 1)\quad \left\{ \frac{p+\epsilon da(aq-bp)}{q+\epsilon db(aq-bp)}\ \middle|\ a, b\ coprime, \epsilon = \pm \right\},$$

$$II.\ (d = 1)\quad \left\{ \frac{p+\epsilon 4a(aq-bp)}{q+\epsilon 4b(aq-bp)}\ \middle|\ a, b\ coprime, \epsilon = \pm \right\},$$

$$III.\ (d = 1)\quad \left\{ \frac{p(b-1)(4ab-4a-2b-1)+\epsilon p'(2ab-2a-b)^2}{q(b-1)(4ab-4a-2b-1)+\epsilon q'(2ab-2a-b)^2},\ \frac{p(1-2a)^2(b-1)+\epsilon p'(2ab-2a-b)^2}{q(1-2a)^2(b-1)+\epsilon q'(2ab-2a-b)^2}\ \middle| \right.$$

$$\left. a, b, p', q'\ integers,\ pq' - p'q = 1, \epsilon = \pm \right\},$$

$$IV.\ (d = 1)\ \left\{ \frac{p(2a-1)(2ab+a-b+1)+\epsilon p'(2ab+a-b)^2}{q(2a-1)(2ab+a-b+1)+\epsilon q'(2ab+a-b)^2},\ \frac{p(2b+1)(2ab+a-b+1)+\epsilon p'(2ab+a-b)^2}{q(2b+1)(2ab+a-b+1)+\epsilon q'(2ab+a-b)^2}\ \middle| \right.$$

$$\left. a, b, p', q'\ integers,\ pq' - p'q = 1, \epsilon = \pm \right\}.$$

The answer to this *which* question is obtained by building on the work of Berge [1] and Gabai [18] (and Moser [26]) on surgeries on knots in solid tori that yield solid tori via the Montesinos technique [25].

Of course, between any two rational tangles, for a given distance, there may exist many possible subtangle replacements if one considers all possible places where these (subtangle) replacements can occur within the given tangle. The second question asks *where* these subtangle replacements can take place.

Theorem 3 ([6]). *The rational-subtangle replacements occur, up to homeomorphism, only at the core arc for the full tangle replacement or, as indicated in Fig. 10, for the subtangle replacements.*

Note that in Fig. 10, we draw the tangles in a slightly different way to show this replacement more clearly. Here, we view B^3 as the one-point compactification of the lower half-space in R^3 and place the four endpoints of the boundary arcs on the x-axis at $(i, 0, 0)$ for $i = 1, 2, 3, 4$. A rational tangle within B^3 can then be arranged so that its z-coordinates have only two local minima and, moreover, it has an open 4-plat form with projection to the xz-plane. The oblong rectangles in Fig. 10 labelled with integers indicate twist regions, where the longer direction of the rectangle gives the twist axis.

The answer to the *where* question for rational-subtangle replacements is more difficult to prove, as, unlike the *which* question above, it does not follow from the Montesinos technique of passing to the double branched cover. (As illustrated in [6], in fact two non-homeomorphic tangles may have homeomorphic branched double covers.) The proofs then have three components: generalising some results of Ernst [16], adapting some results of Paoluzzi [29] and considering tangles as hyperbolic orbifolds. The first relies upon the corresponding knot exteriors being Seifert fibred, the second addresses mutations of tangles and involutions of manifolds with non-trivial JSJ decompositions, and the last relies upon the hyperbolic orbifold surgery theorem. (See [6] for the full proofs.)

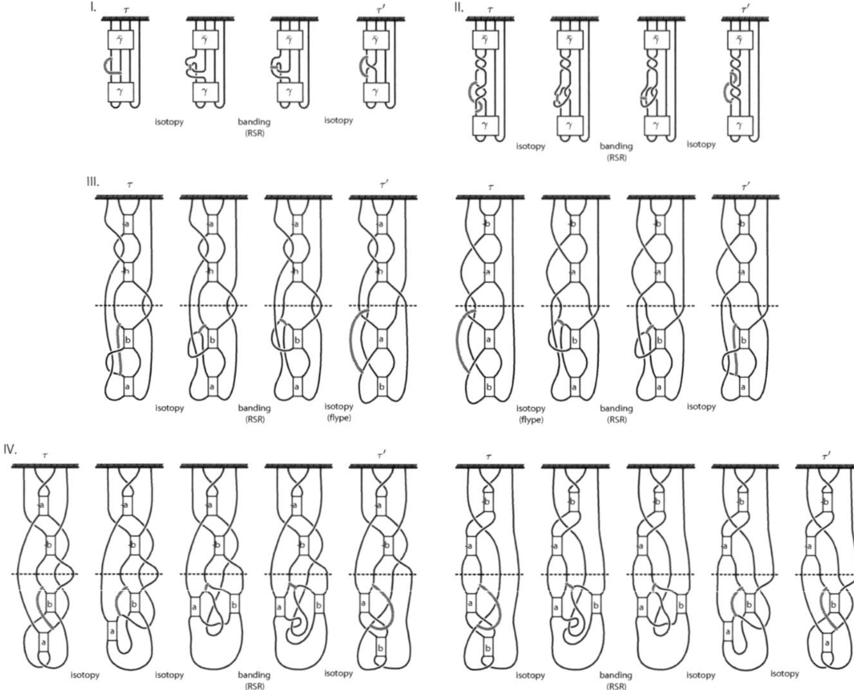

Fig. 10 All rational-subtangle replacements between the tangles τ and τ' occur, up to homeomorphism, only as indicated (From [6])

Finally, we can consider the closure of these rational tangles, which are two-bridge knots or links, and ask the same questions: *which* two-bridge knots are links are related by a rational-subtangle replacement, and *where* can these subtangle replacements occur?

Theorem 4 ([6]). *If the two-bridge link $S(p,q) = [a_1, a_2, \ldots, a_n]$ has an RSR to $S(u,v)$ with distance $d \geq 2$, then there exist integers $c_1, c_2, \ldots, c_k$ such that*

$$S(p,q) = [a_1, a_2, \ldots, a_n, 0, c_1, c_2, \ldots, c_k, 0, -c_k, \ldots, -c_2, -c_1]$$

and

$$S(u,v) = [a_1, a_2, \ldots, a_n, 0, c_1, c_2, \ldots, c_k, \pm d, -c_k, \ldots, -c_2, -c_1].$$

Up to homeomorphism, the site of the RSR is the twist region corresponding to the 0 between the c_k and the $-c_k$ in the plat associated to the continued fraction for $S(p,q)$ and to $\pm d$ in the plat associated to the continued fraction for $S(u,v)$.

To prove this, we have developed the relationship between rational tangles and their closures in the two-bridge classifications by Darcy and Sumners [15] and Torisu [31] in which two-bridge links are related by an RSR of distance at least 2. We then applied Ernst's theorem above [16] to classify where these RSRs occur. (We have also conjectured, building on results of Greene [20] and Lisca [23], which two-bridge links are related to the unknot or unlink by a distance-1 RSR, although the generic case is still open.)

7 Applications

These results can be applied to a variety of protein–DNA interactions to help elucidate local structures and/or the enzyme mechanism. As an illustration, we conclude with two examples below, but emphasise that there are many others.

For site-specific recombination, these theorems can suggest (up to homeomorphism) the DNA configuration of the post-recombinant complex given the synaptic complex (or vice versa), if the protein operation is understood. (This is the case for the resolvase family, which, as discussed above, acts via the subunit rotation mechanism.) Note that the theorems above predict configurations, *independent* of the global topology of the substrate and/or product molecule(s), and so are particularly useful for recombinases such as the tyrosine recombinases and large serine recombinases which act at sites on the same or separate circular or linear DNA molecule(s). Conversely, if there is structural information available for both the pre- and the post-recombinant DNA, then these theorems predict the possible choreographies of the recombinase.

A natural candidate for understanding subrational-tangle replacement is recombinases that utilise additional accessory proteins to trap a fixed conformation near the cleavage/ligation sites. For example, Fig. 11 displays an experimentally derived model of how the rather baroque synaptic complex for the recombinase Sin may look [27]. The two sites on the right are where the cleavage and ligation occur, and the three negative vertical crossings of the DNA axis on the left are trapped by the accessory protein IHF throughout the cleavage/ligation reaction. Recombination mediated by Sin can thus be modelled as a *sub*tangle replacement, as diagrammed in Fig. 12. Within this, by assuming either that the Sin recombinase acts by subunit rotation or the experimentally derived model is correct, one can utilise Theorem 2 above to predict the possible post-recombinant structures. Similarly, one could also use the theorems above to predict whether the subunit rotation mechanism could give rise to a DNA conformation faithful to the experimental evidence (see, e.g., [22]).

A final application considers type II topoisomerases. Their operation, the crossing change, is understood (although the exact details of it are still under discussion). Thus one can use the theorems above to classify all possible *local* structures arising from these crossing changes, as well as predict *where* these crossing changes must occur if the local structures are known (see Fig. 13).

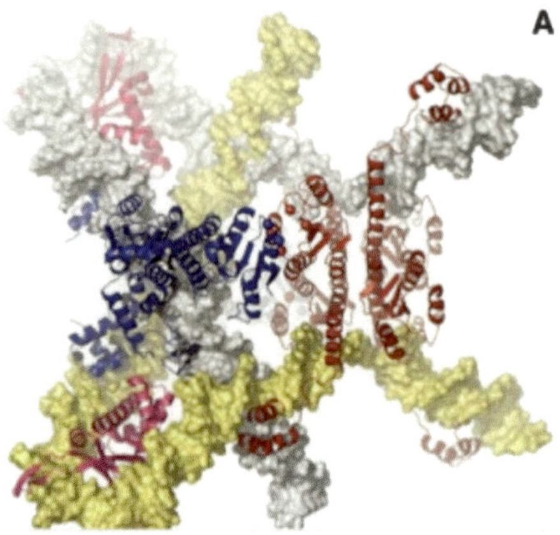

Fig. 11 An experimental model of the complex architecture of the synaptic complex for the Sin recombinase (From [27])

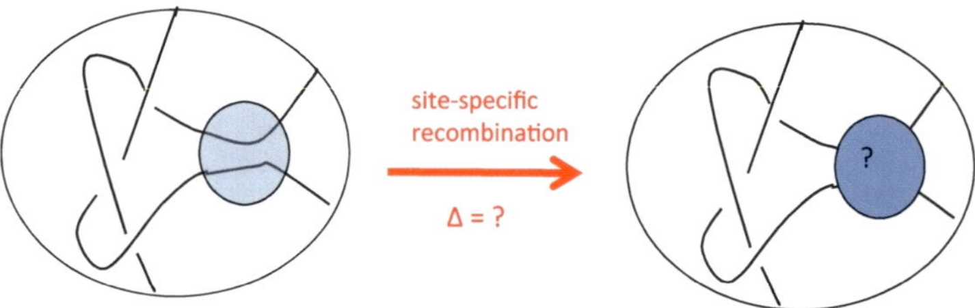

Fig. 12 An application of rational-subtangle replacement theorems to recombination mediated by Sin

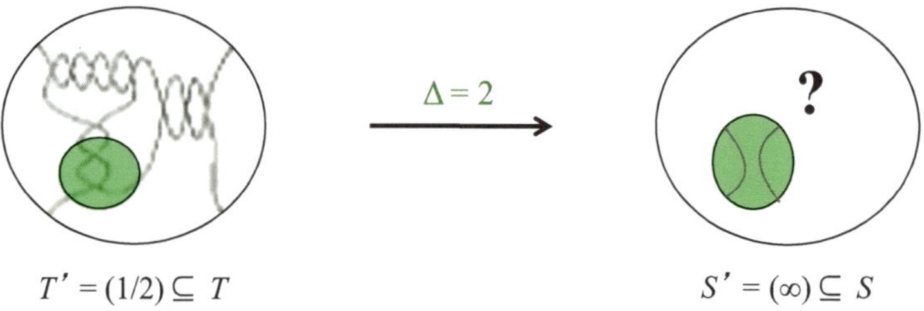

$$T' = (1/2) \subseteq T \qquad\qquad S' = (\infty) \subseteq S$$

Fig. 13 An application of rational-subtangle replacement theorems to a crossing change mediated by type II topoisomerase

8 Conclusions

The applications above illustrate some of the ways that our topological model and results can complement partial experimental knowledge of protein systems that perform localised tranformations. Given structural information (e.g. via X-ray crystallography), these theorems can predict the possibilities for other structures and/or the trajectories of the configurations during the transformation. One subtlety here is understanding how to determine the boundary points of the relevant tangles. Given a larger complex, such as that discussed above for the Sin recombinase [27] or the full λ-Int recombinase–DNA complex structure [2], one needs to establish a natural reference frame for considering subtangle replacement (at the catalytic sites) within the larger rational tangle (composed of accessory sites as well).

This approach should thus complement previous treatments (including the original model of Ernst and Sumners [17]), which also used the closure of the tangles (the global knotting information) to deduce the local structures. Conversely, understanding the mechanism or dynamics of the reaction can help predict some of the static DNA conformations at various points in time (e.g. within the synaptic complex).

We are currently working to implement this computationally, so that experimentalists can easily visualise the putative mechanisms and/or structures. One particularly nice way to see this is via a graph, whose vertices are the structures (the rational tangles or subtangles) and the weighted edges are the distance-Δ replacements between them.

We conclude by noting that these localised DNA transformations are generalisations of crossing changes within knots and links. Given the extraordinary amount of activity to understand which knots/links are related by crossing changes, it is clear that on the purely topological front, as well as the DNA topology front discussed here, there are many tantalising questions to consider within this framework.

Acknowledgements The author would like to express many thanks to Ken Baker, who co-authored the topological proofs that were applied to the biological systems described above, and who provided several of the figures. Thanks also to the reviewers for their careful reading and insightful comments. Also, thanks to the UK's Engineering and Physical Sciences Research Council, which has generously supported the author through grants EP/H0313671, EP/G0395851 and EP/J1075308, and to the London Mathematical Society, which has supported her research through two LMS Scheme 2 Awards. Finally, much gratitude to Natasha Jonoska, Alessandra Carbone, Katarzyna Rejniak, Masahico Saito and Reidun Twarock for organising the stimulating and enjoyable Discrete and Topological Models in Molecular Biology Conference, as well as for inviting this article.

References

1. J. Berge, The knots in $D^2 \times S^1$ which have nontrivial Dehn surgeries that yield $D^2 \times S^1$. Topol. Appl. **38**(1), 1–19 (1991). MR MR1093862 (92d:57005)
2. T. Biswas, H. Aihara, M. Radman-Livaja, D. Filman, A. Landy, T. Ellenberger, A structural basis for allosteric control of DNA recombination by lambda integrase. Nature **435**, 1059–1066 (2005)
3. D. Buck, E. Flapan, Predicting knot or catenane type of site-specific recombination products. J. Mol. Biol. **374**, 1186–1199 (2007)
4. D. Buck, DNA topology, in *Applications of Knot Theory*, ed. by D. Buck, E. Flapan. Proceedings of Symposia in Applied Mathematics, San Diego (AMS, Providence, 2008), pp. 1–43

5. D. Buck, K.L. Baker, Taxonomy of DNA conformations within complex nucleoprotein assemblies. Prog. Theor. Phys. **191**, 55–65 (2011)
6. D. Buck, K.L. Baker, The classification of rational subtangle replacements between rational tangles. Algebr. Geom. Topol. **13**, 1413–1463 (2013)
7. D. Buck, E. Flapan, A topological characterization of knots and links arising from site-specific recombination. J. Phys. A **40**(41), 12377–12395 (2007). MR 2394909 (2010h:92064)
8. D. Buck, M. Mauricio, Connect sum of lens spaces surgeries: application to Hin recombination. Math. Proc. Camb. Philos. Soc. **150**(3), 505–525 (2011). MR 2784772
9. D. Buck, K. Valencia, Characterization of knots and links arising from site-specific recombination on twist knots. J. Phys. A **44**(41), 45002–45038 (2011). MR MR2394909 (2010h:92064)
10. D. Buck, C. Verjovsky Marcotte, Tangle solutions for a family of DNA-rearranging proteins. Math. Proc. Camb. Philos. Soc. **139**(1), 59–80 (2005). MR 2155505 (2006j:57010)
11. D. Buck, C. Verjovsky Marcotte, Classification of tangle solutions for integrases, a protein family that changes DNA topology. J. Knot Theory Ramif. **16**(8), 969–995 (2007). MR 2364885 (2009f:57006)
12. H. Cabrera-Ibarra, D.A. Lizárraga-Navarro, An algorithm based on 3-braids to solve tangle equations arising in the action of Gin DNA invertase. Appl. Math. Comput. **216**, 95–106 (2010)
13. H. Cabrera-Ibarra, D.A. Lizárraga-Navarro, Braid solutions to the action of the Gin enzyme. J. Knot Theory Ramif. **19**, 1051–1074 (2010)
14. M. Culler, C.McA. Gordon, J. Luecke, P.B. Shalen, Dehn surgery on knots. Ann. Math. (2) **125**(2), 237–300 (1987). MR MR881270 (88a:57026)
15. I.K. Darcy, W. De Sumners, Rational tangle distances on knots and links. Math. Proc. Camb. Philos. Soc. **128**(3), 497–510 (2000). MR MR1744106 (2000j:57008)
16. C. Ernst, Tangle equations. J. Knot Theory Ramif. **5**(2), 145–159 (1996). MR MR1395775 (97h:57016)
17. C. Ernst, D.W. Sumners, A calculus for rational tangles: applications to DNA recombination. Math. Proc. Camb. Philos. Soc. **108**(3), 489–515 (1990). MR MR1068451 (92f:92024)
18. D. Gabai, Surgery on knots in solid tori. Topology **28**(1), 1–6 (1989). MR MR991095 (90h:57005)
19. I. Grainge, D. Buck, M. Jayaram, Geometry of site alignment during Int family recombination: antiparallel synapsis by the Flp recombinase. J. Mol. Biol. **298**, 749–764 (2000)
20. J. Greene, The lens space realization problem. `arXiv:1010.6257` `[math.GT]`
21. N.D. Grindley, K.L. Whiteson, P.A. Rice, Mechanisms of site-specific recombination. Annu. Rev. Biochem. **75**, 567–605 (2006)
22. R.A. Keenholtz, S.J. Rowland, M.R. Boocock, W.M. Stark, P.A. Rice, Structural basis for catalytic activation of a serine recombinase. Structure **19**(6), 799–809 (2011)
23. P. Lisca, Lens spaces, rational balls and the ribbon conjecture. Geom. Topol. **11**, 429–472 (2007). MR 2302495 (2008a:57008)
24. Z. Liu, R.W. Deibler, H.S. Chan, L. Zechiedrich, The why and how of DNA unlinking. Nucleic Acids Res. **37**, 661–671 (2009)
25. J.M. Montesinos, Surgery on links and double branched covers of S^3. Knots, groups, and 3-manifolds (Papers dedicated to the memory of R.H. Fox) (Princeton University Press, Princeton, 1975), pp. 227–259. Ann. Math. Stud. No. 84. MR MR0380802 (52 #1699)
26. L. Moser, Elementary surgery along a torus knot. Pac. J. Math. **38**, 737–745 (1971). MR MR0383406 (52 #4287)
27. K.W. Mouw, S.J. Rowland, M.M. Gajjar, M.R. Boocock, W.M. Stark, P.A. Rice, Architecture of a serine recombinase-dna regulatory complex. Mol. Cell **30**(2), 145–155 (2008)
28. F.J. Olorunniji, D.E. Buck, S.D. Colloms, A.R. McEwan, M.C. Smith, W.M. Stark, S.J. Rosser, Gated rotation mechanism of site-specific recombination by ΦC31 integrase. Proc. Natl. Acad. Sci. U.S.A. **109**(48), 19661–19666 (2012)
29. L. Paoluzzi, On hyperbolic type involutions. Rend. Istit. Mat. Univ. Trieste **32**(2001, suppl. 1), 221–256 (2002). Dedicated to the memory of Marco Reni. MR 1893400 (2003a:57014)
30. M.D. Stone, C.D. Hardy, N.J. Crisona, N.R. Cozzarelli, Disentangling DNA during replication: a tale of two strands. Philos. Trans. R. Soc. Lond. B Biol. Sci. **359**, 39–47 (2004)

31. I. Torisu, The determination of the pairs of two-bridge knots or links with Gordian distance one. Proc. Am. Math. Soc. **126**(5), 1565–1571 (1998). MR 1425140 (98j:57020)
32. K. Valencia, D. Buck, Predicting knot and catenane type of products of site-specific recombination on twist knot substrates. J. Mol. Biol. **411**(2), 350–367 (2011)
33. A. Vologodskii, Theoretical models of DNA topology simplification by type IIA DNA topoisomerases. Nucleic Acids Res. **10**, 3125–3133 (2009)
34. J.C. Wang, Cellular roles of DNA topoisomerases: a molecular perspective. Nat. Rev. Mol. Cell Biol. **3**, 430–440 (2002)

Site-Specific Recombination on Unknot and Unlink Substrates Producing Two-Bridge Links

Kenneth L. Baker

Abstract Site-specific recombinases act upon circular DNA, transforming it from one conformation into another. Keeping track of just the axial backbone of the DNA, we may represent the DNA as topological knots and links and model the transformations as localized bandings. As single unknotted loops of DNA and pairs of unlinked unknotted loops of DNA are the typical substrates generated in the laboratory for recombination, and two-bridge links are among the more common conformations arising from recombination, we survey both the classification of which two-bridge links may be thus obtained and the current state of knowledge about where these transformations may occur.

1 Introduction

Site-specific recombinases act upon circular DNA, changing it from one conformation to another. All site-specific recombinases experimentally characterized to date operate by the type of transformation represented in Fig. 1a [7]. Keeping track of just the backbone axis of the DNA rather than its actual double helix, this transformation can be further abstracted and represented as the banding in Fig. 1b.

The single unknotted circle and the pair of unlinked unknotted circles, which we call the *unknot* and *unlink*, respectively, are perhaps the easiest to generate in the laboratory as substrates for recombination. Site-specific recombinases often produce two-bridge links[1] (of one or two components) from DNA molecules in

<hr>

[1] Two-bridge links are defined in Sect. 2.1. Links with one component are commonly called knots. Links with more than one component are often called catenanes by biologists.

K.L. Baker (✉)
Department of Mathematics, University of Miami, Coral Gables, FL 33146, USA
e-mail: k.baker@math.miami.edu

N. Jonoska and M. Saito (eds.), *Discrete and Topological Models in Molecular Biology*, 363
Natural Computing Series, DOI 10.1007/978-3-642-40193-0_17,

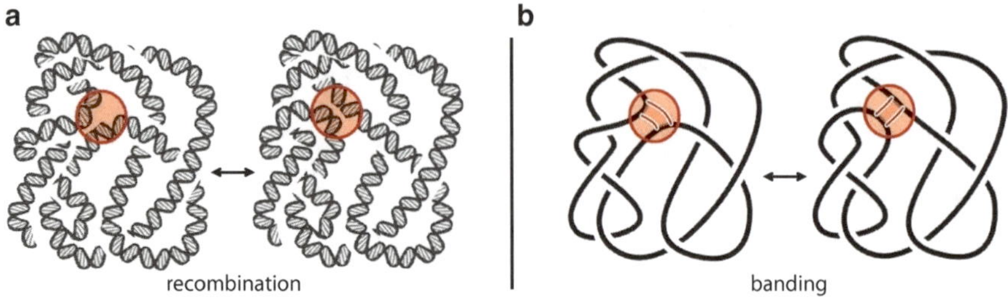

Fig. 1 Recombination may be represented by banding

the conformations of unknots and unlinks [16]. To understand aspects of how recombination may occur, we thus focus our attention here upon bandings (Sect. 2.2) of the unknot and unlink that yield two-bridge links. Two broad questions towards this end may be addressed topologically: *Which* two-bridge links admit a banding from the unknot or unlink? *Where* may these bandings occur?

The purpose of this chapter is to convey the present state of what is known about the answers to these two questions to an audience without much of a background in knot theory or low-dimensional topology. As such, we forgo proofs in favor of only suggesting the overall structure of the results. Nevertheless, we hope that collecting the results and conjectures together here will be of use to the topologists as well.

One further comment. Through taking double branched covers, bandings between links translates to Dehn surgeries between 3-manifolds. Most of the publications, concepts, and results we discuss here we are actually developed in the context of Dehn surgeries on knots between S^3 or $S^1 \times S^2$ and lens spaces. Thus our presentations of these are a translation or adaptation from the sources cited. The correspondence is discussed a bit further in Sect. 2.5.

1.1 Which?

The *which* question has a complete solution. The classification of two-bridge links that admit a banding from the unknot follows from the work of Greene [15]. The classification of those two-bridge links that admit a banding from the unlink follows from the work of Lisca [17] when blended with an observation of Rasmussen recorded in [15] though an oversight is corrected in [4]. Section 2 gives the relevant definitions and sets out the notation $K(p,q)$ for a two-bridge link. (Recall that we also include two-bridge knots among the two-bridge links.)

Theorem 1 (Greene [15]). *A two-bridge link L admits a banding from the unknot if and only if L or its mirror image is isotopic to the two-bridge link $K(p,q)$, where $q \equiv -k^2 \pmod p$ and, for some integers i, d, either*

   I. $p = ik \pm 1$, $\gcd(i,k) = 1$;

  II. $p = ik \pm 1$, $\gcd(i,k) = 2$, *and* $i, k \geq 4$;

III. $\begin{cases} (a)_\pm & p \equiv \pm(2k-1)d \pmod{k^2}, \quad d\,|\,k+1, \ (k+1)/d \ odd; \\ (b)_\pm & p \equiv \pm(2k+1)d \pmod{k^2}, \quad d\,|\,k-1, \ (k-1)/d \ odd; \end{cases}$

IV. $\begin{cases} (a)_\pm & p \equiv \pm(k-1)d \pmod{k^2}, \quad d\,|\,2k+1; \\ (b)_\pm & p \equiv \pm(k+1)d \pmod{k^2}, \quad d\,|\,2k+1; \end{cases}$

  V. $\begin{cases} (a)_\pm & p \equiv \pm(k+1)d \pmod{k^2}, \quad d\,|\,k+1, \ d\,odd; \\ (b)_\pm & p \equiv \pm(k-1)d \pmod{k^2}, \quad d\,|\,k-1, \ d\,odd; \end{cases}$

 VII. $k^2 + k + 1 \equiv 0 \pmod{p}$;

VIII. $k^2 - k - 1 \equiv 0 \pmod{p}$;

  IX. $p = \frac{1}{11}(2k^2 + k + 1)$, $\ k \equiv 2 \pmod{11}$; *or*

   X. $p = \frac{1}{11}(2k^2 + k + 1)$, $\ k \equiv 3 \pmod{11}$.

The numbering of these nine families comes from condensing Berge's list of twelve families [5] as reorganized by Rasmussen [19]. Family VI is contained in family V, and allowing k to run over all integers subsumes families XI and XII into IX and X, respectively.

Theorem 2 (Lisca [17] + Rasmussen via [15], Baker et al. [4]). *A two-bridge link L admits a banding from the unlink if and only if L or its mirror image is isotopic to the two-bridge link $K(p,q)$, where, for some integers m, d, we have $p = m^2$ and either*

   I. $q \equiv md + 1 \pmod{m^2}$, $\gcd(m,d) = 1$;

  II. $q \equiv md + 1 \pmod{m^2}$, $\gcd(m,d) = 2$;

 III. $q \equiv d(m-1) \pmod{m^2}$, $\ d\,|\,k-1$, d *odd; or*

 IV. $q \equiv d(m-1) \pmod{m^2}$, $\ d\,|\,2m+1$.

Family II is missing from the statement of the main result of [17], although it arises in the proof; this detail is unaddressed in [15]. A discussion of this along with a corrected statement is given in [4].

1.2 Where?

The *where* question so far only has a conjectural solution. The conjectural classification of bandings between two-bridge links and the unknot follows from work of Berge (Some knots with surgeries yielding lens spaces, unpublished manuscript) and is described in [1, 2]. A conjectural classification of bandings between two-bridge links and the unlink has more recently been given in [4], building on the ideas of Berge [5]. These two conjectural classifications are unified with the notion of a *doubly primitive band*, which we explain in Sect. 2.3.

Conjecture 1. *Any banding from the unknot or unlink to a two-bridge link is along a doubly primitive band.*

The collections of doubly primitive bands for the unknot and the unlink parallel one another, and can be related by simple changes that will be apparent from our presentation of the known bandings in Sect. 3.

To describe these bandings succinctly, let us first remark that a banding from one link to another confers a dual banding back to the first link. It turns out that the core arc of a banding from a two-bridge link to either the unknot or the unlink that is dual to a doubly primitive banding admits a special presentation as a *simple arc*. The definition of the *simple arc* $K(p,q,k)$ for the two-bridge link $K(p,q)$ is given in Sect. 2.4.

The following theorem may be derived from Berge's work [5] with the aid of Rasmussen's reorganization [19].

Theorem 3 (Berge [5]). *Each two-bridge link $K(p,q)$ with $q \equiv -k^2$ (mod p), as given in Theorem 1, admits a banding to the unknot along the simple arc $K(p,q,k)$.*

Greene's work [15] together with the relationship between simple arcs and doubly primitive bandings of the unknot [5] implies that the bandings given in [1, 2] (and derived from [5]) constitute all the doubly primitive bandings from the unknot to a two-bridge link.

Theorem 4 (Greene [15] + Berge [5]). *If a two-bridge link L is obtained from a doubly primitive banding of the unknot and a is the core arc of the dual banding, then, after perhaps taking a mirror image, there is an isotopy taking L to $K(p,q)$ and a to $K(p,q,k)$, as given in Theorem 3.*

Each arc of Theorem 3 may be labeled according to its associated family as given in Theorem 1. We will relabel them according to their broader groupings. The first five are, collectively, the BERGE–GABAI bandings, which we denote BGI–BGV; the next two are the 3-STRING BRAID types $3SB_{VII}$ and $3SB_{VIII}$, and last two are the SPORADIC types $SPOR_{IX}$ and $SPOR_X$. We have a similar labeling for the bandings between two-bridge links and the unlink.

Theorem 5 (Baker et al. [4]). *Each two-bridge link $K(p,q)$ with integers m, d as given in Theorem 2 (so that $p = m^2$) admits a banding to the unlink along the simple arc $K(p,q,k)$ in the following cases.*

I. $q \equiv md + 1$ (mod m^2), $\gcd(m,d) = 1$, *and either*

    BGI. $k \equiv \pm m$ (mod m^2) *or*
    3SB. $k \equiv \pm dm$ (mod m^2).

II. $q \equiv md + 1$ (mod m^2), $\gcd(m,d) = 2$, *and*

    BGII. $k \equiv \pm m$ (mod m^2).

III. $q \equiv d(m-1) \pmod{m^2}$, $d \mid m-1$, d *odd, and either*

    BGIII. $k \equiv \pm dm \pmod{m^2}$,

     BGV. $k \equiv \pm d \pmod{m^2}$, *or*

    SPOR. $k \equiv \pm 2m \pmod{m^2}$ *and* $m = 1 - 2d$.

IV. $q \equiv d(m-1) \pmod{m^2}$, $d \mid 2m+1$, *and*

    BGIV. $k \equiv \pm m \pmod{m^2}$ *or* $k \equiv \pm dm \pmod{m^2}$.

Conjecture 2. *If a two-bridge link L is obtained from a doubly primitive banding of the unlink and a is the core arc of the dual banding then, after perhaps taking a mirror image, there is an isotopy taking L to $K(p,q)$ and a to $K(p,q,k)$ for some p,q,k given in Theorem 5.*

Our work [4] and Cebanu's dissertation [8] establishes the foundations for the proof of this conjecture. Calculations by Cebanu in [8] confirm the conjecture for two-bridge links in families I and II. He has since claimed proof for families III and IV as well, thereby confirming the conjecture.

2 Two-Bridge Links, Bandings, and Equivalences

To describe these results and indicate their nature, we must develop terminology. Most of these notions should be familiar to experts, although there are variances in the notations. We recommend Cromwell's text [10] for further details.

A link is a finite collection of nonintersecting loops in 3-space. A link of a single component is more commonly called a knot. Two links are considered equivalent if there is an *isotopy* (i.e., deformation) that brings one into the configuration of the other.[2] Links are represented by pictures of their projections onto the plane of the page, and over/under information is represented by breaks in the undercrossing strands; this information is sufficient to reconstruct the link in 3-space up to isotopy.

In our figures, we also describe a set of twists in two strands of a link by an oblong rectangle and an integer coefficient. The longer dimension of the rectangle indicates the axis of the twist. The magnitude of the integer denotes the number of half-twists, and its sign gives the handedness of the twisting, following the right-hand rule. Figure 2a demonstrates this with an example. Frequently, our diagrams will have pairs of twist regions with the same magnitude but opposite handedness.

---

[2]Technically, two links are equivalent if they are related by an *ambient isotopy*, which is a deformation of the space containing the link, but let us just say "isotopy" for simplicity. Using ambient isotopies avoids pathologies that arise if one were to permit any continuous isotopy of a link.

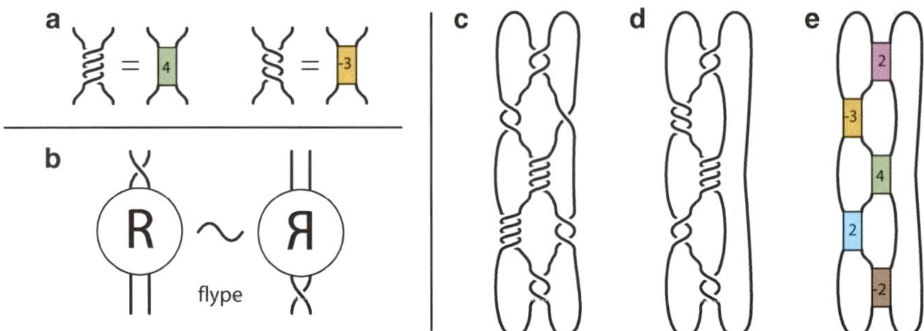

Fig. 2 (**a**) Oblong rectangles with integer coefficients, twist boxes, represent twist regions. (**b**) An illustration of the *flype* isotopy. (**c–e**) A two-bridge link is groomed by flypes and isotopies and then represented with twist boxes

2.1 Two-Bridge Links

A two-bridge link (also called a rational link) is a link of one or two components that may be isotoped into the form of a 4-plat, such as that shown in Fig. 2c. Sequences of flypes and further isotopies may be performed to transfer the twists in the rightmost two strands to the leftmost two strands as in Fig. 2d. (A *flype* is the isotopy illustrated in Fig. 2b that flips a region over, transferring a twist from one side to the other; see [9].) To a two-bridge link in the form of Fig. 2d, we may associate a rational number by putting the numbers in the twist boxes into a continued fraction. In particular, using the coefficients $a_1, a_2, \ldots, a_n$ from the figure, we associate the extended rational number[3] determined by the continued fraction

$$-p/q = [a_1, a_2, \ldots, a_{n-1}, a_n]^- = a_1 - \cfrac{1}{a_2 - \cfrac{1}{\ddots \; a_{n-1} - \cfrac{1}{a_n}}}$$

for coprime integers p, q to the link and denote the link by $K(p, q)$. Note the use of minus signs. Figure 2e (as well as (c) and (d)) shows the two-bridge link $K(-136, 59)$. The two-bridge links $K(p_1, q_1)$ and $K(p_2, q_2)$ are isotopic if and only if $p_1 = \pm p_2$ and either $q_1 \equiv \pm q_2 \mod p_1$ or $q_1 q_2 \equiv \pm 1 \mod p_1$, where the choice of $+$ or $-$ for $\pm$ is used consistently. The link $K(-p, q) = -K(p, q)$ is the mirror image of $K(p, q)$. For example, the link in Fig. 2e may also be denoted

---

[3] Here, "extended" means we also include $\infty = \pm \frac{1}{0}$ among the rational numbers.

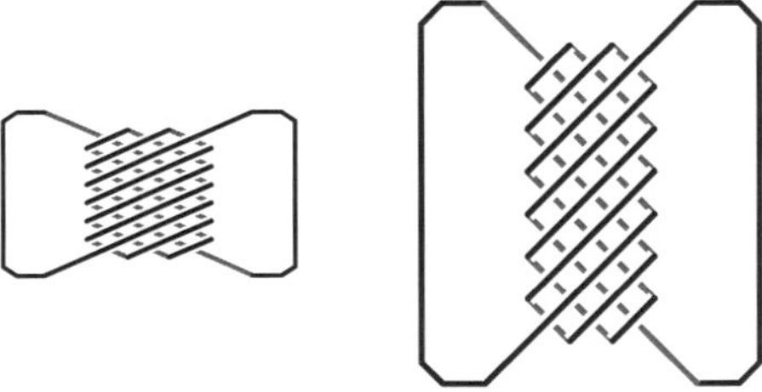

Fig. 3 The square pillowcase model may be rescaled for a cleaner picture

$K(136, 77)$, $-K(136, 59)$, and $K(136, 53)$. The *unknot* is $K(1, 0)$, and the (two-component) *unlink* is $K(0, 1)$.

Two-bridge links also admit a "pillowcase" or "billiards" model. Two rectangles sewn together along their boundaries topologically form a sphere, with four distinguished points coming from the vertices and four arcs connecting these points in a loop coming from the edges. We may view this loop as dividing the sphere into the two rectangles, one in front and one in back. To obtain the knot $K(p, q)$ for coprime integers p and q, we first regard the rectangles of the pillowcase as unit squares lying on top of each other flat in the plane, with a pair of opposite vertices at $(0, 0)$ and $(1, 1)$. Then, starting from $(0, 0)$, we draw a polygonal curve in the square of segments with slopes $\pm q/p$, ricocheting off the edges until another vertex is reached. (This will begin with the segment from $(0, 0)$ to either $(1, |q/p|)$ if $|q| \geq |p| > 0$, $(|q/p|, 1)$ if $|q| \leq |p| > 0$, or $(0, 1)$ if $p = 0$.) The segments with slope $-q/p$ will be on top and those with slope q/p will be on the bottom. A second polygonal curve may then be drawn in a similar manner, connecting the remaining pair of vertices so that it is disjoint from the first on the inflated pillowcase. A flip across the horizontal (if p is even) or vertical (if p is odd) axis of the pillowcase will make the two curves coincide. Finally, we join the vertices $(0, 0)$ and $(0, 1)$ by an arc to the left of the square and the vertices $(1, 0)$ and $(1, 1)$ by an arc to the right of the square. The resulting curves form the two-bridge link $K(p, q)$. This square pillowcase construction of $K(7, -3) = K(7, 4)$ is illustrated in Fig. 3, left. Rescaling the unit square to the $q \times p$ rectangle can make for a cleaner picture in which the slopes of the polygonal curves are all ± 1 (except when $p = 0$ or $q = 0$), as shown in Fig. 3, right.

One may geometrically invoke a generalized Euclidean algorithm to show that the "continued fraction" and the "pillowcase" descriptions of the two-bridge links are equivalent. If $p = a \cdot q - r$, then an isotopy transforms the pillowcase model of $K(p, q)$ into a pillowcase model of $K(q, r)$ with a twists in the pair of strands exiting along one of the "q-sides". Reversing this isotopy stretches $|q| - 1$ pieces of arcs on that q-side of the $K(q, r)$ pillowcase through the a twists to produce the $K(p, q)$ pillowcase, after perhaps another isotopy that "grooms" the arcs on the pillowcase. Since p and q are relatively prime, this may be repeated until the remainder is ± 1, so that the final pillowcase is itself a set of twists. A final isotopy pushes the resulting sequence of twists into a standard 4-plat form that exhibits a

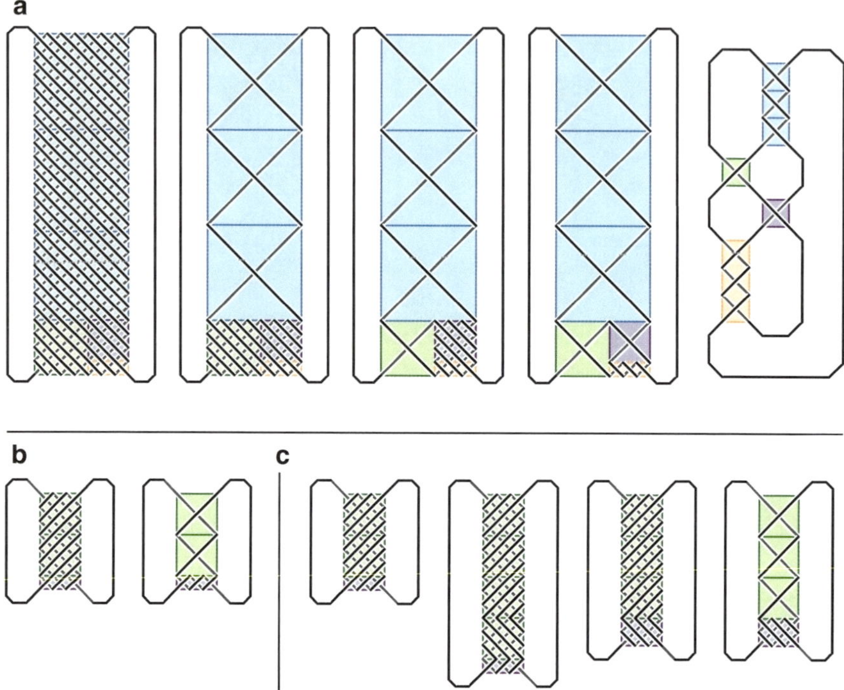

Fig. 4 (**a**) A visual application of the Euclidean algorithm to the pillowcase model provides an isotopy to the continued fraction model. (**b, c**) An illustration of the effects of choices in the generalized Euclidean algorithm

continued fraction expansion of p/q. A straightforward demonstration of the full process is given in Fig. 4a for $K(25, 7)$ using $-25/7 = [-3, 1, -1, 3]^-$ with an underlay of boxes to highlight the relationship, between of the twist regions in the descriptions. Figure 4b, c highlight the two choices $7 = 2 \cdot 3 - (-1)$ and $7 = 3 \cdot 3 - 2$ in the first step of the algorithm for $K(7, 3)$. Observe how the intermediate steps in the isotopy of the latter correspond to the rearrangement $2 \cdot 3 + (3 - 3) - (-1) = (2 \cdot 3 + 3) - (3 + (-1))$.

2.2 Bandings

A *band* is a topological solid rectangle, a disk whose boundary is divided by four points into four edges. If the intersection of a link L and a band b is exactly a pair of opposite edges of b, then the link L' obtained by deleting those edges from L and reconnecting the other pair of edges of b is said to have been obtained from L by a *banding* along b; see Fig. 5. Observe that if b gives a banding from L to L', the same band may be regarded as giving a dual banding from L' to L.

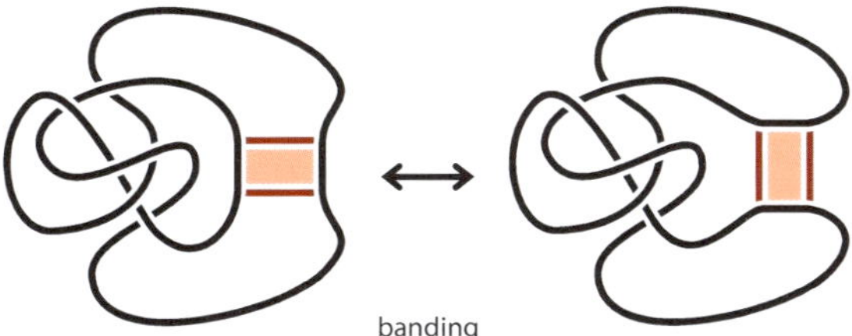

Fig. 5 The transformation between two links caused by a banding

Two bandings b_1 and b_2 on a link L are equivalent if there is an isotopy of $L \cup b_1$ to $L \cup b_2$. Note that in a particular configuration of L, the bands b_1 and b_2 may be on separate components, but an isotopy of L may exchange these components while bringing b_1 to b_2.

The *core arc* of a banding from L to L' along a rectangle b is an arc in b connecting the opposite edges of $b \cap L$. By twisting b, one can see that the same arc may be the core arc of different bandings.

2.3 Three-Bridge Links and Doubly Primitive Bandings

Link may, more generally, be put into a bridge position with many more bridges than two. Those with a two-bridge presentation admit convenient descriptions as discussed above; only the unknot admits a one-bridge presentation. The known bandings from the unknot and unlink to two-bridge links arise from presentations of the unknot and unlink as three-bridge links.

A three-bridge link L, also called a 6-plat, admits a configuration in which there is a plane P transversally intersecting L in six points and dividing L into two sets of three strands, each with the following property: for each of these two sets individually, the three strands can collectively be isotoped, keeping their endpoints fixed, into the dividing plane P without crossing one another and without crossing into the other side of P. Note that there are many such ways to isotope these arcs into P. Moreover, a link L may admit multiple nonisotopic presentations as a three-bridge link. (Observe that this also provides an analogous description of two-bridge links. Compare with the description given above.)

An arc a in that plane P connecting two of the six points of L is called *doubly primitive* if each of the two sets of three strands may be isotoped into the plane so that one of the three strands meets a in only a single endpoint. Such an isotopy may be indicated for each strand with a disk whose boundary is divided into the initial and final positions of the strand and whose interior is filled out with the intermediate stages. Moreover, if the arc a in P is doubly primitive, then each set of the three strands of L to either side of P may be isotoped into P so that one is disjoint from

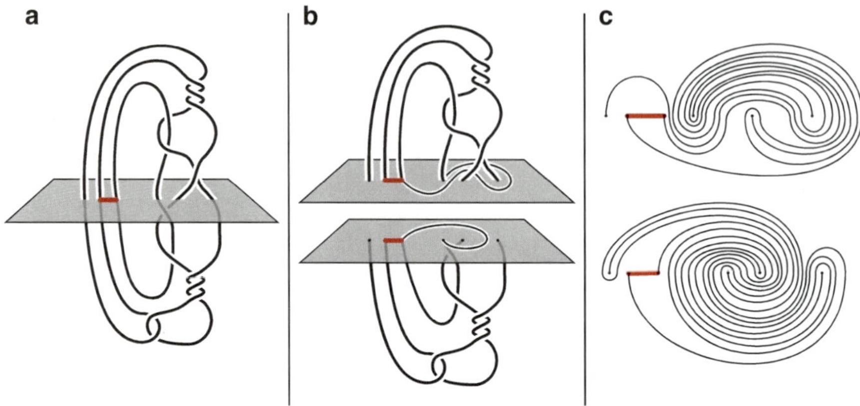

Fig. 6 (**a**) A three-bridge presentation of the unlink with a doubly primitive arc on the bridge plane. (**b**) Isotopies of a strand on each side of the plane into the plane verify that the arc is doubly primitive. Once in the plane, the isotoped strand only meets the arc in one point. (**c**) Further isotopies of the remaining strands into the plane, disjoint from the interior of the arc

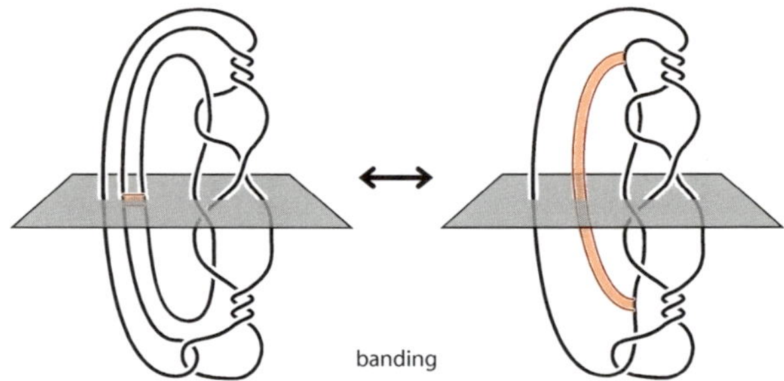

Fig. 7 The banding associated to the doubly primitive arc in Fig. 6a

a and the other two each share an endpoint with a. The point is that a band b for L that meets P in the arc a offers a banding from L to a two-bridge link.

Figure 6a shows the unlink with a plane, giving a three-bridge presentation along with a doubly primitive arc on that plane. Figure 6b demonstrates that this arc is doubly primitive by showing the result of isotoping a strand on each side of the plane into the plane. The remaining strands on each side may be isotoped further onto the plane as in Fig. 6c. Figure 7 shows the banding associated to this doubly primitive arc. It belongs to the SPORADIC family of bandings from the unlink to two-bridge links, as we will see in Sect. 3.3.

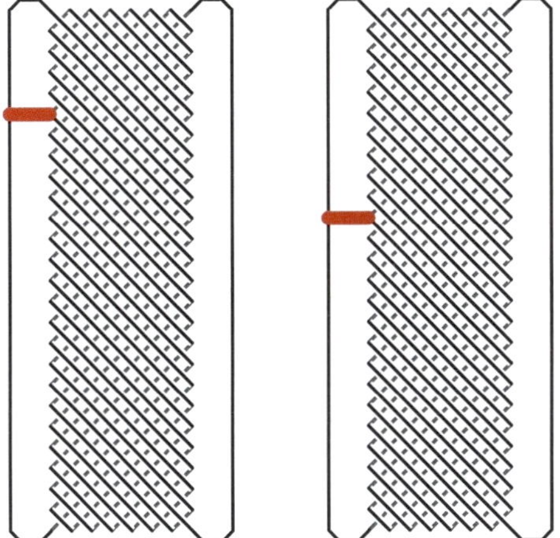

Fig. 8 Examples of the simple arcs $K(25, 7, 5)$ and $K(25, 7, 10)$

2.4 Simple Arcs

Given a two-bridge link $L = K(p, q)$ in its pillowcase presentation as described above, the link intersects the interior of the left edge of the pillowcase in $|p| + 1$ points, two of which are corners. The arc in the plane of the page connecting the kth of these points to the arc joining the left-hand vertices of the square is the *simple arc* $K(p, q, k)$. Figure 8 illustrates the simple arcs $K(25, 7, 5)$ and $K(25, 7, 10)$. By flipping along a horizontal axis, we see that the arc $K(p, q, k)$ is equivalent to the arc $K(p, q, p - k)$. It turns out that the core arcs dual to doubly primitive bandings of the unknot and unlink are actually simple arcs [4, 5].

2.5 Topological Background

The preceding two sections give a description of the "tangle version" of Berge's doubly primitive property that ensures a knot admits a surgery yielding a lens space [5]. To be a little more precise, a knot in a genus-2 Heegaard surface of a 3-manifold is doubly primitive if each genus-2 handlebody to either side of the Heegaard surface contains a compressing disk whose boundary transversally intersects the knot once (Saito provides an account of the main ideas of Berge's unpublished manuscript in his appendix to [20].) Surgery on the knot along the framing induced by the Heegaard surface transforms the manifold and the splitting into a lens space with its genus-1 Heegaard splitting. Our tangle version comes from considering the quotient by the hyperelliptic strong involution induced by the genus-2 splitting. (Montesinos nicely illustrates strong involutions and their tangle quotients in [18].) Under this quotient, the

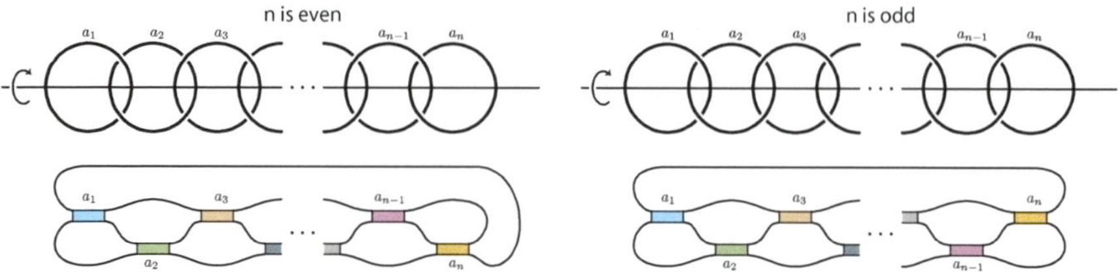

Fig. 9 Strong involutions of surgery descriptions of lens spaces on linear chain links quotient to give our continued fraction models of two-bridge links

manifold is recorded with the link L as its double branched cover, the genus-2 Heegaard surface becomes the dividing plane P that presents L as a three-bridge link, the doubly primitive knot becomes the arc a, and its lens space surgery is given by the banding along b. Here we are actually considering our links L as being in the three-sphere S^3 rather than the three-space $\mathbb{R}^3$. The relevant links L at hand are the unknot, the unlink, and the two-bridge links in general, and their double branched covers are the manifolds S^3, $S^1 \times S^2$, and the lens spaces, respectively.

Through double branched coverings, our diagrams of the two-bridge links shown later in Figs. 11–16 and 19–22 may be regarded as giving presentations of the covering lens spaces as integral surgery on a chain link, or Kirby diagrams for plumbing manifolds bounded by these lens spaces; see Fig. 9. (Among others, Gompf and Stipsicz [13] have provided a nice introduction to Kirby diagrams and their calculus.) In the branched covering, the bands on those two-bridge links lift to framed knots that indicate surgeries from these lens spaces (and in particular, four-dimensional two-handle attachments to these plumbing manifolds) to either S^3 or $S^1 \times S^2$.

The Berge conjecture proposes that any longitudinal surgery on a knot in S^3 that produces a lens space arises from this doubly primitive mechanism [5,14]. In [4], we have extended this conjecture with $S^1 \times S^2$ in place of S^3. Conjecture 1 in Sect. 1.2 is a subconjecture of these two conjectures, as there might be knots that are not strongly invertible yet admit longitudinal lens space surgeries.

3 The Known Bandings Between Two-Bridge Links and the Unknot and Unlink

The doubly primitive bandings of the unknot and unlink constitute the known bandings to two-bridge links from the unknot and unlink. These bandings for the unknot and unlink parallel one another in form and break into three broad families: the Berge–Gabai, sporadic, and 3-string braid families. We present all these bandings up to mirror images and isotopies of the link and band.

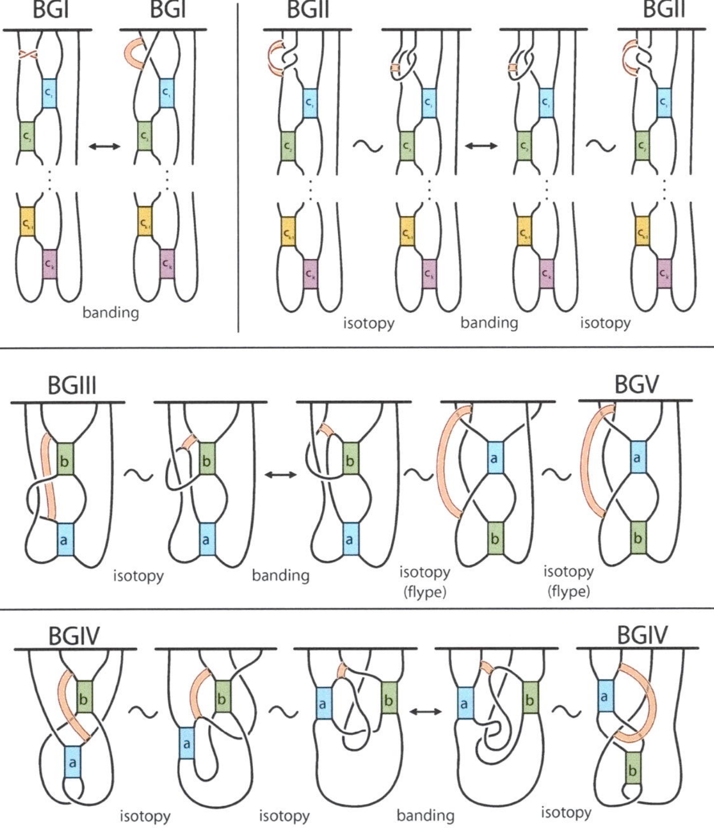

Fig. 10 The four families of bandings between rational tangles up to homeomorphism. The bands of type BGIII and BGV are related by the banding operation, but are generally not homeomorphic

3.1 The Berge–Gabai Bandings

The Berge–Gabai bandings arise from the bandings that transform a rational tangle into a rational tangle. This classification is derived from the work of Berge and Gabai that classified knots in solid tori using longitudinal surgeries yielding solid tori [6, 11, 12]. Baker and Buck [3] have shown that each such knot admits a unique involution whose quotient yields a site in a rational tangle. These sites for bandings between rational tangles partition up to homeomorphism into five types, BGI–BGV, following the numbering of Berge [6]. (Berge also lists a sixth type which is a subfamily of the fifth.) Types BGI, BGII, and BGIV are self-dual, while types BGIII and BGV are dual to one another. These results are summarized in Fig. 10 up to mirror images and isotopy. The tangles are drawn as if in the lower half-space so that the horizontal line represents the boundary plane of the space containing the tangle below. The middle row of Fig. 10 shows the duality between bands of type BGIII

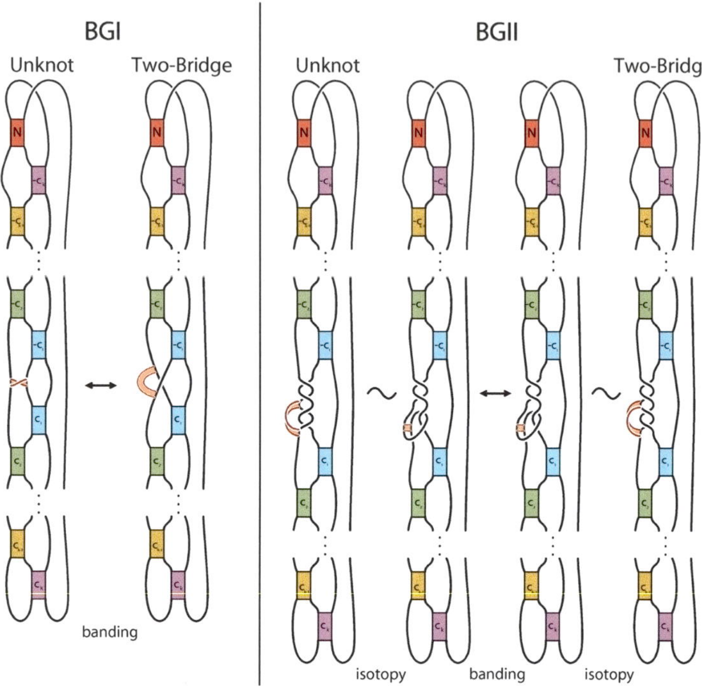

Fig. 11 The BGI and BGII families of bandings between the unknot and a two-bridge knot

and bands of type BGV. For each of the other types in that figure, the tangle together with the band on the right may be isotoped, allowing the endpoints of the strands to move in the boundary into the form of the tangle and band on the left or its mirror, though with different parameters for the twists. We have kept the endpoints of the strands fixed throughout the isotopies and bandings shown in Fig. 10.

To obtain the Berge–Gabai bandings between the unlink or unknot and a two-bridge link, we attach another rational tangle to an input (or output) tangle in Fig. 10 in order to obtain either the unlink or the unknot. To obtain the unlink, only the mirrored tangle suffices; to obtain the unknot, these need only be altered by an integral family. (This follows by considering the double branched covers: among surgeries on the unknot, only 0-surgery yields $S^1 \times S^2$ while $1/n$-surgery for any $n \in \mathbb{Z}$ yields S^3.) The Berge–Gabai bandings from the unknot to a two-bridge link are given in Figs. 11–13. The Berge–Gabai bandings from the unlink to a two-bridge link are given in Figs. 14–16.

3.2 The 3-String Braid Bandings

The 3-string braid bandings arise from presentations of the unknot and the unlink as closed 3-string braids. (In the double branched cover, these bandings correspond to

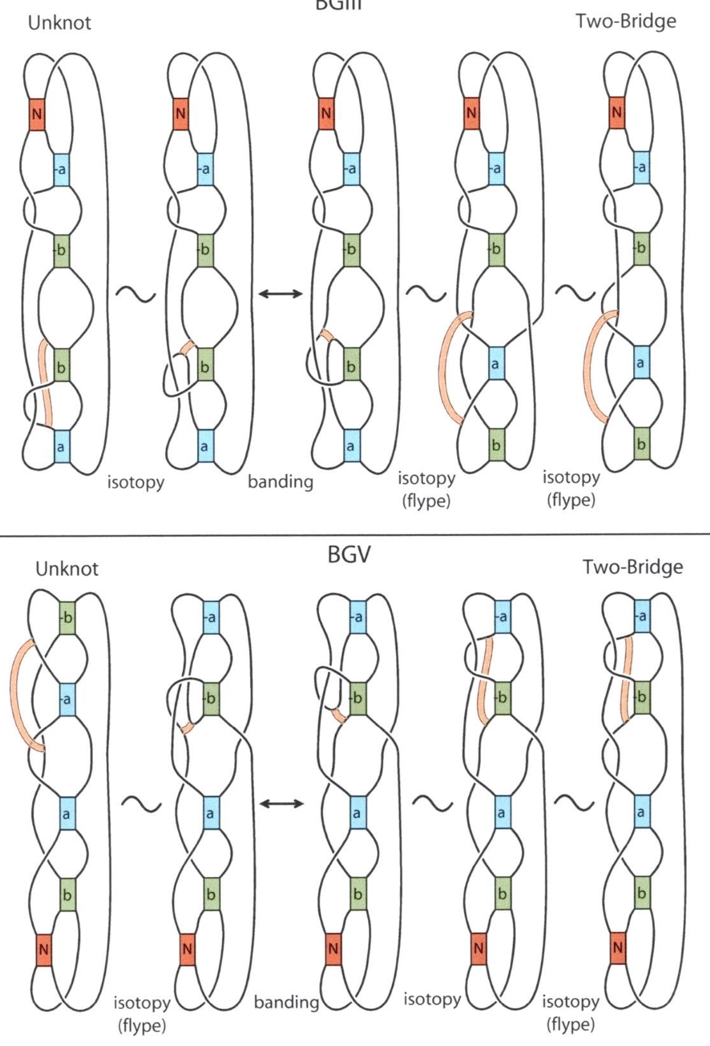

Fig. 12 The BGIII and BGV families of bandings between the unknot and a two-bridge knot

knots that embed on the fiber of a genus-one fibered knot.) Up to conjugation and braid moves, Fig. 17a–c show the three 3-string braid presentations of the unknot, and (d) and (e) show the three 3-string braid presentations of the unlink. Observe that (a) and (c) are mirror images, as are (d) and (e).

Given a 3-string braid γ whose closure is the link L, closures of the various conjugates of γ are all still isotopic to L. Bandings of the arcs that close up these conjugates produce the family of 3-string braid bandings from L to two-bridge links. Figure 18 illustrates such bandings from L to a two-bridge link based on

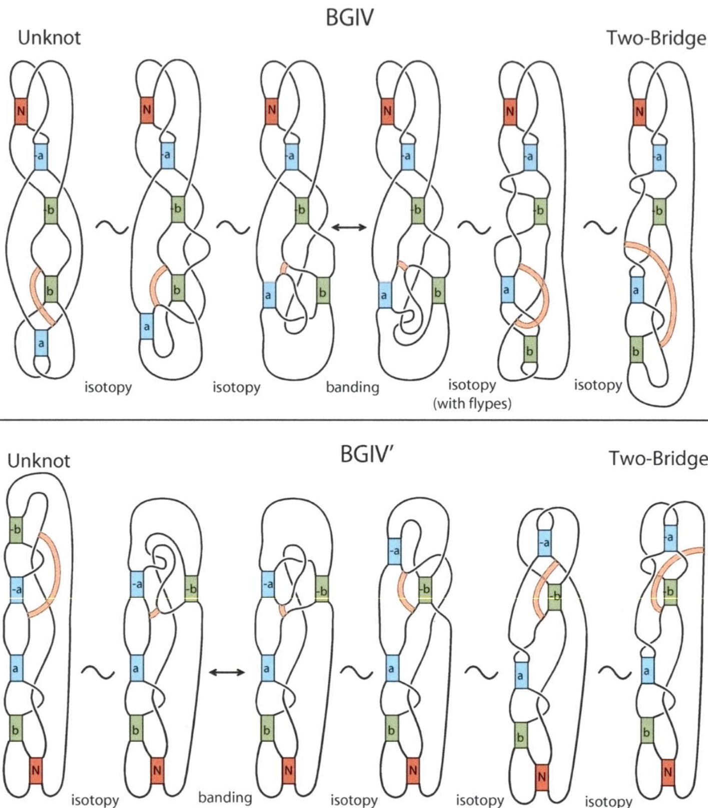

Fig. 13 The BGIV and BGIV' families of bandings between the unknot and a two-bridge knot

the conjugation of γ by another 3-braid β. Using the braids of Fig. 17a, b, d for γ gives the various 3-string braid bandings from the unknot and unlink to a two-bridge knot up to a mirror image. These are shown in Figs. 19 and 20.

3.3 The Sporadic Bandings

The sporadic bandings for the unknot and unlink are depicted in Figs. 21 and 22 up to mirror images. One may regard them as "almost 3-braid" bandings because the clasp at the bottom is the only obstruction to the braiding of the presentation, although the naming refers to them being doubly primitive bandings that do not appear to exhibit as clear a unifying structure as the Berge–Gabai or 3-string braid

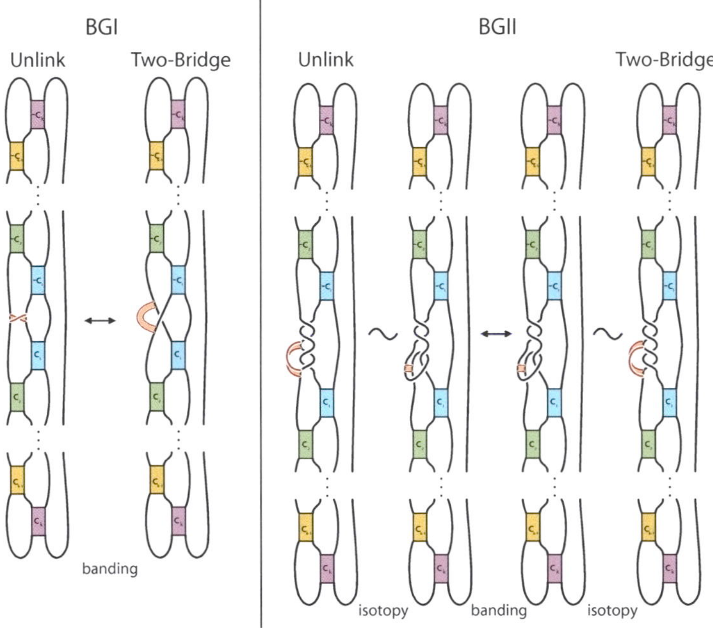

Fig. 14 The BGI and BGII families of bandings between the unlink and a two-bridge knot

bandings do.[4] Greene's work [15] recently established that these were the only remaining doubly primitive bandings of the unknot aside from the Berge–Gabai and the 3-string braid bandings. This remains unconfirmed in general for the unlink; compare the comments following Conjecture 2.

4 Example

We conclude with an illustration of the equivalence between the bandings along simple arcs given in Theorem 2 and their descriptions in Sect. 3. The two-bridge link $K(36, 25)$ is shown at the beginning of Fig. 23 along with the simple arc $K(36, 25, 6)$. The rest of that figure shows a sequence of simplifying isotopies: the link's right-hand side is given a half-twist producing the isotopic presentation as $K(36, -13)$ with the simple arc $K(36, -13, 10)$, then a generalized Euclidean

---

[4]There are doubly primitive bandings of other links that fit neither the Berge–Gabai nor the 3-string braid bandings and yet do not appear to admit similar "almost 3–braid" presentations.

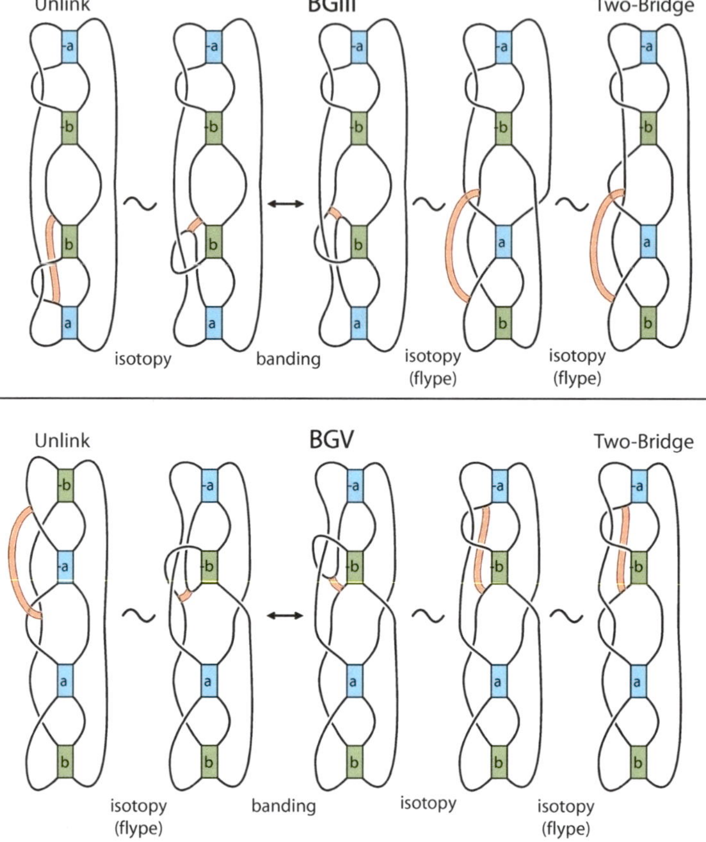

Fig. 15 The BGIII and BGV families of bandings between the unlink and a two-bridge knot

algorithm that fixes the simple arc is applied, and the link is finally isotoped into a 4-plat presentation where the banding for the simple arc is shown. Using $m = 6$ and $d = 5$, Theorem 2 shows that the two-bridge link $K(36, 25)$ of type III admits a banding to the unlink. Taking $k = dm = 30$, Theorem 5 implies there should be a banding from the BGV family along the simple arc $K(36, 25, 30)$. Flipping the arc and link over a horizontal axis shows that this is equivalent to the simple arc $K(36, 25, 6)$ of Fig. 23. Figure 24 demonstrates that the banding does indeed produce the unlink. Figure 25 shows that this banding is isotopic to the BGV banding of Fig. 15 with $a = -2$ and $b = 2$.

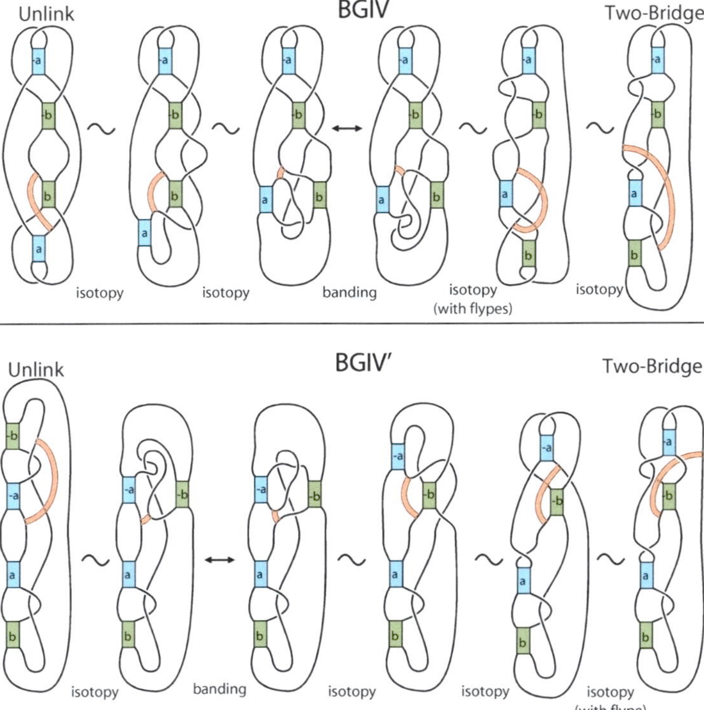

Fig. 16 The BGIV and BGIV′ families of bandings between the unlink and a two-bridge knot

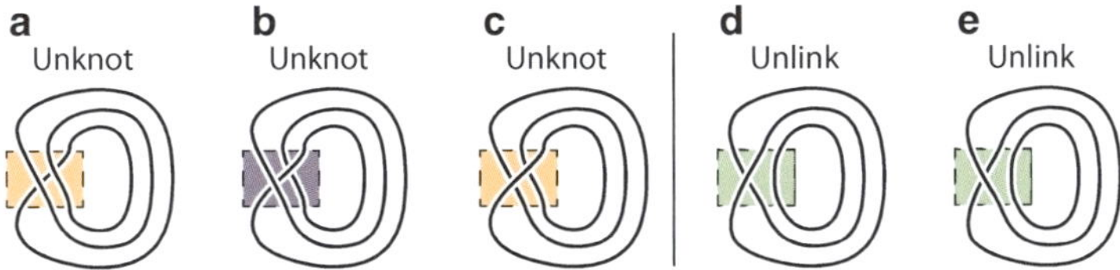

Fig. 17 (**a–c**) There are three closed three-string braids, up to braid isotopy, that represent the unknot. (Note **a** and **c** are orientation reversing homeomorphic.) (**d, e**) There are two closed three-string braids, up to braid isotopy, that represent the unlink (Note these two are orientation reversing homeomorphic)

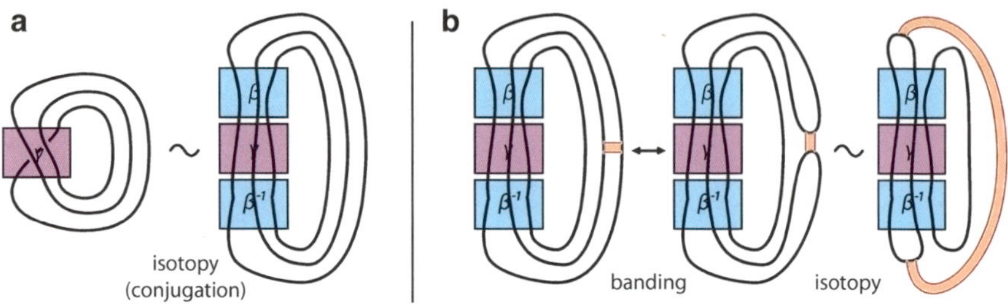

Fig. 18 (**a**) The closure of the three-string braid γ is braid isotopic to the closure of its conjugate $\beta\gamma\beta^{-1}$. (**b**) Each conjugate of γ gives a banding of its braid closure to a two-bridge link

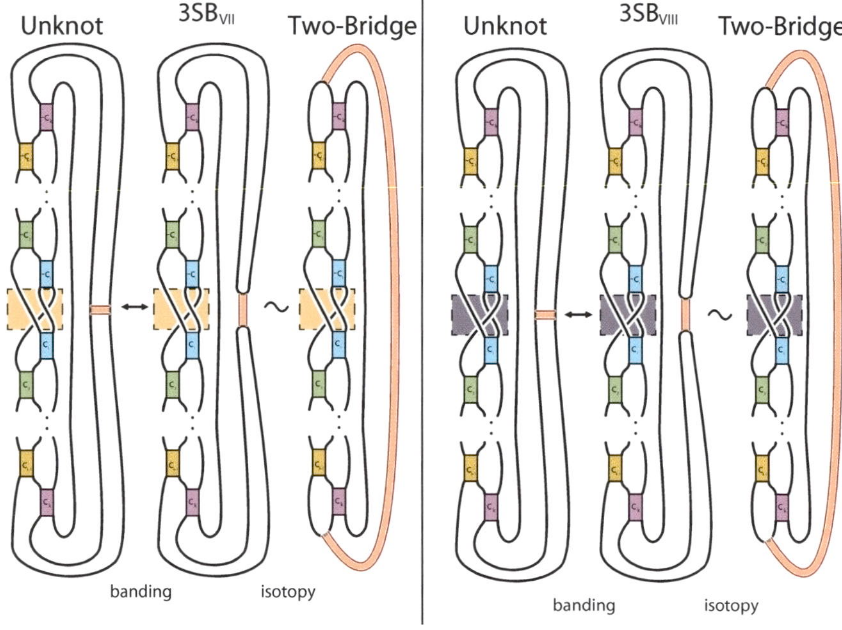

Fig. 19 The two families of three-string braid bandings between the unknot and a two-bridge link

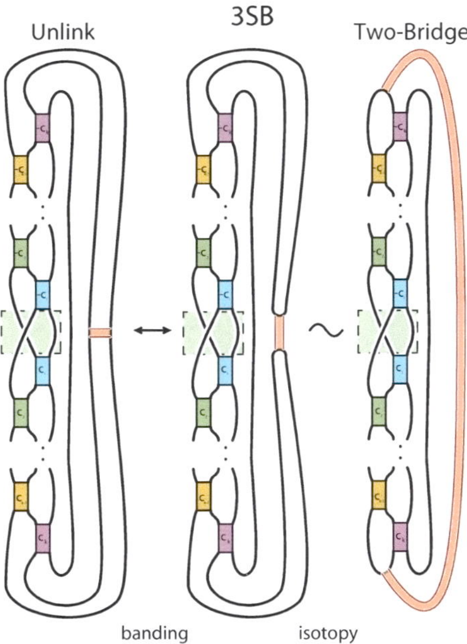

Fig. 20 The family of three-string braid bandings between the unlink and a two-bridge link

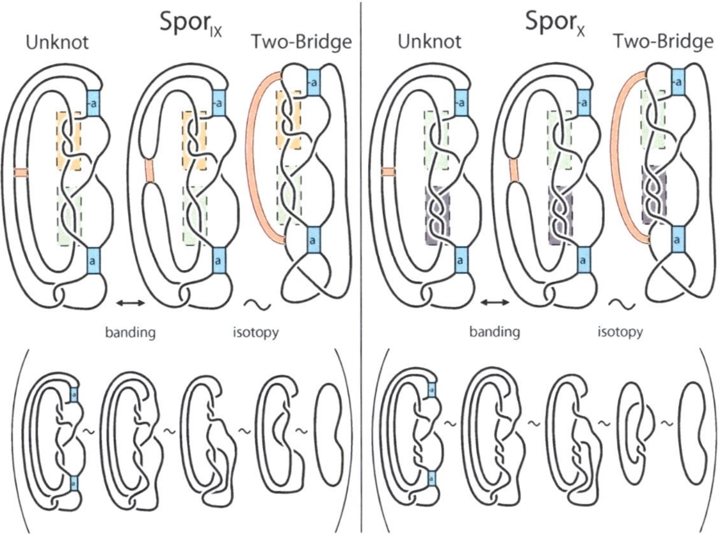

Fig. 21 The two families of sporadic bandings between the unknot and a two-bridge link (Also shown are isotopies verifying the claimed unknots are indeed unknots)

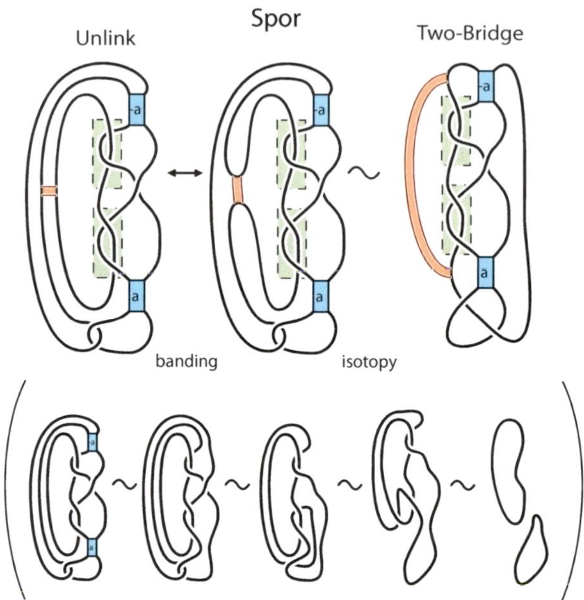

Fig. 22 The family of sporadic bandings between the unknot and a two-bridge link (Also shown are isotopies verifying the claimed unknot is indeed an unknot)

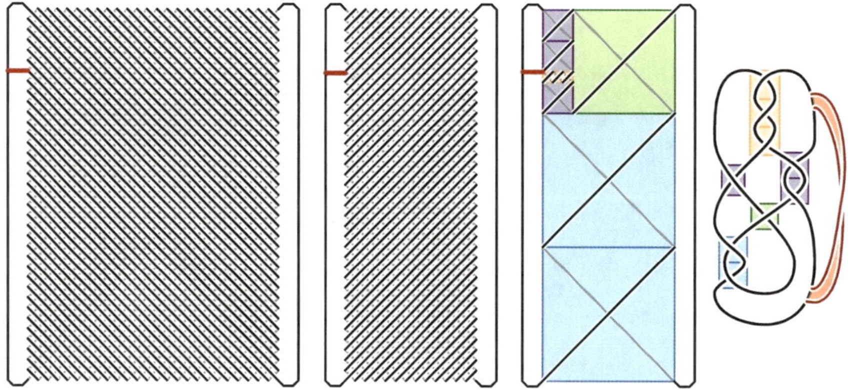

Fig. 23 An isotopy of the pillowcase model of $K(36, 25)$ along with its simple arc $K(36, 25, 6)$ into a cleaner form

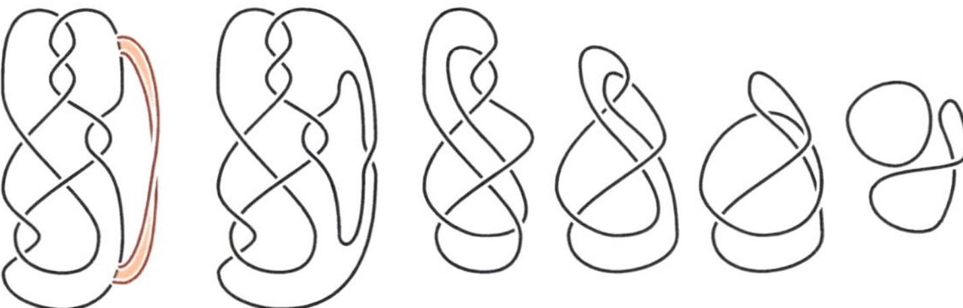

Fig. 24 The result of Fig. 23 is given along with its banding. The result of the banding is then isotoped until it is clearly the unlink

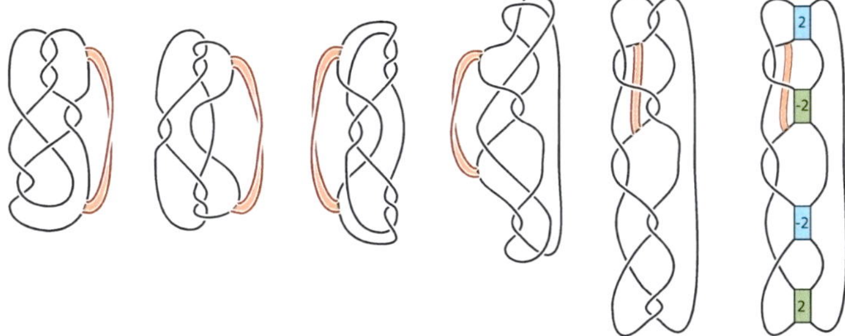

Fig. 25 The result of Fig. 23 is given and followed by an isotopy that recognizes its banding as belonging to family BGV

Acknowledgements We would like to thank Radu Cebanu for sharing his work, John Berge for his discussions about sporadic knots, Neil Hoffman for discussions about involutions and symmetries, Natasha Jonoska and Masahico Saito for organizing the conference and offering me the opportunity to present these results, and my collaborators Dorothy Buck and Ana Lecuona. This work is partially supported by a grant from the Simons Foundation (#209184 to Kenneth L. Baker).

References

1. K.L. Baker, Surgery descriptions and volumes of Berge knots, I: large volume Berge knots. J. Knot Theory Ramif. **17**(9), 1077–1097 (2008). MR 2457837 (2009h:57025)
2. K.L. Baker, Surgery descriptions and volumes of Berge knots, II: descriptions on the minimally twisted five chain link. J. Knot Theory Ramif. **17**(9), 1099–1120 (2008). MR 2457838 (2009h:57026)
3. K.L. Baker, D. Buck, The classification of rational subtangle replacements between rational tangles, Algebr. Geom. Topol. **13**(3), 1413–1463 (2013)
4. K.L. Baker, D. Buck, A.G. Lecuona, Some knots in $S^1 \times S^2$ with lens space surgeries (preprint) arXiv:1302.7011 [math.GT]
5. J. Berge, Some knots with surgeries yielding lens spaces (unpublished manuscript)
6. J. Berge, The knots in $D^2 \times S^1$ which have nontrivial Dehn surgeries that yield $D^2 \times S^1$. Topol. Appl. **38**(1), 1–19 (1991). MR 1093862 (92d:57005)

7. D. Buck, E. Flapan, Predicting knot or catenane type of site-specific recombination products. J. Mol. Biol. **374**, 1186–1199 (2007)

8. R. Cebanu, Une généralisation de la propriété "R". Ph.D. thesis, Université du Québec à Montréal, Nov 2012

9. J.H. Conway, An enumeration of knots and links, and some of their algebraic properties, in *Computational Problems in Abstract Algebra* (Proceedings of a Conference, Oxford, 1967) (Pergamon, Oxford, 1970), pp. 329–358. MR 0258014 (41 #2661)

10. P.R. Cromwell, *Knots and Links* (Cambridge University Press, Cambridge, 2004). MR 2107964 (2005k:57011)

11. D. Gabai, Surgery on knots in solid tori. Topology **28**(1), 1–6 (1989). MR 991095 (90h:57005)

12. D. Gabai, 1-bridge braids in solid tori. Topol. Appl. **37**(3), 221–235 (1990). MR 1082933 (92b:57011)

13. R.E. Gompf, A.I. Stipsicz, 4-*Manifolds and Kirby Calculus*. Graduate Studies in Mathematics, vol. 20 (American Mathematical Society, Providence, 1999). MR 1707327 (2000h:57038)

14. C.McA. Gordon, Dehn surgery on knots, in *Proceedings of the International Congress of Mathematicians, Vol. I, II*, Kyoto, 1990 (Mathematical Society of Japan, Tokyo, 1991), pp. 631–642. MR 1159250 (93e:57006)

15. J.E. Greene, The lens space realization problem. Ann. Math. **177**(2), 449–511 (2013)

16. N.D. Grindley, K.L. Whiteson, P.A. Rice, Mechanisms of site-specific recombination. Annu. Rev. Biochem. **75**, 567–605 (2006)

17. P. Lisca, Lens spaces, rational balls and the ribbon conjecture. Geom. Topol. **11**, 429–472 (2007). MR 2302495 (2008a:57008)

18. J.M. Montesinos, Surgery on links and double branched covers of S^3, in *Knots, Groups, and 3-Manifolds (Papers Dedicated to the Memory of R.H. Fox)*, ed. by R.H. Fox, L.P. Neuwirth. Annals of Mathematics Studies, vol. 84 (Princeton University Press, Princeton, 1975), pp. 227–259. MR 0380802 (52 #1699)

19. J. Rasmussen, Lens space surgeries and L-space homology spheres (preprint). arXiv:0710.2531v1 [math.GT]

20. T. Saito, Dehn surgery and (1,1)-knots in lens spaces. Topol. Appl. **154**(7), 1502–1515 (2007)

Site-Specific Recombination Modeled as a Band Surgery: Applications to Xer Recombination

Kai Ishihara, Koya Shimokawa, and Mariel Vazquez

Abstract The tangle method, first introduced by Ernst and Sumners in the late 1980s, uses tools from knot theory and low-dimensional topology to analyze the topological changes induced by site-specific recombination on a circular DNA substrate. Often, a recombination reaction can be modeled by a band surgery. Here we provide a brief description of the tangle method, followed by an overview of recent applications of Dehn surgeries and band surgeries to the study of XerCD recombination.

1 Introduction

In 1953, Watson and Crick discovered the double-helical structure of DNA [48]. The axis of the double helix can be modeled as a curve in three-dimensional space; when the molecule is circular its topology can be studied. Circular DNA molecules may be knotted or linked. Linear DNA forms are topologically trivial unless the two ends are fixed. In this chapter, we consider enzymes which change the topology of

K. Ishihara
Faculty of Education, Yamaguchi University, 1677-1 Yoshida, Yamaguchi-shi,
Yamaguchi, 753-8511, Japan
e-mail: kisihara@yamaguchi-u.ac.jp

K. Shimokawa (✉)
Department of Mathematics, Saitama University, 255 Shimo-Okubo, Sakura-ku, Saitama-shi,
Saitama, 380-8570, Japan
e-mail: kshimoka@rimath.saitama-u.ac.jp

M. Vazquez
Department of Mathematics, San Francisco State University, 1600 Holloway Ave.,
San Francisco, CA 94132, USA
e-mail: mariel@sfsu.edu

N. Jonoska and M. Saito (eds.), *Discrete and Topological Models in Molecular Biology*,
Natural Computing Series, DOI 10.1007/978-3-642-40193-0_18,

DNA. In particular, we deal with site-specific recombinases and show an application of knot theory to their study.

DNA topology is the study of the geometrical and topological properties of circular DNA. Essentially all reactions involving DNA are influenced by its topology. A good overview of the field of DNA topology can be found in [4]. Some knot theory books include expository chapters on the applications of low-dimensional topology and knot theory to the study of DNA [1, 22, 33]. These books focus on the tangle method introduced by Ernst and Sumners for the analysis of site-specific recombination [21]. In Sects. 2 and 3, we introduce knots, links, and the tangle method. In Sects. 4 and 5, we discuss site-specific recombination as modeled by a rational tangle surgery, which can be interpreted as a Dehn surgery in the covering space. Rational tangle surgeries are divided into band surgeries and nonband surgeries (Sect. 5). The action of tyrosine recombinases is modeled as a band surgery. In Sect. 6, we introduce results on band surgeries which are relevant to the study of site-specific recombination. In Sect. 7, we present an application to the Xer system.

2 Knots and Tangles

A *knot* is a simple closed curve in a three-dimensional space. A disjoint union of knots is called a *link* or *catenane*. Intuitively, a *tangle* is a ball with two strings as illustrated in Fig. 1a. In what follows, we use formal mathematical language to define tangles and related technical terms [33]. A *two-string tangle*, or simply a *tangle*, denoted by $T = (B^3, t)$, is a pair consisting of a three-dimensional ball B^3 and two arcs t. The arcs t are properly embedded in the ball, and the endpoints lie on the boundary of the ball. The endpoints can be mapped to points {NW, NE, SW, SE} on the equatorial circle, thus defining the framing for the tangle (Fig. 1a). A tangle is *rational* if it can be obtained by smooth deformations of the trivial tangle (shown in Fig. 1a, left). More formally, a tangle (B^3, t) is *rational* if there is a homeomorphism of pairs between (B^3, t) and $(D^2 \times [0, 1], \{p_1, p_2\} \times [0, 1])$, where p_1 and p_2 are points in the interior of a two-dimensional disk D^2. Rational tangles are illustrated in Fig. 1a. The class of rational tangles is the simplest class of tangles. There is a one-to-one correspondence between the set of rational tangles and the extended rational numbers $\mathbb{Q} \cup \{\frac{1}{0}\}$ [11]. A rational tangle can be untangled by a finite sequence of horizontal and vertical twists, and can be expressed using a vector representation, called the Conway vector (Fig. 1a, legend).

The *numerator* $N(T)$ of a tangle T is a knot or link obtained by adding arcs outside the tangle ball to connect NW and NE, and SW and SE (Fig. 1b). A knot or link is *rational* if it can be constructed as the numerator $N(T)$ of a rational tangle T. Rational knots and links are the same as 4-plats and 2-bridge knots and links. The tangle sum $S + T$ can be obtained from two tangles S and T by connecting the two east endpoints of S to the two west endpoints of T (Fig. 1c).

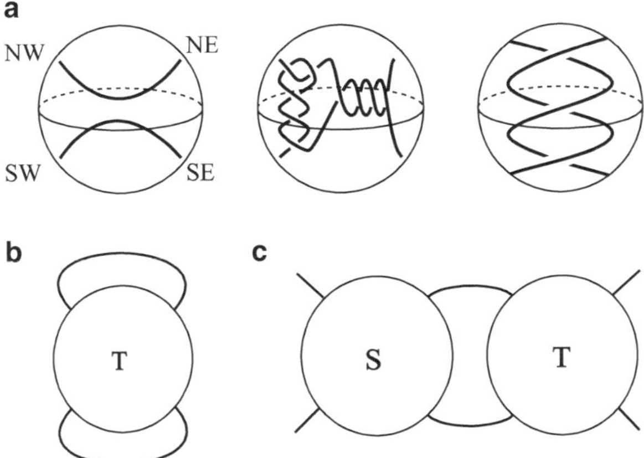

Fig. 1 (a) Rational tangles with Conway vectors (0), $(2, 3, 4)$, and $(-3, 0)$ (*left*, *middle*, and *right*, respectively). The corresponding rational numbers are 0, $\frac{30}{7} = 4 + \frac{1}{3+\frac{1}{2}}$ and $-\frac{1}{3} = 0 + \frac{1}{-3}$, respectively. (b) The numerator $N(T)$ of a tangle T is a knot or a 2-component link. (c) The tangle sum $S + T$ of two tangles S and T

3   Knots, Links, Site-Specific Recombination and the Tangle Method

Circular genomes and naturally occurring plasmids are subject to knotting and linking. Circular DNA forms are common in prokaryotes (e.g. the genome of the bacterium *Escherichia coli* is circular). Even though the DNA in higher-order organisms is commonly linear, it often appears to be subdivided into loops as a consequence of the tight organization of DNA in the cell nucleus, thus justifying its topological study.

Site-specific recombinases are ubiquitous enzymes whose cellular role is to change the genetic code of an organism by integrating a DNA segment into another one, excising a DNA segment, moving a DNA segment to a new location, or inverting a DNA segment within a genome. They belong to one of two families, based on sequence homology and strand-passage mechanism: serine recombinases and tyrosine recombinases [25]. Serine recombinases bind two specific DNA sites, introduce one double-stranded break at each site, recombine the open ends, and reseal the ends. Unlike enzymes in the serine family, tyrosine recombinases act through a Holiday junction intermediate, performing single-stranded cleavage in two steps.

Through the process of cleavage and strand exchange, site-specific recombinases are able to change the topology of their DNA substrates. Changes in topology can be observed experimentally by taking closed circular DNA substrates and incubating them with the enzyme of choice (e.g. [46, 47]). The recombination products are

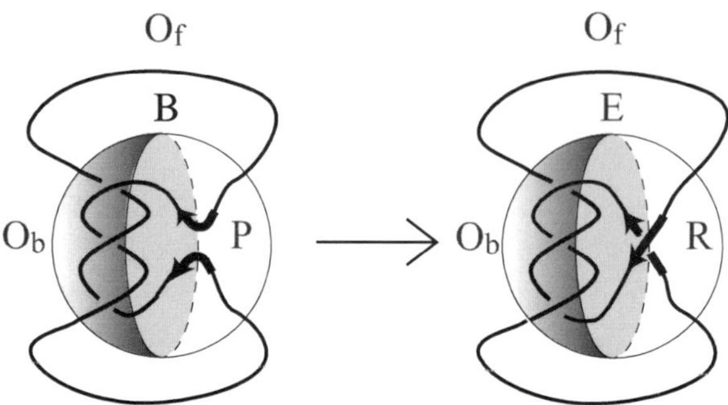

Fig. 2 The tangle B represents two strands of DNA trapped in an enzymatic complex. The tangle O_f represents the DNA outside the complex. B is partioned into the sum of O_b and P. The tangle O_b contains the DNA inside the complex which remains unchanged during recombination. Recombination changes P into another tangle R. We assume P and R are rational

analyzed by gel electrophoresis, electron microscopy, or atomic force microscopy, reviewed in [4]. The enzymatic mechanism can be analyzed using tangles (reviewed in [40]). The tangle method was introduced by Ernst and Sumners [21] and has been used to characterize topologically the action of several site-specific recombinases (e.g. [3, 8, 9, 13, 16–18, 21, 38, 40, 44, 45]).

In the tangle method, the pair consisting of the enzyme (= 3-ball) and the bound DNA (= two strings) is modeled as an two-string tangle B. The tangle O_f is the exterior of B and contains the DNA not bound by the enzyme. The following biologically reasonable assumptions are made [21]:

1. The topological mechanism of recombination is constant and independent of the topology of the substrate. The tangle B is modeled as the sum of two tangles O_b and P. Recombination occurs in the tangle P. The tangle $O = O_f + O_b$ is called the outside tangle, and it remains unchanged during recombination (see Fig. 2).
2. Recombination is modeled by tangle surgery, where the tangle P is changed into another tangle R. The new tangle $O_b + R$ is denoted by E. P can be assumed to be a rational tangle by restricting the ball to enclose two very short DNA regions (5–50 base pairs long) where cleavage takes place. Sometimes R can be proven to be rational (e.g. [21, 44]). In other cases, enough biological evidence is available to assume that R is a tangle with at most one or two crossings.

In the case of the Xer recombination system in *E. coli*, P is defined so that the length of the DNA inside P is short (32 base pairs) [39, 45], so the tangles P and R cannot be very complicated (see assumption 2 above). By pushing any extraneous twists outside P, we can assume that $P = (0)$. In this example, knowledge of the biochemical reaction of cleavage and strand exchange at the local level allows us to assume that $R = (-1), (0, 0)$ or (1) (reviewed in [3, 45]).

Suppose the substrate has a knot or link of type K_1 and that the product is of type K_2. Then site-specific recombination taking K_1 into K_2 is modeled as a system of tangle equations as follows:

$$\begin{cases} N(O + P) = K_1 \\ N(O + R) = K_2 \end{cases} \tag{1}$$

When the tangles involved are rational or sums of rational tangles, solutions to the equations can be obtained using tangle calculus [19–21]. TopoICE-R and TangleSolve are computer programs which can be used to find solutions to such tangle equations [15, 35].

4 Tangle Surgeries and Dehn Surgeries

We refer the reader to [33, 34] for the mathematical terminology used in this section. By the second assumption of the tangle method, a site-specific recombination reaction can be modeled as a tangle surgery, where the tangle P is converted into R (Fig. 3). Such a tangle surgery can be studied at the covering-space level. Consider the case where P and R are rational. The double branched cover of B^3 along two strings of a rational tangle is a solid torus. Hence a rational tangle surgery corresponds to a particular type of surgery in the covering space, called a Dehn surgery.

Dehn surgery is a method of constructing 3-manifolds using knots and links. Let K be a knot in a 3-manifold M, and let $\text{Ext}(K) = M - N(K)$ be the exterior of K, where $N(K)$ is an open regular neighborhood of K. Let γ be a simple closed curve on $\partial\text{Ext}(K)$. Consider the equivalence class of all such curves under ambient isotopy, which we call the *isotopy class* of γ (see [1, 33] for more formal definitions). This isotopy class is called a *slope*. We attach a solid torus $D^2 \times S^1$ to $\text{Ext}(K)$ so that γ bounds a meridian disk of $D^2 \times S^1$. Let $K(\gamma)$ denote the 3-manifold obtained. We say that $K(\gamma)$ *is obtained from M by a Dehn surgery along K*.

Let M be a 3-manifold such that ∂M is a torus. Let γ be an isotopy class of a simple closed curve on ∂M; this isotopy class is also called a slope. Let $M(\gamma)$ denote the closed 3-manifold obtained by attaching a solid torus to ∂M so that γ bounds a meridian disk of the solid torus. This operation is called a *Dehn filling*. A Dehn surgery can be achieved by removing an open regular neighborhood of a knot and applying a Dehn filling to the resulting space.

As stated before, the action of site-specific recombination can be interpreted at the covering-space level in terms of a Dehn surgery. In many cases, the recombination substrates are trivial knots and the products are rational knots and links. The double cover of the 3-sphere branched along the trivial knot is the 3-sphere, and the double cover of the 3-sphere branched along a rational knot or link is a lens space. Hence, the enzymatic action is modeled as a rational tangle

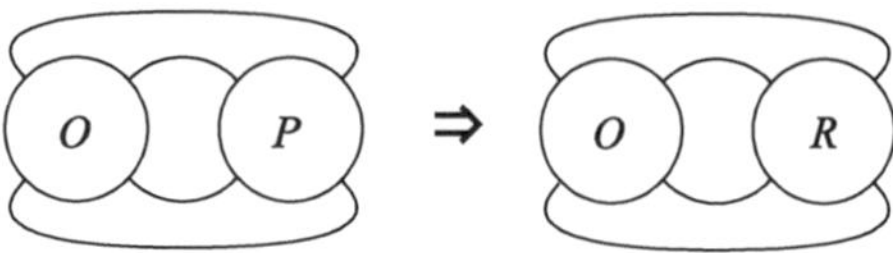

Fig. 3 The tangle method for site-specific recombination. The recombination is modeled by a tangle surgery that changes P into R. The outside tangle O remains unchanged during the recombination reaction

surgery, which corresponds to a Dehn surgery along a knot in the 3-sphere yielding a *lens space*. By construction, the knot is *strongly invertible*. Hence results on Dehn surgeries on strongly invertible knots, such as those in [12, 26], have direct applications to the study of site-specific recombination.

5   Rational Tangle Surgeries

Rational tangle surgeries can be divided into two types: *band surgeries* and *non-band surgeries*. See Sect. 6 for the definition of a band surgery. In the tangle method, the rational tangle surgery taking $P = (0)$ into $R = (1/w)$ for some integer w corresponds to a band surgery. In particular, if $P = (0)$ and $R = (-1), (0, 0)$, or (1), the tangle surgery is a band surgery.

A rational tangle surgery can be converted into a Dehn surgery on a knot by taking the double branched covering. A rational tangle surgery between rational knots and links corresponds to a Dehn surgery on a knot embedded in a lens space, yielding a lens space.

Culler et al. [12] studied Dehn surgeries on knots yielding a 3-manifold M with a cyclic fundamental group (denoted by $\pi_1(M)$). As a lens space has a cyclic fundamental group, we can apply the following result to Dehn surgeries on knots yielding lens spaces.

Theorem 1 ([12] Cyclic surgery theorem). *Let M be a compact, connected, irreducible, orientable 3-manifold such that ∂M is a torus. Suppose M is not a Seifert fibered space. If $\pi_1(M(r))$ and $\pi_1(M(s))$ are cyclic, then $\Delta(r, s) \leq 1$. Hence there are at most three slopes r such that $\pi_1(M(r))$ is cyclic.*

Here, the distance $\Delta(r, s)$ between the slopes r and s is defined by the geometrical intersection number of r and s.

As a corollary to the cyclic surgery theorem, we have the following result about rational tangle surgeries. A nonrational tangle M is called a *Montesinos tangle* if it is homeomorphic to the tangle sum $R_1 + R_2 + \cdots + R_k$ of rational tangles $R_1, R_2, \ldots, R_k$ [19]. The double covering space of S^3, branched along a rational knot or link, is a lens space. Ernst [19] showed that a tangle is either rational or a Montesinos tangle if its double branched covering space branched along two strings is a Seifert fibered space, and obtained the following corollary to the cyclic surgery theorem.

Corollary 1 ([19]). *Suppose $N(O+P)$ and $N(O+R)$ are rational knots and links (including the unknot and the two-component unlink). Assume that $P = (0)$ and R is a rational tangle. If $R \neq (1/w)$ $(w \in \mathbb{Z})$, then O is rational or a Montesinos tangle.*

All such solutions in the case where the tangle O is rational or a Montesinos tangle are calculated in [14, Theorem 3]. Algorithms for computing solutions that are rational or Montesinos, when R is assumed to be integral, were also developed in [20, 43]. Computer programs presented in [15, 35] solve tangle equations under suitable assumptions.

6 Band Surgeries

6.1 Definition of a Band Surgery

Let L be a link in S^3 and let $b : [0, 1] \times [0, 1] \to S^3$ be an embedding such that $b^{-1}(L) = [0, 1] \times \{0, 1\}$. Let L_b be a link obtained from L by replacing a pair of subarcs $b([0, 1] \times \{0, 1\})$ with $b(\{0, 1\} \times [0, 1])$. This operation is called a *band surgery*. For simplicity we use the symbol b to denote the image $b([0, 1] \times [0, 1])$. If L and L_b are oriented and have the same orientation except for the band b, the band surgery is called *coherent*. A coherent band surgery always changes the number of components of the link.

Suppose that two sites on a knot occur in direct repeats. (See the legend of Fig. 4 for the definition of a direct repeat.) We assume that the substrate and the product have orientations inherited from the orientations of the sites.

Observation. *Suppose that a site-specific recombination reaction converting a knot with directly repeated sites to a link, or converting a link to a knot, is modeled by a band surgery. Then the band surgery is coherent with respect to the orientations of the knot and the link induced by the site orientations. (See Fig. 4.)*

6.2 Characterization of Band Surgeries

The following characterizations are known for coherent band surgeries between knots and links.

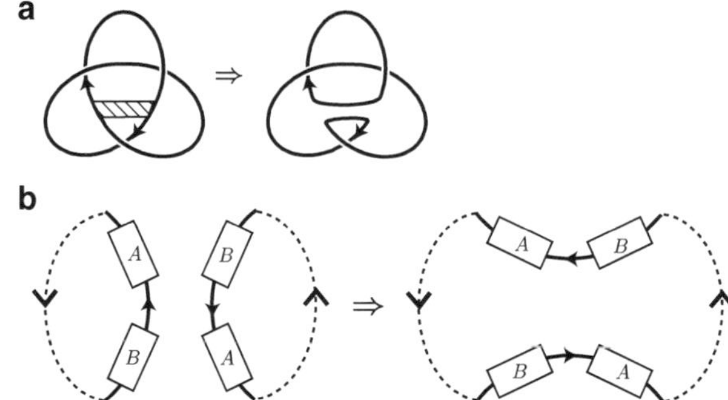

Fig. 4 (**a**) A coherent band surgery from a trefoil knot to a Hopf link (2-cat). The knot and link have the same orientation except at the band. (**b**) The orientation of the site is defined by using its sequence. This orientation induces an orientation of the entire *circle*. A pair of two sites on a knot is called a *direct repeat* if the orientations induced on the entire circle by each of these sites agree. Otherwise, it is called an *inverted repeat*. Site-specific recombination with directly repeated sites corresponds to a coherent band surgery. The part A represents the head part of the site, and B the tail part

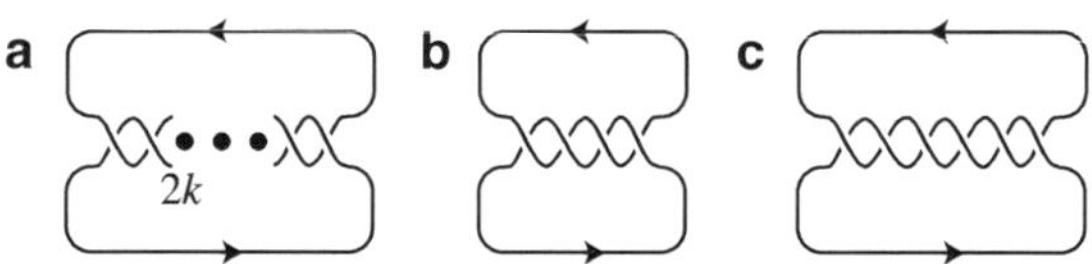

Fig. 5 (**a**) An antiparallel RH (right-handed) $2k$-cat. (**b**) An antiparallel RH 4-cat. (**c**) An antiparallel RH 6-cat

Theorem 2 ([36]). *Let L be a trivial knot. Then L_b is a two-component trivial link if and only if the band is trivial, i.e., there is a disk D bounded by L and $b \subset D$. (See Fig. 6a.)*

Theorem 3 ([26]). *Let L be a trivial knot. Then L_b is a $2k$-cat (see Fig. 5) if and only if b is standard, i.e. there is a disk D bounded by L, $b(\{\frac{1}{2}\} \times [0,1]) \subset D$, and b has $2k$ half-twists with respect to D. (see Fig. 6b.)*

Thompson [42] characterized this band surgery for the case where L_b is a Hopf link. (See Fig. 4a, right.)

Band surgeries on a rational knot $C(2m, 2n)$ yielding a $2k$-cat are characterized in [18]. Note that $C(2m, 2n)$ corresponds to a twist knot if $|m| = 1$ or $|n| = 1$. (See [30] for the definition of twist knots.) See Fig. 5 for pictures of antiparallel right-handed $2k$-cats.

Theorem 4 ([18, 27]). *Let L be a rational knot $C(2m, 2n)$, and let L_b be an antiparallel $2k$-cat. Then b is isotopic to one of the six bands in Fig. 6c, and one of the following holds:*

1. L_b is right-handed and $m = k$, $n = k$, $m + n + 1 = k$, or $m + n - 1 = k$.
2. L_b is left-handed and $m = -k$, $n = -k$, $m + n + 1 = -k$, or $m + n - 1 = -k$.

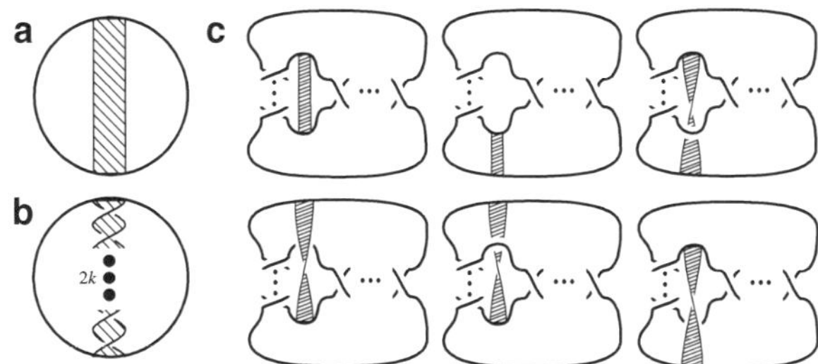

Fig. 6 (**a**) Band surgery from a trivial knot to a two-component trivial link. (**b**) Band surgery from a trivial knot to a $2k$-cat. The band has $2k$ half-twists. (**c**) Band surgeries from $C(2m, 2n)$ to $2k$-cat

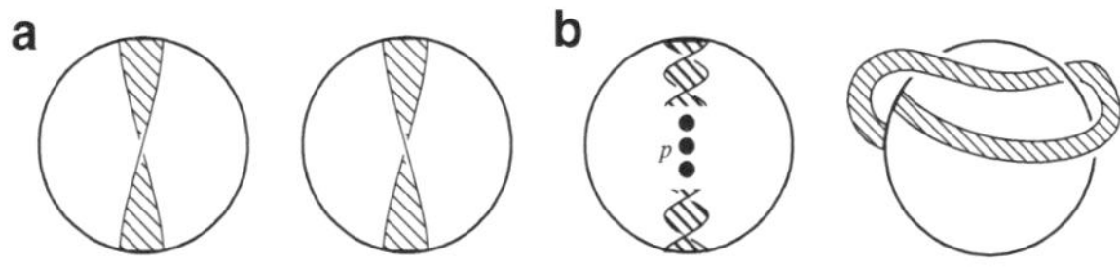

Fig. 7 (**a**) Band surgery from a trivial knot to a trivial knot. (**b**) Band surgery from a trivial knot to a $(2, p)$-torus knot. The band has p half-twists

For noncoherent band surgeries, there are two results on the characterization of band surgeries.

Theorem 5 ([6]). *Let L be a trivial knot. Then L_b is a trivial knot if and only if the band is trivial, i.e., there is a disk D bounded by L, $b(\{\frac{1}{2}\} \times [0, 1]) \subset D$, and b has a half-twist with respect to D. (See Fig. 7a.)*

By using recent results on Dehn surgery of knots [31, 41], the following result was obtained in [27].

Theorem 6 ([27]). *Let L be a trivial knot. Then L_b is a $(2, p)$-torus knot if and only if one of the following holds:*

1. *b is standard, i.e., there is a disk D bounded by L, $b(\{\frac{1}{2}\} \times [0, 1]) \subset D$, and b has p half-twists with respect to D. (See Fig. 7b, left.)*
2. *$p = 5$ and b is isotopic to the band in (Fig. 7b, right).*

6.3   *Band Surgeries and Polynomial Invariants of Links*

In order to show the nonexistence of a band surgery between two given links, link invariants such as the link signature, the Alexander polynomial, and the Jones polynomial are useful.

Let $\sigma(L)$ denote the signature of an oriented link L. See [33] for the definition of $\sigma(L)$.

Theorem 7 ([32]). *Suppose L_b is obtained from L by a coherent band surgery. Then $|\sigma(L_b) - \sigma(L)| \leq 1$.*

It is known that $\sigma(L^*) = -\sigma(L)$, where L^* is the mirror image of L.

Corollary 2. *Suppose a link L is changed into another link L' by a sequence of coherent band surgeries. Let $n = |\sigma(L) - \sigma(L')|$. Then the number of coherent band surgeries in the sequence is at least n.*

Let $\Delta_K(t)$ denote the Alexander polynomial of a knot K.

Theorem 8 ([23]). *Let L be a two-component trivial link. Suppose a knot L_b is obtained by a coherent band surgery from L. Then $\Delta_{L_b}(t) = \pm t^r f(t) f(t^{-1})$ for some integer r and some integral polynomial $f(t)$.*

Kawauchi [29] generalized Theorem 8 to coherent band surgeries on parallel and antiparallel $2k$-cats.

Theorem 9 ([29]). *Let L be a parallel $2k$-cat. Suppose a knot L_b is obtained by a coherent band surgery from L. Then $\Delta_{L_b}(t) \equiv \pm t^r f(t) f(t^{-1})$ mod $(1 - t)(1 - t^{2k})/1 - t^2$ for some integer r and some integral polynomial $f(t)$.*

Theorem 10 ([29]). *Let L be an antiparallel $2k$-cat. Suppose a knot L_b is obtained by a coherent band surgery from L. Then $\Delta_{L_b}(t) \equiv \pm t^r f(t) f(t^{-1})$ mod k for some integer r and some integral polynomial $f(t)$.*

Kanenobu [28] gave a relation between special values of the Jones polynomial and the Q-polynomial of knots and links before and after a band surgery. Here, we include one result on the Jones polynomial. Let $V(L; t)$ denote the Jones polynomial of a link L, and let $\omega = e^{\pi i/3}$.

Theorem 11 ([28]). *Suppose L_b is obtained from L by a band surgery. Then it follows that $V(L; \omega)/V(L_b; \omega) \in \{\pm i, -\sqrt{3}^{\pm 1}\}$.*

7 Applications to Xer Recombination

In this section, we discuss applications of band surgery theory to the tangle analysis of Xer site-specific recombination. The Xer system of E. coli consists of two tyrosine recombinases, XerC and XerD, that act cooperatively at specific recombination sites. The topological mechanism of XerCD has been studied experimentally for unknotted substrates with two *psi* sites in direct repeats [10], and for substrates that are RH torus catenanes with *psi* sites in antiparallel orientations [5]. Xer recombination at *psi* sites requires the presence of two accessory proteins believed to stabilize a specific synapse geometry prior to recombination, thus conferring topological specificity to the system [10]. In vivo the Xer enzymes act on the bacterial chromosome at *dif* sites to resolve chromosome dimers produced by homologous recombination. More recently, it has been proposed that Xer recombination plays a role in the unlinking of replication links (in the absence of topoIV) [24]. In this section, we briefly review the tangle analysis of the various Xer reactions.

Remark 1. The knot table in [34] is commonly used to identify knots and links. However, this table does not provide a consistent way to distinguish a knot from its mirror image, and this is particularly important when dealing with biological objects. We use here the writhe-guided nomenclature for knots reviewed in [7]. This nomenclature offers a way to systematically distinguish a knot from its mirror image based on the mean writhe of an unbiased ensemble of conformations of the given knot type. The writhe is a geometrical property, which provides a way to measure the chain's entanglement complexity and chirality. Here, given a chiral pair (K_1, K_2) of a fixed knot type K, the knot K_i is called K if its mean writhe is positive, otherwise it is called K^*, the mirror image of K.

7.1 *Xer–psi Recombination on an Antiparallel 2k-cat*

In [45], the recombination mechanism of Xer acting on *psi* sites of an unknot to yield an antiparallel RH 4-cat ((2, 4)-torus link) was characterized using the tangle method. There, it was shown that the tangle O from the tangle equation is rational, using a result from [26].

In [5], it was shown that Xer–*psi* recombination on an antiparallel RH $2k$-cat produces a $2k + 1$ crossing knot for $k \geq 3$. In these reactions, the 4-cat was not recombined by XerCD. Using hybrid *psi–loxP* sites and a mutant of the recombinase Cre, the authors of [2] achieved recombination on an antiparallel RH 4-cat, which was converted to a five-noded knot. Here we consider mathematically the case where an antiparallel RH $2k$-cat produces a $(2k + 1)$-crossing knot for $k = 2$ and 3.

The following theorems follow from the arguments in [27] and [18]. See [27] for the details of the proof. In this section, we use the notation for knots given in [34].

Theorem 12 ([18, 27]). *Suppose the substrate $L = N(O + P)$ is an antiparallel RH 4-cat and the product $K = N(O + R)$ is a 5 crossing knot. Suppose further that $P = (0)$ and $R = (1/w)$ for some integer w. Then K is 5_2^*. If $P = (0)$ and $R = (-1)$, then $O = (-4/3)$.*

Proof. First, note that $\sigma(L) = 1$. The product K is either a torus knot 5_1, a twist knot 5_2, or one of their mirror images. As $\sigma(5_1) = 4$ and $\sigma(5_2) = -2$, K must be 5_2^*. The characterization of such a tangle surgery follows from Theorem 4.

For the case where the substrate is an antiparallel RH 6-cat and the product is a seven crossing knot other than 7_7, we have the following theorem.

Theorem 13 ([18, 27]). *Suppose the substrate $L = N(O + P)$ is an anti-parallel RH 6-cat and the product $K = N(O + R)$ is a seven crossing knot other than 7_7 and 7_7^*. Suppose further that $P = (0)$ and $R = (\frac{1}{w})$ for some integer w. Then K is 7_2^* or 7_4^*.*

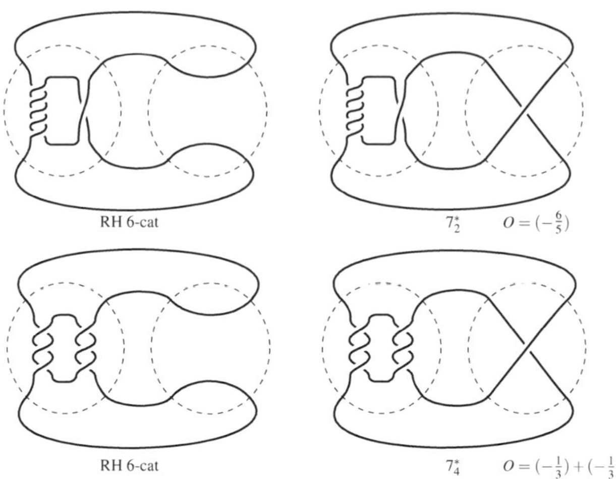

Fig. 8 Tangle models of recombination from antiparallel right-handed 6-cat to 7_2^* and 7_4^*

1. If K is 7_2^*, $P = (0)$ and $R = (-1)$, then $O = (-\frac{6}{5})$.
2. If K is 7_4^*, $P = (0)$ and $R = (-1)$, then $O = (-\frac{1}{3}) + (-\frac{1}{3})$.

(See Fig. 8.)

Proof. As $\sigma(L) = 1$, from Theorem 7 we have $\sigma(K) = 0$ or 2. Then the seven-crossing candidates for K are 7_2^*, 7_4^*, 7_6^*, 7_7, 7_7^* and $3_1^*\#4_1$. As $V(L; \omega) = -\sqrt{3}$ and $V(7_6^*; \omega) = -1$, by applying Theorem 11, we see that there is no coherent band surgery between L and 7_6^*. Theorem 10 gives the proof of nonexistence of a band surgery from L to 7_6^*, as well as to $3_1^*\#4_1$. Band surgeries between antiparallel 6-cats and 7_2^* or 7_4^* are classified in Theorem 4.

We do not know whether a band surgery from an antiparallel RH 6-cat to 7_7 or 7_7^* exists or not.

In [43], we assumed that $R = (k)$ for some integer k, and computed solutions to the above equations that are rational and sums of rational tangles. A solution (which does not correspond to a band surgery) that takes RH 6-cat to the 7_7^* knot was obtained, but in this case $R = (3)$, which is not consistent with the local biochemical mechanism of tyrosine recombinases.

7.2 Unlinking of a Parallel 2k-cat by the Xer–dif–FtsK System

In [24] unlinking of DNA catenanes by the Xer–FtsK system at *dif* sites was reported. In those experiments, parallel $2k$-cats were unlinked gradually. The authors of [24] proposed a *stepwise unlinking model* (Fig. 9) to account for the experimental data. We will characterize the shortest unlinking pathway, as well as the mechanism of each individual step from the trefoil to the unlink. We will also use tangle calculus to find rational and Montesinos tangle solutions to those steps

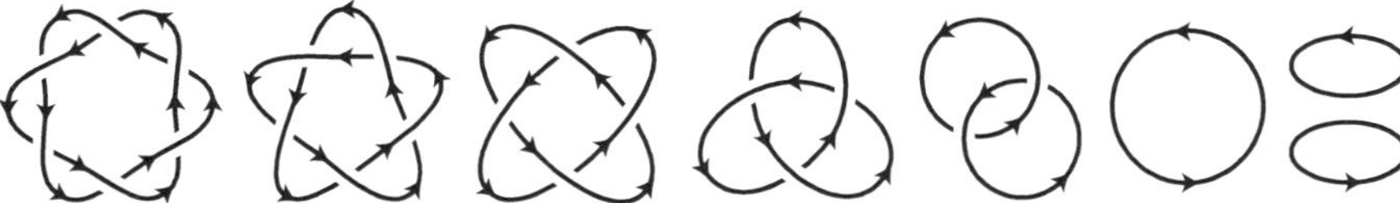

Fig. 9 The *stepwise unlinking model* proposed in [24]. A right-handed 6-cat with sites in parallel alignment is gradually unlinked. The intermediates are an RH five-crossing torus knot, a parallel RH 4-cat, an RH trefoil, a Hopf link, and a trivial knot. In this model, the parallel RH 6-cat is unlinked in six-steps

where a full characterization is not available. The results of this research will be presented in a forthcoming paper [37]. The *iterative recombination model* of this unlinking is also discussed in [38].

Acknowledgements KS would like to thank the conference organizers for giving him the opportunity to talk. KS is supported by JSPS KAKENHI Grant Number 22540066. MV is supported by NSF grants DMS0920887 and DMS1057284.

References

1. C.C. Adams, *The Knot Book: An Elementary Introduction to the Mathematical Theory of Knots* (W. H. Freeman, New York, 1994)
2. A. Akopian, S. Gourlay, H. James, SD Colloms, Communication between accessory factors and the Cre recombinase at hybrid *psi-loxP sites*. J. Mol. Biol. **357**, 1394–1408 (2006)
3. J. Arsuaga, Y. Diao, M. Vazquez, Mathematical methods in DNA topology: application to chromosome organization and site-specific recombination, in *Mathematics of DNA Structure, Function and Interactions*, ed. by C.J. Benham et al. The IMA Volumes in Mathematics and its Applications, vol. 150 (Springer, Dordrecht/New York, 2009), pp. 7–36
4. A.D. Bates, A. Maxwell, *DNA Topology*, 2nd edn. (Oxford University Press, Oxford, 2005)
5. J. Bath, D. Scherratt, S. Colloms, Topology of Xer recombination on catenanes produced by lambda integrase. J. Mol. Biol. **289**, 873–883 (1999)
6. S. Bleiler, M. Scharlemann, A projective plane in $\mathbb{R}^4$ with three critical points is standard. Topology **27**, 519–540 (1988)
7. R. Brasher, R. Scharein, M. Vazquez, New biologically-motivated knot table. Biochem. Soc. Trans. **41**(2), 606–611 (2013)
8. D. Buck, M. Mauricio, Connect sum of lens spaces surgeries: application to Hin recombination. Math. Proc. Camb. Philos. Soc. **150**, 505–525 (2011)
9. D. Buck, C. Verjovsky Marcotte, Tangle solutions for a family of DNA-rearranging proteins. Math. Proc. Camb. Philos. Soc. **139**, 59–80 (2005)
10. S.D. Colloms, J. Bath, D.J. Sherratt, Topological selectivity in Xer site-specific recombination. Cell **88**(6), 855–864 (1997)
11. J.H. Conway, An enumeration of knots band links, and some of their algebraic properties, in *Computational Problems in Abstract Algebra*, ed. by J. Leech (Pergamon Press, Oxford, 1970), pp. 329–358
12. M. Culler, C. Gordon McA, J. Luecke, P.B. Shalen, Dehn surgery on knots. Ann. Math. **125**, 237–300 (1987)

13. I.K. Darcy, Biological distances on DNA knots and links: applications to Xer recombination. J. Knot Theory Ramifi. **10**, 269–294 (2001)
14. I.K. Darcy, Solving unoriented tangle equations involving 4-plats. J. Knot Theory Ramifi. **14**, 993–1005 (2005)
15. I.K. Darcy, R.G. Scharein, TopoICE-R: 3D visualization modeling the topology of DNA recombination. Bioinformatics **22**, 1790–1791 (2006)
16. I.K. Darcy, D.W. Sumners, Rational tangle distances on knots and links. Math. Proc. Camb. Philos. Soc. **128**, 497–510 (2000)
17. I.K. Darcy, J. Luecke, M. Vazquez, Tangle analysis of difference topology experiments: applications to a Mu protein-DNA complex. Algebra. Geome Topol. **9**, 2247–2309 (2009)
18. I.K. Darcy, K. Ishihara, R. Medikonduri, K. Shimokawa, Rational tangle surgery and Xer recombination on catenanes. Algebra. Geome Topol. **12**, 1183–1210 (2012)
19. C. Ernst, Tangle equations. J. Knot Theory Ramifi. **5**, 145–159 (1996)
20. C. Ernst, Tangle equations. II. J. Knot Theory Ramifi. **6**, 1–11 (1997)
21. C. Ernst, D.W. Sumners, A calculus for rational tangles: applications to DNA recombination. Math. Proc. Camb. Philos. Soc. **108**, 489–515 (1990)
22. E. Flapan, *When Topology Meets Chemistry: A Topological Look at Molecular Chirality* (Cambridge University Press, Cambridge, 2000)
23. R.H. Fox, J. Milnor, Singularities of 2-spheres in 4-space and cobordism of knots. Osaka J. Math. **3**, 257–267 (1966)
24. I. Grainge, M. Bregu, M. Vazquez, V. Sivanathan, S. CY Ip, D.J. Sherratt, Unlinking chromosome catenanes in vivo by site-specific recombination. EMBO J. **26**, 4228–4238 (2007)
25. N.D.F. Grindley, K.L. Whiteson, P.A. Rice, Mechanisms of site-specific recombination. Ann. Rev. Biochem. **75**, 567–605 (2006)
26. M. Hirasawa, K. Shimokawa, Dehn surgery on strongly invertible knots which yield lens spaces. Proc. Am. Math. Soc. **128**, 3445–3451 (2000)
27. K. Ishihara, K. Shimokawa, Band surgeries between knots and links with small crossing numbers. Prog. Theor. Phys. Suppl. **191**, 245–255 (2011)
28. T. Kanenobu, Band surgery on knots and links. J. Knot Theory Ramifi. **19**, 1535–1547 (2010)
29. A. Kawauchi, On links by zero-linking twists (in preparation)
30. A. Kawauchi, *A Survey of Knot Theory* (Birkhäuser, Basel, 1995)
31. P. Kronheimer, T. Mrowka, P. Ozsváth, Z. Szabó, Monopoles and lens space surgeries. Ann. Math. **165**, 457–546 (2007)
32. K. Murasugi, On a certain numerical invariant of link types. Trans. Am. Math. Soc. **117**, 387–422 (1965)
33. K. Murasugi, *Knot Theory and Its Applications* (Birkhäuser, Boston, 1996)
34. D. Rolfsen, *Knots and Links* (AMS Chelsea Publishing Vol 346, American Mathematical Society, Providence RI, 2003)
35. Y. Saka, M. Vazquez, TangleSolve: topological analysis of site-specific recombination. Bioinformatics **18**, 1011–1012 (2002)
36. M. Scharlemann, Smooth spheres in $\mathbb{R}^4$ with four critical points are standard. Invent. Math. **79**, 125–141 (1985)
37. K. Shimokawa, K. Ishihara, I. Grainge, D.J. Sherratt, M. Vazquez, (in preparation)
38. K. Shimokawa, K. Ishihara, M. Vazquez, Tangle analysis of DNA unlinking by the Xer/FtsK system. Bussei Kenkyu **92**, 89–92 (2009)
39. D.K. Summers, D.J. Sherratt, Multimerization of high copy number plasmids causes instability: ColE1 encodes a determinant essential for plasmid monomerization and stability. Cell **36**, 1097–1103 (1984)
40. D.W. Sumners, C. Ernst, N.R. Cozzarelli, S.J. Spengler, Mathematical analysis of the mechanisms of DNA recombination using tangles. Q. Rev. Biophys. **28**, 253–313 (1995)
41. M. Tange, Ozsváth Szabó's correction term and lens surgery. Math. Proc. Camb. Philos. Soc. **146**, 119–134 (2009)
42. A. Thompson, Knots with unknotting number one are determined by their complements. Topology **28**, 225–230 (1989)

43. M. Vazquez, Tangle analysis of site-specific recombination: Gin and Xer systems. PhD Dissertation, Florida State University, 2000
44. M. Vazquez, D.W. Sumners, Tangle analysis of Gin site-specific recombination. Math. Proc. Camb. Philos. Soc. **136**, 565–582 (2004)
45. M. Vazquez, S.D. Colloms, D.W. Sumners, Tangle analysis of Xer recombination reveals only three solutions, all consistent with a single three dimensional topology pathway. J. Mol. Biol. **346**, 493–504 (2005)
46. S.A. Wasserman, N.R. Cozzarelli, Determination of the stereostructure of the product of Tn3 resolvase by a general method. Proc. Natl. Acad. Sci. U S A **82**, 1079–1083 (1985)
47. S.A. Wasserman, J.M. Dungan, N.R. Cozzarelli, Discovery of a predicted DNA knot substantiates a model for site-specific recombination. Science **229**, 171–174 (1985)
48. J.D. Watson, F.H.C Crick, A structure for deoxyribose nucleic acid. Nature **171**, 737–738 (1953)

Part V
Dynamics and Kinetics of Molecular Interactions

Understanding DNA Looping Through Cre-Recombination Kinetics

Massa J. Shoura and Stephen D. Levene

Abstract The interior of a cell is a crowded and fluctuating environment where DNA and other biomolecules are both highly constrained and subject to many mechanical forces. The extensive compaction of DNA in living cells is a challenge to many critical biological functions. An evolutionary solution to this challenge may be the juxtaposition of *cis*-acting elements such that multimeric protein complexes simultaneously interact with two or more protein-binding sites. This mode of biological activity involves the formation of looped DNA structures, which, by themselves, are thermodynamically unfavorable. Our knowledge about the roles of DNA bending, twisting, and their respective energetics in DNA looping has come mainly from analyses of ligase-dependent DNA cyclization experiments, which are quantitatively described by the Jacobson–Stockmayer, or J, factor. In this chapter, we discuss a novel quantitative approach to measuring the probability of DNA loop formation in solution using ensemble Förster resonance energy transfer (FRET) measurements of intramolecular and intermolecular Cre-recombination kinetics. Because the mechanism of Cre recombinase does not conform to a simple kinetic scheme, we employ numerical methods to extract rate constants for fundamental steps that pertain to Cre-mediated loop closure.

1 Overview

In order to better understand the physical mechanism of protein-mediated DNA looping, it is necessary to complement both in vitro and in vivo experiments with theoretical approaches that account for the dynamic flexibility of DNA. In the last decade, there has been a series of notable experiments investigating DNA

M.J. Shoura (✉) • S.D. Levene
The University of Texas at Dallas, 800 W. Campbell Rd., Richardson, TX 75080, USA
e-mail: marcela@utdallas.edu; stephen.levene@utdallas.edu

N. Jonoska and M. Saito (eds.), *Discrete and Topological Models in Molecular Biology*, Natural Computing Series, DOI 10.1007/978-3-642-40193-0_19,
© Springer-Verlag Berlin Heidelberg 2014

flexibility and folding that motivated the work described here. We describe a novel analysis of synaptic-complex formation based on the kinetics of the Cre recombination reaction, which yields a quantitative measure of the probability of DNA loop formation, J. Our approach uses time-dependent Förster resonance energy transfer (FRET), a technique that has been used effectively in the analysis of DNA cyclization, to obtain the rates of both intermolecular and intramolecular recombination site synapsis [1]. In addition to providing information about the probability of Cre-mediated loop formation in vitro, our method is potentially applicable to studies of DNA loop formation in living cells.

1.1 *Fluorescence and Förster Resonance Energy Transfer*

Absorption of light (photons) by a population of molecules can induce electronic transitions from a singlet ground level S_0 to an excited state S_1. The excited state can return to the ground state via several different competitive processes. Fluorescence occurs when most of the energy absorbed is emitted as photons while the remaining absorbed energy is nonradiatively dissipated in the surroundings as thermal energy. Thus, the energy of the emitted photons is always lower than that of the absorbed photons. After excitation, a fluorophore (a molecule that emits photons) remains in the excited state for a short time before returning to the ground state; the excited-state, or fluorescence, lifetime ranges from picoseconds to nanoseconds. Fluorescence is characterized by parameters such as the intensity, which equals the number of photons emitted at a given wavelength multiplied by the photon energy, and the quantum yield, the ratio of the number of photons emitted to the number absorbed.

Exogenous molecules added to a fluorescent system can quench the emission and therefore reduce the quantum yield. Furthermore, fluorophores can, under certain conditions, interact through the transfer of energy from an electronically excited fluorophore (the "donor") to a fluorophore in the ground state (the "acceptor"). The excited state of the donor induces an oscillating electric field that excites acceptor electrons, a phenomenon first considered by Theodor Förster [2] and denoted Förster resonance energy transfer. Energy transfer leads to a measurable decrease in both the fluorescence intensity and the fluorescence lifetime of the donor owing to an additional decay pathway from the excited state in the presence of an acceptor. The efficiency of energy transfer, E, is strongly distance-dependent and is given by

$$E = \frac{R_0^6}{R_0^6 + R^6},\tag{1}$$

where R is the distance between the fluorophores and R_0 is the distance at which E is 50 %, which depends on the spectroscopic characteristics of the specific donor–acceptor dye pairs and the spatial relationship between the fluorophores [3–5].

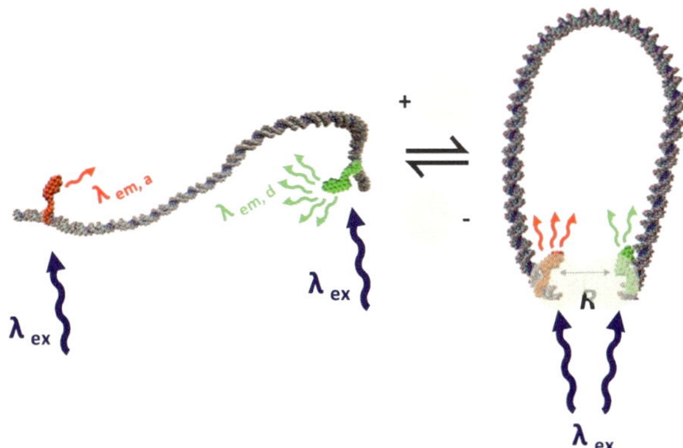

Fig. 1 Principle of Förster resonance energy transfer (*FRET*) applied to DNA loop formation. A linear DNA molecule is labeled with donor (*green*) and acceptor (*red*) modifications near the termini. The FRET detection method described here is based on donor quenching, in which the labeled DNA is excited at a wavelength that primarily excites the donor moiety. In the unlooped state shown on the *left*, the emission intensities from the donor and acceptor are comparable to those observed for the free fluorophores. Addition of a protein that mediates formation of a loop (*right*) brings the DNA ends into close proximity (measured in terms of the end-to-end distance R), giving rise to resonant energy transfer between the fluorophores. Energy transfer in this case is manifested through a decrease in the donor, and corresponding increase in the acceptor, emission intensities

This dependence on R makes FRET a very powerful tool, in principle, for measuring fluorophore–fluorophore distances in the range of 1–10 nm (Fig. 1).

1.1.1 Why FRET?

FRET has been widely used in ensemble and single-molecule studies to characterize both intermolecular interactions and intramolecular conformational transitions. For example, FRET has been used to investigate various protein-protein interactions, such as oligomerization of receptors [6] and transcription factor interactions [7]. Intramolecular FRET can give distance information for dyes tethered to specific atoms within a fluorophore-labeled molecule such as in complex molecular structures of nucleic acids and protein–nucleic acid complexes [8]. Short oligonucleotides can be covalently labeled with dyes and assembled in a specific manner to form structures of interest such as four-way DNA junctions [9]. More recently, single-molecule FRET has been used to probe the dynamics of specifically labeled DNA duplexes [10] and partially mobile four-way DNA junctions and RNA hairpins [11, 12].

Kinetic information about complex biological phenomena such as intrinsic or protein-induced conformational changes can also be obtained from FRET. Chromatin has been studied to examine torsional features of nucleosomal DNA [13]. Nucleosomes with combinations of labeled DNA and labeled histones have been used to test the effects of transcription factor interactions on nucleosome structure [14], and to investigate DNA–histone association or dissociation [15, 16]. Other examples include the nucleoprotein filaments formed by *E. coli* RecA and similar protein assemblies [17]. These bind nonspecifically to DNA and promote strand exchange during homologous recombination. Whereas electron microscopy, crystallographic, and NMR studies have provided important insights into the structure and activity of DNA–recombinase complexes, FRET studies can overcome the deficiencies of those techniques to estimate average fluorophore distances and distance distributions within such protein–DNA complexes [18]. With the advent of advanced single-molecule FRET techniques [17, 19, 20], information can be obtained about the conformational dynamics of macromolecules. The capability to directly address conformational flexibility makes FRET a powerful adjunct to classical structural-biology techniques.

1.2 Cre–loxP Site-Specific Recombination

From viruses to vertebrates, many organisms use transposition and site-specific recombination as efficient mechanisms for rearranging genetic material [21–24]. Site-specific recombination differs from homologous recombination in that DNA is exchanged exclusively at specific sequences [25]. Hallmarks of site-specific recombination systems include target-sequence recognition and specific cleavage/religation chemistry. The integrase system of the bacteriophage λ (λ Int) has served as a paradigm for a large superfamily of conservative site-specific recombinases [26].

The genome of the bacteriophage P1 is a 90-kbp circular molecule; like all circular genomes, daughter molecules must be decatenated after replication [27–29]. This process is facilitated by a protein called Cre recombinase, a phage-encoded member of the λ-Int superfamily of recombinases. The Cre mechanism acts on specific sites, denoted *loxP*, in a multistep reaction scheme. A common feature of the λ-Int superfamily is a catalytic tyrosine residue (in the case of Cre Y324), which cleaves DNA in a mechanism similar to the chemistry of topoisomerase IB [30]. The enzymatic reaction progresses in two distinct stages – an initial round of strand cleavage followed by DNA strand exchange to form a stable recombinase-bound Holliday junction [31, 32]. This junction is resolved by a second set of cleavage and strand-exchange steps, yielding recombinant products.

It has been challenging to study the key steps of such reactions in the context of the entire pathway. The mechanism of Cre recombinase and other members of the integrase family remains controversial owing to limited and contradictory three-dimensional structural information for the protein–DNA synaptic complexes that are central intermediates in these pathways [33]. Although crystallographic structures of Cre [31, 34–36], Flp [37], and λ-Int [38] recombinases with various oligonucleotides are available, the structures of the recombinase–DNA intermediates in solution, such as the Cre–DNA synaptic complex, have remained elusive.

Moreover, the dynamic properties of these intermediates and their relation to the overall kinetics of the reaction pathway remain unclear.

The wild-type *loxP* target site for Cre is a 34-bp DNA sequence that consists of two 13-bp inverted repeats flanking an asymmetric 8-bp core region [39]. The core sequence confers an overall directionality on the *loxP* site. Recombination on directly repeated *loxP* sites leads to the exclusive formation of deletion products, whereas recombination of inversely repeated *loxP* sites results in an inversion of the intervening DNA sequence with respect to the parental substrate [40, 41].

2 Thermodynamics of DNA Looping

The bending and twisting flexibility of DNA has been measured by a number of groups by employing ligase-mediated cyclization [1, 42–45]. The Cyclization efficiencies of DNA fragments can be measured in terms of the Jacobson–Stockmayer factor, J, also termed the J factor. The J factor for a DNA segment of specified length can be thought of as the effective concentration of one end of the DNA in the vicinity of the other. Thermodynamically, J is the ratio of an equilibrium constant for conversion of a linear monomer to a circular molecule, K_c, to the corresponding rate constant for dimerization of a linear monomer, K_b [42, 43, 46]. Dividing K_c by K_b removes the thermodynamic contribution from covalent-bond dissociation and reformation so that the resulting expression for J reflects only the thermodynamic cost of constraining the chain ends. The associated free energy for this process in the standard state is then given by

$$\Delta G^0 = -RT \ln J = -RT \ln \left(K_c / K_b \right), \tag{2}$$

where R is the gas constant and T is the absolute temperature. Theoretically, J can be calculated, in principle, from probability densities computed in Monte Carlo simulations of DNA chains subject to specified boundary conditions imposed on the ends of the chain [47].

Zhang et al. [47] developed a rigorous statistical-mechanical theory for DNA looping based on a generalization of polymer cyclization. In their model, DNA conformations are described by standard base-step parameters of tilt, roll, and twist [48]. Their analysis revealed that there are significant quantitative differences between DNA cyclization and looping. These differences are manifested in the amplitude and phase of J as a function of the helical phasing and are sensitive to both the structure of the protein complex that mediates looping and its intrinsic flexibility. Protein-specific geometry and flexibility can couple to the DNA twist in a loop to give unexpected deviations in the periodicity of J compared with that expected according to results for cyclization. Moreover, unlike the case for cyclization, multiple looped conformations involving the same protein structure but different loop geometries can coexist. These details should be considered when one is analyzing DNA loop formation, both in vitro and in vivo.

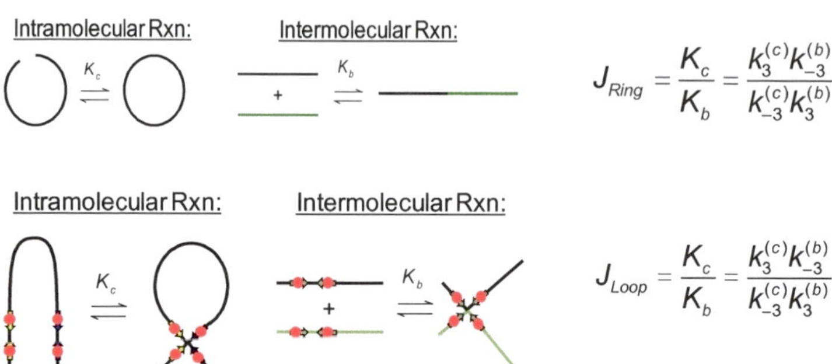

Fig. 2 Comparison between ligase-catalyzed DNA cyclization and DNA looping via Cre recombinase. The cyclization kinetics measures J factors from rates of conversion of linear to circular DNA in the presence of DNA ligase. Rate constants for the intramolecular reaction (a linear to a circular monomer) are then normalized by a factor containing rate constants for the equivalent dimerization reaction for linear molecules (two linear monomers forming a single linear dimer). We extend this scheme to DNA looping performed by Cre recombinase. The intramolecular recombination model shows a linear DNA substrate bearing directly repeated Cre_2loxP sequences (SE$_4$). Formation of the synaptic complex (*SC*) occurs with forward and reverse rate constants $k_3^{(c)}$ and $k_{-3}^{(c)}$. The rate constants $k_4^{(c)}$ and $k_{-4}^{(c)}$ govern the resolution reaction, which leads to product formation. The corresponding model for the intermolecular reaction takes place between two *loxP*-bearing DNA molecules fully occupied by Cre (SE2). The rate constants for the corresponding synapsis and resolution steps are $k_3^{(b)}$, $k_{-3}^{(b)}$, $k_4^{(b)}$, and $k_{-4}^{(b)}$, respectively

3 Measurements of Cre-Mediated DNA Looping

It is not generally possible to measure K_c and K_b directly. The most general method for measuring J experimentally involves determining rate constants separately for intramolecular and intermolecular end-association reactions [46]. We used a similar approach to analyze loop formation in intramolecular Cre-mediated recombination, but this required a number of technical advancements to address the increased complexity of recombinase kinetics relative to ligase-mediated cyclization kinetics (Fig. 2).

3.1 Experimental and Numerical Methods

We used bulk FRET measurements that monitored quenching of a donor dye to quantitate the rate of Cre-mediated site synapsis and recombination. Our experimental design monitored the recombination-mediated exchange of fluorophore-labeled strands on parental duplexes, placing donor and acceptor moieties at adjacent positions on opposing DNA strands in the products (Figs. 3 and 4) Rate constants were obtained by fitting the time-dependent fluorescence signal, $F(t)$, to a system of ordinary differential equations, as described below. This analysis requires knowing

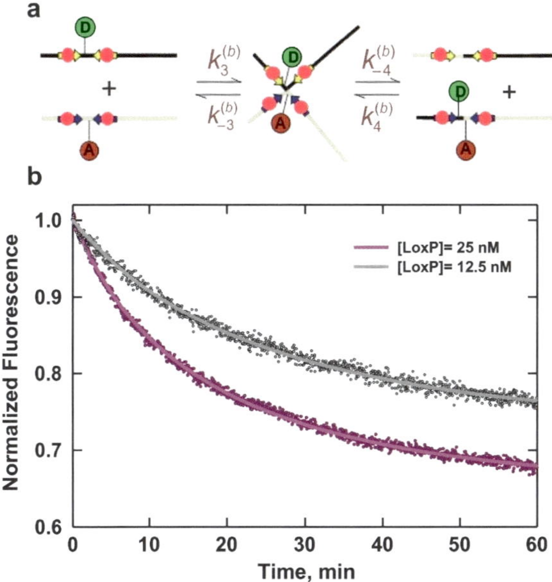

Fig. 3 Time-dependent FRET measurements of intermolecular synapsis and recombination kinetics. (a) Recombination reactions using equimolar ratios of donor-labeled ($T^D B$) and acceptor-labeled (TB^A) duplexes in reactions having total *loxP* concentrations of 12.5 and 25 nM. (b) Time-dependent fluorescence signal, $F(t)$, which monitors recombination-mediated exchange of DNA strands via donor quenching. The recombinant product harbors donor and acceptor moieties at adjacent positions on opposing DNA strands and has a FRET efficiency of 0.99. Rate constants were obtained by fitting each set of $F(t)$ data to the numerical solution of a system of ordinary differential equations (see Fig. 4), which described the time-dependent concentrations of reactants, intermediates, and products along the intermolecular recombination pathway

the extent of donor quenching in the recombination product, which we determined independently from the ratio of the intrinsic donor emission signal from donor-only ($T^D B$) duplexes to that from doubly labeled (donor plus acceptor) duplexes with the probes located at positions identical to those in the recombination product ($T^D B^A$) (see the supplementary information for Ref. [49]). The kinetics of recombination were essentially independent of Cre concentration.

To reduce the number of unknowns that needed to be fitted to the data, we determined the duplex quenching constant, $\phi_{\text{Dup}}^{\text{DA}} = \overline{f}_{T^D B^A}/\overline{f}_{T^D B}$, which is equal to the ratio of the quantum yields for the donor-labeled duplex in the presence and absence of the acceptor. We obtained a value of 0.01, implying that the donor emission is quenched by 99 % in the duplex bearing donor and acceptor fluorophores in the positions expected in the recombinant product. The same value of this parameter was obtained in the presence and absence of Cre protein, indicating that there is negligible excess quenching of donor emission due to Cre binding alone.

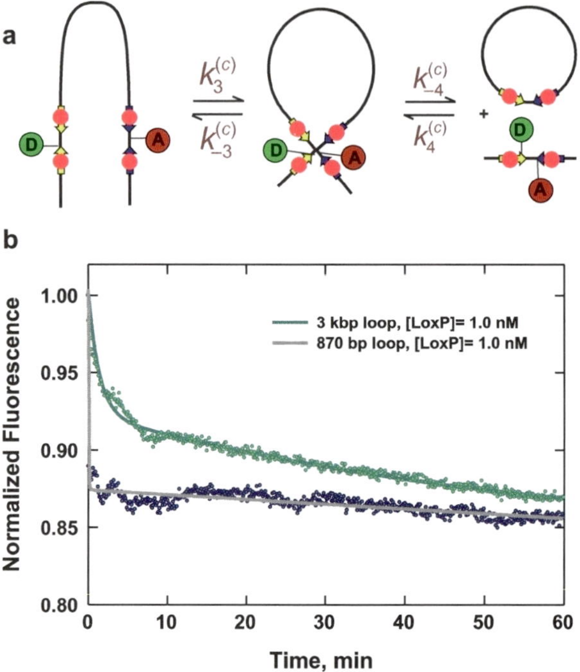

Fig. 4 Intramolecular synapsis and recombination kinetics obtained from time-dependent FRET measurements. (**a**) Schematic of the intramolecular reaction carried out on a DNA fragment bearing donor- and acceptor-labeled *loxP* sites. (**b**) Fluorescence signal, $F(t)$, which monitors donor quenching via FRET during site synapsis and recombination. The Fluorescence decays are for molecules having 3-kbp and 870-bp DNA loops. The positions of the donor and acceptor fluorophores in the labeled product *loxP* sequences are the same as in Fig. 2. Rate constants were obtained by fitting $F(t)$ to a system of ordinary differential equations that described the time-dependent concentrations of reactants, intermediates, and products along the intramolecular recombination pathway (see Fig. 5)

3.1.1 Recombination Pathway for Cre Recombinase: A Mathematical Model

The mechanism of Cre recombinase does not obey a simple Michaelis–Menten scheme; therefore, we employed numerical methods to extract rate constants for fundamental steps in the recombination pathway. The proposed mechanisms of both the intermolecular and the intramolecular recombination pathways and additional details of the curve-fitting procedure are given below and in [49].

The intermolecular and intramolecular recombination pathways share a common set of four rate constants for the elementary recombinase binding and dissociation steps, k_1, k_{-1}, k_2, k_{-2}. The intermolecular and intramolecular site synapsis kinetics are characterized by the apparent rate constants $k_3^{(b)}, k_{-3}^{(b)}, k_4^{(b)}, k_{-4}^{(b)}$ and $k_3^{(c)}, k_{-3}^{(c)}, k_4^{(c)}, k_{-4}^{(c)}$, respectively (Figs. 5 and 6) In our analysis, we fixed

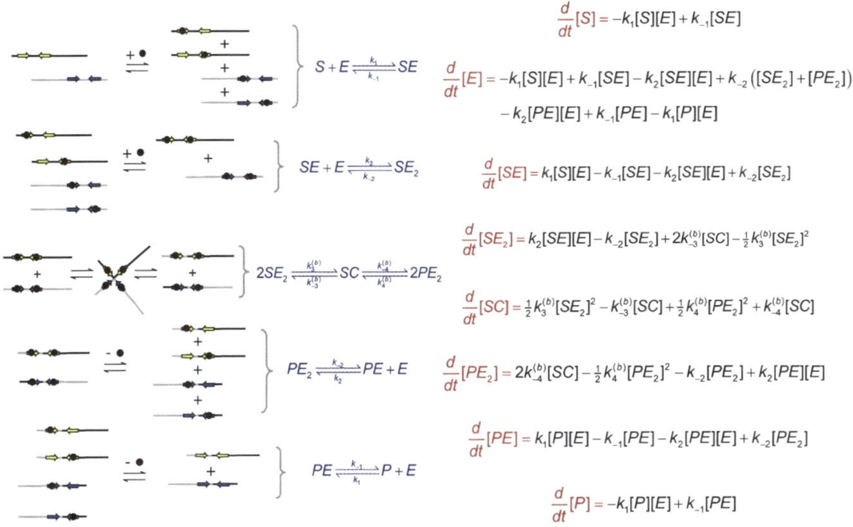

$$\frac{d}{dt}[S] = -k_1[S][E] + k_{-1}[SE]$$

$$S + E \underset{k_{-1}}{\overset{k_1}{\rightleftharpoons}} SE$$

$$\frac{d}{dt}[E] = -k_1[S][E] + k_{-1}[SE] - k_2[SE][E] + k_{-2}([SE_2] + [PE_2])$$
$$- k_2[PE][E] + k_{-1}[PE] - k_1[P][E]$$

$$SE + E \underset{k_{-2}}{\overset{k_2}{\rightleftharpoons}} SE_2$$

$$\frac{d}{dt}[SE] = k_1[S][E] - k_{-1}[SE] - k_2[SE][E] + k_{-2}[SE_2]$$

$$\frac{d}{dt}[SE_2] = k_2[SE][E] - k_{-2}[SE_2] + 2k_{-3}^{(b)}[SC] - \tfrac{1}{2}k_3^{(b)}[SE_2]^2$$

$$2SE_2 \underset{k_{-3}^{(b)}}{\overset{k_3^{(b)}}{\rightleftharpoons}} SC \underset{k_{-4}^{(b)}}{\overset{k_4^{(b)}}{\rightleftharpoons}} 2PE_2$$

$$\frac{d}{dt}[SC] = \tfrac{1}{2}k_3^{(b)}[SE_2]^2 - k_{-3}^{(b)}[SC] + \tfrac{1}{2}k_4^{(b)}[PE_2]^2 + k_{-4}^{(b)}[SC]$$

$$\frac{d}{dt}[PE_2] = 2k_{-4}^{(b)}[SC] - \tfrac{1}{2}k_4^{(b)}[PE_2]^2 - k_{-2}[PE_2] + k_2[PE][E]$$

$$PE_2 \underset{k_2}{\overset{k_{-2}}{\rightleftharpoons}} PE + E$$

$$\frac{d}{dt}[PE] = k_1[P][E] - k_{-1}[PE] - k_2[PE][E] + k_{-2}[PE_2]$$

$$PE \underset{k_1}{\overset{k_{-1}}{\rightleftharpoons}} P + E$$

$$\frac{d}{dt}[P] = -k_1[P][E] + k_{-1}[PE]$$

Fig. 5 Mechanistic steps and kinetic description of the pathway of Cre intermolecular recombination

the parameters k_1, k_{-1}, k_2, k_{-2} at values that were experimentally determined by Ringrose et al. [50] under reaction conditions closely similar to ours. For the intermolecular FRET data, we fitted the time-dependent reduced fluorescence intensity $F(t)$ to five parameters: $k_3^{(b)}, k_{-3}^{(b)}, k_4^{(b)}, k_{-4}^{(b)}$, and ϕ_{SC}^{DA}, an empirical quenching constant for the donor–acceptor pair in the synaptic complex. This parameter is expected to be significantly different from the duplex value, ϕ_{Dup}^{DA}. For intramolecular reactions, we fixed ϕ_{SC}^{DA} at the value obtained for the intermolecular data sets and fitted $F(t)$ using $k_3^{(c)}, k_{-3}^{(c)}, k_4^{(c)}$, and $k_{-4}^{(c)}$ as adjustable parameters. Curve-fitting routines were implemented in MATLAB and used the functions *lsqcurvefit* for nonlinear least-squares fitting and *ode15s* to solve the initial-value problem for the systems of ordinary differential equations.

3.2 Kinetic Characterization of Intermolecular and Intramolecular Cre Recombination

The time-dependent fluorescence of the intermolecular reaction is described very well by the single-intermediate mechanism shown in Fig. 3. The apparent value of ϕ_{SC}^{DA} was virtually constant and equal to 0.12 ± 0.014 for all data sets, independent of [*loxP*] or [Cre]. Detailed studies of the intramolecular reaction were facilitated by a novel fluorophore-labeling method, which places donor and acceptor modifications at specific sites in covalently closed plasmids. We investigated the intramolecular reaction using two linear plasmids that form loops of 3 kbp and

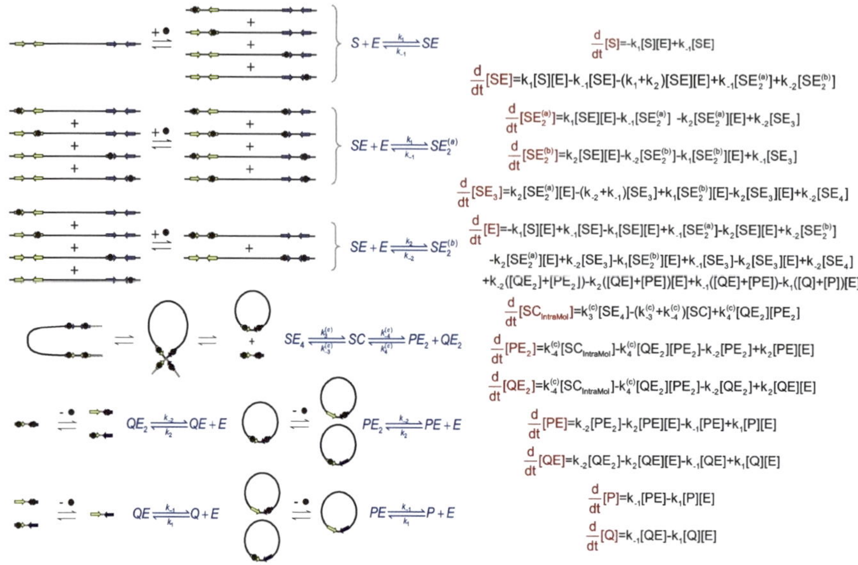

Fig. 6 Mechanistic steps and kinetic description of the pathway of Cre intramolecular recombination

870 bp on synapsis. These substrates had fluorophore-labeled *loxP* sites positioned near the ends of the molecule. In order to avoid interference from the intermolecular reaction, intramolecular-kinetics assays were carried out at substrate concentrations of less than 2 nM ($[loxP] < 4$ nM). This range was below the target-site concentration threshold for intermolecular recombination, which generally requires significantly greater concentrations of the DNA substrate and the recombinase protein. Figure 4 shows that the time-dependent FRET signals were fitted quite well by the solutions of the ordinary differential equations for the intramolecular single-intermediate Cre-recombination pathway. We fixed the value of ϕ_{SC}^{DA} at 0.12, the best-fit value obtained from analysis of the intermolecular reaction, and treated $k_3^{(c)}$, $k_{-3}^{(c)}$, $k_4^{(c)}$, and $k_{-4}^{(c)}$ as adjustable parameters.

The intramolecular recombination reactions showed time-dependent fluorescence signals that were qualitatively different from those observed for the intermolecular reaction. There was a rapid initial decay of the fluorescence emission, followed by a slower, more extended kinetic phase with little change in the fluorescence amplitude beyond 10 min. In the case of the 870-bp substrate, the initial decay phase was faster, and therefore more difficult to characterize, than that for the 3-kbp substrate.

Table 1 Comparison of experimental J-factor values obtained from Cre recombination, J_{exp}, with theoretical J values, J_{theor}, obtained by a transfer matrix method [49]

Loop size (bp)	J_{theor} (nM)	J_{exp} (nM)
3,044	20	18 ($\pm$ 4.7)
870	55	37 ($\pm$ 18)

4 Estimating the Probability of Loop Closure by Recombination Kinetics

Independent measurements of intermolecular and intramolecular recombination rate constants permit quantitative determination of the probability of loop formation, i.e., the J factor. This probability, which is normally expressed in units of concentration, is a generalization of the quantity J defined for cyclization in Sect. 2. Formally, J is a quotient of apparent rate constants for intramolecular and intermolecular synapsis and is given by

$$ J = \frac{K_c}{K_b} = \frac{k_3^{(c)} k_{-3}^{(b)}}{k_{-3}^{(c)} k_3^{(b)}}. \tag{3} $$

Our values of J are based on the kinetics of substrate conversion to synaptic complex and therefore not dependent on the apparent values of $k_4^{(c)}, k_{-4}^{(c)}$ and $k_4^{(b)}, k_{-4}^{(b)}$ determined from the fits to the fluorescence-decay data. Indeed, the latter rate constants are less accurately determined than the other four kinetic parameters because strongly quenched product species contribute little to the total fluorescence signal.

Using (3), we obtained the values of J reported in Table 1. The experimental value J_{exp} for the 3-kbp looped substrate is in excellent agreement with the theoretical J value for both loop conformations. The J_{exp} value for the 870-bp loop falls somewhat below the theoretical value. There is greater uncertainty associated with this measurement, due in large part to the fact that the initial phase of the fluorescence decay is too rapid to be accurately captured in a manual mixing experiment. We note that significant modulations of J due to the helical geometry of DNA are not expected in the range of loop sizes examined here [47]. These modulations were not taken into account in the theoretical calculations.

5 Discussion and Summary

Many methods have been used to directly observe DNA looping in vitro, such as scanning-probe microscopy [51] and electron microscopy [52], and single-molecule techniques [53]. In vivo assays based on helical dependence, in which the DNA length between two protein-binding sites is varied and excess repression or activation of a reporter gene is measured [54, 55], have been a powerful tool in bacterial systems. Several techniques have been developed in the last decade to investigate loop formation in the cells of higher organisms. Chromosome conformation capture (3-C) technology and variants thereof [56] make use of nonspecific protein–protein crosslinking combined with digestion and religation of protein-bound DNA fragments to identify long-range interactions across complex genomes.

However, the number of techniques available for quantifying DNA looping in solution is limited. Extensions of solution-phase in vitro approaches would be valuable complements to multi-C technologies in vivo and may offer improved resolution. We have described a novel approach to characterizing the rate-determining steps for intermolecular and intramolecular synapsis in a Cre site-specific recombination reaction. Cre recombination does not require accessory proteins, DNA supercoiling, or particular metal-ion cofactors and is thus a highly flexible system for quantitatively analyzing DNA loop formation in vitro and in vivo. Our methodology, which could be extended to future in vivo studies, uses time-dependent FRET in conjunction with numerical modeling of the recombination pathway to monitor target site synapsis and the time-dependent yield of Cre-recombination products. The quotient of apparent equilibrium constants for the intramolecular and equivalent intermolecular Cre reactions yields a quantity J that characterizes the thermodynamics of loop-mediated intramolecular site synapsis. Additional information about the synapse geometry is potentially available from FRET, but would require more sophisticated modeling of dye motion within the nucleoprotein complex. The present FRET study was not designed to fully characterize the geometry of the recombination intermediate; however, extensions of this approach could be used in subsequent studies to reveal the dynamic nature of recombination intermediates in solution.

Acknowledgments We thank Andreas Hanke and Stefan Giovan for calculations of J_{theor}. This work was supported by a grant from the NIH/NSF Joint Program in Mathematical Biology (DMS-0800929 from the National Science Foundation) to SDL.

References

1. Y. Zhang, D.M. Crothers, Proc. Natl. Acad. Sci. U. S. A. **100**, 3161–3166 (2003)
2. T. Förster, Z. Naturforsch. A. **4**(7) (1949)
3. R.E. Dale, J. Eisinger, Biopolymers **13**, 1573–1605 (1974)
4. R.E. Dale, J. Eisinger, W.E. Blumberg, Biophys. J. **26**, 161–193 (1979)
5. D. Badali, C.C. Gradinaru, J. Chem. Phys. **134**, 225102 (2011)
6. R.D. Mitra, C.M. Silva, D.C. Youvan, Gene **173**, 13–17 (1996)
7. R. Day, Mol. Endocrinol. **12**, 1410–1419 (1998)
8. B. Treutlein, A. Muschielok, J. Andrecka, A. Jawhari, C. Buchen, D. Kostrewa, F. Hog, P. Cramer, J. Michaelis, Mol. Cell **46**, 136–146 (2012)

9. S.M. Miick, R.S. Fee, D.P. Millar, W.J. Chazin, Proc. Natl. Acad. Sci. U. S. A. **94**, 9080–9084 (1997)
10. A.K. Wozniak, G.F. Schroder, H. Grubmuller, C.A. Seidel, F. Oesterhelt, Proc. Natl. Acad. Sci. U. S. A. **105**, 18337–18342 (2008)
11. S.A. McKinney, A.D. Freeman, D.M. Lilley, T. Ha, Proc. Natl. Acad. Sci. U. S. A. **102**, 5715–5720 (2005)
12. E. Tan, T.J. Wilson, M.K. Nahas, R.M. Clegg, D.M. Lilley, T. Ha, Proc. Natl. Acad. Sci. U. S. A. **100**, 9308–9313 (2003)
13. M. Bussiek, K. Toth, N. Schwarz, J. Langowski, Biochemistry **45**, 10838–10846 (2006)
14. C.L. White, K. Luger, J. Mol. Biol. **342**, 1391–1402 (2004)
15. M. Tomschik, K. van Holde, J. Zlatanova, J. Fluoresc. **19**, 53–62 (2009)
16. C. Bonisch, K. Schneider, S. Punzeler, S.M. Wiedemann, C. Bielmeier, M. Bocola, H.C. Eberl, W. Kuegel, J. Neumann, E. Kremmer, H. Leonhardt, M. Mann, J. Michaelis, L. Schermelleh, S.B. Hake, Nucleic Acids Res. **40**, 5951–5964 (2012)
17. R. Zhou, T. Ha, Methods Mol. Biol. **922**, 85–100 (2012)
18. S.F. Singleton, J. Xiao, Biopolymers **61**, 145–158 (2001)
19. M. Margittai, J. Widengren, E. Schweinberger, G.F. Schroder, S. Felekyan, E. Haustein, M. Konig, D. Fasshauer, H. Grubmuller, R. Jahn, C.A. Seidel, Proc. Natl. Acad. Sci. U. S. A. **100**, 15516–15521 (2003)
20. T. Ha, A.G. Kozlov, T.M. Lohman, Annu. Rev. Biophys. **41**, 295–319 (2012)
21. W.M. Stark, D.J. Sherratt, M.R. Boocock, Cell **58**, 779–790 (1989)
22. S.M. Lewis, Adv. Immunol. **56**, 27–150 (1994)
23. D.J. Sherratt, L.K. Arciszewska, G. Blakely, S. Colloms, K. Grant, N. Leslie, R. McCulloch, Philos. Trans. R. Soc. Lond. B Biol. Sci. **347**, 37–42 (1995)
24. B. Hallet, D.J. Sherratt, FEMS Microbiol. Rev. **21**, 157–178 (1997)
25. R. Oh-McGinnis, M.J. Jones, L. Lefebvre, Brief. Funct. Genomics **9**, 281–293 (2010)
26. A. Landy, Annu. Rev. Biochem. **58**, 913–949 (1989)
27. J.R. Scott, Virology **36**, 564–574 (1968)
28. N. Sternberg, Cold Spring Harb. Symp. Quant. Biol. **43**(Pt 2), 1143–1146 (1979)
29. N. Sternberg, D. Hamilton, S. Austin, M. Yarmolinsky, R. Hoess, Cold Spring Harb. Symp. Quant. Biol. **45**(Pt 1), 297–309 (1981)
30. C.H. Ma, A.H. Kachroo, A. Macieszak, T.Y. Chen, P. Guga, M. Jayaram, PLoS One **4**, e7248 (2009)
31. D.N. Gopaul, F. Guo, G.D. Van Duyne, EMBO J **17**, 4175–4187 (1998)
32. K. Ghosh, G.D. Van Duyne, Methods **28**, 374–383 (2002)
33. A.A. Vetcher, A.Y. Lushnikov, J. Navarra-Madsen, R.G. Scharein, Y.L. Lyubchenko, I.K. Darcy, S.D. Levene, J. Mol. Biol. **357**, 1089–1104 (2006)
34. F. Guo, D.N. Gopaul, G.D. van Duyne, Nature **389**, 40–46 (1997)
35. F. Guo, D.N. Gopaul, G.D. Van Duyne, Proc. Natl. Acad. Sci. U. S. A. **96**, 7143–7148 (1999)
36. E. Ennifar, J.E. Meyer, F. Buchholz, A.F. Stewart, D. Suck, Nucleic Acids Res. **31**, 5449–5460 (2003)
37. Y. Chen, U. Narendra, E.L. Iype, M.M. Cox, A.P. Rice, Mol. Cell **6**, 885–897 (2000)
38. T. Biswas, H. Aihara, M. Radman-Livaja, D. Filman, A. Landy, T. Ellenberger, Nature **435**, 1059–1066 (2005)
39. G.D. Van Duyne, Annu. Rev. Biophys. Biomol. Struct. **30**, 87–104 (2001)
40. R.H. Hoess, A. Wierzbicki, K. Abremski, Nucleic Acids Res. **14**, 2287–2300 (1986)
41. I. Grainge, S. Pathania, A. Vologodskii, R.M. Harshey, M. Jayaram, J. Mol. Biol. **320**, 515–527 (2002)
42. D. Shore, J. Langowski, R.L. Baldwin, Proc. Natl. Acad. Sci. U. S. A. **78**, 4833–4837 (1981)
43. D.M. Crothers, J. Drak, J.D. Kahn, S.D. Levene, Methods Enzymol. **212**, 3–29 (1992)
44. Q. Du, C. Smith, N. Shiffeldrim, M. Vologodskaia, A. Vologodskii, Proc. Natl. Acad. Sci. U. S. A. **102**, 5397–5402 (2005)
45. T.E. Cloutier, J. Widom, Mol. Cell **14**, 355–362 (2004)
46. S.D. Levene, S.M. Giovan, A. Hanke, M.J. Shoura, Biochem. Soc. Trans. **41**, 513–518 (2013)

47. Y. Zhang, A.E. McEwen, D.M. Crothers, S.D. Levene, Biophys. J. **90**, 1903–1912 (2006)
48. R.E. Dickerson, J. Biomol. Struct. Dyn. **6**, 627–634 (1989)
49. M.J. Shoura, A.A. Vetcher, S.M. Giovan, F. Bardai, A. Bharadwaj, M.R. Kesinger, S.D. Levene, Nucleic Acids Res. **40**, 7452–7464 (2012)
50. L. Ringrose, V. Lounnas, L. Ehrlich, F. Buchholz, R. Wade, A.F. Stewart, J. Mol. Biol. **284**, 363–384 (1998)
51. K. Rippe, M. Guthold, P.H. von Hippel, C. Bustamante, J. Mol. Biol. **270**, 125–138 (1997)
52. V.A. Bloomfield, D.M. Crothers, I.J. Tinoco, *Nucleic acids: structures, properties and functions* (University Science Books, Herndon, 2000)
53. L. Finzi, J. Gelles, Science **267**, 378–380 (1995)
54. J. Muller, S. Ochler, B. Muller-Hill, J. Mol. Biol. **257**, 21–29 (1996)
55. T.M. Dunn, S. Hahn, S. Ogden, R.F. Schleif, Proc. Natl. Acad. Sci. U. S. A. **81**, 5017–5020 (1984)
56. E. de Wit, W. de Laat, Genes Dev. **26**, 11–24 (2012)

The QSSA in Chemical Kinetics: As Taught and as Practiced

Casian Pantea, Ankur Gupta, James B. Rawlings, and Gheorghe Craciun

Abstract Chemical mechanisms for even simple reaction networks involve many highly reactive and short-lived species (intermediates), present in small concentrations, in addition to the main reactants and products, present in larger concentrations. The chemical mechanism also often contains many rate constants whose values are unknown a priori and must be determined from experimental measurements of the large species concentrations. A classic model reduction method known as the quasi-steady-state assumption (QSSA) is often used to eliminate the highly reactive intermediate species and remove the large rate constants that cannot be determined from concentration measurements of the reactants and products. Mathematical analysis based on the QSSA is ubiquitous in modeling enzymatic reactions. In this chapter, we focus attention on the QSSA, how it is "taught" to students of chemistry, biology, and chemical and biological engineering, and how it is "practiced" when researchers confront realistic and complex examples. We describe the main types of difficulties that appear when trying to apply the standard ideas of the QSSA, and propose a new strategy for overcoming them, based on *rescaling* the reactive intermediate species.

First, we prove mathematically that the program taught to beginning students for applying the 100-year-old approach of classic QSSA model reduction *cannot*

C. Pantea
Department of Electrical Engineering, Imperial College London, London, UK
e-mail: c.pantea@imperial.ac.uk

A. Gupta · J.B. Rawlings
Department of Chemical and Biological Engineering, University of Wisconsin-Madison, Madison, WI, USA
e-mail: gupta25@wisc.edu; rawlings@engr.wisc.edu

G. Craciun (⊠)
Department of Mathematics and Department of Biomolecular Chemistry, University of Wisconsin-Madison, Madison, WI, USA
e-mail: craciun@math.wisc.edu

N. Jonoska and M. Saito (eds.), *Discrete and Topological Models in Molecular Biology*, 419
Natural Computing Series, DOI 10.1007/978-3-642-40193-0_20,

be carried out for many of the relevant kinetics problems, and perhaps even most of them. By using Galois theory, we prove that the required algebraic equations cannot be solved for as few as five bimolecular reactions between five species (with three intermediates). We expect that many practitioners have suspected this situation regarding nonsolvability to exist, but we have seen no statement or proof of this fact, especially when the kinetics are restricted to unimolecular and bimolecular reactions. We describe algorithms that can test any mechanism for solvability. We also show that an alternative to solving the QSSA equations, the Horiuti–Temkin theory, also does not work for many examples.

Of course, the reduced model (and the full model, for that matter) can be solved numerically, which is the standard approach in practice. The remaining difficulty, however, is how to obtain the values of the large kinetic parameters appearing in the model. These parameters *cannot* be estimated from measurements of the large-concentration reactants and products. We show here how the concept of rescaling the reactive intermediate species allows the large kinetic parameters to be removed from the parameter estimation problem. In general, the number of parameters that can be removed from the full model is less than or equal to the number of intermediate species. The outcome is a reduced model with a set of rescaled parameters that is often identifiable from routinely available measurements. New and freely available computational software (`parest_dae`) for estimating the reduced model's kinetic parameters and confidence intervals is briefly described.

1   Introduction

The quasi-steady-state assumption (QSSA) has become a cornerstone of chemical kinetic modeling and model reduction since its introduction almost a 100 years ago [4,5]. Typical kinetic mechanisms describing any reasonably complex chemical system involve species that have large concentrations, namely the reactants and products, and species that have vanishingly small concentrations, usually referred to as reactive intermediates or simply intermediates. The reactive intermediates have small concentrations because their rates of formation are small compared with their rates of consumption over the range of species concentrations of interest. Models that contain highly reactive intermediates usually display a behavior with two (or more) timescales. The full model exhibits a fast timescale, during which the highly reactive intermediates change from their starting conditions (often zero) to quasi-steady values relative to the reactants and products, and a slow timescale, during which the large-concentration reactants and products evolve. The QSSA is used to remove the highly reactive, low-concentration species from the model and produce a reduced model valid on the slow timescale, which is usually the timescale of interest for analyzing measurements, identifying reduced mechanisms, estimating model parameters, and designing experiments and industrial reactors. As taught in introductory examples, the QSSA reduced model usually contains only the reactants and products. The rate expressions for the production and consumption of reactants and products in terms of only the reactants and products are called the reduced mechanism. Like the QSSA itself, the reduced mechanism is valid over the usual range of species concentrations exhibited by the chemical system of interest.

The quasi-steady-state assumption is widely used in modeling enzymatic reactions, where the intermediates are various enzyme–substrate complexes. The QSSA is at the core of most enzyme kinetics models, such as the Michaelis–Menten kinetics for the basic enzyme reaction [18, 20], the Hill kinetics for cooperative enzymatic models [11, 18], models for enzyme inhibition processes [18], and the Goldbetter–Koshland function in models of futile cycles [9, 24].

The model reduction provides several advantages.

1. *Model validation.* The reduced mechanism allows the model developer to test the structure of the mechanism against experimental measurements. Since the reduced mechanism involves only the more easily measured high-concentration reactants and products, the experimental measurements and therefore the model validation are streamlined.

2. *Parameter estimation.* The full mechanism involving the reactive intermediates involves both large and small rate constants, in which the large rate constants usually correspond to the reactions that consume the intermediates. The large rate constants corresponding to intermediate consumption reactions cannot easily be identified from experimental measurements. To identify these parameters, measurements of the rapidly evolving, low-concentration intermediates are required. In the reduced model, however, these large rate constants often are removed entirely or appear as ratios to other large rate constants. This change in parametrization of the model facilitates estimating the model parameters from slow-timescale measurements of only the high-concentration reactants and products that are typically available.

3. *Model solution.* The evolution of species concentrations is usually described by sets of nonlinear ordinary differential equations (ODEs) that must be solved numerically. Because of the large and small rate constants, the full model's ODEs are often stiff. Even in fortuitous cases in which the large rate constants were somehow available, early ODE solvers often failed to produce accurate solutions for the stiff equations generated by the full model. ODE solvers have improved to the point that even reasonably stiff ODEs corresponding to large, complex chemical mechanisms can be solved accurately. Simplifying the mechanism and reducing the stiffness may, however, lead to a large decrease in the required computation time.

The QSSA method has a long history and a prominent place in the education of chemists, biologists, and chemical and biological engineers. Although the validity of the QSSA model as an approximation to the full mechanism has been studied extensively [6, 10, 20, 21], some fundamental questions about the method have not been addressed. In this chapter, we first explore the following unanswered question: for what class of chemical reactions can the standard procedure for applying the QSSA actually be carried out? The answer to even this basic question is surprising.

We prove in Sect. 2 that the classic QSSA method cannot be carried out for chemical reactions as simple as those with second-order kinetics involving only three intermediates and two reactants and products. In Sect. 3, we consider two simple alternatives to the classical reduction method of the QSSA to overcome this limitation. We show that neither of these simple alternatives is sufficient for model reduction. In Sect. 4, we consider the numerical solution of the full and reduced models, and show how to rescale the species and parameters to obtain a tractable parameter estimation problem. In Sect. 6, we draw conclusions from this study and comment on future research directions.

2   The QSSA and Its Limitations

2.1   The QSSA Method

Reaction networks can involve the formation and consumption of intermediate species, which sometimes are transitory, highly reactive, and unlikely to exist outside the reaction mixture. The quasi-steady-state assumption in based on the rapid equilibration of these species. A slow-timescale model can be derived by setting the net rate of formation of these intermediates to zero. This results in a system of algebraic equations, which, if solved, provides expressions for the concentrations of the intermediates in terms of the reactant and product concentrations. Finally, this permits the construction of reaction-rate expressions for the stable reactants and stable products in terms of the reactant and product concentrations only.

We identify prospective highly reactive intermediates using characteristics such as a high rate of consumption, short lifetime, short induction time, or low concentrations. To see if a model reduction is appropriate for a given chemical mechanism, we also need to see how well the approximate solution obtained using the QSSA describes the exact solution. A detailed discussion of the selection of species in the QSSA and a list of references on this topic are given in [19].

We illustrate the QSSA approach using the following simple reaction network:

$$A \underset{k_{-1}}{\overset{k_1}{\rightleftharpoons}} 2B, \qquad B \xrightarrow{k_2} C. \tag{1}$$

The corresponding system of differential equations is

$$\dot{c}_A = -k_1 c_A + k_{-1} c_B^2,$$

$$\dot{c}_B = 2k_1 c_A - 2k_{-1} c_B^2 - k_2 c_B,$$

$$\dot{c}_C = k_2 c_B. \tag{2}$$

Suppose that $k_2 \gg k_1$, so that the consumption rate of B is large compared with its rate of production; B is a highly reactive intermediate and a candidate for elimination. Setting the production rate of B to zero gives

$$R_B = 2k_1 c_A - 2k_{-1} c_B^2 - k_2 c_B = 0. \tag{3}$$

The reduced model for this example is obtained by solving this equation for c_B and substituting into the differential equations for A and C in (2). Here we select the single nonnegative solution for c_B, but multiple nonnegative roots of QSSA algebraic systems are possible in general. In those cases, one usually selects the solution that leads to the reduced model which best fits the available data [23]. We call this procedure the standard approach to applying the QSSA. We obtain the system of differential equations

$$\dot{c}_A = -k_1 c_A + k_{-1} \left(\frac{-k_2 + \sqrt{k_2^2 + 16 k_1 k_{-1} c_A}}{4 k_{-1}} \right)^2,$$

$$\dot{c}_C = k_2 \cdot \frac{-k_2 + \sqrt{k_2^2 + 16 k_1 k_{-1} c_A}}{4 k_{-1}}, \tag{4}$$

which corresponds to the simple reaction

$$A \xrightarrow{\tilde{r}} 2C, \qquad \tilde{r} = \frac{k_2}{2} \left(\frac{-k_2 + \sqrt{k_2^2 + 16 k_1 k_{-1} c_A}}{4 k_{-1}} \right).$$

Here, the $\tilde{r}$ on top of the reaction arrow denotes a reaction rate function (and not a reaction rate constant). A comparison of the solutions to the differential equations (2) and (4) is depicted in Fig. 1. The rate constants used in the simulation were $k_1 = k_{-1} = 1$, $k_2 = 100$ and the initial conditions were $c_A(0) = 1, c_B(0) = c_C(0) = 0$. The two solutions are very close; the QSSA hypothesis for B is legitimate. Another lesson that is learned from model reduction by the QSSA is that *complex kinetics* (see the c_A dependence in the reaction rate $\tilde{r}$) can emerge from nothing more complicated than a few first- and second-order reactions when reactive intermediates are involved.

A key fact that allowed us to apply the QSSA was the possibility of explicitly solving equation (3) using a finite number of operations of addition, subtraction, multiplication, division, and radicals; explicit solutions of this kind are usually referred to as *solutions expressible by radicals*. While we are always able to solve quadratic equations explicitly, solutions expressible by radicals do not always exist for higher-degree polynomial equations. The next section reviews classical results that address this issue.

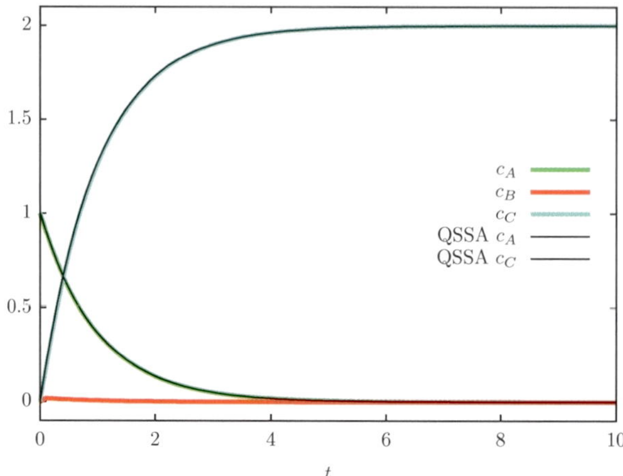

Fig. 1 The concentrations c_A and c_C with and without applying the QSSA to the intermediate B

2.2 Solvability by Radicals

It has been known since 1824, from the work of Niels Abel [1], that there is no formula expressible by radicals for the solution of the general fifth-degree polynomial equation. Around 1830, Évariste Galois produced his celebrated theory that gives a definitive answer to the question of solvability by radicals for any polynomial equation. In particular, this theory shows that the general polynomial equation of degree n cannot be solved using radicals for *any $n \geq 5$.*

Here, we introduce a few notions and theorems that will allow us to use the results of Galois theory. For an excellent overview of the subject, see [15]. The notions and facts about groups and fields presented below are standard and can be found in most textbooks on abstract algebra; see, for example, [14].

Consider the general polynomial equation of degree n,

$$a_n x^n + \ldots + a_1 x + a_0 = 0, \tag{5}$$

with arbitrary coefficients $a_0, \ldots, a_n$ in a field F. For our purposes, F will be the field generated by the coefficients of a polynomial whose roots we want to find. According to the overview of the QSSA in the previous section, we need to solve for the concentrations of intermediates in terms of the concentrations of nonintermediates. Thus our coefficients are not simply elements of the field $\mathbb{Q}$ of rational numbers; they may be polynomial expressions involving variables that represent concentrations of the nonintermediate species. Consequently, our field will be the smallest field that includes all these variables. This field is usually denoted by $\mathbb{Q}(c_1, \ldots, c_l)$, the set of quotients of real polynomials in the variables $c_1, \ldots, c_l$, representing the concentrations of the nonintermediate species. For instance, in the example represented by Eq. (2), to find the concentration c_B we solved $2k_1 c_A - 2k_{-1} c_B^2 - k_2 c_B = 0$, viewed as an equation in c_B with coefficients in $\mathbb{Q}(c_A)$.

If the roots of Eq. (5) are not in F, there always exists a larger field that contains all these roots; let us denote these roots by $x_{01}, \ldots, x_{0n}$. The set of polynomials with variables $x_1, \ldots x_n$ and coefficients in the field F is denoted by $F[x_1, \ldots, x_n]$. A polynomial Q in $F[x_1, \ldots, x_n]$ that vanishes at the point $(x_{01}, \ldots, x_{0n})$ is called a *relation* among the roots of (5). Recall that the *symmetric group S_n* is the group of all permutations of n distinct elements.

Definition 1. The *Galois group* of the algebraic equation (5) over the field F is the subgroup of the symmetric group S_n consisting of the permutations of the roots $(x_{01}, \ldots, x_{0n})$ that preserve all the relations among these roots.

Galois theory introduces the key notion of a *solvable group*; for a detailed description of this notion, see [14]. The following theorem is the main result regarding solvability of algebraic equations by radicals.

Theorem 1. *An algebraic equation of the form (5) is solvable by radicals if and only if its Galois group is solvable.*

Therefore, the question of whether or not the solution of an algebraic equation can be made explicit using radicals is equivalent to checking a certain property of a group involving the coefficients of the equation. Most modern computer algebra software can handle the latter problem; we find the software package Maple [17] particularly suitable for this task.

It can be shown that the Galois group of the general polynomial equation of degree n with arbitrary coefficients is S_n. In view of Theorem 1, the nonsolvability by radicals of the general polynomial equation of degree ≥ 5 is then explained by the following result.

Theorem 2. *The symmetric group S_n is solvable if and only if $n < 5$.*

Before concluding this section, we shall mention another algebraic notion that we will use. Roughly speaking, a *Groebner basis* replaces a system of polynomial equations for n variables by a "nicer" one which has the same set of solutions as the first. For example, if the original system has a finite number of solutions (which will usually be the case for systems arising from chemical kinetics), we can choose a Groebner basis where one of the equations is a univariate polynomial in one of the variables. This way, we are able to address the problem of the explicit solvability of the initial multivariate polynomial system by using Galois theory on the univariate polynomial in the Groebner basis. A standard reference on Groebner bases is [7].

2.3 *Nonsolvable Examples*

As we saw above, a key step in the standard QSSA approach is solving a certain system of polynomial equations. But this is not always possible, and thus we are confronted with an important limitation of this approach to applying the QSSA. In what follows, we present two chemical reaction networks where the standard QSSA approach cannot be used because of the nonsolvability by radicals of the resulting systems of algebraic equations.

The first example consists of the following mechanism:

$$2Y \; \underset{k_{-1}}{\overset{k_1}{\rightleftharpoons}} \; 2B,$$

$$Y + B \; \overset{k_2}{\longrightarrow} \; Z + A,$$

$$Z + B \; \underset{k_{-3}}{\overset{k_3}{\rightleftharpoons}} \; 2X,$$

$$A + X \; \overset{k_4}{\longrightarrow} \; Y + B,$$

$$2Z \; \overset{k_5}{\longrightarrow} \; 2A,$$

whose dynamics is given by the equations

$$\dot{c}_A = k_2 c_B c_Y - k_4 c_A c_X + 2k_5 c_Z^2,$$

$$\dot{c}_B = 2k_1 c_Y^2 - 2k_{-1} c_B^2 - k_2 c_B c_Y - k_3 c_B c_Z + k_{-3} c_X^2 + k_4 c_A c_X,$$

$$\dot{c}_X = 2k_3 c_B c_Z - 2k_{-3} c_X^2 - k_4 c_A c_X,$$

$$\dot{c}_Y = -2k_1 c_Y^2 + 2k_{-1} c_B^2 - k_2 c_B c_Y + k_4 c_A c_X,$$

$$\dot{c}_Z = k_2 c_B c_Y - k_3 c_B c_Z + k_{-3} c_X^2 - 2k_5 c_Z^2. \tag{6}$$

To obtain fast equilibration of X, Y, and Z, we choose k_1, k_{-3}, k_5 to be "large", k_{-1}, k_4 to be "small", and k_2, k_3 to be of order $O(1)$. Then the species X, Y, and Z are fast intermediates, and are candidates for the QSSA approach. Therefore, we need to solve the following system of algebraic equations for c_X, c_Y, and c_Z:

$$2k_3 c_B c_Z - 2k_{-3} c_X^2 - k_4 c_A c_X = 0,$$

$$-2k_1 c_Y^2 + 2k_{-1} c_B^2 - k_2 c_B c_Y + k_4 c_A c_X = 0,$$

$$k_2 c_B c_Y - k_3 c_B c_Z + k_{-3} c_X^2 - 2k_5 c_Z^2 = 0. \tag{7}$$

We have used the software package Maple to generate a Groebner basis of this algebraic system. To facilitate the computation, we chose all the "large" reaction constants to be equal to K, the "small" ones equal to k, and all other reaction constants to be 1. An element of this basis was then found to be the univariate polynomial

$$P(c_Z) = 32k^2 c_B^8 K - 4k^3 c_B^6 c_A^2 - 2k^5 c_B^4 c_A^4 +$$

$$(-k^4 c_A^4 c_B^3 - 2k^2 c_A^2 c_B^5 - 16k^3 K c_B^5 c_A^2) c_Z +$$

$$(32k^3 K^2 c_B^4 c_A^2 + 4k^4 c_A^4 c_B^2 - 64kK^2 c_B^6 + 4k^2 K c_A^2 c_B^4)c_Z^2 +$$

$$(8K^2 k^4 c_A^4 c_B + 48k^2 K^2 c_B^3 c_A^2)c_Z^3 +$$

$$(32K^3 c_B^4 - 16K^3 k^2 c_A^2 c_B^2 + 8K^3 k^4 c_A^4 - 256kK^4 c_B^4)c_Z^4 -$$

$$64K^4 k^2 c_A^2 c_B c_Z^5 +$$

$$(-128K^5 k^2 c_A^2 + 256K^5 c_B^2)c_Z^6 +$$

$$512K^7 c_Z^8.$$

Any c_Z that comes from a solution of the system (7) must also be a root of P. However, using Maple, we found that the Galois group over $\mathbb{Q}(c_A, c_B, k, K)$ of the equation $P(c_Z) = 0$ is S_8. Hence, according to Theorem 1, c_Z cannot be found explicitly in terms of c_A and c_B. Therefore, the QSSA equations are not solvable this case.

Remark 1. To simplify the computations, we set all reaction rates equal to K, k, or 1 in the above calculations. On the other hand, if it were possible to find explicit formulas using radicals for the general set of parameters $k_1, k_{-1}, k_2, k_3, k_{-3}, k_4, k_5$, then the same formulas would apply for the special case of the parameters $K, k, 1$, which, according to the computations above, is impossible.

The second example is a real chemical mechanism for the photochemical decomposition of propanone,

$$CH_3COCH_3 \longrightarrow C_2H_6 + CO.$$

The following mechanism was proposed in [25]:

$$CH_3COCH_3 \underset{k_{-1}}{\overset{k_1}{\rightleftharpoons}} CH_3CO^{\bullet} + CH_3^{\bullet},$$

$$CH_3CO^{\bullet} \overset{k_2}{\longrightarrow} CH_3^{\bullet} + CO,$$

$$CH_3^{\bullet} + CH_3COCH_3 \overset{k_3}{\longrightarrow} CH_4 + {}^{\bullet}CH_2COCH_3,$$

$$CH_3^{\bullet} + CH_3^{\bullet} \overset{k_4}{\longrightarrow} C_2H_6,$$

$$CH_3^{\bullet} + {}^{\bullet}CH_2COCH_3 \overset{k_5}{\longrightarrow} CH_3CH_2COCH_3,$$

$${}^{\bullet}CH_2COCH_3 + {}^{\bullet}CH_2COCH_3 \overset{k_6}{\longrightarrow} CH_3COCH_2CH_2COCH_3,$$

$${}^{\bullet}CH_2COCH_3 \overset{k_7}{\longrightarrow} CH_2CO + CH_3^{\bullet},$$

$$CH_3CO^{\bullet} + CH_3CO^{\bullet} \overset{k_8}{\longrightarrow} CH_3COCOCH_3.$$

To simplify the notation, we write

$$A = CH_3COCH_3, \ B = CO, \ C = C_2H_6, \ D = CH_4, \ E = CH_3CH_2COCH_3,$$

$$F = CH_3COCH_2CH_2COCH_3, \ G = CH_2CO,$$

$$H = CH_3COCOCH_3, \ X = CH_3CO^{\bullet}, \ Y = CH_3^{\bullet}, \ Z = {}^{\bullet}CH_2COCH_3.$$

The corresponding system of differential equations is

$$\dot{c}_A = -k_1 c_A + k_{-1} c_X c_Y - k_3 c_A c_Y, \tag{8}$$

$$\dot{c}_B = k_2 c_X,$$

$$\dot{c}_C = k_4 c_Y^2,$$

$$\dot{c}_D = k_3 c_A c_Y,$$

$$\dot{c}_E = k_5 c_Y c_Z,$$

$$\dot{c}_F = k_6 c_Z^2,$$

$$\dot{c}_G = k_7 c_Z,$$

$$\dot{c}_H = k_8 c_X^2,$$

$$\dot{c}_X = k_1 c_A - k_{-1} c_X c_Y - k_2 c_X - 2k_8 c_X^2,$$

$$\dot{c}_Y = k_1 c_A + k_2 c_X - k_{-1} c_X c_Y - k_3 c_A c_Y - 2k_4 c_Y^2 - k_5 c_Y c_Z + k_7 c_Z,$$

$$\dot{c}_Z = k_3 c_A c_Y - k_5 c_Y c_Z - 2k_6 c_Z^2 - k_7 c_Z.$$

The prospective QSSA intermediates are the radicals X, Y, and Z. The algebraic system in c_X, c_Y, and c_Z is

$$k_1 c_A - k_{-1} c_X c_Y - k_2 c_X - 2k_8 c_X^2 = 0, \tag{9}$$

$$k_1 c_A + k_2 c_X - k_{-1} c_X c_Y - k_3 c_A c_Y - 2k_4 c_Y^2 - k_5 c_Y c_Z + k_7 c_Z = 0,$$

$$k_3 c_A c_Y - k_5 c_Y c_Z - 2k_6 c_Z^2 - k_7 c_Z = 0.$$

If we choose all reaction rate parameters to be 1, then this system has a Groebner basis that contains the following univariate polynomial in c_Z:

$$R(c_Z) = -c_A^5 + c_A^6 + (5c_A^4 - 4c_A^5)c_Z + (-12c_A^3 - 4c_A^5 + c_A^2 + 5c_A^4)c_Z^2 +$$

$$(-16c_A^3 + 16c_A^2 - 2c_A + 8c_A^4)c_Z^3 + (4 + 42c_A^2 - 4c_A - 8c_A^3 + 4c_A^4)c_Z^4 +$$

$$(44c_A^2 + 20)c_Z^5 + (44 + 8c_A + 20c_A^2)c_Z^6 + 48c_Z^7 + 24c_Z^8.$$

As in the previous example, the Galois group of the polynomial $R(c_Z)$ over the field $\mathbb{Q}(c_A)$ is S_8. Therefore c_X, c_Y, and c_Z cannot be expressed in terms of c_A, i.e., the QSSA approach is not solvable in this case.

2.4 Challenges in the Characterization of Systems to Which the Standard QSSA Procedure Can Be Applied

The method described in the previous section can be extended to an algorithm for checking whether the standard QSSA procedure can be applied to a generic system. With all rate constants regarded as symbolic entries, Groebner bases are computed using appropriate monomial orderings, and univariate polynomials are obtained for each intermediate. If all corresponding Galois groups are solvable, we conclude that the standard QSSA procedure may be applied. If at least one such group is not solvable, then the standard QSSA procedure cannot be carried out.

On the other hand, calculations of Groebner bases are combinatorially explosive even for polynomials over real numbers, and are even more computationally expensive over fraction fields with several variables. Therefore, for large enough networks, the computation may not terminate. Moreover, even if the computation terminates and the standard QSSA procedure can be applied, constructing the actual equations of the reduced model may involve further complications, including convoluted algebraic expressions whose signs need to be analyzed, and the possibility of multiple nonnegative roots of the QSSA algebraic system.

Setting some rate constants to simple numerical values, as we have done above, can reduce dramatically the computational size of the problem. However, while this easier problem may be computationally tractable, it is only informative if one of the resulting Galois groups is *not* solvable. In that case we can conclude that the QSSA procedure cannot be applied to the system with general reaction rates. On the other hand, even if all resulting Galois groups are solvable for some fixed reaction rate values, we cannot conclude that the system with general reaction rates is solvable.

3 Alternatives to Solving the QSSA Equations

3.1 Eliminating Reactions Involving Intermediates

The goal in QSSA model reduction is usually not to *evaluate* the concentrations of the intermediates but to *eliminate* them from the reduced model. We next investigate whether there is a general method to remove them short of solving for them. Consider first an example in which this elimination can be done:

$$A \xrightarrow{r_1} B, \qquad B \xrightarrow{r_2} C.$$

We do not assume any simple form for the rate expressions. If B is an intermediate, we set its production rate to zero and obtain

$$R_B = r_1(c_A) - r_2(c_B) = 0.$$

This relationship defines c_B as an implicit function of c_A, but we assume that the algebraic complexity of the function r_2 prevents solving this equation explicitly. For this simple example, knowing only that $r_2(c_B) = r_1(c_A)$ is sufficient. In fact, the production rates of A and C using $r_2 = r_1$ are expressible by

$$R_A = -r_1(A), \qquad R_C = r_1(A),$$

and we have eliminated B and any large rate constants from the model without solving explicitly for c_B as a function of c_A, i.e., $c_B(c_A)$.

But now consider a slightly more complex example:

$$A \xrightarrow{r_1} 2B, \qquad 2B \xrightarrow{r_2} D, \qquad B + C \xrightarrow{r_3} E.$$

If we say B is an intermediate again, we have

$$R_B = 2r_1(c_A) - 2r_2(c_B) - r_3(c_B, c_C) = 0.$$

In general, this equation implicitly defines a function $c_B(c_A, c_C)$, but we again assume that we cannot solve for this function explicitly. The production rates of the reactants and products are as follows:

$$R_A = -r_1(c_A), \qquad R_C = -r_3(c_B, c_C), \qquad R_D = r_2(c_B), \qquad R_E = r_3(c_B, c_C).$$

To eliminate B from the model, we require r_2 and r_3. But the QSSA relation provides only the relation $2r_1 - 2r_2 - r_3 = 0$. We can remove r_2 or r_3 from the model using this equation, but not both. We require the function $c_B(c_A, c_C)$ to substitute into r_2 and r_3 in order to remove B from the reduced model in this example. Eliminating reactions is not a general procedure. If the number of reaction rate expressions in which intermediates appear exceeds the number of intermediates, we do not obtain enough equations to eliminate the intermediates. In the preceding example, we have one intermediate (B), but it appears in two reaction rate expressions (r_2, r_3).

The generalization of this idea of elimination to sets of reactions is known as the Horiuti–Temkin theory [13, 22]. We shall not discuss this generalization further here, because the second simple example shows that the Horiuti–Temkin theory is insufficient in general to eliminate the reaction rates containing intermediates as required to apply the QSSA.

3.2 Solving the Full or Reduced Model Numerically

Since we cannot solve the algebra of the QSSA and we cannot eliminate all reaction rates containing intermediates, we consider next the numerical solution of the full model. A simple approach is to set the rate constants corresponding to consumption of the intermediates large and solve the full model. We assume that the ODE solver can provide accurate solutions given any choice of large rate constants. Alternatively, one can replace the differential equations for the intermediates with the algebraic equations that result from setting their production rates to zero. This procedure produces a set of differential–algebraic equations (DAEs) in place of the ODEs of the full model. The procedure for generating and numerically solving these DAE models is known as computational singular perturbation (CSP). A discussion of the CSP method is provided in [16, 26, 27]. Whether one is numerically solving the full model or the reduced, slow-timescale DAE model, it is generally assumed that all kinetic parameters are known.

To illustrate the issues that arise when the large rate constants are not all known, we consider the following example:

$$A \xrightarrow{k_1} B, \qquad B \xrightarrow{k_2} C, \qquad 2B \xrightarrow{k_3} D, \tag{10}$$

with the full model

$$\dot{c}_A = -k_1 c_A,$$

$$\dot{c}_B = k_1 c_A - k_2 c_B - 2k_3 c_B^2,$$

$$\dot{c}_C = k_2 c_B,$$

$$\dot{c}_D = k_3 c_B^2.$$

The production rate of B is given by

$$R_B = k_1 c_A - k_2 c_B - 2k_3 c_B^2. \tag{11}$$

Setting this production rate to zero and solving for c_B gives the QSSA result,

$$c_B = \frac{2k_1/k_2}{1 + \sqrt{1 + \beta c_A}} c_A, \qquad \beta = 8k_1 k_3/k_2^2.$$

We then express the production rates in terms of the concentrations of only the reactants and products (A, C, and D)

$$R_A = -k_1 c_A, \tag{12}$$

$$R_C = \frac{2k_1}{1 + \sqrt{1 + \beta c_A}} c_A, \tag{13}$$

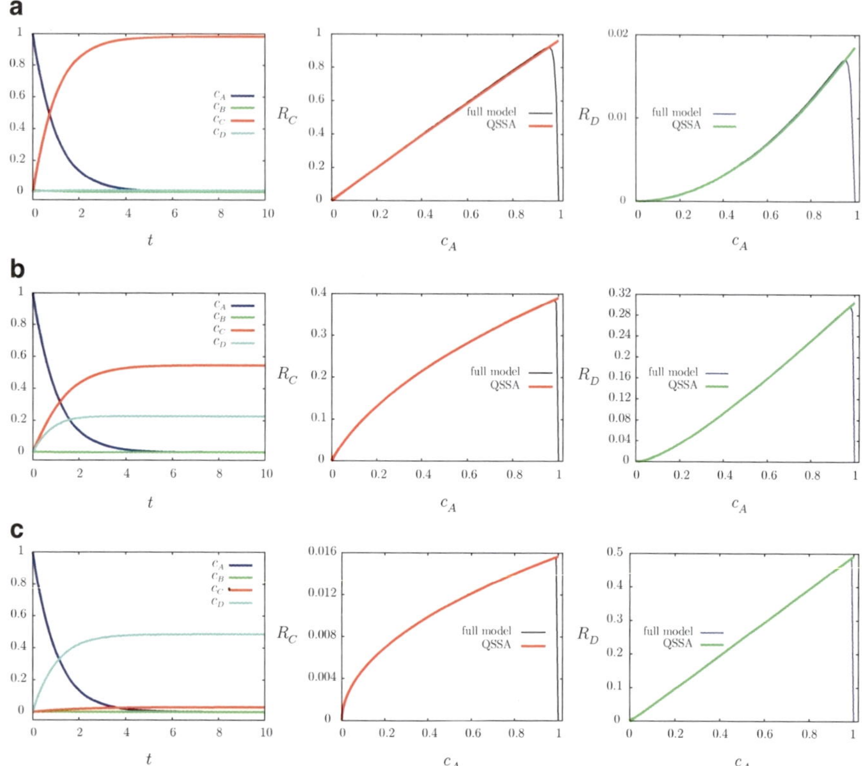

Fig. 2 Numerical solutions of the full model and QSSA approximation. *Left*: c_A, c_B, c_C, c_D versus time. *Middle*: R_C versus c_A. *Right*: R_D versus c_A. (**a**) $k_1 = 1$, $k_2 = 10^2$, $k_3 = 2 \times 10^2$, $\beta = 16 \times 10^{-2}$. R_C versus c_A is linear. (**b**) $k_1 = 1$, $k_2 = 10^2$, $k_3 = 2 \times 10^4$, $\beta = 16$. Neither R_C nor R_D is linear with c_A. (**c**) $k_1 = 1$, $k_2 = 10$, $k_3 = 2 \times 10^5$, $\beta = 16 \times 10^3$. R_D versus c_A is linear

$$R_D = \frac{1}{2} k_1 \left(\frac{-1 + \sqrt{1 + \beta c_A}}{1 + \sqrt{1 + \beta c_A}} \right) c_A. \tag{14}$$

If we were unable to solve the algebra, and instead tried the simple approach of setting all rate constants for consumption of intermediates large, we would set k_2 and k_3 large. But this does not specify the value of β. If we examine the two limiting cases of β large and β small, we obtain the production rates

$$\beta \to 0, \qquad\qquad \beta \to \infty,$$

$$R_A = -k_1 c_A, \qquad R_A = -k_1 c_A,$$

$$R_C = k_1 c_A, \qquad\qquad R_C = 0,$$

$$R_D = 0, \qquad R_D = \frac{1}{2} k_1 c_A$$

corresponding to the two reduced mechanisms

$$A \xrightarrow{k_1} C \quad (\beta \to 0), \qquad A \xrightarrow{k_1} \frac{1}{2} D \quad (\beta \to \infty). \qquad (15)$$

Some relevant numerical results for the example (10) are shown in Fig. 2. Notice that neither the limit $k_2 \gg k_3$ nor $k_3 \gg k_2$ can describe the behavior of the QSSA model for the intermediate range of β shown in Fig. 2b. We see that the QSSA treatment of the full model remains valid, but we cannot obtain the correct behavior by simply setting rate constants large. We need to know the relative sizes of some large rate constants; in this case, we need k_3 / k_2^2. It is a simple matter to estimate the parameters k_1 and β from measurements of C and D, but it is not a simple matter to know the form of the production rates of C and D given in Eqs. (13) and (14). The knowledge of the form of the production rates is the primary benefit of the standard QSSA in this simple example.

4 Rescaling Intermediates, and Parameter Estimation from Data

Since solving for the intermediates in closed form is not always possible, in this section we explore an alternative procedure of rescaling the intermediates. The choice of how to rescale is not unique, but we show in the next section how to automate the procedure for any kinetic model; a primary benefit is that the rescaling procedure often provides a tractable parameter estimation problem as well. For illustrative purposes, consider again the previous example, but assume that we are unable to solve the algebra to determine c_B from setting $R_B = 0$. Instead, we introduce a rescaled B concentration, denoted c_Z, by writing

$$c_Z = k_2 c_B.$$

Note that this choice is not unique. Setting $R_B = 0$ then adds an algebraic equation to the other species' differential equations. The reduced DAE model is

$$\dot{c}_A = -k_1 c_A,$$

$$0 = k_1 c_A - c_Z - 2 K_3 c_Z^2,$$

$$\dot{c}_C = c_Z,$$

$$\dot{c}_D = K_3 c_Z^2.$$

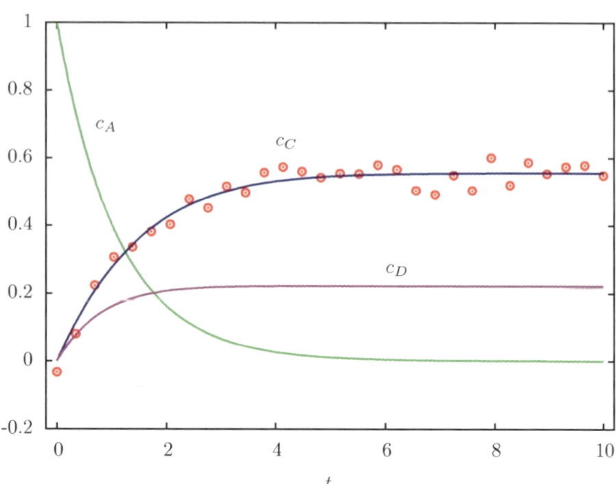

Fig. 3 Measurement of the concentration of C versus time and prediction of the concentrations of A, B, and C from the rescaled model using optimal parameter estimates

Notice that the rescaling removes the two potentially large kinetic parameters k_2 and k_3 and introduces one new scaled parameter, $K_3 = k_3/k_2^2$. Without concentration measurements of reactants and products, we have no idea about the size of K_3; the low concentration of the intermediate B tells us only that k_3 or k_2 or both are large, and it is silent regarding the ratio k_3/k_2^2.

Next, we attempt parameter estimation using both the full and the reduced models. Consider the data set with measurement noise depicted in Fig. 3. To make the parameter estimation problem challenging, we assume that only species C can be measured conveniently at the fairly slow sampling rate ($\Delta = 0.34$) shown in the figure. This data set was generated from the full model using $k_1 = 1.0$, $k_2 = 100$, and $k_3 = 2 \times 10^4$, and the initial conditions $c_{A0} = 1.0$ and $c_{B0} = c_{C0} = c_{D0} = 0.0$. Note that these choices correspond to the intermediate case (b) in Fig. 2.

Normally distributed measurement noise (zero mean, standard deviation $= 0.03$) was added to the c_C of the full model to create the measurement set. Attempting to estimate k_1, k_2, k_3 of the full model from these data is hopeless. An optimizer would find a set of infinitely many estimates with approximately constant k_2/k_3^2, and all points in this set would fit the measurements equally well. This lack of identifiability of parameters plagues all overly complex models using realistic measurement sets.

The estimates obtained for the parameters k_1, K_3 of the reduced model are shown in Fig. 4. Notice that the parameters are well determined and that the approximate 95 % confidence interval is small and contains the "true" parameters used to generate the data.[1] The method used to generate the confidence interval is discussed in standard texts [3, 19]. The values of the estimates and the plus/minus interval (the bounding box of the ellipse shown in Fig. 4) are

---

[1] Note that these are not quite the true parameter values, because we used the full model rather than the reduced model to generate the data.

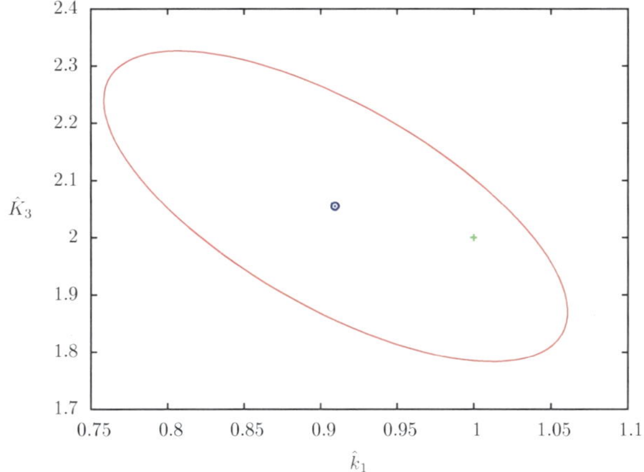

Fig. 4 Optimal parameter estimates $\hat{k}_1$ and $\hat{K}_3$ ($\odot$) based on the measurements of species C in Fig. 3, approximate 95 % confidence interval (*line*), and parameter values used to create the measurements ($+$)

$$\begin{bmatrix} \hat{k}_1 \\ \hat{K}_3 \end{bmatrix} = \begin{bmatrix} 0.910 \\ 2.06 \end{bmatrix} \pm \begin{bmatrix} 0.151 \\ 0.271 \end{bmatrix}, \qquad \begin{bmatrix} k_1 \\ \frac{k_3}{k_2^2} \end{bmatrix} = \begin{bmatrix} 1.0 \\ 2.0 \end{bmatrix}.$$

As we can see, applying the QSSA to the full model provides a validated reduced model that is useful for other scientific studies and for engineering design purposes.

5 A Reparametrization of the QSSA Model

Recall the rescaling of intermediates presented in the previous section, where two kinetic parameters k_2 and k_3 were replaced by a single new parameter K_3. In this section, we show how this rescaling procedure can be performed for any set of QSSA differential–algebraic equations, and highlight some interesting properties of the rescaled model.

Let $(\mathcal{S})$ denote a system of DAEs corresponding to a QSSA model with unknown kinetic parameters and with variables $x_1, \ldots, x_n$. Suppose that the first m variables, $x_1, \ldots, x_m$, correspond to nonintermediate chemical species, and that the variables $x_{m+1}, \ldots, x_n$ correspond to intermediate species. We assume that $(\mathcal{S})$ fulfills the following assumption, which is satisfied by most QSSA systems encountered in practice: for any $t \geq 0$, if $x_1(t), \ldots, x_m(t)$ are known, then the $n - m + 1$ algebraic equations of $(\mathcal{S})$ have unique nonnegative solutions for $x_{m+1}(t), \ldots, x_n(t)$ (note that, as illustrated in Sect. 2.3, these solutions may not be expressible by radicals). As a consequence, a solution of $(\mathcal{S})$ is determined by specifying only the first m coordinates of the initial condition vector.

In what follows, we devise a simpler model $(\bar{\mathcal{S}})$ of $(\mathcal{S})$ with variables $\bar{x}_1, \ldots, \bar{x}_n$ that captures the dynamics of the nonintermediate species $x_1, \ldots, x_m$. More precisely, $(\bar{\mathcal{S}})$ will satisfy the following two properties:

(*i*) $(\bar{\mathcal{S}})$ has fewer parameters than $(\mathcal{S})$;

(*ii*) If $(x_1(t), \ldots, x_n(t))$ and $(\bar{x}_1(t), \ldots, \bar{x}_n(t))$, $t \geq 0$, are solutions of $(\mathcal{S})$ and $(\bar{\mathcal{S}})$ with $x_i(0) = \bar{x}_i(0)$ for all $i \in \{1, \ldots, m\}$, then $x_i(t) = \bar{x}_i(t)$ for all $i \in \{1, \ldots, m\}$ and all $t \geq 0$.

In other words, condition (*ii*) specifies that the outputs $x_1, \ldots, x_m$ from $(\bar{\mathcal{S}})$ and $(\mathcal{S})$ corresponding to nonintermediates are the same.

5.1 Rescaling of the Intermediates

Our strategy is to consider rescalings of intermediates

$$\bar{x}_i = \alpha_i x_i \text{ for } i \in \{m+1, \ldots, n\} \tag{16}$$

that are "optimal" in the sense that the rescaled DAE system has the minimum number of parameters that can be obtained by any reparametrization of the form (16). More precisely, the α_i are chosen such that some monomials with unknown coefficients in $(\mathcal{S})$ are replaced by monomials with coefficient 1 in the reparametrized system $(\bar{\mathcal{S}})$, and this is achieved for as many monomials as possible.

To illustrate, suppose that the DAE system $(\mathcal{S})$ has intermediate variables x, y, z, w, u and suppose that the monomials of $(\mathcal{S})$ involving the intermediate variables are

$$k_1 xy, \ k_2 zw, \ k_3 yz, \ k_4 xw, \ k_5 wu, \text{ and } k_6 yu.$$

Note that these monomials might contain more factors corresponding to nonintermediate variables, but, since these factors do not play a role in our analysis, we will neglect them.

Since the kinetic parameters $k_1, \ldots, k_6$ are unknown positive constants, we treat them as (linearly independent) symbolic indeterminates. Letting

$$\bar{x} = \alpha x, \ \bar{y} = \beta y, \ \bar{z} = \gamma z, \ \bar{w} = \delta w, \ \bar{u} = \eta u,$$

the monomials become

$$\frac{k_1}{\alpha\beta}\bar{x}\bar{y}, \ \frac{k_2}{\gamma\delta}\bar{z}\bar{w}, \ \frac{k_3}{\beta\gamma}\bar{y}\bar{z}, \ \frac{k_4}{\alpha\delta}\bar{x}\bar{w}, \ \frac{k_5}{\delta\eta}\bar{w}\bar{u}, \ \frac{k_6}{\beta\eta}\bar{y}\bar{u}. \tag{17}$$

Let $\mathbf{C}$ denote the vector of the logarithms of the six coefficients in the monomials above, and let $\ln \mathbf{k}$, $\mathbf{v}$, and Ψ denote the 6×1 and 5×1 vectors and the 6×5 matrix, respectively, in the equality

$$
\mathbf{C} = \ln \mathbf{k} - \Psi \mathbf{v} =
\begin{bmatrix} \ln k_1 \\ \ln k_2 \\ \ln k_3 \\ \ln k_4 \\ \ln k_5 \\ \ln k_6 \end{bmatrix}
-
\begin{bmatrix} 1\,1\,0\,0\,0 \\ 0\,0\,1\,1\,0 \\ 0\,1\,1\,0\,0 \\ 1\,0\,0\,1\,0 \\ 0\,0\,0\,1\,1 \\ 0\,1\,0\,0\,1 \end{bmatrix}
\begin{bmatrix} \ln \alpha \\ \ln \beta \\ \ln \gamma \\ \ln \delta \\ \ln \eta \end{bmatrix}
=
\begin{bmatrix} \ln k_1 \\ \ln k_2 \\ \ln k_3 \\ \ln k_4 \\ \ln k_5 \\ \ln k_6 \end{bmatrix}
-
\begin{bmatrix} M_1 \cdot \mathbf{v} \\ M_2 \cdot \mathbf{v} \\ M_3 \cdot \mathbf{v} \\ M_4 \cdot \mathbf{v} \\ M_5 \cdot \mathbf{v} \\ M_6 \cdot \mathbf{v} \end{bmatrix},
\qquad (18)
$$

where M_j denotes row j of Ψ. The maximum number of zero coordinates for the vector $\mathbf{C}$ equals *rank* $\Psi = 4$.

Since M_1, M_2, M_3, and M_5 are linearly independent, we can make the corresponding coordinates of $\mathbf{C}$ equal to zero (and therefore make the coefficients of the first two monomials in (17) equal to 1). We have

$$
\mathbf{C} =
\begin{bmatrix} \ln k_1 \\ \ln k_2 \\ \ln k_3 \\ \ln k_4 \\ \ln k_5 \\ \ln k_6 \end{bmatrix}
-
\begin{bmatrix} M_1 \cdot \mathbf{v} \\ M_2 \cdot \mathbf{v} \\ M_3 \cdot \mathbf{v} \\ (M_1 + M_2 - M_3) \cdot \mathbf{v} \\ M_5 \cdot \mathbf{v} \\ (M_1 + M_5 - M_4) \cdot \mathbf{v} \end{bmatrix}
=
\begin{bmatrix} \ln k_1 \\ \ln k_2 \\ \ln k_3 \\ \ln k_4 \\ \ln k_5 \\ \ln k_6 \end{bmatrix}
-
\begin{bmatrix} \ln k_1 \\ \ln k_2 \\ \ln k_3 \\ \ln k_1 + \ln k_2 - \ln k_3 \\ \ln k_5 \\ \ln k_1 + \ln k_5 - \ln k_4 \end{bmatrix}.
$$

The new monomials

$$
\bar{x}\bar{y}, \ \bar{z}\bar{w}, \ \bar{y}\bar{z}, \ K_1\bar{x}\bar{w}, \ \bar{w}\bar{u}, \ \text{and} \ K_2\bar{y}\bar{u} \qquad (19)
$$

contain only two parameters,

$$
K_1 = \frac{k_4 k_3}{k_1 k_2} \ \text{and} \ K_2 = \frac{k_6 k_4}{k_1 k_5}.
$$

In general, the number of new parameters is equal to the number of old parameters minus the rank of Ψ.

Remarks.

1. The reparametrized system $(\bar{S})$ indeed satisfies the desired properties (i) and (ii). Condition (i) is true, as explained above. Also, the differential equations for $\bar{x}_1, \ldots, \bar{x}_m$ in $(\bar{S})$ are exactly the same as the corresponding differential equations for $x_1, \ldots, x_m$ in (S); therefore, for identical initial conditions, the solutions must coincide, i.e., $\bar{x}_i = x_i$ for $i \in \{1, \ldots, m\}$.

2. Our example shows that the optimal scaling is not always obvious, and that, in general, the number of parameters that can be eliminated may be smaller than the number of intermediates present in the system.
3. Note that, in some cases, even after reducing the number of parameters as described above, we might not obtain a system with uniquely identifiable parameters. This may be the case even if there are no intermediate species; see [8] for examples. In future work, we will analyze this issue in more detail.

5.2 Equivalence of Reparametrizations

As explained above, the reparametrization (19) is optimal with respect to the number of parameters that it removes. However, this reparametrization is not the only one that is optimal. Similarly to the way M_1, M_2, M_3, and M_5 determined an optimal reparametrization in the previous section, any choice of linearly independent rows of Ψ determines another optimal reparametrization.

Recall that the new parameters K_1 and K_2 in (19) are products of powers (positive or negative) of the original parameters $k_1, \ldots, k_6$. This is true for any reparametrization. Interestingly, any two optimal reparametrizations of a system $(\mathcal{S})$, with parameter sets denoted $\mathcal{P}$ and $\tilde{\mathcal{P}}$, are *parameter-set equivalent*, i.e., any element of $\tilde{\mathcal{P}}$ is a product of powers of elements of $\mathcal{P}$ and vice versa.

For example, if we choose $\mathbf{v}$ in (18) such that the second, fourth, fifth, and sixth coordinates of $\mathbf{C}$ are zero (using the fact that M_2, M_4, M_5, and M_6 are linearly independent), we have

$$\mathbf{C} = \begin{bmatrix} \ln k_1 \\ \ln k_2 \\ \ln k_3 \\ \ln k_4 \\ \ln k_5 \\ \ln k_6 \end{bmatrix} - \begin{bmatrix} (M_4 + M_6 - M_5) \cdot \mathbf{v} \\ M_2 \cdot \mathbf{v} \\ (M_2 + M_6 - M_5) \cdot \mathbf{v} \\ M_4 \cdot \mathbf{v} \\ M_5 \cdot \mathbf{v} \\ M_6 \cdot \mathbf{v} \end{bmatrix} = \begin{bmatrix} \ln k_1 \\ \ln k_2 \\ \ln k_3 \\ \ln k_4 \\ \ln k_5 \\ \ln k_6 \end{bmatrix} - \begin{bmatrix} \ln k_4 + \ln k_6 - \ln k_5 \\ \ln k_2 \\ \ln k_2 + \ln k_6 - \ln k_5 \\ \ln k_4 \\ \ln k_5 \\ \ln k_6 \end{bmatrix},$$

and the new monomials are

$$\tilde{K}_1 \bar{x} \bar{y}, \quad \tilde{K}_2 \bar{z} \bar{w}, \quad \bar{y} \bar{z}, \quad \bar{x} \bar{w}, \quad \bar{w} \bar{u}, \text{ and } \bar{y} \bar{u},$$

where

$$\tilde{K}_1 = \frac{k_1 k_5}{k_4 k_6} \text{ and } \tilde{K}_2 = \frac{k_3 k_5}{k_2 k_6}.$$

Note that $\tilde{K}_1 = 1/K_2$ and $\tilde{K}_2 = K_1/K_2$, and the two reparametrizations are indeed parameter-set equivalent.

5.3 *Propanone Example*

We shall now illustrate the usefulness of the reparametrization discussed in this section by estimating parameters for the photochemical decomposition of propanone. A representative data set was generated from the full model described by (8) using $k_1 = 10^{-2}$, $k_{-1} = 10^7$, $k_2 = 10^3$, $k_3 = 10^2$, $k_4 = 5 \times 10^4$, $k_5 = 10^3$, $k_6 = 10^1$, $k_7 = 10^{-1}$, $k_8 = 10^9$, $c_{A0} = 1$, $c_{B0} = c_{C0} = c_{D0} = c_{E0} = c_{F0} = c_{G0} = c_{H0} = c_{X0} = c_{Y0} = c_{Z0} = 0$. These parameter values were chosen so that the major products and reactants A, B, C, D, and H would be present in larger amounts than the minor species E, F, and G, and the reactive intermediates X, Y, and Z would be present in much smaller quantities. Normally distributed measurement noise (zero mean, variance $= 10^{-5}$) was added to c_A, c_B, c_C, c_D, c_E, c_F, c_G, and c_H to create the measurement set. Similarly to the example represented by Eq. (10), estimating all nine parameters of the full model would result in infinitely many parameter sets which fit the data equally well. The reparametrized DAE model had six (rescaled) parameters k_1, $K_{-1} = k_{-1}/(k_2 k_3)$, $K_4 = k_4/k_3^2$, $K_5 = k_5/(k_3 k_7)$, $K_6 = k_6/k_7^2$, and $K_8 = k_8/k_2^2$, which were estimated using the software package `parest_dae` to obtain a good fit to the measurement data and to obtain reasonably small 95 % confidence intervals containing the "true" parameters used to create the measurement data. Figure 5 shows the fit between the measurement data and the estimated solution. The Estimated values of the rescaled parameters and their corresponding confidence intervals are

$$
\begin{bmatrix} \hat{k}_1 \\ \hat{K}_{-1} \\ \hat{K}_4 \\ \hat{K}_5 \\ \hat{K}_6 \\ \hat{K}_8 \end{bmatrix} = \begin{bmatrix} 0.0099 \\ 98.4 \\ 5.03 \\ 101 \\ 993 \\ 996 \end{bmatrix} \pm \begin{bmatrix} 0.0017 \\ 82 \\ 0.39 \\ 5.03 \\ 76 \\ 68 \end{bmatrix}, \qquad \begin{bmatrix} k_1 \\ K_{-1} \\ K_4 \\ K_5 \\ K_6 \\ K_8 \end{bmatrix} = \begin{bmatrix} 0.01 \\ 100 \\ 5 \\ 100 \\ 1000 \\ 1000 \end{bmatrix}.
$$

Note that the confidence interval for the parameter $\hat{K}_{-1}$ is the largest in a relative sense, indicating that this parameter is the least well determined by this experiment. If this uncertainty were deemed too large, one could then apply experimental design methods to determine the optimal experiment to be performed next in order to provide more information about this parameter [2, 19]. This example shows, however, that although estimation of all of the parameters of the full model is not possible, reparametrization of the corresponding DAE model allows us to identify all of the parameters of the reparametrized DAE model. The reduced model is perfectly adequate for predicting the concentrations of all measurable species. If one were interested in identifying all of the full model's rate constants, different experiments using measurements of the QSSA species would be required.

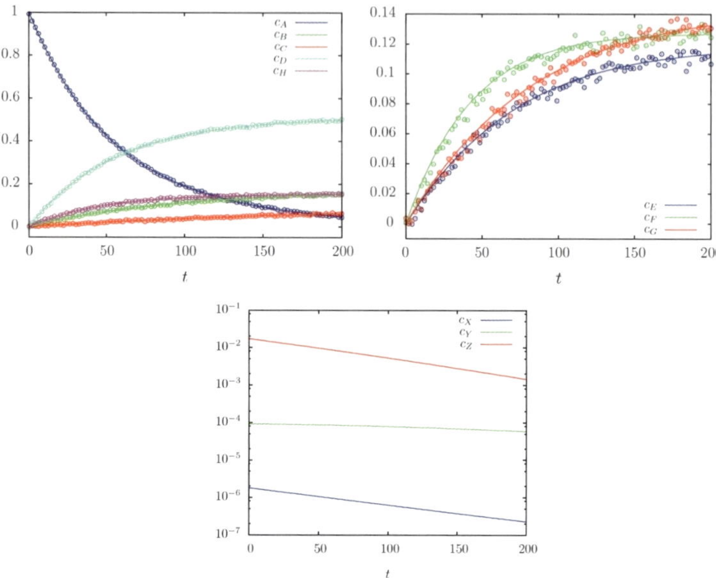

Fig. 5 Measurements of major and minor species and predictions of all species by the reduced DAE model. *Top left*: concentrations of major species c_A, c_B, c_C, c_D, c_H versus time. *Top right*: concentrations of minor species c_E, c_F, c_G versus time. *Bottom*: concentrations of reactive intermediates c_X, c_Y, c_Z versus time

6 Conclusions

The 100-year-old approach of classic QSSA model reduction, as taught in introductory courses, cannot be carried out for many relevant kinetics problems. We have proved that the algebra cannot be solved for even as few as five reactions involving five species (with three intermediates) with nothing more complex than bimolecular mass action kinetics. If any readers can find simpler nonsolvable examples, the authors would like to know about them. We have also analyzed a chemical mechanism taken from the literature and shown that it cannot be reduced by the classical approach. We have described algorithms that can test any mechanism for solvability. We have shown that the alternative approach of the Horiuti–Temkin theory also does not achieve the requirements of model reduction.

The first goal of this chapter was simply to make instructors aware of the limitations of trying to apply the QSSA in this way. Students should probably be told of these limitations when they are introduced to the approach. A second, longer-term goal was to promote the idea of *rescaling* the low-concentration species rather than solving for them. Although the choice of the rescaled parameters is not unique, the minimum number of rescaled parameters remaining in the reduced model *is* unique, and all reduced models with this minimum number of rescaled parameters are equivalent. We expect that in most large enough examples, one cannot solve the QSSA equations explicitly using radicals. Moreover, even when one can solve them explicitly, it may still be preferable to use rescaling in order to identify key combinations of parameters which can be identified from data. The result of the procedure can be a reduced (DAE) model with rescaled parameters that are identifiable from standard measurements. Calculations using the open source software package `parest_dae` that estimates parameters and confidence intervals for DAE models were presented. This package makes use of the recently released SUNDIALS implicit ODE solver IDAS [12].

Looking to the future, as ab initio methods for predicting rate constants become more capable, we may be able to reduce the number of large rate constants and functions of these large constants that must be estimated from measurements. Model validation studies may then be carried out numerically with the full model, rather than by inspecting the reduced model's structural dependence on large-concentration reactants and products. Further research and tool development supporting both rate constant prediction and numerical model validation should prove highly useful to scientists developing and using complex chemical models.

Acknowledgements We thank Andrei Căldăraru and Lev Borisov for useful suggestions and discussions. The work of CP and GC was partially supported by the DOE BACTER Institute, the National Science Foundation, and NIH grant R01GM86881.

References

1. N.H. Abel, Mémoire sur les équations algébriques, où l'on démontre l'impossibilité de la résolution de l'équation générale du cinquième degré in *Oeuvres complètes de Niels Henrik Abel, Édition de Christiana*, vol. 1 (Gröndahl and Son, Oslo, 1881), pp. 28–33
2. A.C. Atkinson, A.N. Donev, *Optimum Experimental Designs* (Oxford University Press, New York, 1992)
3. Y. Bard, *Nonlinear Parameter Estimation* (Academic, New York, 1974)
4. M. Bodenstein, Eine Theorie der photochemischen Reaktionsgeschwindigkeiten. Z. Phys. Chem. **85**, 329–397 (1913)
5. D.L. Chapman, L.K. Underhill, The interaction of chlorine and hydrogen. The influence of mass. J. Chem. Soc. Trans. **103**, 496–508 (1913)
6. A. Ciliberto, F. Capuani, J.J. Tyson, Modeling networks of coupled enzymatic reactions using the total quasi-steady state approximation. PLoS Comput. Biol. **3**(3) e45 (2007)
7. D. Cox, J. Little, D. O'Shea, *Ideals, Varieties, and Algorithms: An Introduction to Computational Algebraic Geometry and Commutative Algebra* (Springer, New York, 2006)
8. G. Craciun, C. Pantea, Identifiability of chemical reaction networks. J. Math. Chem. **44**, 244–259 (2007)
9. A. Goldbeter, D.E. Koshland, An amplified sensitivity arising from covalent modification in biological systems. Proc. Natl. Acad. Sci. U.S.A. **78**(11), 6840–6844 (1981)
10. S. Hanson, S. Schnell, Reactant stationary approximation in enzyme kinetics. J. Phys. Chem. A **112**(37), 8654–8658 (2008)
11. A.V. Hill, The possible effects of the aggregation of the molecules of hemoglobin on its dissociation curves. J. Physiol. **40**, 4–7 (1910)

12. A.C. Hindmarsh, P.N. Brown, K.E. Grant, S.L. Lee, R. Serban, D.E. Shumaker, C.S. Woodward, SUNDIALS: suite of nonlinear and differential/algebraic equation solvers. ACM Trans. Math. Softw. **31**(3), 363–396 (2005)
13. J. Horiuti, T. Nakamura, Stoichiometric number and the theory of steady reaction. Z. Phys. Chem. **11**, 358–365 (1957)
14. M. Isaacs, *Algebra: A Graduate Course* (Brooks Cole, Pacific Grove, 1993)
15. A.G. Khovanskii, On solvability and unsolvability of equations in explicit form. Russ. Math. Surv. **59**(4), 661–736 (2004)
16. S.H. Lam, D.A. Goussis, The CSP method for simplifying kinetics. Int. J. Chem. Kinet. **26**, 461–486 (1994)
17. Maple is a division of Waterloo Maple, Inc. (2007)
18. J.D. Murray, *Mathematical Biology: I. An Introduction* (Springer, New York, 2002)
19. J.B. Rawlings, J.G. Ekerdt, *Chemical Reactor Analysis and Design Fundamentals* (Nob Hill Publishing, Madison, 2004)
20. S. Schnell, P.K. Maini, A century of enzyme kinetics: reliability of the K_M and v_{max} estimates. Comments Theor. Biol. **8**, 169–187 (2003)
21. L.A. Segel, M. Slemrod, The quasi-steady state assumption: a case study in perturbation. SIAM Rev. **31**(3), 446–477 (1989)
22. M.I. Temkin, The kinetics of some industrial heterogeneous catalytic reactions. Adv. Catal. **28**, 173–291 (1979)
23. T. Turanyi, A.S. Tomlin, M.J. Pilling, On the error of the quasi-steady-state approximation. J. Phys. Chem. **97**(1), 163–172 (1993)
24. J.J. Tyson, K.C. Chen, B. Novak, Sniffers, buzzers, toggles and blinkers: dynamics of regulatory and signaling pathways in the cell. Curr. Opin. Cell Biol. **15**, 221–231 (2003)
25. M.R. Wright, *An Introduction to Chemical Kinetics* (Wiley, Chichester/Hoboken, 2004)
26. A. Zagaris, H.G. Kaper, T.J. Kaper, Analysis of the computational singular perturbation reduction method for chemical kinetics. J. Nonlinear Sci. **14**, 59–91 (2004)
27. A. Zagaris, H.G. Kaper, T.J. Kaper, Fast and slow dynamics for the computational singular perturbation method. Multiscale Model. Simul. **2**(4), 613–638 (2004)

Algebraic Models and Their Use in Systems Biology

Reinhard Laubenbacher, Franziska Hinkelmann, David Murrugarra, and Alan Veliz-Cuba

Abstract Progress in systems biology relies on the use of mathematical and statistical models for system-level studies of biological processes. Several different modeling frameworks have been used successfully, including traditional differential-equation-based models, a variety of stochastic models, agent-based models, and Boolean networks, to name some common ones. This chapter focuses on discrete models, and describes a mathematical approach to the construction and analysis of discrete models which relies on combinatorics and computational algebraic geometry. The underlying mathematical concept is that of a polynomial dynamical system over a finite field. Examples are given of the advantages of this approach, and several applications are discussed.

R. Laubenbacher (✉)
Center for Quantitative Medicine, University of Connecticut Health Center,
195 Farmington Ave., Farmington, CT 06030, USA
e-mail: laubenbacher@uchc.edu

F. Hinkelmann
C/o Renzelmann, Kurfürstenstr. 35, 80801 München, Germany
e-mail: franziska.hinkelmann@gmail.com

D. Murrugarra
School of Mathematics, Georgia Tech, Atlanta, GA, USA
e-mail: davidmur@math.gatech.edu

A. Veliz-Cuba
606 PGH, Department of Mathematics, University of Houston, Houston, TX 77004, USA

312 Keck Hall, Department of Biochemistry and Cell Biology, Rice University, Houston,
TX 77251, USA
e-mail: alanavc@math.uh.edu

N. Jonoska and M. Saito (eds.), *Discrete and Topological Models in Molecular Biology*,
Natural Computing Series, DOI 10.1007/978-3-642-40193-0_21,
© Springer-Verlag Berlin Heidelberg 2014

1 Introduction

Molecular systems biology is one of the important developments in the life sciences in the last 20 years, made possible by a succession of new so-called "-omics" technologies (genomics, proteomics, metabolomics, etc.) that allow the simultaneous measurement of many molecular species in a cell, a tissue, or an entire organism. The fundamental paradigm of systems biology is the construction and study of entire networks, or systems, rather than isolated proteins, genes, or metabolites. In the words of [7]:

> The advent of functional genomics has enabled the molecular biosciences to come a long way towards characterizing the molecular constituents of life. Yet, the challenge for biology overall is to understand how organisms function. By discovering how function arises in dynamic interactions, systems biology addresses the missing links between molecules and physiology.

In this description, it is important to note the characterization of "function" as arising through dynamic interactions of many molecular species. This clearly points to mathematical and computational models of dynamical systems as the key enabling technology of systems biology. And, indeed, mathematical and statistical models have been used extensively to study interaction networks of molecular species, ranging from mechanistic biochemical models to statistical phenomeno-logical models. For a sample of such models, see [12] for a model of an iron homeostasis regulatory network in breast epithelial cells, consisting of a system of ordinary differential equations; [31] for a model of the budding yeast cell cycle in the form of a Boolean network; [21] for a Petri net model of systemic iron homeostasis; and [6] for a Bayesian network model of root epidermis cell differentiation in *Arabidopsis*.

The common structure of all these modeling frameworks is as follows. We are given a collection of variables $x_1, \ldots, x_n$, representing biological or biochemical entities, such as the concentration of certain gene products, such as proteins. These variables may also reflect contextual information, such as the location of a protein in the nucleus or the cytosol, and conditions such as a protein being phosphorylated or not, for example in a signaling network. Each variable x_i takes values in a set X_i, which may be finite or infinite, and with or without additional structure. The x_i are related to each other through a collection of relationships

$$R_1(x_1, \ldots, x_n), \ldots, R_n(x_1, \ldots, x_n),$$

where R_i encodes the dependence of x_i on the other variables and itself. For instance, each R_i could be a differential equation describing the change in con-centration of a molecular species represented by variable x_i. Or it could be a logical rule (or a family of such), such as in a Boolean network model. It could also be a stochastic relationship, such as a probability distribution. Finally, there is given a scheme that updates the state of each variable in time. This may happen in continuous or discrete time, synchronously or asynchronously, and in a deterministic or stochastic manner.

In this chapter, the focus is on the special case in which all the value sets X_i are finite sets, and the R_i are given by functions

$$R_i = R_i(x_1, \ldots, x_n) : X_1 \times \cdots \times X_n \longrightarrow X_1 \times \cdots \times X_n.$$

The update rule can be deterministic or stochastic, and synchronous or asynchronous. That is, variables can be updated in parallel or sequentially, and can be determined by a deterministic rule or by a stochastic process. We will call such a structure a *finite dynamical system.*

The goal of this chapter is to look at finite dynamical systems from an algebraic point of view, which makes accessible additional tools for model construction and analysis. Accordingly, we introduce the term "algebraic model" for a finite dynamical system represented as a time-discrete dynamical system over a finite field. We will show how algebraic models can be used to study dynamic molecular networks, and how they can be constructed and analyzed. We will discuss both deterministic and stochastic models. An important tool for model construction within a differential-equation framework is the ability to use parameter estimation methods to fit the model to given experimental data. We will discuss discrete versions of parameter estimation in this chapter. Steady-state analysis of models is a standard problem to be solved. We will discuss both computational and theoretical tools for this purpose. Along the way, we will describe some software tools specific to this framework that are available to the user. Finally, we will briefly discuss stochastic models, as well as a special class of models constructed from logical rules that capture important features of molecular regulatory networks, namely so-called nested canalizing rules. Models constructed from these rules have special dynamic properties that are commonly found in biological systems. The topics selected are intended to represent examples of research questions in this field to provide the reader with snapshots of topics to be explored further. The theme that ties the sections together is the use of a polynomial representation of discrete models and the use of computational and theoretical tools that this representation provides.

2 Background

The work presented in this chapter is a small part of a larger research field that studies and uses finite dynamical systems in biology. In this section, we provide a context by discussing some examples of other work related to finite dynamical systems and, hence, to algebraic models, which may serve as a starting point for the reader who wishes to inquire further. This section is not intended to be comprehensive in any way. The reader might also consult the work presented at the May 2012 workshop "Algebraic Methods in Systems and Evolutionary Biology" at the Mathematical Biosciences Institute (MBI) at The Ohio State University (www.mbi.osu.edu). Video recordings of some lectures are available on the MBI website.

The use of Boolean networks as models for gene regulatory networks goes back to Kauffman [35]. Regulatory mechanisms for each gene are represented by logical functions. This framework was extended in several ways by René Thomas and his collaborators [65,66]. It allows network nodes to take on any finite number of states, and it emphasizes the importance of an asynchronous update scheme for the nodes, with the argument that the molecular processes that underlie the changing states of the different nodes take different lengths of time to complete, so that synchronous updating of the network nodes is not realistic. These so-called logical models have been used very successfully to model a variety of molecular networks, and a Web tool, GINsim, is available for their construction and analysis at http://gin.univ-mrs.fr/. The choice of the update schedule for the network nodes influences the network dynamics, potentially quite strongly. Although the steady states of a model are independent of the choice of the update schedule, the periodic behavior of the model can be affected quite strongly. The analysis of logical models takes into account the effect of all possible update schedules. For a more detailed introduction to logical models and the GINsim software package the reader can consult [51], for example.

An alternative approach is to choose an update schedule at random at each update of the model. It was shown in [11], using a Boolean model of the segment polarity network in *Drosophila melanogaster* [1], that this leads to an improved fit of the model dynamics with known biology. This approach is taken in the software package BoolNet [50], which allows the construction and analysis of Boolean networks using random sequential updating. Thus, effectively, this approach replaces deterministic models with stochastic ones. For more details of how this model type is used in systems biology, see [59].

An alternative approach to making Boolean and multistate discrete models stochastic was introduced in [62] in the form of so-called probabilistic Boolean networks. In this setting, each node has assigned to it a set of possible update rules, endowed with a probability distribution. At each update of the model, an update rule is chosen at random from the node's probability space. A comprehensive introduction to this model type and its use in systems biology can be found in [61].

Another discrete modeling framework that can be used very effectively in systems biology is the language of Petri nets. These have been used in a variety of application areas to describe distributed processes, such as workflow design and software engineering. Another important application area is the modeling of chemical reaction processes. A Petri net is defined as a bipartite graph, with one set of nodes called *places* and the other set called *transitions*. When it is applied to modeling chemical reaction networks, one can think of the places as representing molecular species. In specifying the state of a Petri net, each place is assigned a nonnegative integer, referred to as the *number of tokens* at that place. This can be thought of as the number of molecules present. A transition represents a chemical reaction which moves tokens between places according to a chemical reaction equation. For an introduction to the use of Petri nets in systems biology, see [60]. A software package for the construction and analysis of biological Petri net models is described in [58]. There are also stochastic versions of Petri nets available. A good introduction can be found in [78]. It was shown in [10] that logical models can be translated into Petri nets for further analysis.

Each of these modeling frameworks comes with its own methods for model construction, in particular methods to infer network models from experimental data. Furthermore, each framework has its own methods for model analysis, such as computation of steady states or model reduction.

3 An Example: The *lac* Operon

For the reader not familiar with finite dynamical systems, we present here an example of such a model for a well-known gene regulatory network. The *lac* operon is a set of genes that regulate lactose metabolism in *E. coli*. It has been studied extensively and is one of the earliest gene regulatory networks found to have positive and negative control. In this section, we briefly describe a Boolean model of the *lac* operon proposed in [71]. This model is capable of reproducing known features of this network, including bistability.

3.1 Biological Background

We first describe the key features of the *lac* operon used in the Boolean model. The main components of the *lac* operon network are *lac* mRNA, *lac* permease, *lac* β-galactosidase, lactose, allolactose, CAP (catabolite activator protein), and LacI (repressor protein). Transcription of *lac* mRNA produces the proteins *lac* permease and *lac* β-galactosidase, where *lac* permease is in charge of transporting lactose into the cell and *lac* β-galactosidase is in charge of converting lactose into allolactose. Allolactose deactivates LacI. CAP and LacI regulate the transcription of the *lac* operon genes and, hence, the production of *lac* mRNA; CAP activates transcription, while LacI inhibits it. Glucose is thought to regulate the *lac* operon by inhibiting the activation by CAP and by inhibiting the transport by *lac* permease. This information is summarized in the wiring diagram in Fig. 1, where the following notation is used:

$M = lac$ mRNA;
$P = lac$ permease;
$B = lac$ β-galactosidase;
$C = $ CAP;
$R = $ LacI;
$A = $ allolactose;
$L = $ intracellular lactose;
$L_e = $ extracellular lactose;
$G_e = $ extracellular glucose.

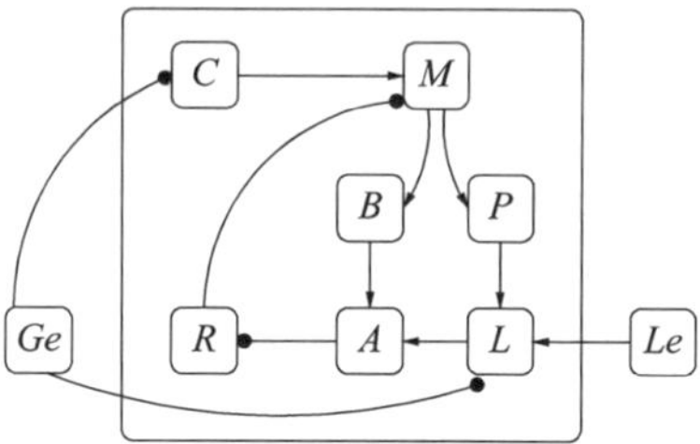

Fig. 1 Wiring diagram of the *lac* operon network

3.2 *Boolean Model*

Figure 1 shows which variables depend on which others. We now provide each variable with a (Boolean) function that describes how it depends on the others. We will use the Boolean operators AND $= \wedge$, OR $= \vee$, and NOT $= \neg$. The AND operator is used when all the variables are required for a process to happen; the OR operator is used when the variables act independently; the NOT operator is used to represent inhibition. For example, since glucose inhibits the activation by CAP, the Boolean function for C is $f_C = \neg G_e$. Since *lac* permease and *lac* β-galactosidase will only be produced when *lac* mRNA is present, we have that $f_P = M$ and $f_B = M$. Since allolactose is produced from lactose by β-galactosidase, we have that $f_A = L \wedge B$. Also, intracellular lactose is present when it is brought into the cell by *lac* permease; however, this process is inhibited in the presence of glucose. Therefore, $f_L = P \wedge L_e \wedge \neg G_e$. The Boolean functions for the other variables are created similarly. The complete Boolean model is given below:

$$
\begin{aligned}
f_M &= C \wedge \neg R \wedge \neg R_m, \\
f_P &= M, \\
f_B &= M, \\
f_C &= \neg G_e, \\
f_R &= \neg A \wedge \neg A_m, \\
f_{R_m} &= (\neg A \wedge \neg A_m) \vee R, \\
f_A &= L \wedge B, \\
f_{A_m} &= L \wedge L_m, \\
f_L &= P \wedge L_e \wedge \neg G_e, \\
f_{L_m} &= ((P \wedge L_{em}) \vee L_e)\neg G_e.
\end{aligned}
$$

In order to have better resolution, the authors of [71] considered some variables to have more than two states, namely, R, A, L, and L_e; the subscript m denotes a medium concentration. This allows us to distinguish between three instead of two concentration levels; for example, for lactose we have low (represented by $(L_m, L) = (0, 0)$), medium (represented by $(L_m, L) = (1, 0)$), and high (represented by $(L_m, L) = (1, 1)$). The quantities L_e and G_e are parameters of the model that can be set to represent different levels of lactose and glucose outside the cell. Computationally, this is accomplished by introducing additional variables, so that all variables in the model remain Boolean.

Given an initial configuration of the variables, the next configuration (time is discrete) is computed by updating the values of all variables using the Boolean functions. That is, if we consider $f = (f_M, f_P, f_B, f_C, f_L, f_{L_m}, f_A, f_{A_m}, f_R, f_{R_m})$, then we have a function $f : \{0, 1\}^{10} \to \{0, 1\}^{10}$ that encodes all the information about the network. The dynamics of the network is given by iteration of f: $x(t + 1) = f(x(t))$, where $x = (M, P, B, C, R, R_m, A, A_m, L, L_m)$. Thus, this model is a finite dynamical system with synchronous update.

3.3 Model Analysis

Now that the model has been constructed, we need to analyze it and compare its dynamics with known features of the *lac* operon, which we review briefly. The *lac* operon is said to be OFF if mRNA, permease, and β-galactosidase are low; it is ON if mRNA, permease, and β-galactosidase are high. The most important features of the dynamics of the *lac* operon are its steady states. It is known that with high levels of extracellular glucose, the *lac* operon will eventually be OFF. On the other hand, with low levels of extracellular glucose, we have the following: with low levels of extracellular lactose, the *lac* operon will eventually be OFF; with high levels of extracellular lactose, the *lac* operon will eventually be ON; and with medium levels of extracellular lactose, the *lac* operon can be ON or OFF (depending on the initial conditions). The latter is a feature called bistability.

Now we analyze the model. We first find the steady states, i.e., those states x for which $f(x) = x$. As explained above, the variables of interest are M, B, and P. If $(M, P, B) = (0, 0, 0)$, then we say that the *lac* operon model is OFF; and if $(M, P, B) = (1, 1, 1)$, then we say that the model is ON. Extracellular glucose can take two values: low ($G_e = 0$) and high ($G_e = 1$). Extracellular lactose can take three values: low (represented by $(L_{em}, L_e) = (0, 0)$), medium (represented by $(L_{em}, L_e) = (1, 0)$), and high (represented by $(L_{em}, L_e) = (1, 1)$). Using the results presented in [30], we find that, for $G_e = 1$, there is one steady state regardless of the value of L_e, namely, $x = 0000110000$ (commas have been omitted). Furthermore, all of the 2^{10} initializations eventually reach this steady state. Since $(M, P, B) = (0, 0, 0)$, we obtain the result that the *lac* operon model is OFF in the presence of glucose. On the other hand, when $G_e = 1$, that is, when extracellular lactose is present, we have the following cases:

- If extracellular lactose is low (that is, $(L_{em}, L_e) = (0, 0)$), there is only one steady state, namely, $x = 0001110000$. That is, the *lac* operon model is OFF. Furthermore, all of the 2^{10} initializations eventually reach this steady state.
- If extracellular lactose is at a medium concentration (that is, $(L_{em}, L_e) = (1, 0)$), there are two steady states, namely, $x = 0001110000$ and $x = 1111000101$. Thus, the *lac* operon model can be either OFF or ON. Furthermore, all of the 2^{10}

initializations eventually reach one of these steady states. That is, the system is bistable.

- If extracellular lactose is high (that is, $(L_{em}, L_e) = (1, 1)$), there is only one steady state, namely, $x = 1111001111$. That is, the *lac* operon model is ON. Furthermore, all of the 2^{10} initializations eventually reach this steady state.

In summary, the behavior of the Boolean model agrees with the biological behavior described above.

4 A General Mathematical Framework: Polynomial Dynamical Systems

For finite dynamical systems that have a small number of nodes (such as the one in the previous section), one can use exhaustive enumeration to analyze them. However, when the number of nodes increases, exhaustive search very quickly becomes unfeasible, as the number of system states grows exponentially with the number of network nodes. In a model with 100 nodes, there are 2^{100} possible network configurations. One current (2011) processor can handle 177,000 million instructions per second (MIPS). Even if we needed only one instruction to test each configuration, it would still take a 100 billion years to check the complete space of configurations of network nodes. We therefore need other tools to replace or supplement model simulation. One way to do this is to frame the problem within a mathematical context that has such tools available, similarly to the introduction of a coordinate system in the plane, which allows the use of algebraic tools to answer geometric questions that could previously only be solved by inspection, such as whether two lines in the plane intersect. With a coordinate system, that question can be answered by computing the solution to a system of two linear equations.

Our objects of study are mappings

$$f = (f_1, \ldots, f_n) : X_1 \times \cdots \times X_n \longrightarrow X_1 \times \cdots \times X_n,$$

where the X_i are finite sets. The "coordinate system" to be introduced is given by an algebraic structure on the X_i that makes them into a finite field. Furthermore, for the purposes of analysis, we would like the X_i all to be the same. Both of these aims are accomplished by adding states to the sets, so that all X_i have the same number of elements, and this number is equal to a prime p. (Typically, the number of states will end up being quite small, such as 3 or 5.) So, we are now dealing with one set X of cardinality p. Then the functions f_i are extended to this larger set in a suitable way that allows the removal of "artificial" network states later on. Finally, we impose the structure of a finite field on X by relabeling the elements as $0, 1, \ldots, p - 1$ and using addition and multiplication in Z/p. We will denote the resulting field by k. We emphasize that the algebraic structure serves as a computational tool only, and there is no biological meaning to it. The last observation that motivates the definition below is that it is well known [45] that any function $k^n \longrightarrow k$ can be represented uniquely as a minimal-degree polynomial in the variables $x_1, \ldots, x_n$ with coefficients in k. The polynomial can be constructed from the function using Lagrange interpolation. (Computationally, of course, this is only feasible in the case where the function depends on only a small number of variables, which is typically the case in the setting we are concerned with.) These observations motivate the framework we propose, polynomial dynamical systems (PDSs) over finite fields.

4.1 Polynomial Dynamical Systems

Definition 1 ([33]). A *polynomial dynamical system* f over a finite field k is a function

$$f = (f_1, \ldots, f_n) : k^n \to k^n,$$

where the coordinate functions f_i are elements of $k[x_1, \ldots, k_n]$, the ring of polynomials in the variables $x_1, \ldots, x_n$, with coefficients in k. Iteration of f results in a time-discrete dynamical system with states in k^n.

In the case of a Boolean network, the corresponding PDS is over $k = \mathbb{F}_2 = (\{0, 1\}, +, \cdot)$. The elements of k can have different biological interpretations for different variables. For example, x_1 may be present at different concentration levels, and x_2 may be in a phosphorylated or nonphosphorylated state; therefore 0 represents "low" when used for x_1 and "nonphosphorylated" when used for x_2.

Example 1.

$$f_x(x, y) = (\neg x \wedge \neg y) \vee (x \wedge y), \tag{1}$$

$$f_y(x, y) = x. \tag{2}$$

The Boolean model described by Eqs. (1) and (2) is equivalent to the following PDS: $f(x, y) = (f_x(x, y), f_y(x, y)) : k^2 \to k^2$ over $k = \mathbb{F}_2$,

$$f_x(x, y) = x + y + 1,$$

$$f_y(x, y) = x.$$

One can easily check that the polynomial system $f(x, y)$ evaluated over $\mathbb{F}_2^2 = \{0, 1\} \times \{0, 1\}$ has the same dynamics as the Boolean model, namely $(1, 1)$ is a steady state, and the other three points in the state space form a limit cycle of length 3. Section 4.2 describes how to construct the desired polynomials.

Example 2. Consider the production of allolactose in the Boolean model of the *lac* operon:

$$f_A = L \wedge B,$$

$$f_{A_m} = L \wedge L_m.$$

Lactose must be present in high concentration for allolactose to be present at a medium level; if, in addition, β-galactosidase is present, allolactose will be present in high concentration. We can express this behavior with the following polynomial over $\mathbb{F}_3$ (using 0, 1, and 2 for *low, medium,* and *high,* respectively):

$$f_A(B, L) = -B^2 * L^2 + B^2 * L - L^2 + L.$$

4.2 Generating a Polynomial from Data Points

Every function in r variables over a finite field k can be uniquely expressed by a polynomial of degree at most $r|k|$. The polynomial is unique. To construct the polynomial that matches the data points, we can use the following interpolation procedure. For simplicity, we will use $k = \mathbb{F}_p$.

Theorem 1. *Given a function* $f : k^r \to k$, *with* $f(s_{j,1}, \ldots, s_{j,r}) = t_j$ *for* $j \in \{1, \ldots, p^r\}$, *the polynomial*

$$\tilde{f}(x_1, \ldots, x_r) = \sum_{j=1}^{p^r} t_j \prod_{i=1}^{r} (1 - (x_i - s_{j,i})^{p-1}) \tag{3}$$

describes the same mapping. Note that k^r is the Cartesian product of r copies of k, and a point in k^r is an r-tuple. The number of points in k^r is p^r, and the index j iterates over all these points.

Proof. For any point $s = (s_1, \ldots, s_r) \in k^r$, $\prod_{i=1}^{r}(1 - (x_i - s_i)^{p-1})$ vanishes for all $x \in k^r$, except for $x = s$. For $x = s$, the product evaluates to 1. Thus, for the index j such that $x = s_j$,

$$\tilde{f}(x) = 0 + \cdots + 0 + t_j + 0 + \cdots + 0 = t_j = f(s_j) = f(x).$$

Thus, $\tilde{f}(x) = f(x)$.

Example 3. Using the dynamics of the Boolean model of Example 1, i.e.,

$$f(0, 0) = (1, 0),$$

$$f(0, 1) = (0, 0),$$

$$f(1, 0) = (0, 1),$$

$$f(1, 1) = (1, 1),$$

Theorem 1 yields the following PDS $f(x, y) = (f_x(x, y), f_y(x, y))$:

$$
\begin{aligned}
f_x(x, y) =& 1 * (1 - (x - 0))(1 - (y - 0)) + \\
& 0 * (1 - (x - 0))(1 - (y - 1)) + \\
& 0 * (1 - (x - 1))(1 - (y - 0)) + \\
& 1 * (1 - (x - 1))(1 - (y - 1)) = \\
& (1 - x)(1 - y) + 0 + 0 + (x)(y) = \\
& 1 - x - y - xy + xy = \\
& x + y + 1, \\
f_y(x) =& 0 * (1 - (x - 0)) + \\
& 1 * (1 - (x - 1)) = \\
& 0 + x = \\
& x.
\end{aligned}
$$

As expected, $f(x, y) = (x + y + 1, x)$ is the PDS that represents the Boolean model of Example 1.

Example 4. The polynomial f_A for allolactose in Example 2 was constructed as follows. The behavior for allolactose as dictated by the Boolean rules for x_A and x_{Am} is described in Table 1. Using the data points in the table and interpolation according to Theorem 1 yields the polynomial f_A over $\mathbb{F}_3$:

$$
\begin{aligned}
f_A(B, L) =& 1 * (1 - (B - 0)^{3-1})(1 - (L - 2)^2) + \\
& 2 * (1 - (B - 1)^2)(1 - (L - 2)^2) + \\
& 2 * (1 - (B - 2)^2)(1 - (L - 2)^2) = \\
& B^2 * L^2 - B^2 * L - L^2 + L + \\
& - B^2 * L^2 + B^2 * L - B * L^2 + B * L \\
& - B^2 * L^2 + B^2 * L + B * L^2 - B * L = \\
& - B^2 * L^2 + B^2 * L - L^2 + L.
\end{aligned}
$$

Note that a different finite field yields different polynomials.

Thus, in principle, any model that can be expressed as a finite dynamical system can also be expressed as a PDS. This procedure was carried out explicitly for the

Table 1 Expression levels of allolactose (A) based on the levels of β-galactosidase (B) and lactose (L), using 0, 1, and 2 for *low, medium,* and *high*, respectively. We assume that medium and high concentrations have the same effect on allolactose, since the Boolean model does not distinguish between those levels

B	L	A
0	0	0
0	1	0
0	2	1
1	0	0
1	1	0
1	2	2
2	0	0
2	1	0
2	2	2

case of logical models and bounded Petri nets in [72]. The Web tool ADAM [30] carries out this conversion for logical models in the GINSim format and Petri nets in the Snoopy format. This shows that for analytical purposes, the PDS framework is universal for a wide range of (deterministic) discrete models. However, the language of polynomials is not very intuitive for the purpose of interpreting model features biologically.

5   Reverse Engineering

The inference of molecular regulatory networks from experimental data is one of the central problems in systems biology, and a very active research area. An annual competition organized by the DREAM (Dialogue for Reverse Engineering and Methods) project provides one focal point of research, and is a good guide to the literature (http://www.the-dream-project.org/). Typically, reverse engineering methods provide as output a directed or undirected graph with genes as nodes and with edges indicating regulatory relationships. A variety of network inference methods are now available, using techniques from statistics [17, 20, 27, 40], dynamical systems [46], metabolic control analysis [16], and parameter estimation [19, 77], to name a few. Different methods have different data requirements and give different output. Some, such as biclustering, give regulatory relationships between modules of coexpressed genes [14], while others infer causal relationships between individual genes [75]. Some can be applied to infer genome-scale networks [20], whereas others are limited to much smaller networks [16, 17, 22], either because of their high data requirements or because of computational limitations. It has been argued (see, e.g., [4]) that a significant improvement in performance can be obtained

Table 2 Partial information for polynomial f_1

x	$f_1(x)$
$s^1 = 01210$	0
$s^2 = 01211$	0
$s^3 = 01214$	1
$s^4 = 30000$	3
$s^5 = 11113$	4

with methods that make use of data that capture the dynamics of the response of genes to perturbations in the form of time course data. This point of view has been adopted in several recent publications presenting new methods that take into account time course data as well as perturbations of the network (see, e.g., [39, 46, 53]).

5.1 Inferring the Wiring Diagram

Here, the goal is to use partial information about the dynamics to construct the wiring diagram of a molecular network, assumed to be described by a PDS. It turns out that the problem can be studied coordinatewise (see, for example, [33, 69]). That is, given partial information about a polynomial f_j, we want to estimate which variables affect it. The set of the variables on which a polynomial depends is called the *support* and is denoted by supp(f_i). Graphically, the support of a polynomial translates into regulatory arrows coming into the corresponding node, which then results in the wiring diagram of the network.

For example, consider a PDS $f = (f_1, f_2, f_3, f_4, f_5) : \mathbb{F}_5^5 \to \mathbb{F}_5^5$, and suppose we are given information about the first coordinate function as listed in Table 2.

5.1.1 Unsigned Wiring Diagram

Given this information, we want to infer the *support* of f_1, that is, the set of variables on which f_1 depends. Of course, this is an underdetermined problem; therefore, we want to find the "simplest" or "minimal" wiring diagrams from among the usually many possible ones. We say that a polynomial is *minimal* if it fits the data and there is no other polynomial that fits the data with strictly smaller support (with respect to inclusion). We say that a wiring diagram is minimal if it is the wiring diagram of a minimal polynomial. For example, consider the polynomials

$$
\begin{aligned}
h_1 &= -2x_1^4 - 2x_5^4 - 2x_5^3 - x_5, \\
h_2 &= -2x_2^4 x_5^4 + x_2^4 x_5^3 + 2x_2^4 x_5^2 + x_2^3 x_5^3 + 2x_2^4 x_5 + 2x_2^3 x_5^2 + x_2^2 x_5^3 + 2x_2^4 + 2x_2^3 x_5 \\
&\quad + 2x_2^2 x_5^2 + x_2 x_5^3 + 2x_5^4 + 2x_2^2 x_5 + 2x_2 x_5^2 + 2x_2 x_5 - 2, \\
h_3 &= -2x_2^4 x_5^4 x_4 + x_2^4 x_5^3 x_4 + 2x_2^4 x_5^2 x_4 + x_2^3 x_5^3 x_4 + 2x_2^4 x_5 x_4 + 2x_2^3 x_5^2 x_4 + x_2^2 x_5^3 x_4 \\
&\quad + 2x_2^4 x_4 + 2x_2^3 x_5 x_4 + 2x_2^2 x_5^2 x_4 + x_2 x_5^3 x_4 + 2x_5^4 x_4 + 2x_2^2 x_5 x_4 + 2x_2 x_5^2 x_4 \\
&\quad + 2x_2 x_5 x_4 - 2.
\end{aligned}
$$

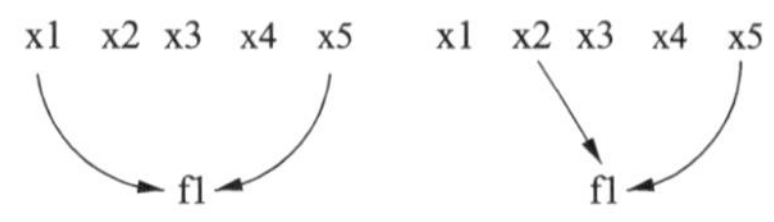

Fig. 2 Two minimal wiring diagrams for Table 2

It is not difficult to show that these polynomials all fit the data. Also, $\text{supp}(h_1) = \{x_1, x_5\}$, $\text{supp}(h_2) = \{x_2, x_5\}$, and $\text{supp}(h_3) = \{x_2, x_4, x_5\}$. Since $\text{supp}(h_2)$ is strictly contained in $\text{supp}(h_3)$, h_3 is not minimal. On the other hand, there is no polynomial with support equal to $\{x_2\}$ that fits the data. Indeed, according to the first row in Table 2, such a polynomial would have to be 0 when evaluated at $x_2 = 1$; and, according to the third row in Table 2, such a polynomial would have to be 1 when evaluated at $x_2 = 1$. Similarly, there is no polynomial with support equal to $\{x_5\}$ that fits the data. Therefore, h_2, with support $\text{supp}(h_2) = \{x_2, x_5\}$, is a minimal polynomial. It can also be shown that h_1 is minimal. Therefore, we have that the wiring diagrams in Fig. 2 are minimal wiring diagrams (these are not necessarily all possible wiring diagrams).

If we were to try to find all the minimal wiring diagrams by exhaustive search, we would need to find all functions that fit the data, then find their support, and, finally, find which polynomials are minimal. This very quickly becomes unfeasible. For example, for Table 2, there are about $6 \times 10^{2,180}$ polynomials that fit the data. Therefore, we need to compute the minimal wiring diagrams using a different approach. In [33], Jarrah et al. proposed a framework and method to compute all minimal unsigned wiring diagrams, using theoretical and computational tools from commutative algebra. First, a monomial ideal is constructed in the polynomial ring that encodes information about the given transitions, namely, which variables must be present in order to fit the partial transition table. It is known that any ideal in a polynomial ring can be written uniquely as an intersection of powers of prime ideals, a generalization of the unique factorization property of integers. Those prime ideals that are minimal with respect to inclusion are called *minimal primes*. The algorithm below is based on the observation that the minimal wiring diagrams are in one-to-one correspondence with the minimal primes:

- For $f_i(s) \neq f_i(s')$, define $P_{s,s'} = \prod_{k:\, s_k \neq s'_k} x_i$.
- Define the ideal $I = \langle P_{s,s'} : f_i(s) \neq f_i(s') \rangle$.
- Compute the *minimal primes* of I.
- The minimal wiring diagrams are given by the generators of the minimal primes.

For example, since $f_1(0, 1, 2, 1, 0) = 0 \neq 1 = f_1(0, 1, 2, 1, 4)$ and $(0, 1, 2, 1, 0)$ differs from $(0, 1, 2, 1, 4)$ in the fifth variable, then $P_{01210,01214} = x_5$. Similarly, since $f_1(0, 1, 2, 1, 0) \neq f_1(3, 0, 0, 0, 0)$, we have $P_{01210,30000} = x_1 x_2 x_3 x_4$. Then the ideal is $I = \langle x_5, x_1 x_2 x_3 x_4, \ldots \rangle$. Using the open source computer algebra system Macaulay2 [26], we have obtained the minimal primes as $\langle x_1, x_5 \rangle$, $\langle x_2, x_5 \rangle$, $\langle x_3, x_5 \rangle$, and $\langle x_4, x_5 \rangle$. Then, by Jarrah et al. [33, Corollary 4], we have the four minimal wiring diagrams shown in Fig. 3.

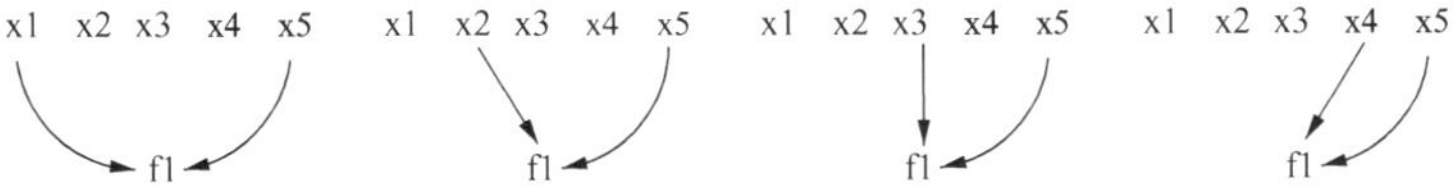

Fig. 3 All the minimal wiring diagrams for Table 2

This result shows the advantage of introducing the algebraic framework of finite fields, since its theoretical and algorithmic tools can then be used to compute more effectively with functions and data than can be done at the set-theoretic level, where the only tool available is enumeration.

5.1.2 Signed Wiring Diagram

Since the polynomials are intended to represent molecular regulation, which can be either positive or negative, one could also require the polynomials to be *unate* functions (also known as biologically meaningful functions [54]); that is, each polynomial in a PDS has to be increasing or decreasing with respect to each variable. For example, in the field $\mathbb{F}_2$, the polynomial $x_1(1 - x_2)$ is increasing with respect to x_1 and decreasing with respect to x_2 (using the order $0 < 1$), so it could represent molecular regulation. On the other hand, the polynomial $x_1 + x_2$ is neither increasing nor decreasing with respect to its variables, so it is less likely to be biologically relevant.

For example, let us revisit Table 2 and the polynomials h_1, h_2, h_3. One can show that h_1 is unate (increasing with respect to x_1 and x_5). However, neither h_2 nor h_3 is unate. So, in addition to keeping track of which polynomials fit the data, we may want to consider unate functions only. Once again, this problem cannot be solved by exhaustive search; so, we need a different approach.

In [69], Veliz-Cuba proposed a framework and method to compute minimal signed wiring diagrams. The mathematical tools are similar to those used in [33]. The algorithm is summarized below:

- For $f_i(s) < f_i(s')$, define $P_{s,s'} = \prod_{k:\, s_k \neq s'_k} (x_i - \text{sign}(s'_k - s_k))$.
- Define the ideal $I = \langle P_{s,s'} : f_i(s) < f_i(s') \rangle$.
- Compute the primary decomposition (or minimal primes) of I.
- The minimal wiring diagrams are given by the generators of each primary ideal, and the signs are given by the constant term.

For example, since $f_1(0, 1, 2, 1, 0) = 0 < 1 = f_1(0, 1, 2, 1, 4)$ and $(0, 1, 2, 1, 0)$ differs from $(0, 1, 2, 1, 4)$ in the fifth variable, then $P_{01210,01214} = (x_5 - \text{sign}(4 - 0)) = x_5 - 1$. Similarly, since $f_1(0, 1, 2, 1, 0) < f_1(3, 0, 0, 0, 0)$, we have $P_{01210,30000} = (x_1 - 1)(x_2 + 1)(x_3 + 1)(x_4 + 1)$. Then, the ideal is $I = \langle x_5 - 1, (x_1 - 1)(x_2 + 1)(x_3 + 1)(x_4 + 1), \ldots \rangle$. Using Macaulay2, we have obtained the result that the primary ideals in the primary decomposition are $\langle x_1 - 1, x_5 - 1 \rangle$

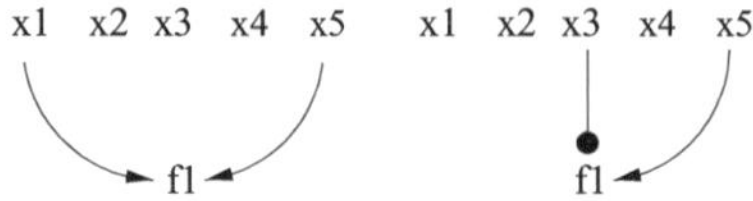

Fig. 4 All the minimal signed wiring diagrams for Table 2. Normal *arrows* indicate activation, and *circles* indicate inhibition

and $\langle x_3 + 1, x_5 - 1 \rangle$. Then, by Veliz-Cuba [69, Theorem 3.6], we have two minimal wiring diagrams, shown in Fig. 4. Notice that we have only two such wiring diagrams, which means that the second and fourth wiring diagrams in Fig. 3 do not correspond to any unate function.

It is important to note that the key advantage of the algebraic framework is that it efficiently finds all minimal wiring diagrams without making any assumptions about the structure of the models, the number of inputs, or the number of states. In practice, this means that given enough data points, perfect recovery of the wiring diagram is guaranteed. For example, in [69], it was shown that as the partial information increases, the estimated wiring diagram quickly approaches the actual wiring diagram. For instance, using the *lac* operon model above, it was shown that only 1.9 % of knowledge was enough to obtain perfect recovery of the wiring diagram. In other words, the algebraic framework provides a systematic approach to studying the problem of *perfect network inference*.

5.2 Inferring the PDS

Here, the goal is to use partial information about the dynamics to estimate the actual polynomials of the PDS. It turns out again that the problem can be studied coordinatewise (see, e.g., [41]). For example, consider a PDS $f = (f_1, f_2, f_3, f_4, f_5)$: $\mathbb{F}_5^5 \to \mathbb{F}_5^5$ and suppose we are given information about the first coordinate function in Table 2. The goal is to find a polynomial that fits the data and is in some sense minimal. It turns out that for any given *monomial ordering*, it is possible to define the concept of a minimal polynomial. We fix a monomial ordering and consider an ideal I. Then there exists a unique polynomial h such that h fits the data and, if $h = h' + g'$, where h' fits the data and $g' \in I$, then we must have that $g' = 0$. We can call this polynomial h the *minimal* polynomial (with respect to the monomial ordering) that fits the data.

Laubenbacher and Stigler [41] gave an algorithm to compute the minimal polynomial that fits a given data set. The algorithm is summarized below. Suppose that we know the values of a polynomial at $s^1, \ldots, s^r$:

- Compute any polynomial that fits the data, h^0.
- Define the ideal $I = \bigcap_{k=1}^{r} \langle x_1 - s_1^k, \ldots, x_n - s_n^k \rangle$.

- Reduce h^0 modulo I.
- The result of the reduction is the minimal polynomial.

The polynomial h^0 can be computed in many ways, for example, using Lagrange interpolation or the formula given in [41]. For our example, let us consider the polynomial $h^0 = -2x_1^4 - 2x_5^4 - 2x_5^3 - x_5$. It is important to mention that the minimal polynomial does not depend on the choice of h^0. Then we define the ideal

$$I = \langle x_1, x_2 - 1, x_3 - 2, x_4 - 1, x_5 \rangle \cap \langle x_1, x_2 - 1, x_3 - 2, x_4 - 1, x_5 - 1 \rangle$$

$$\cap \langle x_1, x_2 - 1, x_3 - 2, x_4 - 1, x_5 - 4 \rangle \cap \langle x_1 - 3, x_2, x_3, x_4, x_5 \rangle$$

$$\cap \langle x_1 - 1, x_2 - 1, x_3 - 1, x_4 - 1, x_5 - 3 \rangle.$$

We now reduce h^0 modulo I (e.g., using the command $h^0 \% I$ in Macaulay2). Using the *graded reverse lexicographic* monomial ordering (with $x_1 > x_2 > x_3 > x_4 > x_5$), we obtain the minimal polynomial $h = -2x_5^2 - x_3 - x_4 + 2x_5 - 2$. Notice that if we change the monomial ordering, then we may obtain a different minimal polynomial. For example, using the *lexicographical* monomial ordering, we obtain the minimal polynomial $h = 2x_4 - x_5^3 - 2x_5^2 - 2x_5 - 2$. The order of the monomials can be chosen so that "important" variables have more weight.

Similarly to Sect. 5.1.2, we may want to restrict the polynomials to unate polynomials. For the Boolean case, an algorithm is given in [28] to find the (Boolean) nested canalizing functions that fit a data set. Nested canalizing functions have been identified as a class of unate functions that appear in many Boolean models of biological systems [47] (see also Sect. 8).

5.3 The Polynome Software Package

The software package Polynome, a freely available Web tool, helps users construct Boolean network models based on experimental data and biological input [18, 19]. For given data, Polynome infers a wiring diagram, polynomial functions, or both, depending on the user's choice. Depending on other choices and the characteristics of the data set, Polynome may invoke different algorithms that lead to (deterministic or probabilistic) PDSs. Polynome presents the user with limited options in order to simplify the process of model inference. More options are available when the algorithms described in [19] are applied manually instead of being invoked through the Polynome interface.

It has been argued (e.g., [9]) that it is advantageous to combine different reverse engineering methods. In [63], the algebraic method described here was combined with a parameter estimation method for systems of ordinary differential equations. An obvious approach to reverse engineering is to set up a generic set of linear ordinary differential equations with unknown parameters and then fit the parameters to given time course data sets. This is the underlying principle in [22], for instance.

The problem is that, if the network has n variables, there will be in excess of n^2 parameters to be fitted (taking into account external-perturbation variables, etc.). Even for relatively small values of n, this becomes a computationally intractable problem. The method presented in [22] deals with this issue by assuming that each node has only a small number of inputs, which dramatically reduces the number of parameters to be estimated. While this assumption is reasonable, there are many cases in which it is not warranted. The approach in [63] is to use the algebraic method as a preprocessing method to identify parameters that can be set equal to 0 in the system of ordinary differential equations. This is done by applying the algebraic method to obtain a collection of minimal wiring diagrams, which are then combined into a consensus network. This network is then biased toward false positives, but will still generally result in a sparse network. The parameters in the system of ordinary differential equations that correspond to nonedges in the consensus network are then set equal to 0. This approach avoids the need to make biological assumptions which may or may not hold.

The combined method, called DICORE (DIscrete and COntinuous Reverse Engineering), was applied in [63] to several time courses of DNA microarray data, collected from experiments to elucidate a transcriptional network that controls the oxidative stress response network in *Saccharomyces cerevisiae*. The benchmark was a network constructed from results in the literature, consisting of 13 genes, which has two transcriptional hubs that control essentially all other genes in the network. In a comparison of DICORE with the methods presented in [8, 22, 80], it performed very favorably. For instance, it was the only method that correctly identified the two regulatory hubs in the network.

6 Steady-State Analysis

Once we have constructed a PDS representing the computational model, we can use algebra to analyze it efficiently. One way in which algebraic theory can be used is in the computation of steady states, or fixed points. The algebraic machinery that allows us to handle complex models too large for a brute force enumeration algorithm is the theory of *Gröbner bases*. We encourage the reader to follow the algebraic calculations; the example code here is written for Macaulay2. For an introduction to varieties, we recommend [13].

Theorem 2 (steady states [30]). *The steady states of a PDS* $f : k^n \to k^n$ *are the points in the variety* $\mathcal{V} = V(I)$, *where the ideal* I *is equal to* $\langle f_1(x) - x_1, \ldots, f_n(x) - x_n \rangle$.

Example 5. Here, we use Theorem 2 to identify the steady states of the model of Example 1, $f(x, y) = (x + y + 1, x)$. The ideal I is generated by $f_x(x, y) - x$ and $f_y(x, y) - y$, i.e., $I = \langle x + y + 1 - x, x - y \rangle = \langle y + 1, x - y \rangle$. It is easy to see that, over $\mathbb{F}_2$, $\mathcal{V}(I) = \{(1, 1)\}$, just as expected.

The following Macaulay2 code computes the variety $\mathscr{V}$:

```
R = ZZ/2[x,y]
F = matrix(R, {{x+y+1, x}})
I = ideal apply( flatten entries F, gens R,
      (f, x) -> f - x )
loadPackage "RationalPoints"
rationalPoints I
```

In Example 1, there are only four possible steady states, since the state space has size 4, and we can easily look at all trajectories to determine the steady state. The Macaulay2 command `rationalPoints` uses a simple-minded brute force approach to find the solutions. For larger systems, however, we would run into the same computational problem as with the computational model. The trick here is to use a Gröbner basis of I in lexicographic order, from which one can easily infer the solutions. For larger models, we see a dramatic decrease in the time it takes for the calculation of steady states by solving polynomial systems as compared with brute force calculation. Even though the complexity of computing a Gröbner basis may be doubly exponential in general, it is fast in practice for ideals representing biological systems [30].

6.1 Analysis of Oscillatory Behavior

A model may also have an attractor that is a set of states between which the system oscillates. We can compute limit cycles of length m by composing f with itself m times, and then solve the system $f^m - x = 0$.

Theorem 3 (limit cycles [30]). *The limit cycles of length m of a PDS $f : k^n \to k^n$ consist of the points in the variety $\mathscr{V} = V(I)$ that are not part of a smaller cycle or a steady state, where the ideal I is equal to $\langle f^m(x) - x \rangle$, and F^m denotes function composition.*

Remark 1. Any point belonging to a cycle of length l is an element in the variety $\mathscr{V}(\langle f^r - x \rangle)$ for r a multiple of l.

Remark 2. A 1-cycle is the same as a steady state.

Example 6. Here, we use Theorem 3 to identify the limit cycles in the model of Example 1, $f(x, y) = (x + y + 1, x)$. We have previously identified $\{(1, 1)\}$ as the steady state of the system. Since there are four different configurations, the longest cycle we may find is a 4-cycle. We first look at 2-cycles. Note that $f^2(x, y) = f(f(x, y))$, i.e., $f(x, y) = (f_x(f_x(x, y), f_y(x, y)), f_y(f_x(x, y), f_y(x, y)))$:

$$f(f(x, y)) = (f_x(x + y + 1, x), f_y(x + y + 1, x))$$

$$= ((x + y + 1) + x + 1, x + y + 1)$$

$$= (y, x + y + 1).$$

Then we compute the variety

$$\mathscr{V}(\langle f(f(x,y)) - (x,y)\rangle) = \mathscr{V}(\langle y - x, x + y + 1 - y\rangle) = \mathscr{V}(\langle y - x, x + 1\rangle) = \{(1,1)\}.$$

We see that the only point in $\mathscr{V}(f^2 - x)$ is $\{(1,1)\}$, a point that we previously identified as a steady state; thus there are no cycles of length 2. We can repeat this exercise for 3- and 4-cycles:

$$\begin{aligned}
f^3(x,y) &= (f_x(F^x(x,y)), f_y(F^x(x,y))) \\
&= (f_x(y, x + y + 1), f_y(y, x + y + 1)) \\
&= (y + (x + y + 1) + 1, y) \\
&= (x, y).
\end{aligned}$$

Thus, the system we need to solve is $x - x, y - y$, or $0 = 0$. Any point in $\mathbb{F}_2^2$ is a solution to this trivial system of equations, and since we have identified $\{(1,1)\}$ as a steady state, the remaining three points must constitute a 3-cycle. We can now compose f once more with itself to calculate the 4-cycles, or, since all points in $\mathbb{F}_2^2$ are part of a steady state or a 3-cycle, we know that there cannot be any 4-cycles.

Again, we can use Macaulay2 to carry out the calculations for us. We use $\mathtt{sub(F,F)}$ to compose f with itself:

```
restart
R = ZZ/2[x,y]
F = matrix(R, {{x+y+1, x}})
F2 = sub(F,F)
I2 = ideal apply( flatten entries F2, gens R,
     (f, x) -> f - x )
loadPackage "RationalPoints"
rationalPoints I2
F3 = sub(F2, F)
I3 = ideal apply( flatten entries F3, gens R,
     (f, x) -> f - x )
rationalPoints ideal gens gb I3
F4 = sub(F3, F)
I4 = ideal apply( flatten entries F4, gens R,
     (f, x) -> f - x )
rationalPoints ideal gens gb I4
```

Composing a system containing many variables with itself several times quickly becomes computationally infeasible, because the polynomials grow too quickly. For large systems, computing f^3 is already impossible, since even for polynomials that involve only a few of the variables, their third iteration can consist of 100,000 monomial terms. To overcome this problem, we use the following computational trick. Instead of solving $f(f(x)) = x$, we introduce an extra set of variables, y, and solve a system that now has twice as many equations:

$$f(x) = y, f(y) = x.$$

Then we compute a Gröbner basis of the ideal generated by $f(x) - y, f(y) - x$, and use the elimination theorem, another theoretical result about Gröbner bases, to delete the unnecessary y variables [13]:

```
restart
R = ZZ/2[x1, x2, y1, y2]
xVars = {(gens R)_0, (gens R)_1}
yVars = {(gens R)_2, (gens R)_3}
F = matrix(R, {{x1+x2+1, x1}})
I1 = ideal apply( flatten entries F, yVars,
    (f, y) -> f - y )
Fy = sub(F, {x1=>y1, x2=>y2})
I2 = ideal apply( flatten entries Fy, xVars,
    (f, x) -> f - x )
I = I1 + I2
I = eliminate({y1, y2}, I)
I = sub(I, ZZ/2[x1, x2])
loadPackage "RationalPoints"
rationalPoints ideal gens gb I
```

6.2 ADAM: Analysis of Dynamic Algebraic Models

The Web-based tool Analysis of Dynamic Algebraic Models (ADAM) provides analysis methods for discrete models [29, 30]. ADAM analyzes discrete models for their dynamical features and provides a graphical representation of the dynamics when possible. ADAM accepts several types of discrete models, including Boolean models, Petri nets, logical models, and PDSs. Internally, the models are transformed to PDSs to allow efficient computation. Once the model has been translated to a PDS, ADAM creates the corresponding system of equations in the appropriate polynomial ring and finds the steady states by using a Gröbner basis calculation, based on Theorems 2 and 3. Furthermore, ADAM constructs the interpolating polynomial for a given set of data points. ADAM's computational engine uses Macaulay2, but can be used without knowledge of the underlying algebra, and, as a Web tool, does not require the installation of any computer algebra system, which makes it accessible to a wide range of users. ADAM can currently compute steady states of models with up to 72 nodes and limit cycles of models with up to 30 nodes.

6.3 Theoretical Results about Model Steady States

As mentioned before, as the number of network nodes grows, it becomes infeasible to analyze discrete dynamical systems through enumerative exploration of the state space. Beyond a certain range, even methods from computational algebra lose their effectiveness. Sometimes, however, it is possible to carry out a theoretical analysis of the network's steady states, in combination with effective model reduction methods. Here, we briefly describe some theoretical approaches that can be helpful in many concrete cases. The results presented here outline a comprehensive program to analyze steady states of Boolean networks and, ultimately, general multistate models.

- *Reduction methods.* The first step in the program is to carry out model reduction while keeping steady-state information. It is possible to transform a Boolean network into another one, with typically many fewer nodes, such that they share dynamical properties [52, 59, 68]. In [68], a reduction method was used to transform a Th-cell differentiation model with 12 nodes into a Boolean network with 2 nodes, and a *lac* operon model with 10 nodes into a Boolean network with only one node. The reduction process is such that one can algorithmically determine the steady states of the original model from the steady states of the reduced model.
- *Conjunctive and disjunctive Boolean networks.* A Boolean network all of whose functions are constructed using only the AND operator is called *conjunctive* (and similarly for *disjunctive* networks). This is a very special class of networks, which can be used to model gene regulatory networks in which all interactions are synergistic activating regulations. For this class of Boolean networks, it has been shown that the number of steady states can be computed directly from topological features of the wiring diagram [34], namely, the *strongly connected components* and the *maximal antichains* in the partially ordered set of strongly connected components. Recall that the nodes of the wiring diagram of a Boolean network are the nodes of the network. There is a directed edge from x_i to x_j if the variable x_i appears in the update function for x_j. For a strongly connected conjunctive Boolean network (that is, one in which every node can be reached by a directed path from any other node), there is a closed formula that computes the number of periodic states of the Boolean network for all possible periods [34]. In particular, this formula allows the determination of all steady states. A similar result holds for disjunctive networks, constructed using only the OR operator [34].
- *AND–NOT networks.* The applicability of conjunctive networks to the modeling of gene regulatory networks is limited in that it does not allow for negative regulation. We can overcome this problem by also allowing the NOT operator, in addition to the AND operator, resulting in AND–NOT networks. Although the nice formula for AND networks giving the number of steady states does not hold anymore, it has been shown for a certain class of AND–NOT networks (*normal AND–NOT* networks) that the steady states can also be computed directly from

the wiring diagram, namely, via the maximal independent sets [70]. This is a generalization of the results about AND–OR networks presented in [3]. Also, it is possible to transform any AND–NOT network into a normal AND–NOT network while preserving steady-state information.

- *Boolean networks as AND–NOT networks.* Finally, in [74], it was shown that any Boolean network can be transformed into an AND–NOT network with a modest increment in the number of nodes. This can make the results about AND–NOT networks accessible to the steady-state analysis of any Boolean network.
- *Relationship with continuous models.* Although it may seem that the algebraic tools described in this chapter can only be used to study discrete models, it has been shown that our approach can also be used to study certain classes of continuous models [73].

7 Stochastic Polynomial Dynamical Systems

In this section, a stochastic modeling framework for gene regulatory networks is presented. Stochastic modeling tools are important because of the inherent stochasticity of gene regulation processes. Accurately modeling this stochasticity is a complex and important goal in molecular system biology. one can follow several different approaches depending on one's level of knowledge about the biological system and the availability of data. For instance, if a gene regulatory network is viewed as a biochemical reaction network, the Gillespie algorithm can be applied to simulate each biochemical reaction separately, generating a random walk corresponding to a solution of the chemical master equation of the system [24, 25]. At an even more detailed level, one could introduce time delays into the Gillespie simulations to account for realistic time delays in activation or degradation, such as in the case of circadian rhythms [5,55,57]. At a higher level of abstraction, stochastic differential equations [67] contain a deterministic approximation of the system and an additional random white noise term. However, all these schemes require all the kinetic rate constants to be known, which could represent a strong constraint owing to the difficulty of measuring kinetic parameters, limiting these approaches to small systems.

As mentioned in Sect. 4, discrete models are an alternative to continuous models, and do not depend on rate constants. To account for stochasticity in this setting, several methods have been considered. In particular, for Boolean networks, stochasticity has been introduced by flipping node states from 0 to 1 or vice versa with some flip probability [2, 15, 56, 79]. However, it has been argued that this way of introducing stochasticity into the system usually leads to overrepresentation of noise [23]. The main criticism of this approach is that it does not take into consideration the correlation between the expression values of input nodes and the probability of flipping the expression of a node due to noise. In fact, this approach models the stochasticity at a node regardless of the susceptibility to noise of the underlying biological function [23]. Probabilistic Boolean networks (PBNs) [42, 61, 62] allow another stochastic method within the discrete-model paradigm. PBNs model a choice among alternative biological functions during the iteration process, rather than modeling the stochasticity of the failure of the function itself. We have adopted a special case of this setting, in which every node has associated to it two functions: a function that governs its evolution over time and the identity function. If the first is chosen, then the node is updated based on its logical rule. When the identity function is chosen, then the state of the node is not updated. The key difference from a PBN is the assignment of probabilities that govern which update is chosen. In our setting, each function has two probabilities assigned to it. More precisely, let x be a variable. We assign to it a probability $p^\uparrow$, which determines the likelihood that x will be updated based on its logical rule if this update leads to an increase or activation of the variable. Likewise, a probability $p^\downarrow$ determines this probability in the case where the variable is decreased or inhibited. The necessity for considering two different probabilities arises because activation and degradation represent different biochemical processes and, even if these two are encoded by the same function, their propensities are different in general. This is very similar to what is considered in modeling by differential equations, where, for instance, the kinetic rate parameters for activation and for degradation or decay are, in principle, different. For other approaches to modeling stochasticity in the Boolean setting, see [23, 44, 56, 64].

7.1　Framework

The framework described below, published in [49], incorporates propensity parameters for activation and degradation. Here, the aim is to model stochasticity at the biological-function level, assuming that even if the expression levels of the input nodes of an update function guarantee activation or degradation, there is a probability that the process will not occur owing to stochasticity. This could happen, for instance, if some of the chemical reactions encoded by the update function fail to occur. This is similar to models based on the chemical master equation. This model type introduces activation and degradation propensities. Since the definition of such models does not depend on a polynomial representation, we discuss them in full generality.

Let $x_1, \ldots, x_n$ be variables which can take values in finite sets $X_1, \ldots, X_n$, respectively. Let $X = X_1 \times \cdots \times X_n$ be the Cartesian product. A *stochastic discrete dynamical system* in the variables $x_1, \ldots, x_n$ is a collection of n triplets

$$F = \{f_i, p_i^\uparrow, p_i^\downarrow\}_{i=1}^n,$$

where

- $f_i : X \to X_i$ is the update function for x_i, for all $i = 1, \ldots, n$;
- $p_i^\uparrow$ is the activation propensity;

- $p_i^{\downarrow}$ is the degradation propensity;
- $p_i^{\uparrow}, p_i^{\downarrow} \in [0, 1]$.

7.2 Dynamics of Stochastic Discrete Dynamical Systems

Let $F = \{f_i, p_i^{\uparrow}, p_i^{\downarrow}\}_{i=1}^n$ be a stochastic discrete dynamical system, and consider $x \in X$. For all i, we define $\pi_{i,x}(x_i \to f_i(x))$ and $\pi_{i,x}(x_i \to x_i)$ by

$$\pi_{i,x}(x_i \to f_i(x)) = \begin{cases} p_i^{\uparrow}, & \text{if } x_i < f_i(x), \\ p_i^{\downarrow}, & \text{if } x_i > f_i(x), \\ 1, & \text{if } x_i = f_i(x), \end{cases}$$

$$\pi_{i,x}(x_i \to x_i) = \begin{cases} 1 - p_i^{\uparrow}, & \text{if } x_i < f_i(x), \\ 1 - p_i^{\downarrow}, & \text{if } x_i > f_i(x), \\ 1, & \text{if } x_i = f_i(x). \end{cases}$$

That is, if the possible future value of the i-th coordinate is larger or smaller, respectively, than the current value, then the activation or degradation propensity determines the probability that the i-th coordinate will increase or decrease its current value. If the i-th coordinate and its possible future value are the same, then the i-th coordinate of the system will maintain its current value with probability 1. Notice that $\pi_{i,x}(x_i \to y_i) = 0$ for all $y_i \notin \{x_i, f_i(x)\}$.

The dynamics of F is given by a weighted graph X, which has an edge from $x \in X$ to $y \in X$ if and only if $y_i \in \{x_i, f_i(x)\}$ for all i. The weight of an edge $x \to y$ is equal to the product

$$w_{x \to y} = \prod_{i=1}^n \pi_{i,x}(x_i \to y_i).$$

By convention, we omit edges with weight zero. The software package ADAM, described in Sect. 6.2, contains algorithms for the construction and analysis of stochastic discrete dynamical systems.

Given a stochastic discrete dynamical system $F = \{f_i, p_i^{\uparrow}, p_i^{\downarrow}\}_{i=1}^n$, it is straightforward to verify that F has the same steady states as the deterministic system $G = \{f_i\}_{i=1}^n$. It is also important to note that the dynamics of F includes the different trajectories that can be generated from G using other common update mechanisms, such as synchronous and asynchronous schemes. Thus, stochastic discrete dynamical systems take advantage of the PDS representation of the regulatory functions $\{f_i\}_{i=1}^n$ for model analysis, for instance to compute the steady states of $F = \{f_i, p_i^\uparrow, p_i^\downarrow\}_{i=1}^n$ as described in Sect. 6.

The propensity parameters $\{p_i^\uparrow, p_i^\downarrow\}_{i=1}^n$ can be interpreted as a measure of relative speeds among the modeled processes, and these can be given a biological meaning. For instance, in [49], a simple degradation model was analyzed and the degradation propensity $p_1^\downarrow$ was related to the degradation rate (in the setting of the Gillespie algorithm) by a linear equation. The propensity parameters are useful when one is trying to distinguish fast from slow processes. Furthermore, two small stochastic biological systems were studied in [49], and it was shown there that stochastic discrete dynamical systems are a suitable framework for studying stochasticity in gene regulatory networks.

8 Nested Canalizing Networks

Finally, we return to a theme touched upon earlier, the nature of the regulatory rules that are biologically most meaningful. An important theoretical principle that has been discussed in evolutionary biology since the 1940s is that of canalization, a property of gene regulatory networks that buffers the functioning of the network against various sources of noise, as well as against the deleterious effects of mutations (see, e.g., [76]). Stuart Kauffman and his collaborators introduced an abstracted notion of canalization to Boolean network models of gene regulation. Taking the concept a step further, the authors of [36, 37] introduced the concept of a *nested canalizing Boolean function*, and they showed that networks constructed from such functions exhibited features associated with the dynamics of gene regulatory networks, in particular robustness with respect to perturbations. Within the general framework described in Sect. 4, a definition of the notion of a *multistate nested canalizing function* was introduced in [47, 48] that reduces to the Kauffman definition in the Boolean case. For a more detailed discussion of this modeling framework, the user is encouraged to consult the references in this section.

8.1 Nested Canalizing Rules

Here we present the general definition of a nested canalizing rule in variables $x_1, \ldots, x_n$, with state space $X = X_1 \times \cdots \times X_n$.

Assume that each X_i is totally ordered; that is, its elements can be arranged in linear increasing order. In the Boolean case, this could be $X_i = \{0 < 1\}$. Let $S_i \subset X_i, i = 1, \ldots, n$, be subsets that satisfy the property that each S_i is a proper, nonempty subinterval of X_i; that is, every element of X_i that lies between two elements of S_i in the chosen order is also in S_i. Furthermore, we assume that the complement of each S_i is also a subinterval; that is, each S_i can be described by a threshold s_i, with all elements of S_i either larger or smaller than s_i.

- The function $f_i : X \to X_i$ is a *nested canalizing rule* in the variable order $x_{\sigma(1)}, \ldots, x_{\sigma(n)}$ with *canalizing input sets* $S_1, \ldots, S_n \subset X$ and *canalizing output values* $b_1, \ldots, b_n, b_{n+1} \in X_i$, with $b_n \neq b_{n+1}$, if it can be represented in the form

$$
f(x_1, \ldots, x_n) = \begin{cases}
b_1 & \text{if } x_{\sigma(1)} \in S_1, \\
b_2 & \text{if } x_{\sigma(1)} \notin S_1, x_{\sigma(2)} \in S_2, \\
b_3 & \text{if } x_{\sigma(1)} \notin S_1, x_{\sigma(2)} \notin S_2, x_{\sigma(3)} \in S_3, \\
\vdots & \\
b_n & \text{if } x_{\sigma(1)} \notin S_1, \ldots, x_{\sigma(n)} \in S_n, \\
b_{n+1} & \text{if } x_{\sigma(1)} \notin S_1, \ldots, x_{\sigma(n)} \notin S_n.
\end{cases}
$$

- The function $f_i : X \to X_i$ is a *nested canalizing function* if it is a nested canalizing function in some variable order $x_{\sigma(1)}, \ldots, x_{\sigma(n)}$ for some permutation σ on $\{1, \ldots, n\}$.

It is straightforward to verify that if $X_i = \{0, 1\}$ for all i, then we recover the definition of a Boolean nested canalizing rule given in [36]. As mentioned in previous sections, several important classes of multistate discrete models can be represented in the form of a dynamical system $f : X \longrightarrow X$, so that the concept of a nested canalizing rule defined in this way has broad applicability.

8.2 *The Dynamics of Nested Canalizing Networks*

Through extensive simulations [47], it has been shown that dynamical systems constructed from nested canalizing rules as coordinate functions have important dynamical properties characteristic of molecular networks, namely very short limit cycles and very few attractors, compared with the set of all possible functions. Thus, aside from incorporating the biological concept of canalization, networks whose nodes are controlled by combinatorial logic expressed by nested canalizing rules have dynamical properties resembling those of biological networks. In particular, they are robust, owing to the fact that they have a small number of attractors, which are therefore large. That is, perturbations are more likely to remain in the same attractor. In addition, limit cycles tend to be very short, compared with random networks, which implies that these networks have very regular behavior.

8.3 *Nested Canalizing Rules Are Biologically Meaningful*

To test the hypothesis that nested canalizing rules are biologically meaningful, a range of published Boolean models have been investigated [36, 37] and their frequency of appearance has been quantified. In the multistate case, it has also been shown that many published models use logical interaction rules whose polynomial form is nested canalizing and that they are very prevalent [47], providing evidence that nested canalization is indeed a common pattern for the regulatory logic in molecular interaction networks. This raises the question of incorporating an additional constraint into network inference algorithms, so they return only nested canalizing functions. In its generality, this problem is unsolved at this time.

8.4 Theoretical Results about Nested Canalizing Functions

The polynomial representation of nested canalizing functions [32, 48] offers a variety of tools to analyze these functions. For instance, in [43], a new characterization for Boolean nested canalizing functions was obtained in terms of the polynomial representation, which provides a finer categorization of nested canalizing functions that takes into account the level of influence of individual variables. The concept of the layer number of a nested canalizing function was introduced in [43] as a measure of the extent of the hierarchy of influence among its variables. It was shown there that layer numbers are good indicators of the stability of networks whose nodes are controlled by nested canalizing functions. Thus the concept of the layer number gives a division of the class of nested canalizing functions into subfamilies with different stability properties; see [43]. However, these results still remain to be extended to the multistage case. The concept of the layer number was also used to evaluate the number of nested canalizing functions, their Hamming weight, the activity of their variables and their average sensitivity. A conjecture about a sharp upper bound for the average sensitivity of nested canalizing functions given in [43] was recently proved in [38].

References

1. R. Albert, H.G. Othmer, The topology of the regulatory interactions predicts the expression pattern of the segment polarity genes in *Drosophila melanogaster*. J. Theor. Biol. **223**, 1–18 (2003)
2. E.R. Álvarez-Buylla, Á. Chaos, M. Aldana, M. Benítez, Y. Cortes-Poza, C. Espinosa-Soto, D.A. Hartasánchez, R.B. Lotto, D. Malkin, G.J. Escalera Santos, P. Padilla-Longoria, Floral morphogenesis: stochastic explorations of a gene network epigenetic landscape. PLoS ONE **3**(11), e3626 (2008)
3. J. Aracena, J. Demongeot, E. Goles, Fixed points and maximal independent sets in AND-OR networks. Discret. Appl. Math. **138**(3), 277–288 (2004)
4. R. Bonneau, Learning biological networks: from modules to dynamics. Nat. Chem. Biol. **4**(11), 658–664 (2008)
5. D. Bratsun, D. Volfson, L.S. Tsimring, J. Hasty, Delay-induced stochastic oscillations in gene regulation. Proc. Natl. Acad. Sci. U.S.A. **102**(41), 14593–14598 (2005)
6. A. Bruex, R.M. Kainkaryam, Y. Wieckowski, Y.H. Kang, C. Bernhardt, Y. Xia, X. Zheng, J.Y. Wang, M.M. Lee, P. Benfey, P.J. Woolf, J. Schiefelbein, A gene regulatory network for

root epidermis cell differentiation in arabidopsis. PLoS Genet. **8**(1), e1002446 (2012). PMID: 22253603

7. F.J. Bruggeman, H.V. Westerhoff, The nature of systems biology. Trends Microbiol. **15**(1), 45–50 (2007)
8. A. Butte, I. Kohane, Mutual information relevance networks: functional genomic clustering using pairwise entropy measurements. Pac. Symp. Biocomput. **5**, 415–426 (2000)
9. I. Cantone, L. Marucci, F. Iorio, M. Ricci, V. Belcastro, M. Bansal, S. Santini, M. di Bernardo, D. di Bernardo, M. Cosma, A yeast synthetic network for in vivo assessment of reverse-engineering and modeling approaches. Cell **137**(1), 172–181 (2009)
10. C. Chaouiya, E. Remy, P.R.D. Thieffry, Qualitative modeling of genetic networks: from logical regulatory graphs to standard Petri nets. Springer Lect. Notes Comput. Sci. **3099**, 137–156 (2004)
11. M. Chaves, E. Sontag, R. Albert, Methods of robustness analysis for Boolean models of gene control networks. IET Syst. Biol. **153**, 154–167 (2006)
12. J. Chifman, A. Kniss, P. Neupane, I. Williams, B. Leung, Z. Deng, P. Mendes, V. Hower, F.M. Torti, S.A. Akman, S.V. Torti, R. Laubenbacher, The core control system of intracellular iron homeostasis: a mathematical model. J. Theor. Biol. **300**, 91–99 (2012). PMID: 22286016
13. D. Cox, J. Little, D. O'Shea, *Ideals, Varieties, and Algorithms*, 2nd edn. (Springer, New York, 1997)
14. P. Dao, R. Colak, R. Salari, F. Moser, E. Davicioni, A. Schonhuth, M. Ester, Inferring cancer subnetwork markers using density-constrained biclustering. Bioinformatics **26**(18), 625–631 (2010)
15. M.I. Davidich, S. Bornholdt, Boolean network model predicts cell cycle sequence of fission yeast. PLoS One **3**(2), e1672 (2008)
16. A. de la Fuente, P. Brazhnik, P. Mendes, Linking the genes: inferring quantitative gene networks from microarray data. Trends Genet. **18**(8), 395–398 (2002)
17. A. de la Fuente, N. Bing, I. Hoeschele, P. Mendes, Discovery of meaningful associations in genomic data using partial correlation coefficients. Bioinformatics **20**(18), 3565–3574 (2004)
18. E. Dimitrova, L.D. Garcia-Puente, F. Hinkelmann, A.S. Jarrah, R. Laubenbacher, B. Stigler, M. Stillman, P. Vera-Licona, Polynome (2010). Available at http://polymath.vbi.vt.edu/polynome/
19. E. Dimitrova, L.D. Garcìa-Puente, F. Hinkelmann, A.S. Jarrah, R. Laubenbacher, B. Stigler, M. Stillman, P. Vera-Licona, Parameter estimation for Boolean models of biological networks. Theor. Comput. Sci. **412**(26), 2816–2826 (2011)
20. J. Faith, B. Hayete, J. Thaden, I. Mogno, J. Wierzbowski, G. Cottarel, S. Kasif, J. Collins, T. Gardner, Large-scale mapping and validation of Escherichia coli transcriptional regulation from a compendium of expression profiles. PLoS Biol. **5**(1), e8 (2007)
21. D. Formanowicz, A. Sackmann, P. Formanowicz, J. Błazewicz, Petri net based model of the body iron homeostasis. J. Biomed. Inform. **40**(5), 476–485 (2007). PMID: 17258508
22. T. Gardner, D. di Bernardo, D. Lorenz, J. Collins, Inferring genetic networks and identifying compound mode of action via expression profiling. Science **301**(5629), 102–105 (2003)
23. A. Garg, K. Mohanram, A. Di Cara, G. De Micheli, I. Xenarios, Modeling stochasticity and robustness in gene regulatory networks. Bioinformatics **25**(12), i101–i109 (2009)
24. D.T. Gillespie, Exact stochastic simulation of coupled chemical reactions. J. Phys. Chem. **81**(25), 2340–2361 (1977)
25. D. Gillespie, Stochastic simulation of chemical kinetics. Annu. Rev. Phys. Chem. **58**, 35–55 (2007)
26. D.R. Grayson, M.E. Stillman, Macaulay2, a software system for research in algebraic geometry (1992). Available at http://www.math.uiuc.edu/Macaulay2/
27. A. Haury, F. Mordelet, P. Vera-Licona, J. Vert, TIGRESS: trustful inference of gene regulation using stability selection. BMC Syst. Biol. **6**, 145 (2012)
28. F. Hinkelmann, A.S. Jarrah, Inferring biologically relevant models: nested canalyzing functions. ISRN Biomath. **2012**, 7 (2012)

29. F. Hinkelmann, M. Brandon, B. Guang, R. McNeill, A. Veliz-Cuba, G. Blekherman, R. Laubenbacher, ADAM: analysis of analysis of dynamic algebraic models (2010). Available at http://adam.vbi.vt.edu/

30. F. Hinkelmann, M. Brandon, B. Guang, R. McNeill, G. Blekherman, A. Veliz-Cuba, R. Laubenbacher, ADAM: Analysis of discrete models of biological systems using computer algebra. BMC Bioinform. **12**(1), 295 (2011)

31. C. Hong, M. Lee, D. Kim, D. Kim, K.-H. Cho, I. Shin, A checkpoints capturing timing-robust Boolean model of the budding yeast cell cycle regulatory network. BMC Syst. Biol. **6**(1), 129 (2012). PMID: 23017186

32. A. Jarrah, B. Raposa, R. Laubenbacher, Nested canalyzing, unate cascade, and polynomial functions. Physica D **233**, 167–174 (2007)

33. A.S. Jarrah, R. Laubenbacher, B. Stigler, M. Stillman, Reverse-engineering of polynomial dynamical systems. Adv. Appl. Math. **39**(4), 477–489 (2007)

34. A. Jarrah, R. Laubenbacher, A. Veliz-Cuba, The dynamics of conjunctive and disjunctive Boolean network models. Bull. Math. Biol. **72**, 1425–1447 (2010)

35. S.A. Kauffman, The large-scale structure and dynamics of gene control circuits: an ensemble approach. J. Theor. Biol. **44**, 167 (1973)

36. S. Kauffman, C. Peterson, B. Samuelsson, C. Troein, Random Boolean network models and the yeast transcriptional network. Proc. Natl. Acad. Sci. **100**(25), 14796–14799 (2003)

37. S. Kauffman, C. Peterson, B. Samuelsson, C. Troein, Genetic networks with canalyzing Boolean rules are always stable. Proc. Natl. Acad. Sci. **101**(49), 17102–17107 (2004)

38. J.G. Klotz, R. Heckel, S. Schober, Bounds on the average sensitivity of nested canalizing functions. PLoS ONE **8**(5), e64371 (2013)

39. N. Kramer, J. Schafer, A. Boulesteix, Regularized estimation of large-scale gene association networks using graphical Gaussian models. BMC Bioinform. **10**, 384 (2009)

40. R. Küffner, T. Petri, P. Tavakkolkhah, L. Windhager, R. Zimmer, Inferring gene regulatory networks by ANOVA. Bioinformatics **28**(10), 1376–1382 (2012)

41. R. Laubenbacher, B. Stigler, A computational algebra approach to the reverse engineering of gene regulatory networks. J. Theor. Biol. **229**, 523–537 (2004)

42. R. Layek, A. Datta, R. Pal, E.R. Dougherty, Adaptive intervention in probabilistic Boolean networks. Bioinformatics **25**(16), 2042–2048 (2009)

43. Y. Li, J.O. Adeyeye, D. Murrugarra, B. Aguilar, R. Laubenbacher, Boolean nested canalizing functions: a comprehensive analysis. Theor. Comput. Sci. **481**(0), 24–36 (2013)

44. J. Liang, J. Han, Stochastic Boolean networks: an efficient approach to modeling gene regulatory networks. BMC Syst. Biol. **6**(1), 113 (2012)

45. R. Lidl, H. Niederreiter, *Finite Fields* (Cambridge University Press, New York, 1997)

46. A. Madar, A. Greenfield, E. Vanden-Eijnden, R. Bonneau, DREAM3: network inference using dynamic context likelihood of relatedness and the Inferelator. PLoS ONE **5**(3), e9803 (2010)

47. D. Murrugarra, R. Laubenbacher, Regulatory patterns in molecular interaction networks. J. Theor. Biol. **288**(0), 66–72 (2011)

48. D. Murrugarra, R. Laubenbacher, Multi-states nested canlyzing functions. Phys. D Nonlinear Phenom. **241**, 921–938 (2012)

49. D. Murrugarra, A. Veliz-Cuba, B. Aguilar, S. Arat, R. Laubenbacher, Modeling stochasticity and variability in gene regulatory networks. EURASIP J. Bioinform. Syst. Biol. **2012**, 5 (2012)

50. C. Müssel, M. Hopfensitz, H.A. Kestler, BoolNet – an R package for generation, reconstruction and analysis of Boolean networks. Bioinformatics **26**(10), 1378–1380 (2010)

51. A. Naldi, D. Berenguier, A. Fauré, F. Lopez, D. Thieffry, C. Chaouiya, Logical modelling of regulatory networks with GINsim 2.3. Biosystems **97**(2), 134–139 (2009)

52. A. Naldi, E. Remy, D. Thieffry, C. Chaouiya, A reduction of logical regulatory graphs preserving essential dynamical properties, in *Computational Methods in Systems Biology*, ed. by P. Degano, R. Gorrieri. Volume 5688 of Lecture Notes in Computer Science (Springer, Berlin/Heidelberg, 2009), pp. 266–280

53. R. Porreca, E. Cinquemani, J. Lygeros, G. Ferrari-Trecate, Identification of genetic network dynamics with unate structure. Bioinformatics **26**(9), 1239–1245 (2010)

54. L. Raeymaekers, Dynamics of Boolean networks controlled by biologically meaningful functions. J. Theor. Biol. **218**(3), 331–341 (2002)
55. A.S. Ribeiro, Stochastic and delayed stochastic models of gene expression and regulation. Math. Biosci. **223**(1), 1–11 (2010)
56. A.S. Ribeiro, S.A. Kauffman, Noisy attractors and ergodic sets in models of gene regulatory networks. J. Theor. Biol. **247**(4), 743–755 (2007)
57. A. Ribeiro, R. Zhu, S.A. Kauffman, A general modeling strategy for gene regulatory networks with stochastic dynamics. J. Comput. Biol. **13**(9), 1630–1639 (2006)
58. C. Rohr, W. Marwan, M. Heiner, Snoopy – a unifying Petri net framework to investigate biomolecular networks. Bioinformatics **26**(7), 974–975 (2010)
59. A. Saadatpour, I. Albert, R. Albert, Attractor analysis of asynchronous Boolean models of signal transduction networks. J. Theor. Biol. **266**(4), 641–656 (2010)
60. A. Sackmann, M. Heiner, I. Koch, Application of Petri net based analysis techniques to signal transduction pathways. BMC Bioinform. **7**(1), 482 (2006)
61. I. Shmulevich, E.R. Dougherty, *Probabilistic Boolean Networks: The Modeling and Control of Gene Regulatory Networks* (SIAM, Philadelphia, 2010)
62. I. Shmulevich, E.R. Dougherty, S. Kim, W. Zhang, Probabilistic Boolean networks: a rule-based uncertainty model for gene regulatory networks. Bioinformatics **18**(2), 261–274 (2002)
63. B. Stigler, D. Camacho, A. Martins, W. Sha, E.S. Dimitrova, P. Vera-Licona, V. Shulaev, P. Mendes, R. Laubenbacher, Reverse engineering a yeast oxidative stress response network. Under review (2013)
64. S. Teraguchi, Y. Kumagai, A. Vandenbon, S. Akira, D.M. Standley, Stochastic binary modeling of cells in continuous time as an alternative to biochemical reaction equations. Phys. Rev. E Stat. Nonlinear Soft Matter Phys. **84**(6 Pt 1), 062903 (2011)
65. D. Thieffry, R. Thomas, Qualitative analysis of gene networks. Pac. Symp. Biocomput. **3**, 77–88 (1998)
66. R. Thomas, Regulatory networks seen as asynchronous automata: a logical description. J. Theor. Biol. **153**, 1–23 (1991)
67. T. Toulouse, P. Ao, I. Shmulevich, S. Kauffman, Noise in a small genetic circuit that undergoes bifurcation. Complexity **11**(1), 45–51 (2005)
68. A. Veliz-Cuba, Reduction of Boolean network models. J. Theor. Biol. **289**, 167–172 (2011)
69. A. Veliz-Cuba, An algebraic approach to reverse engineering finite dynamical systems arising from biology. SIAM J. Appl. Dyn. Syst. **11**(1), 31–48 (2012)
70. A. Veliz-Cuba, R. Laubenbacher, On the computation of fixed points in Boolean networks. J. Appl. Math. Comput. accepted (2011)
71. A. Veliz-Cuba, B. Stigler, Boolean models can explain bistability in the *lac* operon. J. Comput. Biol. **18**(6), 783–794 (2011)
72. A. Veliz-Cuba, A.S. Jarrah, R. Laubenbacher, Polynomial algebra of discrete models in systems biology. Bioinformatics **26**(13), 1637–1643 (2010)
73. A. Veliz-Cuba, J. Arthur, L. Hochstetler, V. Klomps, E. Korpi, On the relationship of steady states of continuous and discrete models arising from biology. Bull. Math. Biol. accepted (2012)
74. A. Veliz-Cuba, K. Buschur, R. Hamershock, A. Kniss, E. Wolff, R. Laubenbacher, AND-NOT logic framework for steady state analysis of Boolean network models (2012). arXiv:1211.5633
75. M. Vignes, J. Vandel, D. Allouche, N. Ramadan-Alban, C. Cierco-Ayrolles, T. Schiex, B. Mangin, S. de Givry, Gene regulatory network reconstruction using Bayesian networks, the Dantzig selector, the Lasso and their meta-analysis. PLoS ONE **6**(12), e29165 (2011)
76. C.H. Waddington, Canalisation of development and the inheritance of acquired characters. Nature **150**, 563–564 (1942)
77. H. Wang, L. Qian, E. Dougherty, Inference of gene regulatory networks using S-system: a unified approach. IET Syst. Biol. **4**(2), 145–156 (2010)
78. D. Wilkinson, *Stochastic Modeling for Systems Biology* (Chapman and Hall/CRC, Boca Raton, 2006)

79. K. Willadsen, J. Wiles, Robustness and state-space structure of Boolean gene regulatory models. J. Theor. Biol. **249**(4), 749–765 (2007)
80. P. Zoppoli, S. Morganella, M. Ceccarelli, TimeDelay-ARACNE: reverse engineering of gene networks from time-course data by an information theoretic approach. BMC Bioinform. **11**(1), 154 (2010)

Deconstructing Complex Nonlinear Models in System Design Space

Michael A. Savageau and Jason G. Lomnitz

Abstract Achieving predictive understanding of complex nonlinear systems, such as those manifested at various levels of biological organization, represents an enormous challenge. The task would be facilitated if such systems could be generically decomposed into a series of tractable subsystems and the results of their analysis reassembled to provide insight into the original system. In this chapter, we describe an approach in which subsystems are integrated into a system design space that allows qualitatively distinct phenotypes of a complex system to be rigorously defined and counted, their relative fitness to be analyzed and compared, their global tolerance to be measured, and their biological design principles to be identified. We then illustrate the approach in the context of the "genotype–phenotype" question for a couple of simple, well-studied systems. Finally, we discuss the extent to which this approach might be generalized to other classes of nonlinear models. Although this effort has been the focus of our recent biological work, we believe that this methodology has application beyond biology. It also raises a number of mathematical issues that need to be explored further and extended.

1 Introduction

Throughout the pregenomic era, there was sustained interest in the relationship between genotype and phenotype [1]. Although we now have a generic concept of "genotype" provided by the detailed DNA sequence, there is no corresponding

M.A. Savageau (✉)
Biomedical Engineering Department and Microbiology Graduate Group, University of California, One Shields Avenue, Davis, CA 95616, USA
e-mail: masavageau@ucdavis.edu

J.G. Lomnitz
Biomedical Engineering Department, University of California, One Shields Avenue, Davis, CA 95616, USA
e-mail: jlomn@ucdavis.edu

N. Jonoska and M. Saito (eds.), *Discrete and Topological Models in Molecular Biology*, Natural Computing Series, DOI 10.1007/978-3-642-40193-0_22,
© Springer-Verlag Berlin Heidelberg 2014

generic concept of "phenotype". Without a generic concept of a phenotype, there can be no rigorous framework for a deep understanding of the complex nonlinear systems that link genotype to phenotype. The concept of "phenotype" must ultimately be grounded in the underlying biochemistry. The vast majority of biochemical models are represented in terms of the power functions of chemical kinetics or the rational functions of biochemical kinetics, which result from chemical kinetics plus constraints.

Although simple models within these formalisms can occasionally be treated analytically, this is seldom possible when the models become even moderately complex. For this reason, the more complex models are typically analyzed by linearization about fixed points and simulated by numerical methods. Finding the fixed points of nonlinear models is a challenging problem in its own right [2]. Indeed, there is no method that is guaranteed to find all the roots of a complex nonlinear model. Numerical simulation also has limitations because it is impractical to sample all combinations of parameter values; as a result, important behaviors can be missed because of failure to sample certain regions of parameter space.

One is often interested in comparing alternative models as a way of testing hypotheses. If such comparisons are to be well controlled, one must take pains to ensure that the portions of the alternative models that should remain identical do in fact operate around the same states. This is a subtle issue that only comes up in nonlinear models. If the comparisons are not well controlled in this sense, then the resulting differences that one would like to attribute to the specific alternatives being tested might actually result from nonspecific consequences attributed to a change in state elsewhere in the model.

Because of these and other difficulties that arise in the analysis of nonlinear models [3], it would be highly desirable if there were a method to decompose a complex model into a collection of simpler nonlinear models that are analytically tractable, and then to reassemble the results to provide understanding of the original complex nonlinear model. One of the so-called Grand Challenges of modern biology is to understand the relationship between the information in our genes (the genotype) and the expression of that information (the phenotype) as the structure, function, and behavior of the organism in its environmental context [1]. It is hard to overestimate the magnitude of this challenge, and for this reason it manifests all of the difficulties mentioned above. Thus, it provides an ideal context in which to address these difficulties.

In this chapter, we review the elements of a method, developed in the context of this "genotype to phenotype" problem, that has shown promise for accomplishing the desirable goal of systematic deconstruction mentioned above [4]. It involves a concept of *qualitatively distinct phenotypes* that is rigorously defined and applies to systems at levels from the molecular to the organismal. We have provided examples elsewhere to demonstrate how this concept is manifested within a *system design space*, in which qualitatively distinct phenotypes can be identified and counted, their relative fitness analyzed and compared, their tolerance to global change measured, and biological design principles identified. In this approach, the link between genotype and phenotype is a mathematical model whose parameter values are genetically determined, whose independent (input) variables are environmentally determined, and whose dependent (output) variables determine the phenotype.

First, we show here how chemical and biochemical models can be recast into a generic nonlinear form called a *generalized mass action system,* or GMA system, within a power-law formalism. Second, we introduce the concept of *dominant processes* and define a *phenotype* as a valid combination of dominant processes characterized by a class of tractable nonlinear subsystems, called S-systems. Third, we show how the results of analyzing these subsystems yield a repertoire of qualitatively distinct phenotypes, which, when integrated into a system design space, provide important understanding of the original model. Finally, we discuss the extent to which this approach might be generalized to other classes of nonlinear models, and raise a number of mathematical issues that need to be explored further and extended.

2 Recasting Chemical and Biochemical Models into a Generic Nonlinear Representation

Although chemical change can be described in a variety of ways [5–7], the most useful ways deal with its quantitative characteristics: to what extent a reaction normally takes place, and how fast it proceeds. These thermodynamic and kinetic aspects have had extensive mathematical development [8–10]. Not only do these approaches provide useful information about the mechanism of individual processes, but the information they provide is in such a form that it can also be incorporated into an appropriate description of systems containing many such processes. Hence, in the search for a formalism that is appropriate to the analysis of organizationally complex systems, one naturally looks to kinetic approaches.

Among the kinetic approaches, there are several complementary ways to describe chemical change. The most common in biology are stochastic [11], deterministic [12], and Boolean [13] rate laws. One biologist might say, "I am interested in knowing when a chemical bond breaks, and this involves a probability distribution function," which is a discrete stochastic description. Another biologist might say, "I am interested in large numbers of such events in a given increment of time, and for this I would like to know the rate law function," which is a continuous deterministic description. Finally, a developmental biologist might say, "I am only interested in knowing whether or not some gene gets turned on or off during development, and for this purpose it is sufficient to use a Boolean function," which is a discrete deterministic description.

The rate law description has the advantage of permitting mathematical analysis in addition to computer simulation. It is particularly relevant when one is more interested in population means than in individuals. It also facilitates the elucidation of biological design principles and leads to qualitative predictions that can be readily tested experimentally. From a pragmatic point of view, testable predictions that are experimentally verified provide the ultimate justification for this approach. Nevertheless, there are inherent limitations, involving small numbers and short time scales, that must be kept in mind; adding stochastic variation to the deterministic description in these cases can lead to additional insights [14].

2.1   Rate Law Representation

The *rate law* for a process is defined as the mathematical function that represents the instantaneous rate of the process as an explicit function of all the state variables that have a direct influence on the rate of the process. In general, the rate law is a function of n state variables and can be written in *functional form* as

$$v = v\,(X_1, X_2, \cdots, X_n)\,. \tag{1}$$

The state variables and flux variables in biological systems can nearly always be represented as positive quantities. Therefore, the rate law can be represented equivalently in a logarithmic space, i.e., a space in which the logarithm of the rate is a function of the logarithms of the state variables [15, 16]. Indeed, this is a natural representation for biological systems because the logarithm of a concentration (the "activity") is the thermodynamic potential (or "across variable") in a chemical system. The functional form of the general rate law in logarithmic coordinates is given by

$$\ln v = f\,(\ln X_1, \ln X_2, \cdots \ln X_n)\,. \tag{2}$$

Rate laws exhibit a wide variety of forms when they are expressed explicitly. It is necessary to find a convenient generic representation for rate laws in explicit form if we are to develop systematically structured methods of representation and analysis for systems composed of many variables. Otherwise, the representation and analysis of such systems can only be treated in an ad hoc fashion, and our ability to develop general principles is greatly diminished.

2.2   Taylor Series in Logarithmic Space

One generic representation of functions that has been enormously useful, and that has formed the basis for much of classical mathematical analysis, is the Taylor series [17]. When a state variable $\ln X$ deviates only slightly from a reference value $\ln X_0$, a rate law can be approximated by the first two terms of the Taylor series, and all higher-order terms in $(\ln X - \ln X_0)$ are negligible. The equation for this linear approximation in logarithmic space is

$$\ln v\,(\ln X) = \ln v\,(\ln X_0) + [d\,(\ln v)\,/d\,(\ln X)]_0\,(\ln X - \ln X_0)\,, \tag{3}$$

or

$$\ln v = \ln \alpha + g \ln X, \tag{4}$$

where the coefficients have been redefined as $\ln \alpha = \ln v_0 - [d(\ln v)/d(\ln X)]_0 \ln X_0$ and $g = [d(\ln v)/d(\ln X)]_0$. The subscript 0 signifies that the appropriate function or variable is evaluated at the nominal operating point. Exponentiating both sides of (4) gives the equivalent expression in Cartesian space,

$$v(X) = \alpha X^g. \tag{5}$$

A natural extension of this approach to processes influenced by several variables yields the equation

$$\ln v = \ln \alpha + g_1 \ln X_1 + g_2 \ln X_2 + \cdots + g_n \ln X_n \tag{6}$$

and the equivalent expression in Cartesian space,

$$v(X) = \alpha X_1^{g_1} X_2^{g_2} \cdots X_n^{g_n}. \tag{7}$$

The two types of parameters in this rate law will be referred to as *multiplicative parameters* (α) and *exponential parameters* (g_i). They also will be referred to as *rate constants* and *kinetic orders*, respectively, since these are the conventional terms in the context of chemical and biochemical kinetics.

2.3 Power-Law Formalism

The power-law function above has the mathematical simplicity desired. It also has the ability to conform approximately to a variety of nonlinearities. However, the full repertoire of nonlinear behavior capable of representation by power-law functions only becomes evident when they are combined into a system of differential equations, in what is called the power-law formalism [15]. The two most commonly used representations within this formalism are the *synergistic (S-system) representation*

$$\frac{dX_i}{dt} = \alpha_i \prod_{j=1}^{n+m} X_j^{g_{ij}} - \beta_i \prod_{j=1}^{n+m} X_j^{h_{ij}}, \quad X_i(0) = X_{i0} \quad i = 1, 2, \cdots, n, \tag{8}$$

and the *generalized mass action system* representation,

$$\frac{dX_i}{dt} = \sum_{k=1}^{r} \alpha_{ik} \prod_{j=1}^{n+m} X_j^{g_{ijk}} - \sum_{k=1}^{r} \beta_{ik} \prod_{j=1}^{n+m} X_j^{h_{ijk}}, \quad X_i(0) = X_{i0} \quad i = 1, 2, \cdots, n, \tag{9}$$

where the X's are nonnegative real variables that typically represent chemical concentrations in the system, the α's and β's are non-negative real parameters that represent rate constants for production and loss, the g's and h's are parameters that represent kinetic orders for production and loss, m is the number of independent variables whose values are determined by the natural environment of the system or by direct experimental manipulation, and n is the number of dependent variables whose values are dependent upon the values of the independent variables and parameters.

A focus on the elementary chemical reactions in a system leads to a special case of (9) known as traditional mass action kinetics. The representation in this special case involves rate laws that are simple power-law functions with integer exponents. That is, the g and h parameters in (9) have only small integer values (<4).

2.4 Rational-Function Representation

A focus on the biochemical reactions in a system typically leads to a mathematical representation in terms of rational-function kinetics [18, 19],

$$\frac{dX_i}{dt} = \frac{\sum_{k=1}^{r} \alpha_{ik} \prod_{j=1}^{n+m} X_j^{g_{ijk}}}{\sum_{k=1}^{r} \gamma_{ik} \prod_{j=1}^{n+m} X_j^{p_{ijk}}} - \frac{\sum_{k=1}^{r} \beta_{ik} \prod_{j=1}^{n+m} X_j^{h_{ijk}}}{\sum_{k=1}^{r} \delta_{ik} \prod_{j=1}^{n+m} X_j^{q_{ijk}}}, \quad X_i(0) = X_{i0} \quad i = 1, 2, \cdots, n,$$

$$(10)$$

where again the X's are nonnegative real variables that represent chemical concentrations in the system, the multiplicative (Greek) parameters are nonnegative real numbers, and the exponential parameters have small integer values.

Rational-function kinetics arise when constraints are imposed on a mass action system. The simplest and most familiar examples are provided by the Michaelis–Menten expressions of enzyme kinetics. For example, the equation for a reversible mechanism for a single enzymatic reaction,

$$\frac{dP}{dt} = \frac{V_{\max f}\frac{S}{K_f}}{1 + \frac{S}{K_f} + \frac{P}{K_r}} - \frac{V_{\max r}\frac{P}{K_r}}{1 + \frac{S}{K_f} + \frac{P}{K_r}} = \frac{V_{\max f}\frac{S}{K_f}\left(1 - \frac{1}{K_{eq}\Gamma}\right)}{1 + \frac{S}{K_f} + \frac{P}{K_r}}, \quad P(0) = P_0, \quad (11)$$

is obtained when the sum of the concentrations for the various forms of the enzyme (free, substrate- bound, and product- bound) is constrained to be a constant. The net flux of production for the product P in the steady state is the difference between the forward rate (synthesis of P from the substrate S) and the reverse rate (synthesis of S from the product P). The four kinetic parameters represent the maximum velocities in the forward ($V_{\mathrm{max,f}}$) and reverse ($V_{\mathrm{max,r}}$) directions, the concentrations of substrate for half the maximum velocity in the forward direction in the absence of product (K_{f}), and the concentrations of product for half the maximum velocity in the reverse direction in the absence of substrate (K_{r}). Only three of the four kinetic parameters are independent [20–22], since the equation for the overall equilibrium constant ($K_{\mathrm{eq}} = V_{\mathrm{max,f}}K_{\mathrm{r}}/(V_{\mathrm{max,r}}K_{\mathrm{f}})$) must be satisfied and the displacement from thermodynamic equilibrium is given by $\Gamma = S/P$.

2.5 A Generic Nonlinear Representation via Recasting

We showed some time ago that a broad class of nonlinear functions and systems of ordinary nonlinear differential equations can be recast exactly into the power-law formalism [23]. The recasting procedure consists of a few simple steps repeated a finite number of times to yield the generic representation in (8) or (9). Equation (9) differs from the traditional mass action representation in two ways: (a) the exponential parameters need not have small integer values but can have real values (positive or negative), and (b) there are specific algebraic constraints among the initial conditions for the equations. Although it is perhaps less obvious, the rational-function representation can also be considered a special case of (9) as a result of recasting (see Sect. 2.6).

The recasting steps can be summarized as follows:

1. *Reorganization of the original equations.* Rewrite the equations as sums of products of factors:

$$\frac{dX_i}{dt} = \sum_{j} \prod_{k} f_{ijk}, \qquad i = 1, 2, \cdots, n, \qquad (12)$$

   where the factors f_{ijk} can be any nested set of elementary functions.
2. *Translocation to the positive orthant.* If any variable has the potential to become negative, transform it in such a manner that the new variable is always positive and then return to Step 1. If all the variables are always positive, then proceed to Step 3.
3. *Decomposition by the chain rule of differentiation.* Any f_{ijk} that is not already a power-law function is replaced by a new variable. An additional differential equation is then generated by differentiating the new variable. The chain rule of differentiation results in the decomposition of complex composite functions into products of simpler functions. After all the original functions have been treated in this fashion, return to Step 1. When all f_{ijk} are in the form of power-law functions, the procedure is complete and the resulting equations have a GMA system representation within the power-law formalism.

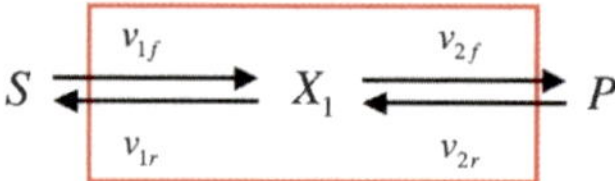

Fig. 1 System consisting of a two-step pathway of reversible enzymatic reactions governed by the Michaelis–Menten mechanism. See the text following (11) for a discussion of this mechanism

4. *Reduction of sums by the product rule of differentiation.* If the maximum number of terms with the same sign in a given equation is r, then let the corresponding dependent variable be replaced by a product of r new variables. Differentiation of this product then generates r derivatives in the new variables that can be equated to a single positive term and a single negative term. When this operation has been completed for all equations in the GMA system, the resulting set of differential equations will be in the S-system form.

In what follows, we need only be concerned with recasting into the GMA system representation. The recasting of rational-function models into GMA models is a particularly simple process. Since all the concentrations are nonnegative variables by definition, there is no need to consider Step 2. Although there are alternative strategies for Step 3, the simplest is to define each polynomial in the denominator as a new variable. The differentiation of each new variable generates a new differential equation that is already in the GMA form. A simple example will illustrate the recasting procedure.

2.6 An Example of Recasting

Consider the example of two reversible reactions governed by Michaelis–Menten mechanisms shown in Fig. 1. In this example, the substrate and product concentrations are environmentally determined independent variables, there is a chemically determined equilibrium for each reaction, and three of the four kinetic parameters for each reaction are determined genetically (see (11)). The resulting equation for the dependent variable X_1, whose values are determined by the values of the independent variables, the parameters, and the initial conditions, is

$$\frac{dX_1}{dt} = \frac{V_{1f}\frac{S}{K_{1f}}\left(1 - \frac{X_1}{K_{e1}S}\right)}{1 + \frac{S}{K_{1f}} + \frac{X_1}{K_{1r}}} - \frac{V_{2f}\frac{X_1}{K_{2f}}\left(1 - \frac{P}{K_{e2}X_1}\right)}{1 + \frac{X_1}{K_{2f}} + \frac{P}{K_{2r}}}, \quad X_1(0) = X_{10}, \tag{13}$$

where $V = V_{max}$, $K_e = K_{eq}$, and the additional subscript signifies the reaction in question.

The recast GMA version of (13) is obtained by the procedure described in Sect. 2.5, in which two new variables ($X_2 = 1 + S/K_{1f} + X_1/K_{1r}$ and $X_3 = 1 + P/K_{2r} + X_1/K_{2f}$) are defined and differentiated:

$$\frac{dX_1}{dt} = \frac{V_{1f}}{K_{1f}} S X_2^{-1} + \frac{V_{2f}}{K_{2f} K_{e2}} P X_3^{-1} - \frac{V_{1f}}{K_{1f} K_{e1}} X_1 X_2^{-1} - \frac{V_{2f}}{K_{2f}} X_1 X_3^{-1}, \tag{14}$$

$$\frac{dX_2}{dt} = \frac{V_{1f}}{K_{1f} K_{1r}} S X_2^{-1} + \frac{V_{2f}}{K_{2f} K_{e2} K_{1r}} P X_3^{-1} - \frac{V_{1f}}{K_{1f} K_{e1} K_{1r}} X_1 X_2^{-1} - \frac{V_{2f}}{K_{2f} K_{1r}} X_1 X_3^{-1}, \tag{15}$$

$$\frac{dX_3}{dt} = \frac{V_{1f}}{K_{1f} K_{2f}} S X_2^{-1} + \frac{V_{2f}}{K_{2f} K_{e2} K_{2f}} P X_3^{-1} - \frac{V_{1f}}{K_{1f} K_{e1} K_{2f}} X_1 X_2^{-1} - \frac{V_{2f}}{K_{2f} K_{2f}} X_1 X_3^{-1}, \tag{16}$$

with initial conditions $X_1(0) = X_{10}$, $X_2(0) = 1 + S/K_{1f} + X_{10}/K_{1r}$, and $X_3(0) = 1 + P/K_{2r} + X_{10}/K_{2f}$.

Alternatively, the result can be expressed as a set of differential–algebraic equations in the GMA representation,

$$\frac{dX_1}{dt} = \frac{V_{1f}}{K_{1f}} S X_2^{-1} + \frac{V_{2f}}{K_{2f} K_{e2}} P X_3^{-1} - \frac{V_{1f}}{K_{1f} K_{e1}} X_1 X_2^{-1} - \frac{V_{2f}}{K_{2f}} X_1 X_3^{-1},$$
$$X_1(0) = X_{10}, \tag{17}$$

$$X_2 = 1 + \frac{S}{K_{1f}} + \frac{X_1}{K_{1r}}, \tag{18}$$

$$X_3 = 1 + \frac{P}{K_{2r}} + \frac{X_1}{K_{2f}}. \tag{19}$$

The fixed points for this GMA system can then be obtained from the following set of nonlinear algebraic equations:

$$0 = \frac{V_{1f}}{K_{1f}} S X_2^{-1} + \frac{V_{2f}}{K_{2f} K_{e2}} P X_3^{-1} - \frac{V_{1f}}{K_{1f} K_{e1}} X_1 X_2^{-1} - \frac{V_{2f}}{K_{2f}} X_1 X_3^{-1}, \tag{20}$$

$$0 = X_2 - 1 - \frac{S}{K_{1f}} - \frac{X_1}{K_{1r}}, \tag{21}$$

$$0 = X_3 - 1 - \frac{P}{K_{2r}} - \frac{X_1}{K_{2f}}. \tag{22}$$

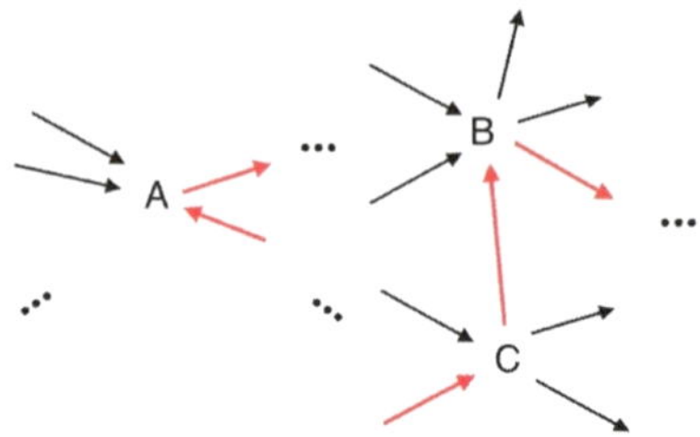

Fig. 2 Three selected metabolites in an arbitrary biochemical system

3 Dominant Processes

In any realistic biochemical system there are typically several processes that contribute to the influx into or efflux from the pool of molecules for any given biochemical entity. This corresponds to a GMA system with several positive terms and several negative terms in each equation. In general, such nonlinear systems are very intractable. However, in any given condition, it is highly probable that one of the terms in each sum is larger than the others in that sum; in other words, that one process dominates the net influx and another dominates the net efflux of each entity. For example, suppose Fig. 2 represents an arbitrary biochemical system in which three metabolites have been singled out. Metabolite A has three processes contributing to its net influx and one to its net efflux, metabolite B has three processes contributing to its net influx and three to its net efflux, and metabolite C has two processes contributing to its net influx and three to its net efflux. The red arrows indicate a possible combination of dominant processes for these metabolites.

3.1 *A Collection of Dominant Terms Corresponds to an S-System*

The intuitive picture in the previous paragraph corresponds in general to a reduction of the GMA system in (23) below to a particular S-system in (24) that is considerably more tractable. (We will have more to say about S-systems in Sect. 5.) These equations are as follows:

$$\frac{dX_i}{dt} = \sum_{k=1}^{r} \alpha_{ik} \prod_{j=1}^{n+m} X_j^{g_{ijk}} - \sum_{k=1}^{r} \beta_{ik} \prod_{j=1}^{n+m} X_j^{h_{ijk}}, \qquad X_i(0) = X_{i0}, \quad i = 1, 2, \cdots, n, \tag{23}$$

$$\frac{dX_i}{dt} = \alpha_{ip} \prod_{j=1}^{n+m} X_j^{g_{ijp}} - \beta_{iq} \prod_{j=1}^{n+m} X_j^{h_{ijq}}, \qquad X_i(0) = X_{i0}, \quad i = 1, 2, \cdots, n, \tag{24}$$

There are 162 combinations of potentially dominant terms for the simple example illustrated in Fig. 2; there are 36 combinations of potentially dominant terms for the system in Fig. 1, as can be seen from (20) to (22). However, not all combinations of dominant terms are valid. To be valid, a particular combination must meet two requirements. First, the resulting S-system must have a steady-state solution. Second, given that solution, all of the other terms in each sum must be smaller than the presumed dominant term. In general, many combinations of potentially dominant terms are not valid. Finding the valid combinations is a tractable linear programming problem that involves solving the S-system equations (a set of linear equations in log space) along with the dominance conditions (a set of linear inequalities in log space) [4].

3.2 Valid Combinations of Dominant Terms

An example of a valid combination of dominant terms (or processes) for the model illustrated in Fig. 1 is the following. Selecting the second positive term in (20), and the first, second, and third negative terms in (20), (21), and (22), respectively yields the following S-system in the steady state:

$$0 = \frac{V_{2f}}{K_{2f}K_{e2}}PX_3^{-1} - \frac{V_{1f}}{K_{1f}K_{e1}}X_1X_2^{-1}, \tag{25}$$

$$0 = X_2 - \frac{S}{K_{1f}}, \tag{26}$$

$$0 = X_3 - \frac{X_1}{K_{2f}}. \tag{27}$$

This is a system of linear equations in logarithmic coordinates; it can easily be solved for the logarithm of the dependent concentration variable and converted back to Cartesian coordinates:

$$X_1 = \sqrt{\frac{V_{2f}K_{e1}SP}{V_{1f}K_{e2}}}. \tag{28}$$

Thus, the intermediate concentration increases with an increase in the square root of the substrate concentration; it also decreases with a decrease in the square root of the product concentration.

The corresponding dominance conditions that must be satisfied are

$$\frac{V_{1f}}{K_{1f}}SX_2^{-1} < \frac{V_{2f}}{K_{2f}K_{e2}}PX_3^{-1}, \qquad \frac{V_{1f}}{K_{1f}K_{e1}}X_1X_2^{-1} > \frac{V_{2f}}{K_{2f}}X_1X_3^{-1}, \tag{29}$$

$$\frac{S}{K_{1f}} > \frac{X_1}{K_{1r}}, \qquad 1 < \frac{S}{K_{1f}}, \tag{30}$$

$$\frac{P}{K_{2r}} < \frac{X_1}{K_{2f}} \,, \qquad\qquad 1 < \frac{X_1}{K_{2f}}. \tag{31}$$

When the steady-state solution is substituted into this system of inequalities, which is also linear in logarithmic coordinates, one obtains the boundaries for the parameter values within which this dominant S-system is valid:

$$\frac{V_{1f}K_{2f}}{V_{2f}K_{1f}K_{e1}} > 1, \qquad\qquad \frac{V_{1f}K_{2f}}{V_{2f}K_{1f}K_{e1}} < \frac{1}{\Gamma K_{eq}}, \tag{32}$$

$$\frac{V_{1f}K_{2f}}{V_{2f}K_{1f}K_{e1}} \left(\frac{K_{1r}^2}{K_{1f}K_{2f}K_{e1}} \right) > \frac{1}{K_{eq}\Gamma}, \qquad\qquad \frac{S}{K_{1f}} > 1, \tag{33}$$

$$\frac{V_{2f}K_{1f}K_{e1}}{V_{1f}K_{2f}} \left(\frac{K_{2r}^2}{K_{1f}K_{2f}K_{e2}K_{eq}} \right) > \frac{1}{K_{eq}\Gamma}, \qquad\qquad \frac{V_{2f}K_{1f}K_{e1}}{V_{1f}K_{2f}} \left(\frac{SP}{K_{1f}K_{2f}K_{e2}} \right) > 1, \tag{34}$$

where the two additional parameters represent the overall thermodynamic equilibrium between S and P ($K_{eq} = K_{e1}K_{e2}$) and the displacement from equilibrium ($\Gamma = S/P$).

The values for the net flux in the steady state can also be obtained by substituting the steady-state solution for the dependent concentration variable into the rate law for the influx or efflux in the S-system equation. Thus,

$$V_{net} \approx -\sqrt{\frac{V_{1f}V_{2f}K_{e1}}{K_{eq}\Gamma}}. \tag{35}$$

The flux in the reverse direction increases with an increase in the square root of the product concentration; it also decreases with an increase in the square root of the substrate concentration.

We define a *phenotype* as the manifestation of a valid combination of dominant processes. Each system has a finite number of *qualitatively distinct phenotypes*, whose characteristics and interrelationships are made evident in the construction of the system design space, which is described in Sect. 4.

3.3 Invalid Combinations of Dominant Terms

An example of an invalid combination of dominant terms for the model illustrated in Fig. 1 is the following. Selecting the first positive term in (20), and the second, second, and third negative terms in (20), (21), and (22), respectively yields the following S-system in the steady state:

$$0 = \frac{V_{1f}}{K_{1f}} S X_2^{-1} - \frac{V_{2f}}{K_{2f}} X_1 X_3^{-1}, \tag{36}$$

$$0 = X_2 - \frac{S}{K_{1f}}, \tag{37}$$

$$0 = X_3 - \frac{X_1}{K_{2f}}. \tag{38}$$

This is an underdetermined system of linear equations in logarithmic coordinates. It is consistent if $V_{1f} = V_{2f}$, but there is no unique solution. Instead, we have

$$\max \left\{ \begin{array}{c} K_{2f} \\ \frac{K_{2f}}{K_{2r}} P \end{array} \right\} < X_1 < \frac{K_{1r}}{K_{1f}} S. \tag{39}$$

4 System Design Space

The process of constructing the system design space is best illustrated by a two simple examples, one chemical and the other biochemical. These examples are already well understood, so our purpose here is not to show that the system design space methodology leads to new information. Rather, these examples have been selected so that the results from our methodology can be compared easily with well-known results and intuition.

4.1 A Chemical Example

Interconversion of the two cyclic forms of glucose involves two reversible reactions and an acyclic intermediate in a chemical process that has been studied for decades [24]. This simple system is represented in Fig. 3, and the equation describing the interconversion is

$$\frac{dX_1}{dt} = [k_{1f}S - k_{1r}X_1] - [k_{2f}X_1 - k_{2r}P], \quad X_1(0) = X_{10}. \tag{40}$$

The intermediate X_1 is the one dependent variable. The forward rate constants k_{if} and the reverse rate constants k_{ir} are related through the equilibrium constants $K_{ei} = k_{if}/k_{ir}$, and, as a result, only two of the four parameters are independent. Thus, the behavior of the system (its phenotypes) is determined by two equilibrium constants (which are fixed thermodynamic quantities), two kinetic parameters (which are subject to change with the design of a catalyst), and two independent concentration variables, the substrate S and product P (which are subject to direct experimental manipulation of the environment).

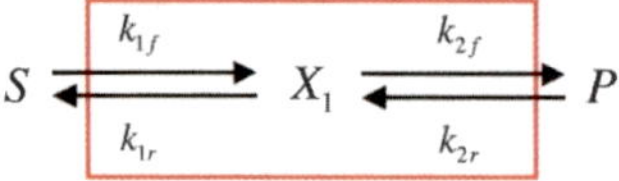

Fig. 3 System consisting of a two-step pathway of reversible chemical reactions. See text

4.1.1 Qualitatively Distinct Phenotypes Based on Dominant Terms

There are four qualitatively distinct phenotypes based on the dominance among the positive and negative terms of (40):

Case 1: $\frac{dX_1}{dt} \approx [k_{1f}S - k_{1r}X_1]$ $\qquad$ when $\frac{k_{1r}}{k_{2f}} > \frac{1}{K_{eq}\Gamma}$ and $\frac{k_{1r}}{k_{2f}} > 1,$

Case 2: $\frac{dX_1}{dt} \approx [k_{1f}S - k_{2f}X_1]$ $\qquad$ when $\frac{k_{1r}}{k_{2f}} > \frac{1}{K_{eq}\Gamma}$ and $\frac{k_{1r}}{k_{2f}} < 1,$

Case 3: $\frac{dX_1}{dt} \approx [k_{2r}P - k_{1r}X_1]$ $\qquad$ when $\frac{k_{1r}}{k_{2f}} < \frac{1}{K_{eq}\Gamma}$ and $\frac{k_{1r}}{k_{2f}} > 1,$

Case 4: $\frac{dX_1}{dt} \approx [k_{2r}P - k_{2f}X_1]$ $\qquad$ when $\frac{k_{1r}}{k_{2f}} < \frac{1}{K_{eq}\Gamma}$ and $\frac{k_{1r}}{k_{2f}} < 1,$

where the two additional parameters again represent the overall thermodynamic equilibrium between S and P ($K_{eq} = K_{e1}K_{e2}$) and the displacement from equilibrium ($\Gamma = S/P$).

This method of selecting one dominant positive term and one dominant negative term generates, in general, a set of *nonlinear* equations known as an S-system, whose solution in the steady state reduces to a linear problem for which one can obtain a solution explicitly. This method of selecting dominant terms from the differential equation has several advantages. First, finding the steady-state solution of the dominant differential equations is much simpler than finding an analytical solution of the original system. Second, having the differential equations based on dominant positive and negative terms means that we also have access to the local dynamic behavior for each of the phenotypes. However, will this method tell us how many distinct phenotypes the system is capable of exhibiting? By integrating information from all the steady-state solutions and their corresponding conditions for validity, we can address this question in the context of the system design space.

4.1.2 Mathematically Defined Boundaries in System Design Space

The phenotypes corresponding to the steady-state solutions only make sense if the solutions also satisfy the set of inequalities required to justify the assumption of dominance. Furthermore, determining if the set of inequalities is satisfied somewhere in the design space is equivalent to solving for the feasibility of a special class of geometric programming problems. These geometric programming problems involve the solution of the steady-state equations, which are linear equations in logarithmic coordinates, along with the corresponding set of dominance conditions, which are linear inequalities in logarithmic coordinates. This class of problems includes linear programming problems in logarithmic coordinates for

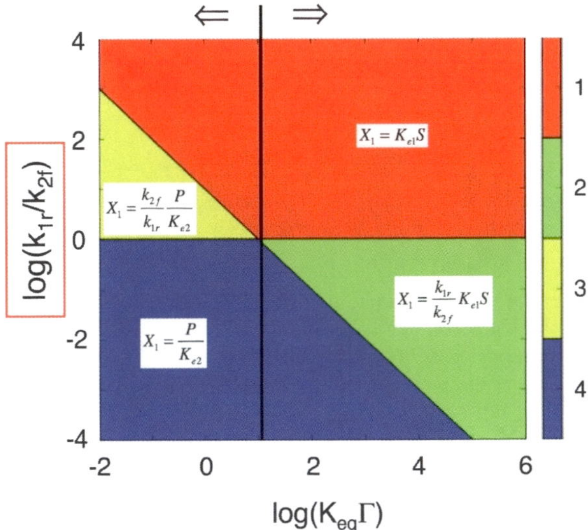

Fig. 4 System design space for the chemical mechanism illustrated in Fig. 3. The environmentally determined variables are plotted on the horizontal axis, and the chemically determined parameters are plotted on the vertical axis. The color bar indicates the case number, and the steady-state solution of the concentration of the intermediate is shown for each qualitatively distinct phenotype. The *vertical black line* represents thermodynamic equilibrium and the *arrows* indicate the direction of the net flux. All values are logarithms to base 10. Nominal parameter values: $k_{e2} = 0.1$, $S = 1,000$, and $k_{1f} = k_{2f} = k_{e1} = P = 1$

which feasibility can be readily determined [25, 26]. The result is a set of linear boundaries in logarithmic space delimiting valid regions, which define *qualitatively distinct phenotypes;* these can be visualized graphically in the *system design space,* as shown in Fig. 4.

4.1.3   S-System Characterization of Phenotypes in System Design Space

The representation of the system within each phenotypic region is always a simple S-system, for which determination of the local nonlinear behavior reduces to conventional linear analysis [15, 16]. Thus, the phenotypes involving local (small) variations are completely determined, and their relative performance can be compared on the basis of relevant performance criteria. These criteria can be quantified using the logarithmic gain, parameter sensitivity (local robustness), and response time (see Sect. 5). The boundaries that delineate a given phenotype can be used to quantify the global tolerance to large changes in the genotype and the environment [27]. An example involving the steady-state solution in each phenotypic region is shown in Fig. 5.

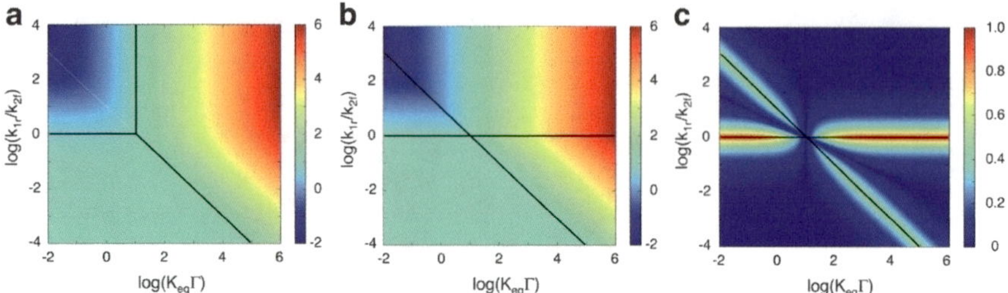

Fig. 5 Concentration of the intermediate normalized with respect to a fixed concentration of product, plotted in the z-direction as a heat map in the system design space of Fig. 4. (**a**) Actual solution with intuitive boundaries between qualitatively distinct regions. (**b**) Solution by means of dominant S-systems, with mathematically defined boundaries between qualitatively distinct regions. (**c**) Quantitative differences, showing errors on the boundaries between regions. See text for discussion

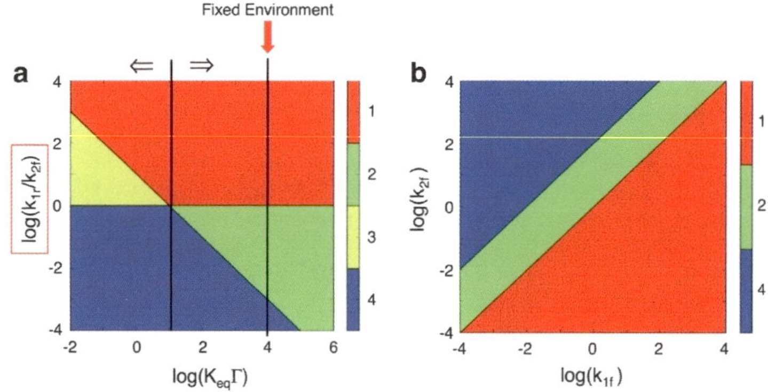

Fig. 6 A fixed environment corresponding to a particular slice through the system design space. (**a**) The system design space of Fig. 5 with an environment fixed at a value of $K_{eq}\Gamma = 10{,}000$. (**b**) A slice through the design space exhibiting only the independent chemically determined parameters k_{1f} and k_{2f}

The intuitive boundaries in Fig. 5a do not capture the real significance of the differences. However, the mathematically defined boundaries in Fig. 5b do. These show that the behavior in the upper right region depends only on the environmentally determined independent variables, the behavior in the upper left region depends only on the chemically determined parameters, the behavior in the lower right region depends on both the chemically determined parameters and the environmentally determined independent variables, and the behavior in the lower left region is influenced by neither.

A fixed environment corresponds to a slice through the design space involving only the chemically determined parameters of the system. For example, a slice in which the net flux to the right is fixed is shown in Fig. 6a along with the corresponding view involving only the kinetic parameters in Fig. 6b. This slice

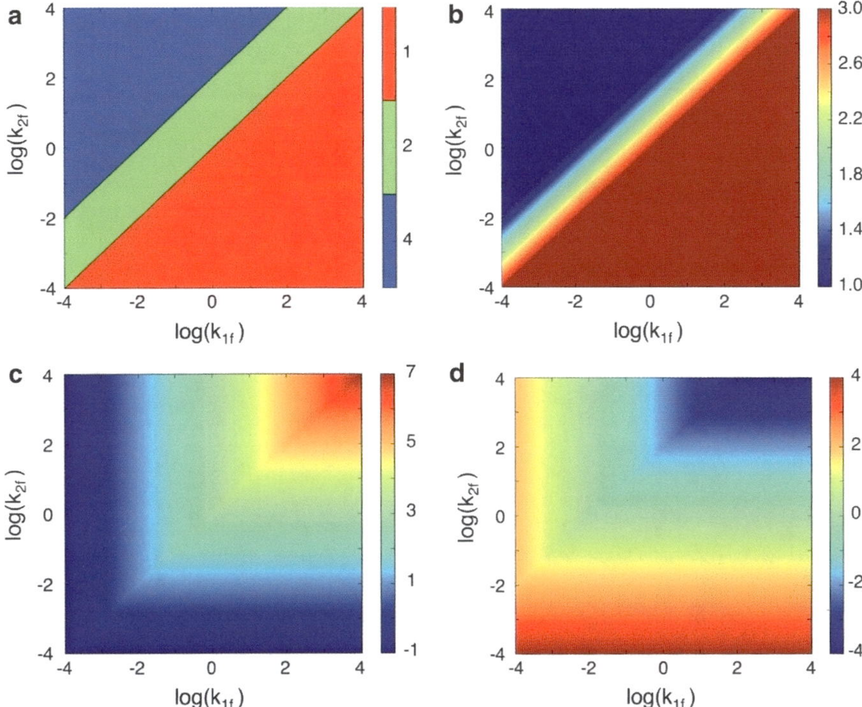

Fig. 7 Phenotypic characteristics of the chemical mechanism plotted as a heat map in the z-direction of the system design space. (**a**) The design space as represented in Fig. 6b, (**b**) intermediate concentration X_1, (**c**) net flux V_{net} and (**d**) toxicity $T = X_1/V_{net}$. All values are logarithms to base 10

is also shown in Fig. 7a for comparisons with the intermediate concentration X_1, net flux V_{net} and the toxicity (defined here as the intermediate concentration for a given net flux $T = X_1/V_{net}$), which are plotted in the z-direction as heat maps. The intermediate concentration (Fig. 7b) is an average of the values resulting from quasi-equilibrium with the substrate or with the product. When the rate of the second reaction is greater than that of the first, the intermediate concentration approaches its minimum; when the rate of the second reaction is less than that of the first, the intermediate concentration approaches its maximum.

The corresponding net flux (Fig. 7c) is limited by the rate of the slower of the two reactions. For a given net flux, an increase in the forward rate constant of the first reaction results in an amplified increase in the intermediate concentration. On the other hand, an increase in the forward rate constant of the second reaction results in an amplified decrease in the intermediate concentration. This is consistent with our intuition. It also shows that specifying only the net flux ignores the importance of the intermediate concentration. In many cases, intermediates are highly reactive, and in high concentrations are toxic to cells. The toxicity (Fig. 7d) decreases as the net flux increases, particularly when the flux is increased by increasing the rate of the second reaction. Thus, other things being equal, it is an advantage if the forward rate constant of the first reaction is less than that of the second.

4.2   A Biological Example

Although the example in Fig. 3 illustrates the basic concepts in a straightforward analytical fashion, it is a linear system, for which a method of deconstruction is not really required. Nor are the values of the chemically determined parameters subject to biological selection; rather, they were determined by the particular chemistry during the chemical evolution of the universe [28]. More complex biological mechanisms are nonlinear and typically involve recasting systems of rational-function equations into GMA systems. These systems, which are subject to biological selection based on their relative fitness as manifested through their phenotype, are of greater interest for us. The *fitness* of a biological system, as defined here, refers to how well the system performs its function in a given context, measured according to some performance criteria. For example, in the context of bacterial growth, this criterion might refer to reproductive success as measured by growth rate. In another context, it might simply be survival over some period of time (e.g., a dry season) as measured by survival rate. In the context of a given intracellular pathway it might refer to the maximum flux that the pathway can deliver for cellular growth, or the maximum concentration of a highly reactive intermediate, as measured by the steady-state value of that dependent variable.

The model illustrated in Fig. 1 appears to be very similar to that in Fig. 3; however, the mechanisms represented are very different. The nonlinear equation describing the biological system in Fig. 1 involves rational functions [(13)]. Fixed points can be determined from the recast GMA system [(20), (21), and (22)]. A bound on the possible number of qualitatively distinct phenotypes is given by the number of combinations of potentially dominant terms; in this case, the number is 36. However, as noted in Sect. 3.3, some of these are invalid; in this case, the number of invalid combinations is 4. Thus, this system exhibits a total of 32 qualitatively distinct phenotypes. The fitness of the phenotypes can be characterized and compared according to putative selection criteria by analyzing the S-system in each case.

Thus, we can consider some characteristics of the phenotypes in addition to the intermediate concentration, net flux, and toxicity. For example, two additional measures of system fitness might be the metabolic cost and the efficiency of producing a given net flux. The cost can be defined as the sum of the two enzyme concentrations or, for simplicity, as $C = V_{1f} + V_{2f}$, and the efficiency can be defined as the net flux for a given cost, $E = V_{net}/C$. Selection might well tend to minimize the toxicity T, minimize the cost C, and maximize the efficiency E, subject to constraints and trade-offs. Other characteristics might also be considered. Our purpose here is not to be exhaustive or to capture the evolutionary path for any specific system;

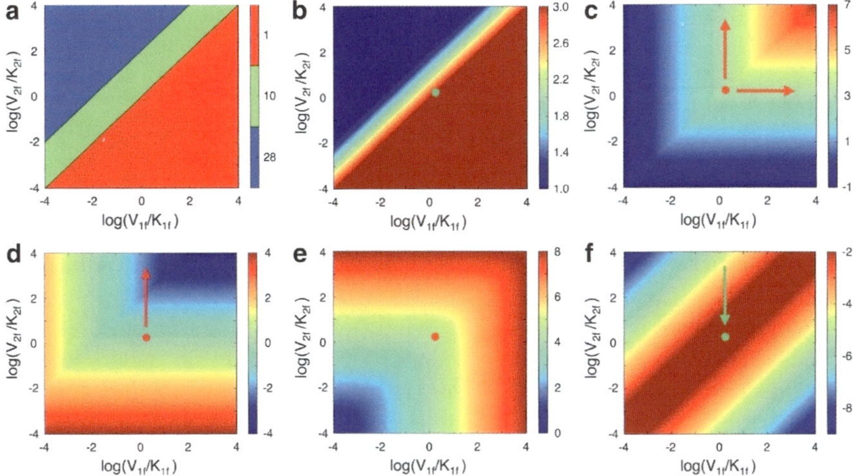

Fig. 8 Characterization of the qualitatively distinct phenotypes in the system design space for the enzymatic mechanism illustrated in Fig. 1 within its linear range of operation. (**a**) A slice through the design space with V_{1f}/K_{1f} on the horizontal axis and V_{2f}/K_{2f} on the vertical axis. (**b**) Intermediate concentration, (**c**) net flux, (**d**) toxicity, (**e**) cost (only relevant when there are changes in enzyme levels), and (**f**) efficiency, plotted in the z-direction as heat maps. All values are logarithms to base 10. Nominal parameter values: $K_{e1} = 1$, $K_{e2} = 0.1$, $S = 1,000$, $P = 1$, and $V_{1f} = V_{2f} = K_{1f} = K_{2f} = K_{1r} = K_{2r} = 10,000$. See text for discussion

rather, it is to show how hypotheses can be formulated and their consequences explored using the system design space approach. Given this objective, we shall examine this particular example briefly from three different perspectives.

4.2.1 Characterization of Phenotypes for the System in Its Linear Range

As our first example, consider the situation in which the system is in a fixed environment and operating in its linear range. This corresponds to a particular slice through the system design space, as shown in Fig. 8a, and the resulting phenotypes are characterized by the independent kinetic parameters that are genetically determined. In this context, the ratio of the maximum velocity V_{if} to the associated Michaelis constant K_{if} is mathematically equivalent to the corresponding rate constant of the chemical system, k_{if}. Thus, the results plotted with the ratio V_{1f}/K_{1f} on the x-axis and V_{2f}/K_{2f} on the y-axis (Fig. 8a–d) are essentially identical to those in Fig. 7.

However, the phenotypic consequences of changes in the different classes of genetically determined parameters can be very different. If the net flux (Fig. 8c) were to increase because of a coordinated increase in the amounts of the two enzymes, there would be a corresponding increase in the metabolic cost to the cell (Fig. 8e). Alternatively, if the same increase in net flux were to occur because of a coordinated decrease in the two Michaelis constants, there would be no increase in the cost (data not shown). However, there is a limit to the extent of such a decrease; for example, if K_{1f} were to become less then S (or K_{1f} were to become less then X_1), then the system would no longer be operating in its linear range.

If the system were under strong balancing selection to maintain a given net flux, there are options for the further evolution of the system. For example, starting from the operating point marked by the dot, mutations that lead to an increase in V_{1f} alone leave the net flux (Fig. 8c) and toxicity (Fig. 8d) unchanged; however, they lead to an increase in the intermediate concentration (Fig. 8b) and the cost (Fig. 8e), with a loss in efficiency (Fig. 8f). These results suggest a decrease in fitness. On the other hand, mutations that lead to an increase in V_{2f} alone would lead to similar results for the net flux (Fig. 8c), the cost (Fig. 8e), and the efficiency (Fig. 8f); however, they would lead to a decrease in the intermediate concentration (Fig. 8b) and the toxicity (Fig. 8d). This result suggests an increase in fitness. Thus, if the net flux were the primary criterion for selection, these results suggest that the system could evolve an appropriate trade-off between increased efficiency and reduced toxicity by selection for a preferential increase in the amount of the second enzyme.

Alternatively, if the changes were to involve a decreases in K_{1f} alone or K_{2f} alone, the results would be the same, except that there would be no change in the cost. The conclusions would then suggest that an appropriate trade-off between increased efficiency and decreased toxicity could be achieved by a reduction in K_{2f} alone.

4.2.2 Characterization of Phenotypes in the Nonlinear Range with Saturation

Another view of the system design space is shown in Fig. 9a where the maximum velocity of the first reaction (normalized with respect to the corresponding Michaelis constant, i.e., V_{1f}/K_{1f}) is on the horizontal axis, the Michaelis constant of the second reaction with respect to the intermediate (normalized with respect to the fixed equilibrium constant of the second reaction and the constant product concentration, i.e., $K_{2f}K_{e2}/P$) is on the vertical axis, and the geometrical landmarks are represented by the other genetically determined parameters: the maximum velocity of the second reaction V_{2f}, and the Michaelis constants K_{1r} and K_{2r}.

Give an initial state of the system corresponding to a point in the upper left portion of this design space, with a high value of $K_{2f}K_{e2}/P = 1{,}000$ and a low value of V_{1f}, mutations that lead to an increase in the maximum velocity for the first reaction would causes the intermediate concentration (Fig. 9b), the net flux (Fig. 9c), and the cost (Fig. 9e) to increase monotonically. On the other hand, the efficiency would increase, reaches a maximum and then decrease (Fig. 9f), whereas the toxicity would decrease and then remain essentially constant (Fig. 9d). This behavior, which is equivalent to a transition of V_{1f} from left to right across the tops of the panels in Fig. 8, is expected, since the system is then operating in its linear range.

The results associated with the lower portion of Fig. 9a reflect the nonlinear behavior that comes into play when K_{2f} approaches the concentration of the intermediate. Given an initial state of the system in the lower portion of this design

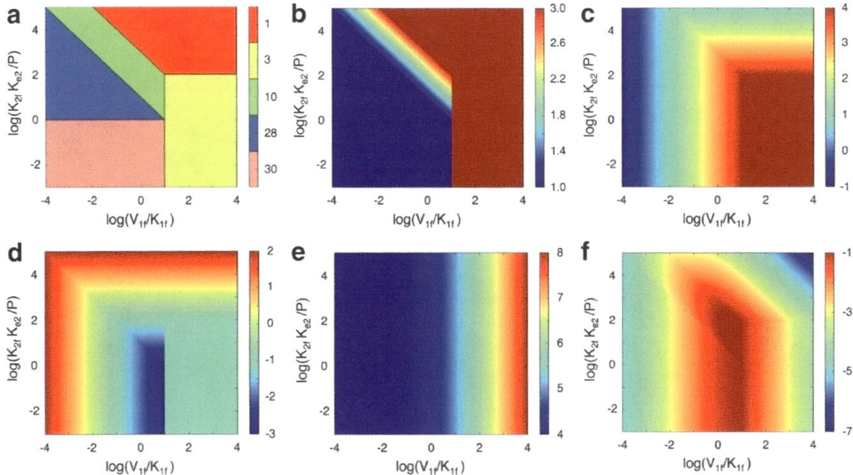

Fig. 9 Characterization of the qualitatively distinct phenotypes in the system design space for the enzymatic mechanism illustrated in Fig. 1 within its unrestricted nonlinear range of operation. Regions 1, 10, and 28 correspond to linear operation, and regions 3 and 30 represent nonlinear operation involving saturation of the second enzyme. (**a**) A slice through the design space with V_{1f}/K_{1f} on the horizontal axis and $K_{2f}K_{e2}/P$ on the vertical axis. (**b**) Intermediate concentration, (**c**) net flux, (**d**) toxicity, (**e**) cost, and (**f**) efficiency, plotted in the z-direction as heat maps. All values are logarithms to base 10. See text for discussion

space, with a value of $K_{2f}K_{e2}/P < 1$ and a low value of V_{1f}, mutations that lead to an increase in the maximum velocity for the first reaction would cause the intermediate concentration (Fig. 9b), net flux (Fig. 9c), and cost (Fig. 9e) to increase sharply when V_{1f} exceeds the fixed value of V_{2f} and the rate of the second reaction saturates. On the other hand, the efficiency would increase, reach a maximum and then decrease (Fig. 9f), whereas the toxicity would decrease, reach a minimum, and then increase (Fig. 9d).

Similar results are obtained when the first reaction is saturated by the substrate; the results plotted with K_{1f}/S on the y-axis and V_{1f}/K_{1f} on the x-axis are, essentially, flipped horizontally compared with those in Fig. 9 (data not shown). In the linear region of operation, with K_{1f} increasing, the critical values are now shifted to higher values of V_{1f} to compensate for the saturation of the first enzyme.

A few other scenarios are the following. If mutations were to result in a simultaneous increase in V_{1f} and K_{1f} (or V_{2f} and K_{2f}), there would be no change in the intermediate concentration, the net flux, or the toxicity, but there would be an increase in the cost and a decrease in the efficiency. If V_{1f} and V_{2f} were to increase simultaneously, there would be no change in either the intermediate concentration or the efficiency, but the cost and the net flux would increase, whereas the toxicity would decrease. Alternatively, if K_{1f} and K_{2f} were to increase simultaneously, there would be no change in either the intermediate concentration or the cost, but the net flux and efficiency would decrease whereas toxicity would increase.

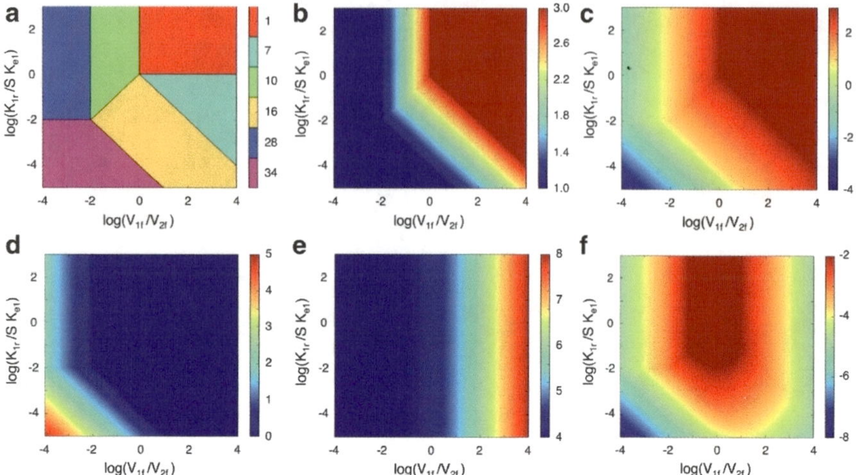

Fig. 10 Characterization of the qualitatively distinct phenotypes in the system design space for the enzymatic mechanism illustrated in Fig. 1 within its unrestricted nonlinear range of operation. Regions 1, 10, and 28 correspond to linear operation, and regions 7, 16, and 34 represent nonlinear operation involving inhibition of the first enzyme. (**a**) A slice through the design space, with V_{1f}/V_{2f} on the horizontal axis and $K_{1r}/(SK_{e1})$ on the vertical axis. (**b**) Intermediate concentration, (**c**) net flux, (**d**) toxicity, (**e**) cost, and (**f**) efficiency, plotted in the z-direction as heat maps. All values are logarithms to base 10. See text for discussion

4.2.3 Characterization of Phenotypes in the Nonlinear Range with Inhibition

Another type of nonlinear behavior is exhibited in the system design space shown in Fig. 10a. The maximum velocity of the first reaction (normalized with respect to the maximum velocity of the second reaction, i.e., V_{1f}/V_{2f}) is on the horizontal axis, the Michaelis constant of the first reaction with respect to the intermediate (normalized with respect to the fixed equilibrium constant of the first reaction and the constant substrate concentration, i.e., $K_{1r}/(SK_{e1})$) is on the vertical axis, and the geometrical landmarks are represented by the other genetically determined parameters: the Michaelis constants K_{1f}, K_{2f} and K_{2r}.

Again, given an initial state of the system in the upper left portion of this design space with a high value of $K_{1f}/(SK_{e1}) = 10$ and a low value of V_{1f}, mutations that result in a systematic increase in the maximum velocity of the first reaction would lead to a monotonic increase in the intermediate concentration (Fig. 10b), net flux (Fig. 10c), and cost (Fig. 10e). On the other hand, the efficiency would increase, reach a maximum and then decrease (Fig. 10f); whereas the toxicity would decrease and then remain essentially constant (Fig. 10d). This behavior, which is equivalent to a transition of V_{1f} from left to right across the tops of the panels in Fig. 8, is expected since the system is then operating in its linear range.

The results associated with the lower portion of Fig. 10a reflect the nonlinear behavior that comes into play when K_{1r} approaches the concentration of the intermediate. Given an initial state of the system represented in the lower portion of this design space with a value of $K_{1f}/(SK_{e1}) < 0.01$ and a low value of V_{1f}, mutations that result in a systematic increase in the maximum velocity of the first reaction would cause the intermediate concentration (Fig. 10b), net flux (Fig. 10c), and cost (Fig. 10e) to increase monotonically. However, the value of V_{1f} at which these increases commence shifts to higher values as K_{1r} decreases. This is because a decrease in K_{1r} leads to an increase in the competitive inhibition of the first reaction; this causes a decrease in the effective activity of the first enzyme, which must be compensated by an increase in the amount of the first enzyme V_{1f}. The efficiency would increase, reach a maximum and then decrease (Fig. 10f); whereas the toxicity would decrease and then remain essentially constant (Fig. 10d). Thus, the results are nearly the same as in the linear range of operation, except for a shift due to the competitive inhibition of the first enzyme.

Similar results are obtained when the second reaction is inhibited by the product; the results plotted with K_{2r}/P on the y-axis and V_{1f}/V_{2f} on the x-axis are, essentially, flipped horizontally compared with those in Fig. 10 (data not shown). In the nonlinear region of operation, when K_{2r} decreases, the critical values are now shifted to higher values of V_{2f} to compensate for the competitive inhibition of the second enzyme.

The full system design space for the mechanism illustrated in Fig. 1 is an eight-dimensional space filled with 32 irregular polytopes. A slice through each of the qualitatively distinct phenotypes (or polytopes) can always be obtained, and at least one set of parameter values can be found to identify a point that lies within that phenotypic region [29]. Although we could continue the analysis in the same vein if space permitted, our purpose here is not to treat this particular example exhaustively. Rather, our goal is only to suggest how the function, design, and evolution of such biochemical systems can be analyzed to achieve greater understanding by using the system design space approach.

5 S-Systems

As suggested in Sect. 3, the power-law formalism has within it a special class of nonlinear equations known as S-systems, which have properties that make them tractable. Their tractability derives in large part from the fact that much of their analysis can be reduced to linear analysis in a logarithmic space. This class of equations was first introduced in the late 1960s. There are many applications of these equations in the literature [30], so we need only summarize a few of the key properties here.

The analytical methods of the theory of linear systems provide powerful techniques for predicting the behavior of systems that are well represented by the linear formalism [31]. However, most complex systems of biological interest are highly nonlinear, and the S-system equations are still nonlinear. How, then, are we to proceed when faced with the task of analyzing such nonlinear systems? The task of extracting an understanding of systemic behavior from the equations that characterize the system is conventionally broken down into two steps. In the first, one determines the steady-state behavior. This is the long-term behavior that remains after all the initial transients have died away. In the second, one determines the transient behavior that occurs whenever there is an abrupt change in the state of the system.

This separation into steady-state and transient responses is an idealization. Such idealizations are important because they provide a conceptual basis for understanding the behavior of complex systems. However, steady-state analysis also has practical significance, because many systems in nature operate in a quasi-steady state. When the differences in time scale are sufficiently great, these systems can be treated as if they were in a bona fide steady state.

5.1 Explicit Steady-State Solution

In general, the S-system representation in the power-law formalism has the following form:

$$\frac{dX_i}{dt} = \alpha_i \prod_{j=1}^{n+m} X_j^{g_{ij}} - \beta_i \prod_{j=1}^{n+m} X_j^{h_{ij}}, \quad X_i(0) = X_{i0}, \quad i = 1, 2, \cdots, n, \tag{41}$$

where n is the number of dependent state variables and m is the number of independent state variables. The steady-state equations are obtained by setting $dX_i/dt = 0$ for all i in (41). When none of the X_j's and none of the rate constants are equal to zero, these equations can be divided by $\alpha_i \prod_{j=1}^{n+m} X_j^{h_{ij}}$, resulting in the following linear system [16]:

$$[A]\ y] = b], \tag{42}$$

where $[A]$ is a matrix with elements representing differences in kinetic orders, $a_{ij} = (g_{ij} - h_{ij})$; $y]$ is a column vector with elements representing logarithms of variables, $y_i = \ln X_i$; and $b]$ is a column vector with elements representing differences in logarithms of rate constants $b_i = \ln \beta_i - \ln \alpha_i$.

The arrays in (42) can be partitioned into dependent and independent elements,

$$\begin{bmatrix} A_d & \vdots & A_i \end{bmatrix} \begin{bmatrix} y_d \\ \cdots \\ y_i \end{bmatrix} = b], \tag{43}$$

and separated as follows:

$$[A]_d \, y]_d = -[A]_i \, y]_i + b], \tag{44}$$

where the subscript "d" signifies that the matrix $[A]_d$ contains only kinetic orders with respect to dependent state variables and the vector $y]_d$ contains only logarithms of dependent state variables. The subscript "i" has a similar interpretation, but for the independent state variables.

For all matrices $[A]_d$ with nonzero determinant, there exists an inverse operator, $[A]_d^{-1} = [M]$ [16]. The inverse operator allows one to solve (44) and obtain the dependent state variables *explicitly* in terms of the independent state variables and the parameters of the system. The explicit steady-state solution for the S-system in (41) can then be written [16]

$$y]_d = [L] \, y]_i + [M] \, b], \atop \uparrow \qquad \uparrow \atop \text{slope} \qquad \text{intercept} \tag{45}$$

where $[L] = -[M] \, [A]_i$. This solution for the logarithms of the dependent concentrations $y]_d(y_j, \ j = 1, \cdots, n)$ is divided into two parts. The first exhibits the linear dependence on the logarithms of the independent state variables $y]_i(y_j, \ j = n+1, \cdots, n+m)$; the second exhibits the linear dependence on the logarithms of the rate constants $b](b_j = \ln(\beta_j/\alpha_j), \ j = 1, \cdots, n)$.

The flux through any pool X_k in the steady state is obtained by a simple secondary calculation involving the aggregate rate law for the influx or efflux of X_k and the known values for the state variables in the steady state. For example, starting with the rate law

$$V_{+k} = \alpha_k \prod_{j=1}^{n+m} X_j^{g_{ij}}, \quad k = 1, 2, \cdots, n, \tag{46}$$

taking logarithms and expressing the results in matrix notation yields

$$(\ln V_+)] = (\ln \alpha)] + [G] \, y]. \tag{47}$$

The solution for the flux variables is also a linear function of the independent state variables. This is readily demonstrated by separating the dependent and independent components of the solution, substituting the explicit solution for the dependent state variables, and regrouping terms; for example,

$$(\ln V_+)] = \left\{ [G]_i + [G]_d \, [L] \right\} \, y]_i + \left\{ (\ln \alpha)] + [G]_d \, [M] \, b] \right\}. \atop \uparrow \qquad\qquad \uparrow \atop \{\text{slope}\} \qquad\qquad \{\text{intercept}\} \tag{48}$$

Thus, the explicit solution presented in (45) and (48) gives the complete relationship between the steady-state values of the dependent state variables on the one hand and the values of the independent state variables and the parameters of the system on the other.

The behavior of the dependent variables in (45) and (48) was separated deliberately into two components and represented by distinct symbols to emphasize the influence of the independent state variables and the parameters of the system. The independent state variables may be thought of as those that are determined by factors outside the system of interest, i.e., as the environment of the system. The parameters, which characterize the relatively fixed aspects of the system itself, may be thought of as physically and genetically determined. To use an analogy, the music emanating from a CD player is a function of the externally supplied stimuli (the disks), which are variables, and of the system parameters, which are relatively fixed and determined by the optical–electromechanical components of the player. The separation of these two types of influences is important for a clear understanding of the behavior of the system.

5.2 Systemic Measures of Signal Propagation: Logarithmic Gain

The behavior of a complex system in response to its environment is characterized by the responses of the dependent variables to changes in the independent variables. The explicit solution obtained with the S-system representation provides a complete characterization of the local steady-state behavior about any operating state. However, it is also useful to characterize the properties of the system in terms of standard factors that relate its outputs (dependent variables) to its inputs (independent variables), namely *logarithmic-gain factors* [16]. The elements L_{ik} of the $m \times n$ matrix $[L] = -[M]\,[A]_i$ are analogous to the conventional gain or amplification factors of linear network theory, and they are referred to as logarithmic-gain factors in the power-law formalism because the changes are characterized in a logarithmic rather than a Cartesian space.

The magnitudes of the logarithmic-gain factors provide a quantitative measure of the influence exerted by a given independent variable over a particular dependent variable. These magnitudes allow one to determine the distribution of total influence either over independent variables or over dependent variables. Because the logarithmic-gain factors can be expressed explicitly in terms of the kinetic orders of the component processes, they also are important for relating systemic behavior of the system to the underlying determinants of the system.

5.3 Systemic Measures of Local Robustness: Parameter Sensitivity

The global behavior of a given system in response to changes in its underlying structure is characterized by the responses of the dependent variables to changes in the *parameters* internal to the system. The explicit solution in the power-law formalism provides a complete characterization of the local steady-state behavior about any operating state of the system. However, it also is useful to characterize the system's behavior in terms of standard factors that relate its outputs (dependent variables) to its parameter values (rate constants and kinetic orders); these factors are referred to as *parameter sensitivities* [16, 32]. The elements M_{ik} of the $n \times n$ matrix $[M]$, which is the inverse of the $n \times n$ system matrix $[A]_d$, are identical to the traditional parameter sensitivities [33–36]. In the power-law formalism, they are referred to as *rate-constant sensitivities*. The sensitivities with respect to the expo- nential parameters have also been presented and a variety of relationships among this class of parameter sensitivities have been summarized elsewhere [16, 32].

By means of the parameter sensitivities described in this section one can characterize the systemic response to change in each parameter of the system, and, because these parameter sensitivities can be expressed explicitly in terms of the kinetic orders of the component processes, they also are important for relating the systemic behavior to the underlying determinants of the system.

If every quantity that can change is treated as a *variable*, then the parameters considered in this section are actually fixed quantities within a given representation. The changes considered in this section would then be virtual changes, and the corresponding systemic responses would be virtual responses. What meaning can be attached to such virtual changes? The answer has to do with the interpretation of experimental error. The actual values of the parameters are unknown to us, and they must instead be estimated from appropriate experimental data. These estimates are inevitably corrupted to some degree by error and, consequently the values that we assign to the parameters of our models are altered from their true values. If the parameter sensitivities are small, then the use of an erroneous parameter value will have little affect on the systemic behavior that is predicted by the model. In this case, the model is said to be *robust*. On the other hand, if the parameter sensitivities are large, then even a small error will produce major changes in the expected systemic behavior. In this case, the model is said to be *fragile*. In general there will be a distribution of parameter sensitivities, and one can use a knowledge of these sensitivities to direct experimental effort toward the most efficient determination of the parameter values. Effort should be directed toward careful measurement of those parameters that have the largest sensitivities. However, if the sensitivity is sufficiently large, then the parameter may be impossible to determine with the current experimental error. In this situation, it may be possible to estimate the value of the parameter by adjusting its value in the model until the systemic response of the model matches that of the actual system. By using the model itself as an amplifier of changes in the sensitive parameter, one may achieve a more accurate estimate by indirect means.

5.4 Dynamic Behavior of S-Systems

One of the principal goals of biochemical systems theory is to relate the behavior of the system (the dependent variables) to that of its environment (the independent variables) and that of its genetic determinants (the parameters). The algebraic methods discussed in Sects. 5.1, 5.2, and 5.3 are very useful for analyzing the steady-state behavior of systems described by S-system equations, but they tell us nothing about the system's dynamic properties. To gain insight into these properties, the differential equations of the system must be examined. In this subsection, we shall consider the simplest forms of dynamic behavior, which are associated with responses to small disturbances about a nominal steady state. The more complex forms of dynamic behavior that are associated with the nonlinearities of the S-system representation [16] are beyond the scope of this chapter.

"Local dynamics" refers to the behavior of a system in response to small perturbations or disturbances. If the perturbation is sufficiently small or sufficiently localized about the normal operating point, then the behavior of a nonlinear system will be identical to that of its linearized representation. Thus, the analysis of local dynamics reduces to the analysis of the dynamics of linear systems [31, 37].

The general S-system equations for an n- variable problem (41) can be linearized as follows [16]:

$$\frac{du}{dt} = F^{\mathrm{T}}\mathbf{A}\,u, \tag{49}$$

where $u_i = y_i - y_{i0} = (X_i - X_{i0})/X_{i0}$ (i.e., the percentage variation in X_i), and $a_{ij} = g_{ij} - h_{ij}$. The elements of the premultiplier are

$$F_i = \alpha_i \prod_{j=1}^{n+m} X_{j0}^{g_{ij}} / X_{i0} = V_{i0}/X_{i0}, \quad i = 1, 2, \cdots, n, \tag{50}$$

and thus F_i may be viewed as a pseudo-first-order rate constant, or the reciprocal of the turnover time for the pool X_i.

The eigen values of such systems can be determined by well-known methods, and the local stability is reflected in the character of their complex values: the system response is locally stable if all of the eigen values have negative real parts, locally unstable if one or more have a positive real part, and marginally stable if one or more has a zero real part and all others have negative real parts.

5.5 Telescopic Properties

Algebraic constraints among the dependent variables X_i in (41) can also be expressed readily in the power-law formalism [16, 38]. When the redundant equations in (41) are eliminated and the algebraic dependencies substituted into the remaining equations, the resulting set has exactly the same form as the original set in (41), but with fewer variables.

For example, if p of the n dependent variables in (41) are temporally dominant in the system (and these are renumbered to be the first p variables) so that the remaining $(n-p)$ differential equations reduce to the algebraic equations

$$\alpha_i \prod_{j=1}^{n+m} X_j^{g_{ij}} = \beta_i \prod_{j=1}^{n+m} X_j^{h_{ij}}, \qquad i = p+1, \cdots, n, \tag{51}$$

then this algebraic system can be solved explicitly for the "fast" variables in terms of the "slow" variables:

$$X_k = \gamma_k \prod_{j=1}^{p+m} X_j^{f_{kj}}, \qquad k = p+1, \cdots, n. \tag{52}$$

When the explicit expression for the "fast" variables is substituted into the p equations for the "slow" variables, the result is an S-system with different parameters but lower dimension:

$$\frac{dX_i}{dt} = \alpha_i' \prod_{j=1}^{p+m} X_j^{g_{ij}'} - \beta_i' \prod_{j=1}^{p+m} X_j^{h_{ij}'}, \qquad i = 1, \cdots, p. \tag{53}$$

This is called the "telescopic property" of the representation [38]. It allows an integrated approach to the modeling and analysis of nonlinear systems at various hierarchical levels of organization. The linear representation has this important property, but very few nonlinear representations do. Once the algebraic dependencies have been taken into account in this manner, the representation and subsequent analysis are identical to that given in Sects. 5.1, 5.2, 5.3, and 5.4.

6 Discussion

The task of characterizing complex nonlinear systems could be facilitated if such systems could be generically decomposed into a series of tractable subsystems and the results of the analysis of those subsystems reassembled to provide insight into the original system. This may sound reminiscent of the traditional notion of identifying modular structures in a biological system. It is therefore important to emphasize that the subsystems identified in the system design space approach do not correspond to the conventional notion of "modularity". The S-systems of our approach are rigorously defined by the underlying mechanisms of the system itself, and all of the parameters are involved in determining the geometrical landmarks in the design space. Moreover, these S-systems typically involve distributed aspects of the entire system and not a localized module clearly separable from the remainder of the system.

We have attempted to present the system design space methodology in terms of simple didactic examples. More complex realistic examples would require a great deal more space, in order to provide the biological background as well as details of the method. This approach has been applied to a number of biological systems, including the induction of prophage lambda [39], the NADPH redox cycle in human erythrocytes [40], anaerobic–aerobic transitions regulated by the global transcription factor FNR [41, 42], lactose operon induction [43], growth phase transitions in *E. coli* [44], and the coupling of toxin–antitoxin systems in persistence [14]. These are still relatively small systems.

Applying the system design space methodology to larger systems is still a considerable challenge. This is probably true of any methodology for dealing with large systems. Some of the challenges are obvious. The number of possible phenotypes grows rapidly with the number of combinations, although the analysis of any given phenotype is a manageable problem. This is an "embarrassingly parallelizable" problem [45], so this challenge will soon be addressed. Visualization in dimensions greater than three is another challenge [46]. Other types of nonlinear models, outside the range of the mass action and rational-function models typically found in biology, have yet to be examined systematically. There are some trivial examples, such as

$$\frac{dx_1}{dt} = \gamma_1 + \alpha_1 x_2 - \beta_1 \frac{x_1^2}{\sqrt{x_1^2 + x_2^2}}, \tag{54}$$

$$\frac{dx_2}{dt} = \gamma_2 + \alpha_2 x_1 - \beta_2 \frac{x_2^2}{\sqrt{x_1^2 + x_2^2}}, \tag{55}$$

which can be written as the following differential–algebraic GMA system:

$$\frac{dx_1}{dt} = \gamma_1 + \alpha_1 x_2 - \beta_1 x_1^2 x_3^{-1}, \tag{56}$$

$$\frac{dx_2}{dt} = \gamma_2 + \alpha_2 x_1 - \beta_2 x_2^2 x_3^{-1}, \tag{57}$$

$$x_3^2 = x_1^2 + x_2^2. \tag{58}$$

Dynamics in system design space is an interesting topic that has only been touched upon, and the treatment of partial differential equations is another important challenge. Clearly, there are many avenues that need to be explored before we can know the full potential of this approach. It is hoped that this chapter will stimulate others to take up some of these challenges.

Acknowledgments We thank Pedro Coelho and Dean Tolla for fruitful discussions, and Rick Fasani for advice on the construction of the system design spaces. This work was supported in part by U.S. Public Health Service Grant R01-GM30054 and by a Stanislaw Ulam Distinguished Scholar Award from the Center for Non-Linear Studies at the Los Alamos National Laboratory.

References

1. S. Brenner, Genomics: the end of the beginning. Science **287**, 2173–2174 (2000)
2. D. Lazard, Thirty years of polynomial system solving, and now? J. Symb. Comput. **44**, 222–231 (2009)
3. W.S. Hlavacek, How to deal with large models? Mol. Syst. Biol. **5**, 240–242 (2009)
4. M.A. Savageau, P.M.B.M. Coelho, R. Fasani, D. Tolla, A. Salvador, Phenotypes and tolerances in the design space of biochemical systems. Proc. Natl. Acad. Sci. U. S. A. **106**, 6435–6440 (2009)
5. J. Barriol, *Elements of quantum mechanics with chemical applications* (Barnes & Noble, New York, 1971)
6. F. Daniels, R.A. Alberty, *Physical chemistry* (Wiley, New York, 1967)
7. W. Forst, *Theory of unimolecular reactions* (Academic, New York, 1973)
8. C. Capellas, B.H.J. Bielski, *Kinetic systems* (Wiley, New York, 1972)
9. I.M. Klotz, *Chemical thermodynamics* (Benjamin, New York, 1972)
10. R. Kopelman, Rate processes on fractals: theory, simulations, and experiments. J. Stat. Phys. **42**, 185–200 (1986)
11. M.A. Gibson, J. Bruck, Efficient exact stochastic simulation of chemical systems with many species and many channels. J. Phys. Chem. A **104**, 1876–1889 (2000)
12. K.A. Connors, *Chemical kinetics, the study of reaction rates in solution* (Wiley, New York, 1990)
13. S.A. Kauffman, *The origins of order: Self-organization and selection in evolution* (Oxford University Press, New York, 1993)
14. R.A. Fasani, M.A. Savageau, Molecular mechanisms of multiple toxin-antitoxin systems are coordinated to govern the persister phenotype. Proc. Natl. Acad. Sci. U. S. A. **110**, E2528–E2537 (2013). doi:10.1073/pnas.1301023110, Early Edition published 17 June 2013
15. M.A. Savageau, Design principles for elementary gene circuits: elements, methods, and examples. Chaos **11**, 142–159 (2001)
16. M.A. Savageau, *Biochemical systems analysis: a study of function and design in molecular biology*, 40th anniversary edn. [http://www.amazon.com/Biochemical-Systems-Analysis-Function-Molecular/dp/1449590764/] (2009) [a reprinting of the original edition published by Addison-Wesley, Reading, Mass. (1976)]
17. G.B. Thomas Jr., R.L. Finney, *Calculus and analytic geometry*, 9th edn. (Addison Wesley, Reading, 1996)
18. D.E. Koshland, K.E. Neet, The catalytic and regulatory properties of enzymes. Annu. Rev. Biochem. **39**, 359–410 (1968)
19. J. Monod, J. Wyman, J.-P. Changeux, On the nature of allosteric transitions: a plausible model. J. Mol. Biol. **12**, 88–118 (1965)

20. R.G. Duggleby, Quantitative analysis of the time courses of enzyme-catalyzed reactions. Methods **24**, 168–174 (2001)
21. I.H. Segel, *Enzyme kinetics: behavior and analysis of rapid equilibrium and steady-state enzyme systems* (Wiley, New York, 1993)
22. J.T. Wong, C.S. Hanes, Kinetic formulations for enzymic reactions involving two substrates. Can. J. Biochem. Physiol. **40**, 763–804 (1962)
23. M.A. Savageau, E.O. Voit, Recasting nonlinear differential equations as S-systems: a canonical nonlinear form. Math. Biosci. **87**, 83–115 (1987)
24. W. Pigman, H.S. Isbell, Mutarotation of sugars in solution. Part 1. History, basic kinetics, and composition of sugar solutions. Adv. Carbohydr. Chem. Biochem. **23**, 11–57 (1968)
25. G.B. Dantzig, *Linear programming and extensions* (Princeton University Press, Princeton, 1963)
26. R.J. Vanderbei, *Linear programming: foundations and extensions* (Springer, New York, 2008)
27. P.M.B.M. Coelho, A. Salvador, M.A. Savageau, Global tolerance of biochemical systems and the design of moiety-transfer cycles. PLOS Comput. Biol. **5**(3), e1000319 (2009)
28. D.D. Clayton, *Principles of stellar evolution and nucleosynthesis* (University of Chicago Press, Chicago, 1983)
29. J.G. Lomnitz, M.A. Savageau, Phenotypic deconstruction of gene circuitry. Chaos **23**(2), 025108 (2013). http://dx.doi.org/10.1063/1.4809776
30. E.O. Voit, Biochemical systems theory: a review. Int. Scholarly Res. Network (ISRN – Biomathematics), Article 897658, 1–53 (2013)
31. J.P. Hespanha, *Linear systems theory* (Princeton University Press, Princeton, NJ, 2009)
32. M.A. Savageau, A. Sorribas, Constraints among molecular and systemic properties: implications for physiological genetics. J. Theor. Biol. **141**, 93–115 (1989)
33. H.W. Bode, *Network analysis and feedback amplifier design* (Van Nostrand, Princeton, 1945)
34. J.B. Cruz (ed.), *System sensitivity analysis* (Dowden, Hutchinson and Ross, Stroudsburg, 1973)
35. S.J. Mason, Feedback theory – some properties of signal flow graphs. Proc. I.R.E **41**, 1144–1156 (1953)
36. J.G. Truxal, *Automatic feedback control system synthesis* (McGraw-Hill, New York, 1955)
37. A.M. Lyapunov, *The general problem of the stability of motion (translation)* (Taylor & Francis, London, 1992)
38. M.A. Savageau, Growth of complex systems can be related to the properties of their underlying determinants. Proc. Natl. Acad. Sci. U. S. A. **76**, 5413–5417 (1979)
39. M.A. Savageau, R.A. Fasani, Qualitatively distinct phenotypes in the design space of biochemical systems. FEBS Lett. **583**, 3914–3922 (2009)
40. P.M.B.M. Coelho, A. Salvador, M.A. Savageau, Relating genotype to phenotype via the quantitative behavior of the NADPH redox cycle in human erythrocytes: mutant analysis. PLoS ONE **5**(9), e13031 (2010)
41. D.A. Tolla, M.A. Savageau, Regulation of aerobic-to-anaerobic transitions by the FNR cycle in *Escherichia coli*. J. Mol. Biol. **397**, 893–905 (2010)
42. D.A. Tolla, M.A. Savageau, Phenotypic repertoire of the FNR regulatory network in *Escherichia coli*. Mol. Microbiol. **79**, 149–165 (2011)
43. M.A. Savageau, Design of the lac gene circuit revisited. Math. Biosci. **231**, 19–38 (2011)
44. A. Martínez-Antonio, J.G. Lomnitz, S. Sandoval, M. Aldana, M.A. Savageau, Regulatory design governing progression of population growth phases in bacteria. PLoS ONE **7**(2), e30654 (2012). doi:10.1371/journal.pone.0030654
45. I. Dutra, D. Page, V.S. Costa, J. Shavlik, M. Waddell, Toward automatic management of embarrassingly parallel applications. Lecture Notes Comput. Sci. **2790**, 509–516 (2003)
46. U. Fayyad, G. Grinstein, A. Wierse, *Information visualization in data mining and knowledge discovery*, 1st edn. (Morgan-Kaufmann Publishers, San Francisco, 2001)

IBCell Morphocharts: A Computational Model for Linking Cell Molecular Activity with Emerging Tissue Morphology

Katarzyna A. Rejniak

Abstract Despite the fact that many genes and signaling pathways responsible for early carcinoma lesions have been identified, it is not yet fully understood how these molecular defects produce emergent abnormal morphologies. Nonetheless, morphological changes in three-dimensional multicellular cultures provide the first insight into invasive potential of cells and phenotypic/genotypic changes when compared with normal nontumorigenic morphologies. Thus, a quantitative tool for matching the molecular and morphological scales can be useful for designing testable experimental hypotheses. The computational model presented here can capture both the spatial and the temporal dynamics of developing multicellular structures and produces multidimensional charts (morphocharts) of various acinar and mutant morphologies arising from simultaneously varied model parameters. These morphocharts allow us to map experimental morphologies onto the model parameter space to identify deregulated molecular processes that can be subjected to further experimental validation. This approach provides mechanistic information on intermediate scales between molecular pathways and multicellular organization.

1 Introduction

Three-dimensional (3D) multicellular cultures are widely used in cancer biology to characterize the proliferative potential of cells in in-vivo-like conditions, in which the cells need to compete for space and resources and respond to local cell–cell

K.A. Rejniak (✉)
Integrated Mathematical Oncology, H. Lee Moffitt Cancer Center and Research Institute, Tampa, FL, USA

Department of Oncologic Sciences, College of Medicine, University of South Florida, Tampa, FL, USA
e-mail: Kasia.Rejniak@moffitt.org

N. Jonoska and M. Saito (eds.), *Discrete and Topological Models in Molecular Biology*, Natural Computing Series, DOI 10.1007/978-3-642-40193-0__23,

interactions and microenvironmental pressures. Many non–tumorigenic cells grown in 3D in vitro cultures can preserve the form and, to some extent, the function of their organs of origin. For instance, epithelial cells derived from breast (MCF-10A), kidney (MDCK), prostate (RWPE-1), and ovary (OVAR-5) each produce cyst-like structures with an epithelial shell enclosing a hollow lumen [1–11] when grown in 3D conditions. On the other hand, when the epithelial cells are mutated so that certain oncogenes are overexpressed, the growing cell clusters lose their well-organized epithelial architecture and form either multicellular spheroids with filled lumens or irregularly shaped condensed cell masses [12–22]. It is widely believed that the aggressiveness of any particular cancer cell line correlates with the degree to which its 3D structure departs from the normal epithelial-like architecture [23–26]. However, there is still limited understanding of the cellular steps that link a specific precancerous molecular alteration to a specific pattern of disruption in the epithelial architecture. This is a challenging problem because it requires integration across several scales: from genes to molecules, cellular phenotypes, and multicellular organization. For example, alterations in an oncogene-related molecular pathway may change proliferation rates, which may lead to disorganized glandular morphologies. Computational modeling can provide an overall theoretical framework and quantitative links between these different scales.

In this chapter, we describe a computational model, IBCell, which represents a two-dimensional central cross section through a 3D cell culture. We do not model the whole 3D structure, in order to reduce the computational cost of simulations. However, the model can recapitulate quite faithfully the development of a normal epithelial acinus and various acinar mutants. We describe the construction of the IBCell morphocharts, a tool for the analysis and classification of different multicellular morphologies. We give examples of morphocharts generated by altering three different cellular traits (3D morphocharts), but, in principle, morphocharts of higher dimensions can be generated. We also discuss how experimental morphologies obtained from a 3D culture system can be mapped onto the IBCell morphochart space in order to identify alterations in cell properties that lead to specific acinar mutant morphologies.

2 The IBCell Model

To address questions about morphological and molecular changes in a growing cell cluster, we developed the two-dimensional IBCell model (Immersed Boundary model of the Cell [27]), which can recapitulate morphological changes in individual cells and in the emerging multicellular architectures, based on the local cell–cell and cell–microenvironment interactions governed by the molecular state of cells. This model is based on the immersed boundary method [28–30], a classic method for studying fluid–structure interactions, designed to capture interactions between elastic bodies (here, cells) and a surrounding viscous incompressible fluid (here, the cytoplasm inside the cells, the extracellular matrix (ECM) outside the tissue,

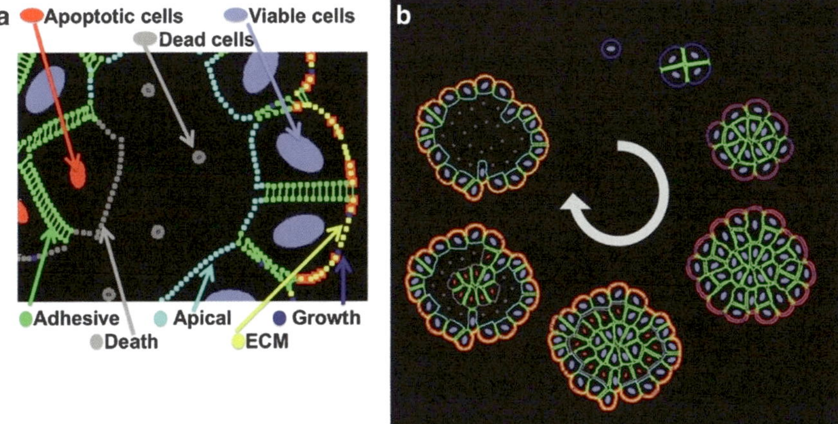

Fig. 1 (**a**) Schematic illustration of IBCell receptors and nuclear staining. (**b**) A sequence of consecutive stages of acinar development from a single cell to a monolayer of epithelial cells enclosing a hollow lumen. Receptor and nuclear staining as in (**a**)

and the lumen inside the hollow acinar structure). In IBCell, all cells are modeled as two-dimensional fully deformable bodies. The cell structure includes an elastic plasma membrane, represented by a network of linear springs that define the cell shape and enclose a viscous, incompressible fluid, representing the cytoplasm and providing cell mass. Individual cells can interact with one another and with the environment via a set of discrete membrane receptors located on the cell boundaries. These receptors are used to sense signals from other cells and from the microenvironment, as well as to secrete various ECM molecules. By integrating these external signals, the host cell can initiate certain cell life processes, such as proliferation, division, apoptotic death, and epithelial polarization. The decision to enter or continue a specific process depends on the cell's molecular signature (a distribution of growth, death, apical, cell–cell adhesion, and cell–ECM adhesion receptors along the cell boundary; see Fig. 1a). In IBCell, we prescribe the receptor signature thresholds for each of the cell life process (proliferation, apoptosis, cell–cell adhesion, cell–ECM adhesion, and cell polarization), but the model has no built-in knowledge about the shape of the tissue structures that the individual cells may construct together. Instead, the form of the multicellular architecture emerges from local interactions between cells, and between the cells and their environment. By examining various receptor thresholds (see IBCell morphocharts in Sect. 3), we can identify the combinations of thresholds that lead to the self-assembly of hollow mammary acini, and those that result in acinar mutants.

A sequence of consecutive stages in the development of a hollow acinus is shown in Fig. 1b. We initiated these simulations with a single cell, which upon consecutive divisions gives rise to a cluster of randomly oriented cells that adhere to their immediate neighbors (green receptors). All cells secrete ECM proteins in the vicinity of those membrane receptors that are not engaged in cell–cell adhesion, that is, initially along the whole cell boundary. However, with the increased number of adhesive receptors, the ECM secretion remains confined to the outer part of the cell membrane that is in direct contact with the external medium. Moreover, upon accumulation of the ECM (increased intensity of staining, from pink to red, around cell receptors), the free growth receptors (blue) change their function to ECM receptors (yellow), subsequently leading to the differentiation of all outer cells and their epithelial polarization. The polarized cells possess three different membrane domains: lateral domains in contact with other cells (green receptors), basal domain in contact with the ECM (yellow receptors), and apical domains (cyan receptors), which eventually develop as a result of breaking all adhesive connections with the inner cells. This subsequently increases the number of death receptors (gray) on the boundaries of the inner cells and triggers their apoptotic death.

The following equations describe the mathematical basis of the IBCell model and couple the interactions between the cells, the fluid, and the ECM molecules:

$$\rho \left(\frac{\partial \mathbf{u}(\mathbf{x}, t)}{\partial t} + (\mathbf{u}(\mathbf{x}, t) \cdot \nabla)\mathbf{u}(\mathbf{x}, t) \right) = -\nabla p(\mathbf{x}, t) + \mu \Delta \mathbf{u}(\mathbf{x}, t) + \frac{\mu}{3\rho} \nabla s(\mathbf{x}, t) + \mathbf{f}(\mathbf{x}, t),$$

$$\tag{1}$$

$$\rho \nabla \cdot \mathbf{u}(\mathbf{x}, t) = s(\mathbf{x}, t), \tag{2}$$

$$\mathbf{f}(\mathbf{x}, t) = \sum_{i=1,\ldots,N} \int_{\Gamma_i} \mathbf{F}(l_i, t) \, \delta(\mathbf{x} - \mathbf{X}(l_i, t)) \, dl, \tag{3}$$

$$\mathbf{F}(l_i, t) = \mathscr{F}^* \frac{\|\mathbf{X}(l_i^*, t) - \mathbf{X}(l_i, t)\| - \mathscr{L}^*}{\|\mathbf{X}(l_i^*, t) - \mathbf{X}(l_i, t)\|} \, (\mathbf{X}(l_i^*, t) - \mathbf{X}(l_i, t)), \tag{4}$$

$$s(\mathbf{x}, t) = \sum_{i=1,\ldots,N} \left(\sum_{k \in \Xi_i^+} S^+(\mathbf{Y}_{i,k}, t) \, \delta(\mathbf{x} - \mathbf{Y}_{i,k}) + \sum_{m \in \Xi_i^-} S^-(\mathbf{Z}_{i,m}, t) \, \delta(\mathbf{x} - \mathbf{Z}_{i,m}) \right), \tag{5}$$

$$\forall i \quad \sum_{k \in \Xi_i^+} S^+(\mathbf{Y}_{i,k}, t) + \sum_{m \in \Xi_i^-} S^-(\mathbf{Z}_{i,m}, t) = 0, \tag{6}$$

$$\frac{\partial \mathbf{X}(l_i, t)}{\partial t} = \mathbf{u}(\mathbf{X}(l_i, t), t) = \int_{\Omega} \mathbf{u}(\mathbf{x}, t) \, \delta(\mathbf{x} - \mathbf{X}(l_i, t)) \, d\mathbf{x}, \tag{7}$$

$$\frac{\partial \gamma(\mathbf{X}(l_i, t))}{\partial t} = \kappa_1 \mathbf{X}(l_i, t) - \kappa_2 \gamma(\mathbf{X}(l_i, t)). \tag{8}$$

In this system, Eq. (1) is the Navier–Stokes equation of a viscous, incompressible fluid with a velocity $\mathbf{u}$ defined on a Cartesian grid $\mathbf{x} = (x_1, x_2)$, where p is the fluid pressure, μ is the fluid viscosity, ρ is the fluid density, s is the local fluid expansion, and f is the external-force density. Equation (2) is the law of mass balance. The fluid flow is influenced both by the force density $\mathbf{F}(l_i, t)$, defined at the material points $\mathbf{X}(l_i, t)$ that form the boundary Γ_i of the ith cell (l_i is the position along the cell boundary), and by point sources $\mathbf{Y}_{i,k}$ and sinks $\mathbf{Z}_{i,m}$ that are nonzero if the ith cell is either growing or dying. The source and sink points are defined in the local microenvironment of the ith cell, i.e., $\mathbf{Y}_{i,k}, \mathbf{Z}_{i,m} \in \Theta_{\Gamma_i}^{\varepsilon} = \bigcup_{\mathbf{X} \in \Gamma_i} \{ \mathbf{x} : \| \mathbf{x} - \mathbf{X} \| < \varepsilon \}$. The interactions between the fluid and the material points on cell boundaries are defined in Eqs. (3)–(7). Here, both the force density $\mathbf{F}(l_i, t)$ and the sources $S^+(\mathbf{Y}_{i,k}, t)$ and sinks $S^-(\mathbf{Z}_{i,m}, t)$ are applied to the fluid using the two-dimensional Dirac delta function δ, and all material boundary points $\mathbf{X}(l_i, t)$ are carried along with the fluid. The boundary forces $\mathbf{F}(l_i, t)$ arise from the elastic properties of the cell membranes, from cell–cell adhesion, and from contractile forces that split the cell during its division, and are represented by short linear Hooke springs in Eq. (4), where $\mathscr{F}^*$ is the stiffness of the spring, $\mathscr{L}^*$ is the resting length, and $\mathbf{X}(l_i^*, t)$ is the adjacent, opposite, or neighboring point for the elastic, contractile, or adhesive force, respectively. The strengths of the individual sources $S^+(\mathbf{Y}_{i,k}, t) = \mathscr{S}^+$ and sinks $S^-(\mathbf{Z}_{i,m}, t) = \mathscr{S}^-$ are chosen such that they balance around every cell separately. The kinetics of the ECM protein concentration γ is described at the material points $\mathbf{X}(l_i, t)$ on the cell boundaries in Eq. (8). It includes a constant rate of ECM secretion at the boundary points (receptors) that are not engaged in cell–cell and cell–ECM adhesion (i.e., free growth receptors) and a decay proportional to the local concentration at each boundary point. Since we are modeling a generic ECM protein, an arbitrary small value is chosen for the receptor secretion rate κ_1. The decay rate κ_2 is defined in such a way as to allow the natural decay of ECM around receptors in which protein secretion has been stopped owing to a change in receptor function (such as in the case of adhesive or apical receptors), but still allow the steady accumulation of ECM around free growth receptors. All physical parameters used in the model are listed in Table 1. Computational implementation of this model requires a discrete version of the Dirac function and periodic boundary conditions on the domain boundaries in order to solve the immersed boundary equations. These numerical procedures are described in detail in [27, 35].

All cell life processes in the IBCell model, such as growth, division, death, cell–cell adhesion, and cell–ECM adhesion, depend entirely upon cues sensed from neighboring cells and from the ECM. This takes place through the cell membrane receptors located on the boundary of each cell (color-coded in Fig. 1a). The default state of the receptors is *growth* (blue). A cell–cell *adhesive* receptor (green) is activated when some boundary points of two distinct cells are sufficiently close to one another. An *ECM* receptor (yellow) is specified by a boundary point that is always in contact with the extracellular medium and is activated when the concentration of ECM in its vicinity exceeds a prescribed density. An *apical* sensor (cyan) is activated in an outer polarized cell when existing cell–cell adhesive links

Table 1 Physical parameters of the IBCell model of acinar morphogenesis

Model parameter	Value	Units	References
Fluid density	$\rho = 1.35$	g/cm^3	[31, 32]
Fluid viscosity	$\mu = 100$	g/(cm $\cdot$ s)	[31, 32]
Elastic-membrane force stiffness	$\mathscr{F}_{\text{elastic}} = 500$	g/(cm $\cdot$ s^2)	[32]
Adhesive-force stiffness	$\mathscr{F}_{\text{adhesive}} = \mathscr{F}_{\text{elastic}}$	g/(cm $\cdot$ s^2)	[33]
Contractile-force stiffness	$\mathscr{F}_{\text{contractile}} = 50 \times \mathscr{F}_{\text{elastic}}$	g/(cm $\cdot$ s^2)	[33]
Fluid sources and sinks	$\mathscr{S}_+ = \mathscr{S}_- = 5 \times 10^{-10}$	g/s	[33]
Elastic-spring resting length	$\mathscr{L}_{\text{elastic}} = 0.5$	μm	[33]
Adhesive-spring resting length	$\mathscr{L}_{\text{adhesive}} = 0.5$	μm	[33]
Contractile-spring resting length	$\mathscr{L}_{\text{contractile}} = 2.5$	μm	[33]
ECM secretion rate (at the growth receptors)	$\kappa_1 = 1 \times 10^{-10}$	mM/s	[34]
ECM decay rate	$\kappa_2 = 0.05 \times \kappa_1$	1/s	[34]

with inner cells are disassembled. A *death* receptor (gray) is created in an inner cell upon its detachment from a polarized cell or from another dying cell.

The activation of a growth receptor causes a certain amount of extracellular fluid to move into the cell, so the overall size of the cell increases. When the cell doubles in size, the division process is triggered. Division itself is modeled by the recruitment of boundary points to internal receptors that attract each other across the cell diameter, forming a furrow (the contractile ring) and eventually separating the mother cell into two daughter cells. The state of a cell is modulated by the percentage of recruited receptors and by the critical receptor thresholds. By varying these thresholds, one can explore the spectrum of model behavior (via IBCell morphocharts). For example, the cell can grow only if it can sense sufficient space in its vicinity, as defined by a threshold level of growth receptors. If this threshold is not met, the cell is considered to be resting. Similarly, cell death is triggered if the host cell accumulates a certain percentage of death receptors. If this threshold is not met, the cell is viable, but quiescent. The dynamics of cell membrane receptors is summarized in a flowchart in Fig. 2a, and the rules for their emergence and activation, as well as their links to the model equations, are listed in Table 2.

Each simulation in our model is initiated with a single, circular cell, and each of its membrane receptors is activated as a growth receptor. This results in a uniform distribution of sources and sinks of fluid along the cell boundary and, as a result, the growth of this first cell is uniform. After that, the cell receptors are activated depending on the local interactions of the host cell with other cells and the microenvironment. The receptor thresholds provide a decision mechanism that modifies the state of a cell by initiating certain life processes based on the cell receptor distribution. Figure 2b depicts a flowchart showing the evolution of five phenotypically distinct subpopulations of cells: resting, growing, polarized, apoptotic, and dead. Each cell phenotype (except the dead cells) can make decisions based on the configuration of cell membrane receptors (growth, death,

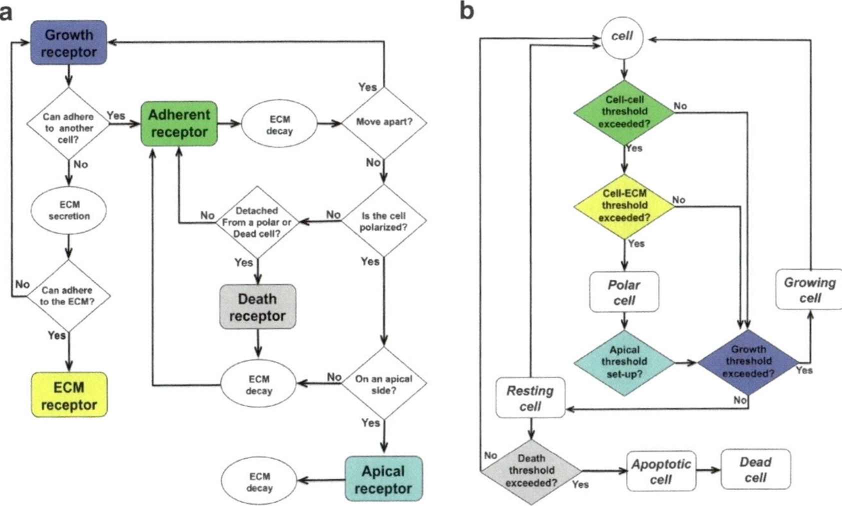

Fig. 2 (**a**) Flowchart describing the dynamics of cell membrane receptors. The default state is growth; transition to another state (ECM, adherent, apical, or death) depends on cues sensed from other cells and from the microenvironment. (**b**) Flowchart describing the evolution of phenotypically different subpopulations of resting, growing, polarized, apoptotic, and dead cells. The initiation of cell life processes depends on the distribution of all cell membrane receptors and the predefined receptor thresholds (Reproduced from [34], Fig. 1, with permission)

Table 2 Rules for the emergence and activation of cell membrane receptors in the IBCell model and their links to the model equations. Cell membrane receptors are color-coded as in Fig. 1

Cell receptors	Receptor activation/emergence	Model equations
Adhesion (green)	Activated when two receptors from distinct cells are close enough	Adhesive forces $\mathbf{F}(l_i, t)$ in Eqs. (3) and (4)
ECM (yellow)	Activated when the expression of ECM proteins exceeds the threshold	ECM $\gamma(\mathbf{X}(l_i, t))$ kinetics in Eq. (8)
Growth (blue)	All receptors free from any cell–cell or cell–ECM contacts (a default state)	Sources $S^+(\mathbf{Y}_{i,k}, t)$ inside and sinks $S^-(\mathbf{Z}_{i,m}, t)$ outside the growing cell, Eqs. (5) and (6)
Death (gray)	Emerge after detachment from a polarized or another dying cell	Sinks $S^-(\mathbf{Z}_{i,m}, t)$ inside and sources $S^+(\mathbf{Y}_{i,k}, t)$ outside the dying cell, Eqs. (5) and (6)
Apical (cyan)	Emerge during the development of a cell's apical membrane domain	Adhesive forces $\mathbf{F}(l_i, t)$ in Eqs. (3) and (4) are disassembled; all extrinsic signals are blocked

apical, cell–cell, and cell–ECM) to enter the associated cell life processes (growth, polarization, apoptotic death, and rest) that are responsible for movement through the flowchart.

3   The IBCell Morphocharts

In order to connect molecular defects in individual cells with altered morphologies
of emerging multicellular structures, we can explore the whole parameter space
of the IBCell thresholds. This can be done by producing multidimensional charts
(morphocharts) that plot collections of final morphologies arising from IBCell
simulations that were done with distinct combinations of receptor thresholds in the
initial cells, inherited thereafter by every newborn daughter cell. This establishes a
direct computational link between morphology and cell properties, enabling us to
classify the final morphologies into similarity classes with respect to the imposed
cell sensitivities (receptor thresholds). The IBCell morphocharts then provide a tool
for exploring what morphological structures will be produced when individual cell
responses (as defined by the receptor thresholds) are combined simultaneously in
the same simulation.

Moreover, the IBCell morphocharts can integrate multiple cell processes triggered by the receptor thresholds. These thresholds are defined as the percentages
of receptors needed to initiate certain cell life processes. That is, in the case of
proliferation or apoptosis, the cell must accumulate a predefined percentage of
particular cell membrane receptors, namely growth or death, respectively. In the case
of the ECM receptor threshold, we included an external inhibitory growth signal
that proliferating cells may receive from the ECM they secrete. This inhibitory
growth signal is directly proportional to the density of ECM accumulated on the
outer boundary of the cells, establishing a negative feedback between proliferation
and ECM density. Therefore, as the density of the matrix increases in the vicinity
of a cell, the receptors in contact with the ECM cross the critical threshold of ECM
density and are converted to ECM receptors, thus inhibiting growth.

Figure 3 shows an example of an IBCell morphochart that integrates the
three individual processes of cell growth, death, and ECM adhesion, resulting in
the formation of normal epithelial acini (red subspace), acini with filled lumens
(blue subspace), acini with degenerate shapes (yellow subspace), or nonstabilized
growing acinar structures (green subspace). Figure 3b shows the whole 3D parameter space, with representative morphologies from each class. Figure 3a shows one
plane of the 3D morphochart (a middle horizontal cross section), with morphologies
for all combinations of growth and death receptor thresholds and a fixed threshold
for the ECM receptors. Note that in this case, for a fixed growth receptor threshold
(with a value in the range from 5 to 25 %), the acinar structure changes from
hollow to partially occupied (by inner cells) and then to completely filled, as
the death receptor threshold increases from 5 to 35 % and then to 55 %. With
an increased death receptor threshold, the cells need to accumulate more death
receptors in order to start the apoptotic process, and thus they become more resistant
to the death signals. For a growth threshold of 0 %, all structures are nonstabilized
(still growing), since cell growth stabilization requires that all outer cells become
epithelially polarized. When the cells cannot acquire this phenotype, they initiate
the proliferation process, since there is no constraint on the cell growth receptor
threshold.

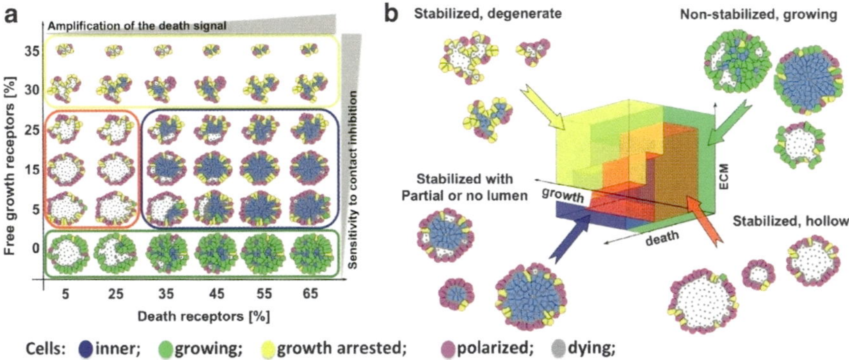

Fig. 3 (**a**) 2D cross section through the center of the 3D IBCell morphochart obtained from (**b**), generated by varying the thresholds for death, growth, and ECM receptors that trigger the corresponding cell life processes of proliferation, apoptosis, and ECM adhesion. The final morphologies are divided into four classes: hollow acini (*red*), filled cysts (*blue*), degenerate forms (*yellow*), and nonstabilized structures (*green*). Cell phenotypes are color-coded (Adapted from [34], Fig. 3, with permission)

On the other hand, for a fixed death receptor threshold and an increasing growth receptor threshold (from 5 to 35 %), the structures become more deformed as a result of growth arrest of individual cells. In such cases, the cells cannot accumulate enough growth receptors to initiate cell proliferation. Note that when the receptors are engaged in either cell–cell or cell–ECM adhesion, the percentage of growth receptors is diminished. Thus, the growing capability of the cell in our model is correlated with its sensitivity to contact inhibition. The more contacts the cell makes with other cells or with the ECM, the less likely it is to start to grow. Thus, the lower the growth receptor threshold is set, the less sensitive the host cell is to contact inhibition. For example, in the case when the growth receptor threshold is equal to 5 %, the cell may have up to 95 % of its receptors engaged in adhesion either to other cells or to the ECM, and the cell will still be able to initiate the proliferation process.

Finally, when the ECM threshold is varied, the model generates acinar structures of various sizes. This threshold regulates the time at which cell receptors become engaged in ECM adhesion. For low values of the ECM threshold, the ECM receptors emerge early, leading to cell growth suppression and the formation of smaller stable (growth-arrested) structures, either hollow or filled. For higher values of the ECM threshold, the acini grow larger before stabilizing, and in extreme of high values, cases the structures never stabilize, and instead continue to proliferate. Therefore, ECM adhesion contributes to cell growth arrest and to the stabilization of acinar structures, and the higher the ECM threshold, the larger the final acinar morphology.

The subspaces for each morphological class in Fig. 3 are quite broad, showing how robust the emerging acinar morphologies are to perturbations (here, in the growth, death, and ECM adhesion sensitivities). This may also be interpreted as adaptability and plasticity of the cells in response to extrinsic cues, or as a measure of the robustness of the acinus-building algorithm with respect to fluctuations of cellular traits.

4   The IBCell Morphological Classification

In Fig. 3b, we identified four broad classes of acinar morphologies. However, each class can be subdivided into subcategories containing significantly different structures. We will discuss each of the four subspaces separately here in order to identify the cellular traits and molecular components responsible for each morphological class.

The subspace of normal, hollow acini comprises a quarter of all simulated acini (Fig. 4). These structures, composed of one layer of epithelially polarized cells enclosing a hollow lumen, arise from fully filled multicellular clusters in which, first, all outer cells in contact with the ECM become epithelially polarized, and second, the inner cells are triggered to die by apoptosis, forming the hollow lumen. The final sizes of these acinar structures depend on when the outer cells became growth-arrested because of their polarization. This, in turn, is modulated by the ECM threshold. Thus, small acini are observed for small values of the ECM threshold (pale pink region in Fig. 4), moderate-sized acini (red region in Fig. 4) are generated for medium values of the ECM threshold, and large acini (brown region in Fig. 4) occur at the top threshold values. For a fixed ECM threshold value, the final size of the acinar structures diminishes with increased growth receptor threshold (a step-wise change in the color-coded parameter subspace, that is, a change from brown to red and from red to pink in the horizontal plane). In contrast, there is no change in the final acinar size classification for a fixed growth receptor threshold and an increased death receptor threshold. Thus, our model indicates that the normal, hollow acinar structures have much greater plasticity in adapting to death signals than to proliferation signals or ECM-related cues. The results of changes in the growth and ECM thresholds show that the cells in our model are more resistant to variation in apoptotic signals and more sensitive to extrinsic ECM-related cues.

The transition from the normal acinar space to the space of stabilized but nonhollow structures is quite sharp across the whole parameter space. In our model, a death receptor threshold of 35 % separates the hollow acini from the acini with cells that are able to survive inside the luminal compartment. The subspace of filled acini comprises about 20 % of all simulated acini (Fig. 4) and is divided into two categories, completely filled structures (light blue region in Fig. 4) and structures with small microlumens (blue region in Fig. 4). The subspace of partially filled acini is limited to structures of moderate size, with the death receptor threshold at the lower end of the nonhollow-acini space. Small structures (with small ECM thresholds) do not produce lumens, because they stabilize quickly and the inner cells have no time to accumulate enough death receptors to trigger the apoptotic process. In the structures emerging when the ECM threshold is set high, the epithelial polarization process occurs late, since the outer cells proliferate continuously. But once the polarization takes place, it happens in all outer cells almost simultaneously,

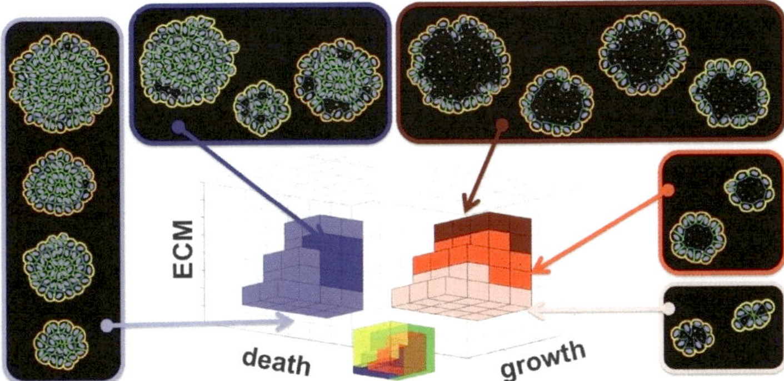

Fig. 4 Parameter subspace of hollow and filled cysts, subdivided into subcategories of small (*pale pink*), medium (*red*), and large (*brown*) normal acinar structures, and partially filled (*blue*) and fully filled (*pale blue*) architectures resembling DCIS (*ductal carcinoma in situ*). The *inset* shows the whole 3D parameter space divided into four classes. Receptors are color-coded as in Fig. 1

preventing the inner cells from initiating the death process, owing to inadequate numbers of expressed death receptors. However, in acinar structures of moderate size, the process of epithelial polarization is prolonged, which enables the inner cells to collect larger numbers of death receptors upon the disassembly of adhesion contacts with multiple outer cells. Thus, these acini contain small luminal areas near the layer of epithelial cells but with an intact core cluster of cells. In our model, we assume that the inner cells are not capable of reentering the proliferation process, even if the dying nearby cells create free space. If this assumption is not made, the inner microlumens can be repopulated, as we simulated in [36].

When the growth receptor threshold is increased, some of the outer cells become growth-arrested while their neighbors may still be able to proliferate. This results in irregular, degenerate final acinar morphologies (Fig. 5). This subspace comprises about 30 % of all simulated acini and includes structures with hollow lumens (yellow region in Fig. 5) and structures with partially or fully filled luminal compartments (orange region in Fig. 5). The transition from one subcategory to the other depends on the death receptor threshold. Multicellular patterns resembling shapes from this subspace may emerge when cells are resistant to growth signals or sensitive to contact inhibition.

Finally, when either the growth receptor threshold is set up to 0 % (meaning that the cells are insensitive to contact inhibition) or the ECM receptor threshold is very high, the final acinar structures remain nonstabilized and contain cells that proliferate continuously (Fig. 5). The subspace of nonstabilized acinar structures comprises about 30 % of all simulated cases, but these structures are highly localized on the boundary of the parameter space considered here. Among these structures, we can identify dense growing clusters (forest green region in Fig. 5), growing clusters with microlumens (dark green region in Fig. 5), and empty acini with cells growing within the outer monolayer (green region in Fig. 5). Again, the fully filled structures arise when the death receptor threshold is high, and the structures with empty

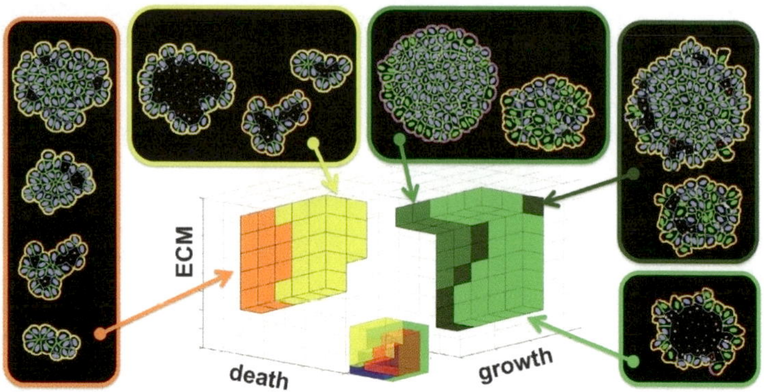

Fig. 5 Parameter subspace of degenerate and nonstabilized structures subdivided into subcategories of degenerate filled (*orange*) and empty (*yellow*) architectures, and still-growing hollow (*green*), partially filled (*dark green*), and fully filled (*forest green*) morphologies. The *inset* shows the whole 3D parameter space divided into four classes. Receptors are color-coded as in Fig. 1. *Green* nuclei represent growing cells, and the nuclei of resting cells are *blue*

lumens are produced when the death receptor threshold is low. The non-stabilized morphologies with microlumens form a transition zone between the structures with empty and filled lumens, but they arise sporadically. Morphologies corresponding to the shapes generated in this subspace may arise either from high resistance to the ECM adhesion signals or from elevated cell proliferation accompanied by changes in the response of cells to accumulated ECM.

The ten specific morphological classes identified by the IBCell model can be used to map the experimentally observed 3D morphologies onto the model parameter space in order to delineate the cell processes that are deregulated and to design laboratory experiments to confirm or rule out the predictions of the model.

5   The IBCell Morphochart Mapping

Many of the morphological structures identified by our computational model have been observed experimentally or clinically, indicating that IBCell has the potential to reproduce and analyze the formation of epithelial acini and acinar mutants.

We have shown that our model is capable of generating hollow acinar structures of various sizes (Fig. 4, red subspace). This is consistent with results reported in [13], where nontumorigenic breast cells (MCF10A) were grown in 3D Matrigel cultures and produced hollow acini that reached a diameter of $67.5 \pm 13.4\,\mu$m over a period of 10–20 days. Similar results have also been reported in [37], where 3D Matrigel cultures of lung alveolar cells (AT II) formed hollow cysts with diameters varying from 20 to 65 μm depending on the initial seeding density. Moreover, we have shown previously [36] that our model parameters can be tuned to qualitatively and quantitatively match experimental data collected from the MCF10A cell line and a particular MCF10A-derived mutant cell line, MCF10A-Her2^{YVMA}.

The parameter space of nonhollow acinar structures (Fig. 4, blue regions) contains morphologies that resemble early carcinoma lesions, such as breast ductal carcinoma in situ and prostate intraepithelial neoplasias, in which luminal filling is often observed [38–41]. Similarly, filled DCIS-like structures have been observed in 3D in vitro cultures of tumorigenic cell lines, such as prostate RWPE-1 cells [19] and breast MCF10CA1a cells [20]. Moreover, the acinar structures with multiple microlumens have been detected in 3D cultures of mammary cells derived from transgenic mice exposed to doxycycline treatment [42], in the breast tumor cells 21 MT and 21 MTΔ187, with the DEAR1 mutation [43], and in the kidney cell line MDCK [44, 45].

Furthermore, the transition from the hollow to filled acini that emerges naturally in our model (the border between the red and blue subspaces in Fig. 4) has been observed experimentally [13]. It was shown there that normal hollow acinar morphologies could be preserved even if either inner-cell apoptosis was inhibited (by overexpressing the antiapoptotic proteins Bcl-2 and Bcl-XL) or the cell proliferation rate was increased (by overexpressing cyclin D1 or the oncoprotein HPV16E7). However, when both of these perturbations were combined, the hollow morphology was disrupted, resulting in filled lumens. This phenomenon is recapitulated by our morphocharts, which show that a simultaneous change in two receptor thresholds (growth and apoptosis) is needed for the lumen to become fully filled (Fig. 3).

Our model has also identified a subspace of acinar morphologies that deviate from the normal regular spherical architecture and become stabilized as irregular, often tortuous structures (Fig. 5, yellow subspace). Such irregular morphologies have also been noticed in experimental cultures. MCF10A cells with an activated ErbB2 oncogene [13] or overexpressed Mek2-DD [14] grow in structures composed of multiple acini merged together. Similarly, the premalignant MCF10AT cell line [20] and the iFGFR1-activated HC11 cell line [15] both form highly distorted architectures in 3D cultures.

The last acinar subspace established by our computational model contains nonstabilized structures with actively growing cells even at a time corresponding to 30 days in culture (Fig. 5, green region). Morphologies of this type have been observed experimentally when MCF10A cells were overexpressing various oncogenes (Her2, HER2-Bcl2, Her2-E7, Her2^{YVMA}, and Ras-Scribble [16, 18, 36]). These experimental structures are characterized by various levels of cell death. In most cases, only small empty (luminal) spaces of a size corresponding to one to two cells were visible. However, we are not aware of any experimental examples in which the cells form an outer monolayer and are capable of persistent proliferation. Thus, this acinar phenotype may represent a combination of model parameters that do not coexist in nature.

In each experimental case mentioned above, the particular cell line was grown in a 3D in vitro culture that reconstructed some of the cell–cell and cell–microenvironment interactions typical of in vivo situations. These cultures were used to investigate the impact that certain intrinsic and extrinsic factors might have on multicellular growth and on anticancer treatments. However, it is very difficult to quantify how these factors alter individual cell behavior, especially in terms of multiple processes. By utilizing our computational morphocharts, we can inspect multiple processes in a systematic manner and tease out which ones have been altered and how. This can be done by mapping a particular experimental morphology onto a set of semiquantitative model parameters in order to quantify the range of possible changes in cell sensitivities defined by the particular cell receptor thresholds.

We have previously applied the IBCell morphochart mapping technique to estimate changes in the proliferation, death, and ECM rates in the MCF10A-Her2^{YVMA} mutant compared with the parental MCF10A cell line [36]. Our approach consisted of several steps. (1) A generic two-dimensional model of the development of hollow acini was tuned to qualitatively and quantitatively match the data acquired from central cross sections through MCF10A three-dimensional cultures. These data included the size, diameter, and shape of the central cross section through the developing 3D acinus at several time points, starting with the day of seeding and ending on the 28th day, when the acini were stabilized. We also used the counts and locations of all viable, proliferating, and dying cells in each acinar cross section. (2) The MCF10A-tuned model, that is, the model with a parameter set that reproduced an average MCF10A acinus, was tuned again to quantitatively reproduce a sequence of data from the developing MCF10A-Her2^{YVMA} mutant, and all parameter changes were recorded. (3) The modified parameters were interpreted biologically in order to highlight which cellular features drove the morphology of the mutant acini compared with that of the normal MACF10A acini and to formulate experimentally testable hypotheses. In the case of the MCF10A-Her2^{YVMA} mutant, the morphocharts showed that alterations in the proliferation and apoptosis sensitivities, no matter how extreme, were not able to account for the observed mutant morphologies until a third process had been considered. We showed that after changing the ECM thresholds, we were able to recover both the mutant 3D architecture and the quantitative metrics in terms of cell counts and structure size. Thus, we hypothesized that the lack of structure stabilization in the MCF10A-Her2^{YVMA} mutant is a result of its altered interactions with the microenvironment – particularly with ECM proteins that in vivo would account for the formation of the stabilizing basement membrane, but in vitro are manifested as a band of ECM proteins accumulating on the edge of the acinus.

6 Conclusions

The IBCell model (Immersed Boundary model of the Cell) was introduced in [46] to model abnormal bending of the human trophoblast bilayer tissue and was subsequently applied to simulate early tumor growth [27, 47] and the emergence of several types of pathological ductal tumors [48]. It was the first mathematical model used to investigate how epithelial acini and their mutants are formed [33, 49], and how their morphologies can be compared through computational simulations and morphocharts in order to delineate molecular or genetic differences between them

[34, 36]. In comparison with other recently developed models of acinus formation [50–52], which are based on hexagonal cellular automata, the IBCell model is not defined on a fixed grid, and each cell has a variable number of neighbors and communicates with other cells and with the microenvironment via a set of cell membrane receptors that, taken together, account for cell plasticity and the dynamic molecular status.

We have used IBCell to address the changes in cellular traits underlying the transformation of normal epithelial acinar morphologies into mutant-like cancerous structures. In principle, such fundamental insights into tissue organization and cancer progression could be derived by measuring changes in real time for multiple traits, as cells in a tissue respond to perturbations. However, this task is technically impractical or, at least, exceedingly labor-intensive. Thus, we developed an in silico acinus culture that integrates multiple traits of individual cells in a tissue and, by systematically varying model parameters, explores the morphogenetic space relative to trait combinations (resulting in a morphochart). Morphocharts provide a tool to integrate experimental data on a given cancer type and to quantify the relative impact of each of the cellular core traits that are modified in relation to the normal cell line. However, it is important to realize that IBCell must initially be tuned to the normal cell line for a given cancer in terms of both morphology and quantitative data (e.g., cell counts, acinar diameters, cellular density, and lumen/cellular ratio) in order to generate relevant morphocharts. Subsequently, the experimental morphologies can be mapped onto the morphocharts to estimate, from the position of the mutant acini, the changes in the proliferation, death, and extracellular matrix secretion rates (or other cellular traits of interest) that must have occurred in the mutant cells to produce the given morphology. Thus, IBCell effectively links genetic mutations to cell traits and can guide further experimentation to identify relationships between cancer mutations and tissue lesions, especially in early lesions that are driven by the mutation itself.

The application of the IBCell model that we have presented in this chapter is a tangible example of the interactions between theory and experimentation that lead to new biology. The morphocharts provide an analytical predictive tool in an area of biology (i.e., tissue morphogenesis) where a multiscale understanding of observed phenomena is especially difficult. The individual scales of this process (e.g., genetics, signaling, and cell trait behavior) continue to be extensively studied and have produced a massive amount of information. The task of integrating this information across scales is, however, daunting and easily escapes human intuition. Nowhere is this more evident than in the case of neoplastic diseases: there is understanding of the correlations between genetic mutations and tissue structure aberrations, but causal links remain elusive. Computational and mathematical modeling approaches, such as the use of IBCell morphocharts, can provide the necessary link bridging different biological scales.

Acknowledgements This work was partially supported by the NIH-NCI Integrative Cancer Biology Program (ICBP), U54CA113007.

References

1. C. Hagios, A. Lochter, M.J. Bissell, Tissue architecture: the ultimate regulator of epithelial function? Philos. Trans. R. Soc. Lond. B **353**, 857–870 (1998)
2. J. Debnath, S.K. Muthuswamy, J.S. Brugge, Morphogenesis and oncogenesis on MCF-10A mammary epithelial acini grown in three-dimensional basement membrane culture. Methods **30**, 256–268 (2003)
3. K.R. Mills-Shaw, C.N. Wrobel, J.S. Brugge, Use of three-dimensional basement membrane cultures to model oncogene-induced changes in mammary epithelial morphogenesis. J. Mammary Gland Biol. Neoplasia **9**(4), 297–310 (2004)
4. J. Debnath, J.S. Brugge, Modeling glandular epithelial cancers in three-dimensional cultures. Nat. Rev. Cancer **5**, 675–688 (2005)
5. C.M. Nelson, M.J. Bissell, Modeling dynamic reciprocity: engineering three-dimensional culture models of breast architecture, function, and neoplastic transformation. Semin. Cancer Biol. **15**, 342–352 (2005)
6. M.J. Reginato, S.K. Muthuswamy, Illuminating the center: mechanisms regulating lumen formation and maintenance in mammary morphogenesis. J. Mammary Gland Biol. Neoplasia **11**, 205–211 (2006)
7. D.R. Tyson, J. Inokuchi, T. Tsunoda, A. Lau, D.K. Ornstein, Culture requirements of prostatic epithelial cell lines for acinar morphognesis and lumen formation in vitro: role of extracellular calcium. Prostate **67**, 1601–1613 (2007)
8. A.A. Mailleux, M. Overholtzer, J.S. Brugge, Lumen formation during mammary epithelial morphogenesis. Cell Cycle **7**, 57–62 (2008)
9. D.M. Bryant, K.E. Mostov, From cells to organs: building polarized tissue. Nat. Rev. Mol. Cell Biol. **9**, 887–901 (2008)
10. R. Xu, A. Boudreau, M.J. Bissell, Tissue architecture and function: dynamic reciprocity via extra- and intra-cellular matrices. Cancer Metastasis Rev. **28**, 167–176 (2009)
11. F. Xu, J. Celli, I. Rizvi, S. Moon, T. Hasan, U. Demirci, A three-dimensional in vitro ovarian cancer coculture model using a high-throughput cell patterning platform. Biotechnol. J. **6**, 204–212 (2011)
12. D. Bello-DeOcampo, H.K. Kleinman, N.D. DeOcampo, M.M. Webber, Laminin-1 and alpha6-beta1 integrin regulate acinar morphogenesis of normal and malignant human prostate epithelial cells. Prostate **46**, 142–153 (2001)
13. J. Debnath, K.R. Mills, N.L. Collins, M.J. Reginato, S.K. Muthuswamy, J.S. Brugge, The role of apoptosis in creating and maintaining luminal space within normal and oncogene-expressing mammary acini. Cell **111**, 29–40 (2002)
14. M.J. Reginato, K.R. Mills, E.B.E. Becker, D.K. Lynch, A. Bonni, S.K. Muthuswamy, J.S. Brugge, Bim regulation of lumen formation in cultured mammary epithelial acini is targeted by oncogenes. Mol. Cell. Biol. **25**, 4591–4601 (2005)
15. W. Xian, K.L. Schwertfegen, T. Vargo-Gogola, J.M. Rosen, Pleiotropic effects of FGFR1 on cell proliferation, survival, and migration in a 3D mammary epithelial cell model. J. Cell Biol. **171**, 663–673 (2005)
16. S.E. Wang, A. Narasanna, M. Perez-Torres, B. Xiang, F.Y. Wu, S. Yang, G. Carpenter, A.F. Gazdar, S.K. Muthuswamy, C.L. Arteaga, HER2 kinase domain mutation results in constitutive phosphorylation and activation of HER2 and EGFR and resistance to EGFR tyrosine kinase inhibitors. Cancer Cell **10**, 25–38 (2006)
17. S.J. Sequeira, A.C. Ranganathan, A.P. Adam, B.V. Iglesias, E.F. Farias, J.A. Aquirre-Ghiso, Inhibition of proliferation by PERK regulates mammary acinar morphogenesis and tumor formation. PLoS ONE **7**, e615 (2007)
18. L.E. Dow, I.A. Elsum, C.L. King, K.M. Kinross, H.E. Richardson, P.O. Humbert, Loss of human Scribble cooperates with H-Ras to promote cell invasion through deregulation of MAPK signalling. Oncogene **27**(46), 5988–6001 (2008)

19. J. Inokuchi, A. Lau, D.R. Tyson, D.K. Ornstein, Loss of annexin A1 disrupts normal prostate glandular structure by inducing autocrine IL-6 signaling. Carcinogenesis **30**, 1082–1088 (2009)
20. A.K. Imbalzano, I. Tatarkova, A.N. Imbalzano, J.A. Nickerson, Increasingly transformed MCF-10A cells have a progressively tumor-like phenotype in three-dimensional basement membrane culture. Cancer Cell Int. **9**, 7 (2009)
21. A. Hockla, D.C. Radisky, E.S. Radisky, Mesotrypsin promotes malignant growth of breast cancer cells through shedding of CD109. Breast Cancer Res. Treat **124**, 27–38 (2010)
22. M.A. Cichon, V.G. Gainullin, Y. Zhang, D.C. Radisky, Growth of lung cancer cells in three-dimensional microenvironments reveals key features of tumor malignancy. Integr. Biol. **4**, 440–448 (2012)
23. V.M. Weaver, S. Lelievre, J.N. Lakins, M.A. Chrenek, J.C.R. Jones, F. Giancotti, Z. Werb, M.J. Bissell, Beta4 integrin-dependent formation of polarized three-dimensional architecture confers resistance to apoptosis in normal and malignant mammary epithelium. Cancer Cell **2**, 205–216 (2002)
24. M.J. Bissell, A. Rizki, I.S. Mian, Tissue architecture: the ultimate regulator of breast epithelial function. Curr. Opin. Cell Biol. **15**, 753–762 (2003)
25. S.E. Seton-Rogers, Y. Lu, L.M. Hines, M. Koundinya, J. LaBaer, S.K. Muthuswamy, J.S. Brugge, Cooperation of the ErbB2 receptor and transforming growth factor beta induction of migration and invasion in mammary epithelial cells. Proc. Natl. Acad. Sci. U.S.A. **101**, 1257–1262 (2004)
26. P.A. Kenny, G.Y. Lee, C.A. Myers, R.M. Neve, J.R. Semeiks, P.T. Spellman, K. Lorenz, E.H. Lee, M.H. Barcellos-Hoff, O.W. Petersen, J.W. Gray, M.J. Bissell, The morphologies of breast cancer cell lines in three-dimensional assays correlate with their profiles of gene expression. Mol. Oncol. **1**, 84–96 (2007)
27. K.A. Rejniak, An immersed boundary framework for modelling the growth of individual cells: an application to the early tumour development. J. Theor. Biol. **247**, 186–204 (2007)
28. C.S. Peskin, Flow patterns around heart valves: a numerical method. J. Comput. Phys. **10**, 252–271 (1972)
29. C.S. Peskin, Numerical analysis of blood flow in the heart. J. Comput. Phys. **25**, 220–252 (1977)
30. C.S. Peskin, The immersed boundary method. Acta Numer. **11**, 479–517 (2002)
31. M. Dembo, F. Harlow, Cell motion, contractile networks, and the physics of interpenetrating reactive flow. Biophys. J. **50**, 109–121 (1986)
32. V.M. Laurent, E. Planus, R. Fodil, D. Isabey, Mechanical assessment by magnetocytometry of the cytosolic and cortical cytoskeletal compartments in adherent epithelial cells. Biorheology **40**, 235–240 (2003)
33. K.A. Rejniak, A. Anderson, A computational study of the development of epithelial acini. I. Sufficient conditions for the formation of a hollow structure. Bull. Math. Biol. **70**, 677–712 (2008)
34. K.A. Rejniak, V. Quaranta, A.R.A. Anderson, Computational investigation of intrinsic and extrinsic mechanisms underlying the formation of carcinoma. Math. Med. Biol. **29**, 67–84 (2012)
35. K.A. Rejniak, Modelling the development of complex tissues using individual viscoelastic cells, in *Single-Cell-Based Models in Biology and Medicine*, ed. by A.R.A. Anderson, M.A.J. Chaplain, K.A. Rejniak (Birkhauser, Basel, 2007)
36. K.A. Rejniak, S.E. Wang, N.S. Bryce, H. Chang, B. Parvin, J. Jourquin, L. Estrada, J.W. Gray, C.L. Arteaga, A.M. Weaver, V. Quaranta, A.R.A. Anderson, Linking changes in epithelial morphogenesis to cancer mutations using computational modeling. PLoS Comput. Biol. **6**(8), e10009000 (2010)
37. W. Yu, X. Fang, A. Ewald, K. Wong, C.A. Hunt, Z. Werb, M.A. Matthay, K. Mostov, Formation of cysts by alveolar type II cells in three-dimensional culture reveals a novel mechanism for epithelial morphogenesis. Mol. Biol. Cell **18**, 1693–1700 (2007)
38. F.A. Tavassoli, Ductal intraepithelial neoplasia of the breast. Vichrows Arch. **438**, 221–227 (2001)

39. D.P. Winchester, J.M. Jeske, R.A. Goldschmidt, The diagnosis and management of ductal carcinoma in-situ of the breast. Cancer J. Clin. **50**, 184–200 (2000)
40. D.G. Bostwick, M.B. Amin, P. Dundore, W. Marsh, D.S. Schultz, Architectural patterns of highgrade prostatic intraepithelial neoplasia. Hum. Pathol. **24**, 298–310 (1993)
41. M. Che, D. Grignon, Pathology of prostate cancer. Cancer Matastasis Rev. **21**, 381–395 (2002)
42. M. Jechlinger, K. Podsypanina, H. Varmus, Regulation of transgenes in three-dimensional cultures of primary mouse mammary cells demonstrates oncogene dependence and identifies cells that survive deinduction. Genes Dev. **23**, 1677–1688 (2009)
43. S.T. Lott, N. Chen, D.S. Chandler, Q. Yang, L. Wang, M. Rodriguez, H. Xie, S. Balasenthil, T.A. Bucholz, A.A. Sahin, K. Chaung, B. Zhang, S.-E. Olufemi, J. Chen, H. Adams, V. Band, A.K. El-Naggar, M.L. Frazier, K. Keyomarsi, K.K. Hunt, S. Sen, B. Haffty, S.M. Hewitt, R. Krahe, A.M. Killary, DEAR1 is a dominant regulator of acinar morphogenesis and an independent predictor of local recurrence-free survival in early-onset breast cancer. PLoS Med. **6**, e1000068 (2009)
44. A.E. Rodriguez-Fraticelli, S. Vergarajauregui, D.J. Eastburn, A. Datta, M.A. Alonso, K. Mostov, F. Martin-Belmonte, The Cdc42 GEF intersectin 2 controls mitotic spindle orientation to form the lumen during epithelial morphogenesis. J. Cell Biol. **189**, 725–738 (2010)
45. F. Martin-Belmonte, W. Yu, A.E. Rodriguez-Fraticelli, A. Ewald, Z. Werb, M.A. Alonso, K. Mostov, Cell-polarity dynamics controls the mechanism of lumen formation in epithelial morphogenesis. Curr. Biol. **18**, 507–513 (2008)
46. K.A. Rejniak, H.J. Kliman, L.J. Fauci, A computational model of the mechanics of growth of the villous trophoblast bilayer. Bull. Math. Biol. **66**(2), 199–232 (2004)
47. K.A. Rejniak, A single-cell approach in modeling the dynamics of tumor regions. Math. Biosci. Eng. **2**, 643–655 (2005)
48. K.A. Rejniak, R.H. Dillon, A single cell based model of the ductal tumor microarchitecture. Comput. Math. Methods Med. **8**(1), 51–69 (2007)
49. K.A. Rejniak, A. Anderson, A computational study of the development of epithelial acini. II. Necessary conditions for structure and lumen stability. Bull. Math. Biol. **70**, 1450–1479 (2008)
50. S.H.J. Kim, J. Debnath, K. Mostov, S. Park, C.A. Hunt, A computational approach to resolve cell level contributions to early glandular epithelial cancer progression. BMC Syst. Biol. **3**, 122 (2009)
51. J.A. Engelberg, A. Datta, K.E. Mostov, C.A. Hunt, MDCK cystogenesis driven by cell stabilization within computational analogues. PLoS Comput. Biol. **7**(4), e1002030 (2011)
52. J. Tang, H. Enderling, S. Becker-Weimann, Ch. Pham, A. Polyzos, Ch.-Y. Chena, S.V. Costes, Phenotypic transition maps of 3D breast acini obtained by imaging-guided agent-based modeling. Integr. Biol. **3**, 408–421 (2011)

Printed by Printforce, the Netherlands